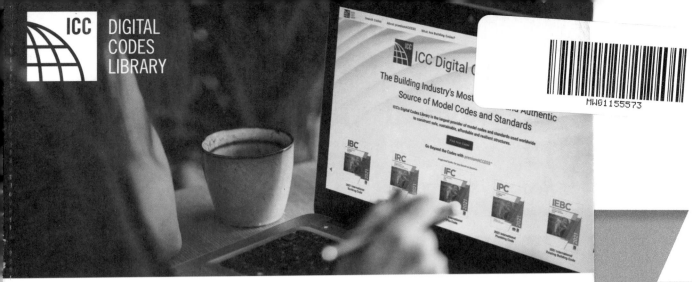

Telework with ICC Digital Codes Library
An essential online platform for all code users

ICC's Digital Codes Library is the most trusted and authentic source of model codes and standards, which conveniently provides access to the latest code text from across the United States. With a *premiumACCESS* subscription, users can further enhance their code experience with powerful features such as concurrent access, bookmarks, notes, errata and revision history, and more.

- Never miss a code update
- Available anywhere 24/7
- Use on any mobile or digital device
- View 800+ code titles and standards

Learn more at codes.iccsafe.org

2021 International Existing Building Code®

Date of First Publication: December 28, 2020

First Printing: December 2020

ISBN: 978-1-60983-969-7 (soft-cover edition)
ISBN: 978-1-60983-970-3 (loose-leaf edition)

COPYRIGHT © 2020
by
INTERNATIONAL CODE COUNCIL, INC.

ALL RIGHTS RESERVED. This 2021 *International Existing Building Code®* is a copyrighted work owned by the International Code Council, Inc. ("ICC"). Without advance written permission from the ICC, no part of this book may be reproduced, distributed or transmitted in any form or by any means, including, without limitation, electronic, optical or mechanical means (by way of example, and not limitation, photocopying or recording by or in an information storage retrieval system). For information on use rights and permissions, please contact: ICC Publications, 4051 Flossmoor Road, Country Club Hills, IL 60478. Phone 1-888-ICC-SAFE (422-7233).

Trademarks: "International Code Council," the "International Code Council" logo, "ICC," the "ICC" logo, "International Existing Building Code," "IEBC" and other names and trademarks appearing in this book are registered trademarks of the International Code Council, Inc., and/or its licensors (as applicable), and may not be used without permission.

PRINTED IN THE USA

PREFACE

Introduction

The *International Existing Building Code®* (IEBC®) establishes minimum requirements for existing buildings using prescriptive and performance-related provisions. It is founded on broad-based principles intended to encourage the use and reuse of existing buildings while requiring reasonable upgrades and improvements. This 2021 edition is fully compatible with all of the International Codes® (I-Codes®) published by the International Code Council® (ICC®), including the *International Building Code®* (IBC®), *International Energy Conservation Code®* (IECC®), *International Fire Code®* (IFC®), *International Fuel Gas Code®* (IFGC®), *International Green Construction Code®* (IgCC®), *International Mechanical Code®* (IMC®), *International Plumbing Code®* (IPC®), *International Private Sewage Disposal Code®* (IPSDC®), *International Property Maintenance Code®* (IPMC®), *International Residential Code®* (IRC®), *International Swimming Pool and Spa Code®* (ISPSC®), *International Wildland-Urban Interface Code®* (IWUIC®), *International Zoning Code®* (IZC®) and *International Code Council Performance Code®* (ICCPC®).

The I-Codes, including the IEBC, are used in a variety of ways in both the public and private sectors. Most industry professionals are familiar with the I-Codes as the basis of laws and regulations in communities across the US and in other countries. However, the impact of the codes extends well beyond the regulatory arena, as they are used in a variety of nonregulatory settings, including:

- Voluntary compliance programs such as those promoting sustainability, energy efficiency and disaster resistance.
- The insurance industry, to estimate and manage risk, and as a tool in underwriting and rate decisions.
- Certification and credentialing of individuals involved in the fields of building design, construction and safety.
- Certification of building and construction-related products.
- US federal agencies, to guide construction in an array of government-owned properties.
- Facilities management.
- "Best practices" benchmarks for designers and builders, including those who are engaged in projects in jurisdictions that do not have a formal regulatory system or a governmental enforcement mechanism.
- College, university and professional school textbooks and curricula.
- Reference works related to building design and construction.

In addition to the codes themselves, the code development process brings together building professionals on a regular basis. It provides an international forum for discussion and deliberation about building design, construction methods, safety, performance requirements, technological advances and innovative products.

Development

This 2021 edition presents the code as originally issued, with changes reflected in the 2006 through 2018 editions and further changes approved by the ICC Code Development Process through 2019. A new edition such as this is promulgated every 3 years.

This code is founded on principles intended to encourage the use and reuse of existing buildings that adequately protect public health, safety and welfare; provisions that do not unnecessarily increase construction costs; provisions that do not restrict the use of new materials, products or methods of construction; and provisions that do not give preferential treatment to particular types or classes of materials, products or methods of construction.

Maintenance

The IEBC is kept up to date through the review of proposed changes submitted by code enforcement officials, industry representatives, design professionals and other interested parties. Proposed changes are carefully considered through an open code development process in which all interested and affected parties may participate.

The ICC Code Development Process reflects principles of openness, transparency, balance, due process and consensus, the principles embodied in OMB Circular A-119, which governs the federal government's use of private-sector standards. The ICC process is open to anyone; there is no cost to participate, and people can participate without travel cost through the ICC's cloud-based app, cdpAccess®. A broad cross section of interests are represented in the ICC Code Development Process. The codes, which are updated regularly, include safeguards that allow for emergency action when required for health and safety reasons.

In order to ensure that organizations with a direct and material interest in the codes have a voice in the process, the ICC has developed partnerships with key industry segments that support the ICC's important public safety mission. Some code development committee members were nominated by the following industry partners and approved by the ICC Board:

- American Institute of Architects (AIA)
- National Association of Home Builders (NAHB)

The code development committees evaluate and make recommendations regarding proposed changes to the codes. Their recommendations are then subject to public comment and council-wide votes. The ICC's governmental members—public safety officials who have no financial or business interest in the outcome—cast the final votes on proposed changes.

The contents of this work are subject to change through the code development cycles and by any governmental entity that enacts the code into law. For more information regarding the code development process, contact the Codes and Standards Development Department of the ICC.

While the I-Code development procedure is thorough and comprehensive, the ICC, its members and those participating in the development of the codes disclaim any liability resulting from the publication or use of the I-Codes, or from compliance or noncompliance with their provisions. The ICC does not have the power or authority to police or enforce compliance with the contents of this code.

Code Development Committee Responsibilities (Letter Designations in Front of Section Numbers)

In each code development cycle, proposed changes to this code are considered at the Committee Action Hearings by the International Existing Building Code Development Committee. Proposed changes to a code section having a number beginning with a letter in brackets are considered by a different code development committee. For example, proposed changes to code sections that are preceded by the designation [F] (e.g., [F] 1504.2 are considered by the International Fire Code Development Committee at the Committee Action Hearings.

The content of sections in this code that begin with a letter designation is maintained by another code development committee in accordance with the following:

[A] = Administrative Code Development Committee;

[BE] = IBC—Means of Egress Code Development Committee;

[BG] = IBC—General Code Development Committee;

[BS] = IBC—Structural Code Development Committee;

[E] = International Commercial Energy Conservation Code Development Committee or International Residential Energy Conservation Code Development Committee;

[F] = International Fire Code Development Committee;

[FG] = International Fuel Gas Code Development Committee;

[M] = International Mechanical Code Development Committee; and

[P] = International Plumbing Code Development Committee.

For the development of the 2024 edition of the I-Codes, there will be two groups of code development committees and they will meet in separate years, as shown in the following Code Development Hearings Table.

Code change proposals submitted for code sections that have a letter designation in front of them will be heard by the respective committee responsible for such code sections. Because different committees hold Committee Action Hearings in different years, it is possible that some proposals for this code will be heard by a committee in a different year than the year in which the primary committee for this code meets. In the case of the IEBC, the primary committees that maintain this code will meet in 2022.

For instance, Section 1501.3 is designated as the responsibility of the IBC—General Code Development Committee. This committee will conduct its code development hearings in 2021 to consider all code change proposals to the *International Building Code* and any portions of other codes that it is responsible for, including Section 1501.3 of the IEBC and other general provisions of the IEBC (designated with [BG] in front of those sections). Therefore, any proposals received for Section 1501.3 will be considered in 2021 by the IBC—General Code Development Committee.

It is very important that anyone submitting code change proposals understand which code development committee is responsible for the section of the code that is the subject of the code change proposal. For further information on the Code Development Committee responsibilities, please visit the ICC website at www.iccsafe.org/current-code-development-cycle.

CODE DEVELOPMENT HEARINGS

Group A Codes (Heard in 2021, Code Change Proposals Deadline: January 11, 2021)	Group B Codes (Heard in 2022, Code Change Proposals Deadline: January 10, 2022)
International Building Code – Egress (Chapters 10, 11, Appendix E) – Fire Safety (Chapters 7, 8, 9, 14, 26) – General (Chapters 2–6, 12, 27–33, Appendices A, B, C, D, K, N)	Administrative Provisions (Chapter 1 of all codes except IECC, IRC and IgCC, IBC Appendix O, the appendices titled "Board of Appeals" for all codes except IECC, IRC, IgCC, ICCPC and IZC, administrative updates to currently referenced standards, and designated definitions)
International Fire Code	**International Building Code** – Structural (Chapters 15–25, Appendices F, G, H, I, J, L, M)
International Fuel Gas Code	**International Existing Building Code**
International Mechanical Code	**International Energy Conservation Code—Commercial**
International Plumbing Code	**International Energy Conservation Code—Residential** – IECC—Residential – IRC—Energy (Chapter 11)
International Property Maintenance Code	**International Green Construction Code** (Chapter 1)
International Private Sewage Disposal Code	**International Residential Code** – IRC—Building (Chapters 1–10, Appendices AE, AF, AH, AJ, AK, AL, AM, AO, AQ, AR, AS, AT, AU, AV, AW)
International Residential Code – IRC—Mechanical (Chapters 12–23) – IRC—Plumbing (Chapters 25–33, Appendices AG, AI, AN, AP)	
International Swimming Pool and Spa Code	
International Wildland-Urban Interface Code	
International Zoning Code	

Note: Proposed changes to the ICCPC will be heard by the code development committee noted in brackets [ ] in the text of the ICCPC.

Marginal Markings

Solid vertical lines in the margins within the body of the code indicate a technical change from the requirements of the 2018 edition. Deletion indicators in the form of an arrow (➡) are provided in the margin where an entire section, exception or table has been deleted or an item in a list of items or row of a table has been deleted.

A single asterisk [*] placed in the margin indicates that text or a table has been relocated within the code. A double asterisk [**] placed in the margin indicates that the text or table immediately following it has been relocated there from elsewhere in the code. The following table indicates such relocations in the 2021 edition of the IEBC.

RELOCATIONS

2021 LOCATION	2018 LOCATION
303	1106
307	502.6, 503.14, 803.4.3, 1104
308	502.7, 503.15, 804, 1105
1009.1	809.1
1501.2.1	705.2

Coordination of the International Codes

The coordination of technical provisions is one of the strengths of the ICC family of model codes. The codes can be used as a complete set of complementary documents, which will provide users with full integration and coordination of technical provisions. Individual codes can also be used in subsets or as stand-alone documents. To make sure that each individual code is as complete as possible, some technical provisions that are relevant to more than one subject area are duplicated in some of the model codes. This allows users maximum flexibility in their application of the I-Codes.

Italicized Terms

Terms italicized in code text, other than document titles, are defined in Chapter 2. The terms selected to be italicized have definitions that the user should read carefully to better understand the code. Where italicized, the Chapter 2 definition applies. If not italicized, common-use definitions apply.

Adoption

The ICC maintains a copyright in all of its codes and standards. Maintaining copyright allows the ICC to fund its mission through sales of books, in both print and electronic formats. The ICC welcomes adoption of its codes by jurisdictions that recognize and acknowledge the ICC's copyright in the code, and further acknowledge the substantial shared value of the public/private partnership for code development between jurisdictions and the ICC.

The ICC also recognizes the need for jurisdictions to make laws available to the public. All I-Codes and I-Standards, along with the laws of many jurisdictions, are available for free in a nondownloadable form on the ICC's website. Jurisdictions should contact the ICC at adoptions@iccsafe.org to learn how to adopt and distribute laws based on the IEBC in a manner that provides necessary access, while maintaining the ICC's copyright.

To facilitate adoption, several sections of this code contain blanks for fill-in information that needs to be supplied by the adopting jurisdiction as part of the adoption legislation. For this code, please see:

Section 101.1 Insert: **[NAME OF JURISDICTION]**

Section 103.1 Insert: **[NAME OF DEPARTMENT]**

Effective Use of the International Existing Building Code

The IEBC is a model code in the International Code family of codes intended to provide requirements for repair and alternative approaches for alterations, changes of occupancy and additions to existing buildings. A large number of existing buildings and structures do not comply with the current building code requirements for new construction. Although many of these buildings are potentially salvageable, rehabilitation is often cost-prohibitive because compliance with all the requirements for new construction could require extensive changes that go well beyond the value of the building or the original scope of the alteration. At the same time, it is necessary to regulate construction in existing buildings that undergo additions, alterations, extensive repairs or change of occupancy. Such activity represents an opportunity to ensure that new construction complies with the current building codes and that existing conditions are maintained, at a minimum, to their current level of compliance or are improved as required to meet basic safety levels. To accomplish this objective, and to make the alteration process easier, this code allows for options for controlled departure from full compliance with the International Codes dealing with new construction, while maintaining basic levels for fire safety, structural and life safety features of the rehabilitated building.

This code provides three main options for a designer in dealing with alterations of existing buildings. These are laid out in Section 301 of this code:

OPTION 1: Work for alteration, change of occupancy or addition of all existing buildings shall be done in accordance with the Prescriptive Compliance Method given in Chapter 5. It should be noted that this method originates from the former Chapter 34 of the *International Building Code* (2012 and earlier editions).

OPTION 2: Work for alteration, change of occupancy or addition of all existing buildings shall be done in accordance with the Work Area Compliance Method given in Chapters 6 through 12.

OPTION 3: Work for alteration, change of occupancy or addition of all existing buildings shall be done in accordance with the Performance Compliance Method given in Chapter 13. It should be noted that this option was also provided in the former Chapter 34 of the *International Building Code* (2012 and earlier editions).

Under limited circumstances, a building alteration can be made to comply with the laws under which the building was originally built, as long as the accessibility requirements are met, there has been no substantial structural damage and there will be limited structural alteration. Flood hazard provisions also must still be addressed where there is a substantial improvement.

Note that all repairs must comply with Chapter 4 and all relocated buildings are addressed by Chapter 14.

ARRANGEMENT AND FORMAT OF THE 2021 IEBC

Before applying the requirements of the IEBC, it is beneficial to understand its arrangement and format. The IEBC, like other codes published by ICC, is arranged and organized to follow logical steps that generally occur during a plan review or inspection.

The following table shows how the IEBC is divided. The ensuing chapter-by-chapter synopsis details the scope and intent of the provisions of the IEBC.

CHAPTER TOPICS

Chapter	Subjects
1–2	Administrative Requirements and Definitions
3	Provisions for all Compliance Methods
4	Repairs
5	Prescriptive Compliance Method for Existing Buildings
6–12	Work Area Compliance Method for Existing Buildings
13	Performance Compliance Method for Existing Buildings
14	Relocated Buildings
15	Construction Safeguards
16	Referenced Standards
Appendix A	Guidelines for Seismic Retrofit of Existing Buildings
Appendix B	Supplementary Accessibility Requirements for Existing Buildings
Appendix C	Guidelines for Wind Retrofit of Existing Buildings
Appendix D	Board of Appeals
Resource A	Guidelines on Fire Ratings of Archaic Materials and Assemblies

Chapter 1 Scope and Administration

This chapter contains provisions for the application, enforcement and administration of subsequent requirements of the code. In addition to establishing the scope of the code, Chapter 1 identifies which buildings and structures come under its purview. Chapter 1 is largely concerned with maintaining "due process of law" in enforcing the regulations contained in the body of the code. Only through careful observation of the administrative provisions can the code official reasonably expect to demonstrate that "equal protection under the law" has been provided.

Chapter 2 Definitions

All defined terms in the code are provided in Chapter 2. While a defined term may only be used in one chapter or another, the meaning provided in Chapter 2 is applicable throughout the code.

Where understanding of a term's definition is especially key to or necessary for understanding of a particular code provision, the term is shown in italics wherever it appears in the code. This is true only for those terms that have a meaning that is unique to the code. In other words, the generally understood meaning of a term or phrase might not be sufficient or consistent with the meaning prescribed by the code; therefore, it is essential that the code-defined meaning be known.

Guidance regarding tense, gender and plurality of defined terms as well as guidance regarding terms not defined in this code is also provided.

Chapter 3 Provisions for All Compliance Methods

This chapter serves several purposes. The main role is to explain the three compliance options available in the code. Clarification is provided as to how provisions in other I-Codes related to repairs, alterations, additions, relocation and changes in occupancy must also be addressed unless they conflict with the IEBC. In that case, the IEBC takes precedence. In addition, this chapter also lays out the methods to be used for seismic design and evaluation throughout the IEBC. There are also several issues that are addressed globally for all methods for consistent application including storm shelters, accessibility, smoke alarms, carbon monoxide detection and exterior wall coverings.

Chapter 4 Repairs

Chapter 4 governs the repair of existing buildings. The provisions define conditions under which repairs may be made using materials and methods like those of the original construction or the extent to which repairs must comply with requirements for new buildings.

This chapter, like Chapter 14 related to relocated or moved buildings, is independent from the three methods presented by this code.

Chapter 5 Prescriptive Compliance Method

This chapter provides one of the three main options of compliance available in the IEBC for buildings and structures undergoing alteration, addition or change of occupancy.

Chapter 6 Classification of Work

This chapter provides an overview of the Work Area Method available as an option for rehabilitation of a building. The chapter defines the different classifications of alterations and provides general requirements for alterations, change of occupancy, additions and historic buildings. Detailed requirements for all of these are given in subsequent Chapters 7 through 12.

Chapter 7 Alterations—Level 1

This chapter provides the technical requirements for those existing buildings that undergo Level 1 alterations as described in Section 503, which includes replacement or covering of existing materials, elements, equipment or fixtures using new materials for the same purpose. This chapter, similar to other chapters of this code, covers all building-related subjects, such as structural, mechanical, plumbing and electrical as well as the fire and life safety issues when the alterations are classified as Level 1. The purpose of this chapter is to provide detailed requirements and provisions to identify the required improvements in the existing building elements, building spaces and building structural system. This chapter is distinguished from Chapters 8 and 9 by only involving replacement of building components with new components. In contrast, Level 2 alterations involve more space reconfiguration and Level 3 alterations involve more extensive space reconfiguration, exceeding 50 percent of the building area.

Chapter 8 Alterations—Level 2

Like Chapter 7, the purpose of this chapter is to provide detailed requirements and provisions to identify the required improvements in the existing building elements, building spaces and building structural system when a building is being altered. This chapter is distinguished from Chapters 7 and 9 by involving space reconfiguration that could be up to and including 50 percent of the area of the building. In contrast, Level 1 alterations (Chapter 7) do not involve space reconfiguration, and Level 3 alterations (Chapter 9) involve extensive space reconfiguration that exceeds 50 percent of the building area. Depending on the nature of alteration work, its location within the building and whether it encompasses one or more tenants, improvements and upgrades could be required for the open floor penetrations, sprinkler system or the installation of additional means of egress such as stairs or fire escapes.

Chapter 9 Alterations—Level 3

This chapter provides the technical requirements for those existing buildings that undergo Level 3 alterations. The purpose of this chapter is to provide detailed requirements and provisions to identify the required improvements in the existing building elements, building spaces and building structural system. This chapter is distinguished from Chapters 7 and 8 by involving alterations that cover 50 percent of the aggregate area of the building. In contrast, Level 1 alterations do not involve space reconfiguration and Level 2 alterations involve extensive space reconfiguration that does not exceed 50 percent of the building area. Depending on the nature of alteration work, its location within the building and whether it encompasses one or more tenants, improvements and upgrades could be required for the open floor penetrations, sprinkler system or the installation of additional means of egress such as stairs or fire escapes. At times and under certain situations, this chapter also intends to improve the safety of certain building features beyond the work area and in other parts of the building where no alteration work might be taking place.

Chapter 10 Change of Occupancy

The purpose of this chapter is to provide regulations for the circumstances when an existing building is subject to a change of occupancy or a change of occupancy classification. A change of occupancy is not to be confused with a change of occupancy classification. The IBC defines different occupancy classifications in Chapter 3 and special occupancy requirements in Chapter 4. Within specific occupancy classifications, there can be many different types of actual activities that can take place. For instance, a Group A-3 occupancy classification deals with a wide variation of different types of activities, including bowling alleys and courtrooms, indoor tennis courts and dance halls. When a facility changes use from, for example, a bowling alley to a dance hall, the occupancy classification remains A-3, but the different uses could lead to drastically different code requirements. Therefore, this chapter deals with the special circumstances that are associated with a change in the use of a building within the same occupancy classification as well as a change of occupancy classification.

Chapter 11 Additions

Chapter 11 provides the requirements for additions, which correlate to the code requirements for new construction. There are, however, some exceptions that are specifically stated within this chapter. An "*Addition*" is defined in Chapter 2 as "an extension or increase in the floor area, number of stories or height of a building or structure." Chapter 11 contains the minimum requirements for an addition that is not separated from the existing building by a fire wall.

Chapter 12 Historic Buildings

This chapter provides some exceptions from code requirements when the building in question has historic value. The most important criterion for application of this chapter is that the building must be essentially accredited as being of historic significance by a state or local authority after careful review of the historical value of the building. Most, if not all, states have such authorities, as do many local jurisdictions. The agencies with such authority can be located at the state or local government level or through the local chapter of the American Institute of Architects (AIA). Other considerations include the structural condition of the building (i.e., is the building structurally sound), its proposed use, its impact on life safety and how the intent of the code, if not the letter, will be achieved.

Chapter 13 Performance Compliance Methods

This chapter allows for existing buildings to be evaluated so as to show that alterations or a change of occupancy, while not meeting new construction requirements, will improve the current existing situation. Provisions are based on a numerical scoring system involving 21 various safety parameters and the degree of code compliance for each issue.

Chapter 14 Relocated or Moved Buildings

Chapter 14 is applicable to any building that is moved or relocated. This chapter, like the chapter on repairs, is independent from the three methods presented in this code.

Chapter 15 Construction Safeguards

The building construction process involves a number of known and unanticipated hazards. Chapter 15 establishes specific regulations in order to minimize the risk to the public and adjacent property. Some construction failures have resulted during the initial stages of grading, excavation and demolition. During these early stages, poorly designed and installed sheeting and shoring have resulted in ditch and embankment cave-ins. Also, inadequate underpinning of adjoining existing structures or careless removal of existing structures has produced construction failures.

There are also several fire safety and means of egress issues addressed by this chapter. This chapter is also consistent with Chapter 33 of the IBC and IFC.

Chapter 16 Referenced Standards

The code contains numerous references to standards that are used to regulate materials and methods of construction. Chapter 16 contains a comprehensive list of all standards that are referenced in the code. The standards are part of the code to the extent of the reference to the standard. Compliance with the referenced standard is necessary for compliance with this code. By providing specifically adopted standards, the construction and installation requirements necessary for compliance with the code can be readily determined. The basis for code compliance is, therefore, established and available on an equal basis to the building code official, contractor, designer and owner.

Chapter 16 is organized in a manner that makes it easy to locate specific standards. It lists all of the referenced standards, alphabetically, by acronym of the promulgating agency of the standard. Each agency's standards are then listed in either alphabetical or numeric order based upon the standard identification. The list also contains the title of the standard; the edition (date) of the standard referenced; any addenda included as part of the ICC adoption; and the section or sections of this code that reference the standard.

Appendix A Guidelines for the Seismic Retrofit of Existing Buildings

Appendix A provides guidelines for upgrading the seismic resistance capacity of different types of existing buildings. It is organized into separate chapters which deal with buildings of different types, including unreinforced masonry buildings, reinforced concrete and reinforced masonry wall buildings, and lightframe wood buildings. This appendix includes its own referenced standards.

Appendix B Supplementary Accessibility Requirements for Existing Buildings and Facilities

Chapter 11 of the IBC contains provisions that set forth requirements for accessibility to buildings and their associated sites and facilities for people with physical disabilities. Section 306 addresses accessibility provisions and alternatives permitted in existing buildings. Appendix B was added to address accessibility in construction for items that are not typically enforceable through the traditional building code enforcement process.

Appendix C Guidelines for the Wind Retrofit of Existing Buildings

This appendix is intended to provide guidance for retrofitting existing structures to strengthen their resistance to wind forces. This appendix is similar in scope to Appendix A which addresses seismic retrofits for existing buildings except that the subject matter is related to wind retrofits. These retrofits are voluntary measures that serve to better protect the public and reduce damage from high wind events for existing buildings.

The purpose of the appendix is to provide prescriptive alternatives for addressing retrofit of buildings in high-wind areas. Currently there are two chapters which deal with the retrofit of gable ends and the fastening of roof decks, Appendix Chapters C1 and C2, respectively. This appendix includes its own referenced standards.

Appendix D Board of Appeals

Section 112 of Chapter 1 requires the establishment of a board of appeals to hear appeals regarding determinations made by the code official. Appendix D provides qualification standards for members of the board as well as operational procedures of such board.

Resource A Guidelines on Fire Ratings of Archaic Materials and Assemblies

In the process of repair and alteration of existing buildings, based on the nature and the extent of the work, the IEBC might require certain upgrades in the fire-resistance rating of building elements, at which time it becomes critical for the designers and the code officials to be able to determine the fire-resistance rating of the existing building elements as part of the overall evaluation for the assessment of the need for improvements. This resource document provides a guideline for such an evaluation for fire-resistance rating of archaic materials that is not typically found in the modern model building codes.

TABLE OF CONTENTS

CHAPTER 1 SCOPE AND ADMINISTRATION .. 1-1

PART 1—SCOPE AND APPLICATION 1-1

Section
101 Scope and General Requirements 1-1
102 Applicability 1-1

PART 2—ADMINISTRATION AND ENFORCEMENT 1-2
103 Code Compliance Agency 1-2
104 Duties and Powers of Code Official 1-2
105 Permits 1-4
106 Construction Documents 1-5
107 Temporary Structures and Uses 1-7
108 Fees 1-7
109 Inspections 1-7
110 Certificate of Occupancy 1-8
111 Service Utilities 1-9
112 Means of Appeals 1-9
113 Violations 1-9
114 Stop Work Order 1-10
115 Unsafe Structures and Equipment 1-10
116 Emergency Measures 1-10
117 Demolition 1-11

CHAPTER 2 DEFINITIONS 2-1
Section
201 General 2-1
202 General Definitions 2-1

CHAPTER 3 PROVISIONS FOR ALL COMPLIANCE METHODS 3-1
Section
301 Administration 3-1
302 General Provisions 3-1
303 Storm Shelters 3-2
304 Structural Design Loads and Evaluation and Design Procedures 3-2
305 In-Situ Load Tests 3-3
306 Accessibility for Existing Buildings 3-3
307 Smoke Alarms 3-5
308 Carbon Monoxide Detection 3-6
309 Additions and Replacements of Exterior Wall Coverings and Exterior Wall Envelopes 3-6

CHAPTER 4 REPAIRS 4-1
Section
401 General 4-1
402 Building Elements and Materials 4-1
403 Fire Protection 4-1
404 Means of Egress 4-1
405 Structural 4-1
406 Electrical 4-2
407 Mechanical 4-2
408 Plumbing 4-3

CHAPTER 5 PRESCRIPTIVE COMPLIANCE METHOD 5-1
Section
501 General 5-1
502 Additions 5-1
503 Alterations 5-2
504 Fire Escapes 5-4
505 Windows and Emergency Escape Openings 5-5
506 Change of Occupancy 5-5
507 Historic Buildings 5-6

CHAPTER 6 CLASSIFICATION OF WORK 6-1
Section
601 General 6-1
602 Alteration—Level 1 6-1
603 Alteration—Level 2 6-1
604 Alteration—Level 3 6-1
605 Change of Occupancy 6-1
606 Additions 6-1
607 Historic Buildings 6-1

CHAPTER 7 ALTERATIONS—LEVEL 1 7-1
Section
701 General 7-1
702 Building Elements and Materials 7-1

TABLE OF CONTENTS

703	Fire Protection	7-2
704	Means of Egress	7-2
705	Reroofing	7-2
706	Structural	7-3
707	Electrical	7-4
708	Energy Conservation	7-4

CHAPTER 8 ALTERATIONS—LEVEL 2 8-1
Section

801	General	8-1
802	Building Elements and Materials	8-1
803	Fire Protection	8-3
804	Means of Egress	8-5
805	Structural	8-9
806	Electrical	8-10
807	Mechanical	8-10
808	Plumbing	8-11
809	Energy Conservation	8-11

CHAPTER 9 ALTERATIONS—LEVEL 3 9-1
Section

901	General	9-1
902	Special Use and Occupancy	9-1
903	Building Elements and Materials	9-1
904	Fire Protection	9-2
905	Means of Egress	9-3
906	Structural	9-3
907	Energy Conservation	9-3

CHAPTER 10 CHANGE OF OCCUPANCY 10-1
Section

1001	General	10-1
1002	Special Use and Occupancy	10-1
1003	Building Elements and Materials	10-2
1004	Fire Protection	10-2
1005	Means of Egress	10-2
1006	Structural	10-2
1007	Electrical	10-2
1008	Mechanical	10-3
1009	Plumbing	10-3
1010	Other Requirements	10-3
1011	Change of Occupancy Classification	10-3

CHAPTER 11 ADDITIONS 11-1
Section

1101	General	11-1
1102	Heights and Areas	11-1
1103	Structural	11-1
1104	Energy Conservation	11-2

CHAPTER 12 HISTORIC BUILDINGS 12-1
Section

1201	General	12-1
1202	Repairs	12-1
1203	Fire Safety	12-2
1204	Change of Occupancy	12-2
1205	Structural	12-3
1206	Relocated Buildings	12-3

CHAPTER 13 PERFORMANCE COMPLIANCE METHODS 13-1
Section

1301	General	13-1

CHAPTER 14 RELOCATED OR MOVED BUILDINGS 14-1
Section

1401	General	14-1
1402	Requirements	14-1

CHAPTER 15 CONSTRUCTION SAFEGUARDS 15-1
Section

1501	General	15-1
1502	Protection of Adjoining Property	15-2
1503	Temporary Use of Streets, Alleys and Public Property	15-3
1504	Fire Extinguishers	15-3
1505	Means of Egress	15-3
1506	Standpipes	15-3
1507	Automatic Sprinkler System	15-4
1508	Accessibility	15-4
1509	Water Supply for Fire Protection	15-4

CHAPTER 16 REFERENCED STANDARDS 16-1

TABLE OF CONTENTS

APPENDIX A GUIDELINES FOR THE SEISMIC RETROFIT OF EXISTING BUILDINGS .... APPENDIX A-1

CHAPTER A1 SEISMIC STRENGTHENING PROVISIONS FOR UNREINFORCED MASONRY BEARING WALL BUILDINGS .... APPENDIX A-1

Section
A101 Purpose .... APPENDIX A-1
A102 Scope .... APPENDIX A-1
A103 Definitions .... APPENDIX A-1
A104 Symbols and Notations .... APPENDIX A-2
A105 General Requirements .... APPENDIX A-3
A106 Materials Requirements .... APPENDIX A-4
A107 Quality Control .... APPENDIX A-5
A108 Design Strengths .... APPENDIX A-6
A109 Analysis and Design Procedure ... APPENDIX A-6
A110 General Procedure .... APPENDIX A-7
A111 Special Procedure .... APPENDIX A-8
A112 Analysis and Design .... APPENDIX A-11
A113 Detailed Building System Design Requirements .... APPENDIX A-13
A114 Walls of Unburned Clay, Adobe or Stone Masonry ..... APPENDIX A-14

CHAPTER A2 EARTHQUAKE HAZARD REDUCTION IN EXISTING REINFORCED CONCRETE AND REINFORCED MASONRY WALL BUILDINGS WITH FLEXIBLE DIAPHRAGMS ...... APPENDIX A-15

Section
A201 Purpose .... APPENDIX A-15
A202 Scope .... APPENDIX A-15
A203 Definitions .... APPENDIX A-15
A204 Symbols and Notations .... APPENDIX A-15
A205 General Requirements .... APPENDIX A-15
A206 Analysis and Design .... APPENDIX A-16
A207 Materials of Construction ...... APPENDIX A-17

CHAPTER A3 PRESCRIPTIVE PROVISIONS FOR SEISMIC STRENGTHENING OF CRIPPLE WALLS AND SILL PLATE ANCHORAGE OF LIGHT, WOOD-FRAME RESIDENTIAL BUILDINGS .... APPENDIX A-19

Section
A301 General .... APPENDIX A-19
A302 Definitions .... APPENDIX A-19
A303 Structural Weaknesses .... APPENDIX A-20
A304 Strengthening Requirements ..... APPENDIX A-20

CHAPTER A4 EARTHQUAKE RISK REDUCTION IN WOOD-FRAME RESIDENTIAL BUILDINGS WITH SOFT, WEAK OR OPEN FRONT WALLS ..... APPENDIX A-37

Section
A401 General .... APPENDIX A-37
A402 Definitions .... APPENDIX A-37
A403 Analysis and Design .... APPENDIX A-37
A404 Prescriptive Measures for Weak Story .... APPENDIX A-39
A405 Materials of Construction ...... APPENDIX A-40
A406 Construction Documents .... APPENDIX A-40
A407 Quality Control .... APPENDIX A-41

CHAPTER A5 REFERENCED STANDARDS .... APPENDIX A-43

Section
A501 Referenced Standards .... APPENDIX A-43

APPENDIX B SUPPLEMENTARY ACCESSIBILITY REQUIREMENTS FOR EXISTING BUILDINGS AND FACILITIES .... APPENDIX B-1

Section
B101 Qualified Historic Buildings and Facilities .... APPENDIX B-1
B102 Fixed Transportation Facilities and Stations .... APPENDIX B-1
B103 Dwelling Units and Sleeping Units .... APPENDIX B-2
B104 Referenced Standards .... APPENDIX B-2

TABLE OF CONTENTS

APPENDIX C GUIDELINES FOR THE WIND RETROFIT OF EXISTING BUILDINGS .... APPENDIX C-1

CHAPTER C1 GABLE END RETROFIT FOR HIGH-WIND AREAS .... APPENDIX C-1

Section

C101 General .... APPENDIX C-1
C102 Definitions.................. APPENDIX C-1
C103 Materials of Construction .... APPENDIX C-2
C104 Retrofitting Gable End Walls to Enhance Wind Resistance ...... APPENDIX C-4

CHAPTER C2 ROOF DECK FASTENING FOR HIGH-WIND AREAS .... APPENDIX C-21

Section

C201 General .... APPENDIX C-21
C202 Roof Deck Attachment for Wood Roofs............. APPENDIX C-21

CHAPTER C3 REFERENCED STANDARDS .... APPENDIX C-23

Section

C301 Referenced Standards .... APPENDIX C-23

APPENDIX D BOARD OF APPEALS .... APPENDIX D-1

Section

D101 General .... APPENDIX D-1

RESOURCE A GUIDELINES ON FIRE RATINGS OF ARCHAIC MATERIALS AND ASSEMBLIES .... RESOURCE A-1

Section

1 Fire-related Performance of Archaic Materials and Assemblies ...... RESOURCE A-2
2 Building Evaluation............. RESOURCE A-3
3 Final Evaluation and Design Solution .... RESOURCE A-6
4 Summary .... RESOURCE A-13
   Appendix .... RESOURCE A-15
   Resource A Table of Contents.... RESOURCE A-15
   Bibliography .... RESOURCE A-134

INDEX INDEX-1

CHAPTER 1
SCOPE AND ADMINISTRATION

User note:

About this chapter: *Chapter 1 establishes the limits of applicability of the code and describes how the code is to be applied and enforced. Chapter 1 is in two parts: Part 1—Scope and Administration (Sections 101–102) and Part 2—Administration and Enforcement (Sections 103–117). Section 101 identifies which buildings and structures come under its purview and references other I-Codes® as applicable.*

This code is intended to be adopted as a legally enforceable document, and it cannot be effective without adequate provisions for its administration and enforcement. The provisions of Chapter 1 establish the authority and duties of the code official appointed by the authority having jurisdiction and also establish the rights and privileges of the registered design professional, contractor and property owner.

PART 1—SCOPE AND APPLICATION

SECTION 101
SCOPE AND GENERAL REQUIREMENTS

[A] 101.1 Title. These regulations shall be known as the *Existing Building Code* of **[NAME OF JURISDICTION]**, hereinafter referred to as "this code."

[A] 101.2 Scope. The provisions of this code shall apply to the *repair*, *alteration*, *change of occupancy*, *addition* to and relocation of *existing buildings*.

> **Exception:** Detached one- and two-family dwellings and townhouses not more than three stories above grade plane in height with a separate means of egress, and their accessory structures not more than three stories above grade plane in height, shall comply with this code or the *International Residential Code*.

101.2.1 Application of fire code. Where work regulated by this code is also regulated by the construction requirements for *existing buildings* in Chapter 11 of the *International Fire Code,* such work shall comply with applicable requirements in both codes.

[A] 101.3 Purpose. The intent of this code is to provide flexibility to permit the use of alternative approaches to achieve compliance with minimum requirements to provide a reasonable level of safety, health, property protection and general welfare insofar as they are affected by the *repair*, *alteration*, *change of occupancy*, *addition* and relocation of *existing buildings*.

[A] 101.4 Applicability. This code shall apply to the *repair*, *alteration*, *change of occupancy*, *addition* and relocation of *existing buildings*, regardless of occupancy, subject to the criteria of Sections 101.4.1 and 101.4.2.

> **[A] 101.4.1 Buildings not previously occupied.** A building or portion of a building that has not been previously occupied or used for its intended purpose, in accordance with the laws in existence at the time of its completion, shall be permitted to comply with the provisions of the laws in existence at the time of its original permit unless such permit has expired. Subsequent permits shall comply with the *International Building Code* or *International Residential Code*, as applicable, for new construction.

[A] 101.4.2 Buildings previously occupied. The legal occupancy of any building existing on the date of adoption of this code shall be permitted to continue without change, except as is specifically covered in this code, the *International Fire Code*, or the *International Property Maintenance Code*, or as is deemed necessary by the *code official* for the general safety and welfare of the occupants and the public.

[A] 101.5 Safeguards during construction. Construction work covered in this code, including any related demolition, shall comply with the requirements of Chapter 15.

[A] 101.6 Appendices. The *code official* is authorized to require retrofit of buildings, structures or individual structural members in accordance with the appendices of this code if such appendices have been individually adopted.

[A] 101.7 Correction of violations of other codes. *Repairs* or *alterations* mandated by any property, housing or fire safety maintenance code, or mandated by any licensing rule or ordinance adopted pursuant to law, shall conform only to the requirements of that code, rule or ordinance, and shall not be required to conform to this code unless the code requiring such *repair* or *alteration* so provides.

SECTION 102
APPLICABILITY

[A] 102.1 General. Where there is a conflict between a general requirement and a specific requirement, the specific requirement shall be applicable. Where, in any specific case, different sections of this code specify different materials, methods of construction or other requirements, the most restrictive shall govern.

[A] 102.2 Other laws. The provisions of this code shall not be deemed to nullify any provisions of local, state or federal law.

[A] 102.3 Application of references. References to chapter or section numbers or to provisions not specifically identified by number shall be construed to refer to such chapter, section or provision of this code.

[A] 102.4 Referenced codes and standards. The codes and standards referenced in this code shall be considered part of the requirements of this code to the prescribed extent of each such reference and as further regulated in Sections 102.4.1 and 102.4.2.

> **Exception:** Where enforcement of a code provision would violate the conditions of the listing of the equipment or appliance, the conditions of the listing shall govern.

[A] 102.4.1 Conflicts. Where conflicts occur between provisions of this code and referenced codes and standards, the provisions of this code shall apply.

[A] 102.4.2 Conflicting provisions. Where the extent of the reference to a referenced code or standard includes subject matter that is within the scope of this code, the provisions of this code, as applicable, shall take precedence over the provisions in the referenced code or standard.

[A] 102.5 Partial invalidity. In the event that any part or provision of this code is held to be illegal or void, this shall not have the effect of making void or illegal any of the other parts or provisions.

PART 2—ADMINISTRATION AND ENFORCEMENT

SECTION 103
CODE COMPLIANCE AGENCY

[A] 103.1 Creation of agency. The [INSERT NAME OF DEPARTMENT] is hereby created, and the official in charge thereof shall be known as the *code official*. The function of the agency shall be the implementation, administration and enforcement of the provisions of this code.

[A] 103.2 Appointment. The *code official* shall be appointed by the chief appointing authority of the jurisdiction.

[A] 103.3 Deputies. In accordance with the prescribed procedures of this jurisdiction and with the concurrence of the appointing authority, the *code official* shall have the authority to appoint a deputy *code official*, other related technical officers, inspectors and other employees. Such employees shall have powers as delegated by the *code official*.

SECTION 104
DUTIES AND POWERS OF CODE OFFICIAL

[A] 104.1 General. The *code official* is hereby authorized and directed to enforce the provisions of this code. The *code official* shall have the authority to render interpretations of this code and to adopt policies and procedures in order to clarify the application of its provisions. Such interpretations, policies and procedures shall be in compliance with the intent and purpose of this code. Such policies and procedures shall not have the effect of waiving requirements specifically provided for in this code.

[A] 104.2 Applications and permits. The *code official* shall receive applications, review construction documents and issue permits for the *repair, alteration, addition*, demolition, *change of occupancy* and relocation of buildings; inspect the premises for which such permits have been issued; and enforce compliance with the provisions of this code.

[A] 104.2.1 Determination of substantially improved or substantially damaged *existing buildings* and structures in flood hazard areas. For applications for reconstruction, rehabilitation, repair, *alteration, addition* or other improvement of *existing buildings* or structures located in *flood hazard areas*, the building official shall determine where the proposed work constitutes *substantial improvement* or *repair* of *substantial damage*. Where the building official determines that the proposed work constitutes *substantial improvement* or *repair* of *substantial damage*, and where required by this code, the building official shall require the building to meet the requirements of Section 1612 of the *International Building Code*, or Section R322 of the *International Residential Code*, as applicable.

[A] 104.2.2 Preliminary meeting. When requested by the permit applicant or the *code official*, the *code official* shall meet with the permit applicant prior to the application for a construction permit to discuss plans for the proposed work or *change of occupancy* in order to establish the specific applicability of the provisions of this code.

> **Exception:** *Repairs* and Level 1 *alterations*.

[A] 104.2.2.1 Building evaluation. The *code official* is authorized to require an *existing building* to be investigated and evaluated by a registered design professional based on the circumstances agreed on at the preliminary meeting. The design professional shall notify the *code official* if any potential noncompliance with the provisions of this code is identified.

[A] 104.3 Notices and orders. The *code official* shall issue necessary notices or orders to ensure compliance with this code.

[A] 104.4 Inspections. The *code official* shall make the required inspections, or the *code official* shall have the authority to accept reports of inspection by *approved* agencies or individuals. Reports of such inspections shall be in writing and be certified by a responsible officer of such *approved* agency or by the responsible individual. The *code official* is authorized to engage such expert opinion as deemed necessary to report on unusual technical issues that arise, subject to the approval of the appointing authority.

[A] 104.5 Identification. The *code official* shall carry proper identification when inspecting structures or premises in the performance of duties under this code.

[A] 104.6 Right of entry. Where it is necessary to make an inspection to enforce the provisions of this code, or where the *code official* has reasonable cause to believe that there exists in a structure or on a premises a condition that is contrary to or in violation of this code that makes the structure or premises *unsafe, dangerous* or hazardous, the *code official* is authorized to enter the structure or premises at reasonable times to inspect or to perform the duties imposed by this code, provided that if such structure or premises be occupied that credentials be presented to the occupant and entry requested. If such structure or premises be unoccupied, the *code official* shall first make a reasonable effort to locate the owner, the owner's authorized agent or other person having charge or control of the structure or premises and request entry. If entry is refused, the *code official* shall have recourse to the remedies provided by law to secure entry.

[A] 104.7 Department records. The *code official* shall keep official records of applications received, permits and certificates issued, fees collected, reports of inspections, and notices and orders issued. Such records shall be retained in the official records for the period required for retention of public records.

[A] 104.8 Liability. The *code official*, member of the Board of Appeals or employee charged with the enforcement of this code, while acting for the jurisdiction in good faith and without malice in the discharge of the duties required by this code or other pertinent law or ordinance, shall not thereby be rendered civilly or criminally liable personally and is hereby relieved from personal liability for any damage accruing to persons or property as a result of any act or by reason of an act or omission in the discharge of official duties.

> **[A] 104.8.1 Legal defense.** Any suit or criminal complaint instituted against an officer or employee because of an act performed by that officer or employee in the lawful discharge of duties and under the provisions of this code shall be defended by legal representatives of the jurisdiction until the final termination of the proceedings. The *code official* or any subordinate shall not be liable for cost in any action, suit or proceeding that is instituted in pursuance of the provisions of this code.

[A] 104.9 Approved materials and equipment. Materials, equipment and devices *approved* by the *code official* shall be constructed and installed in accordance with such approval.

> **[A] 104.9.1 Used materials and equipment.** The use of used materials that meet the requirements of this code for new materials is permitted. Used equipment and devices shall be permitted to be reused subject to the approval of the *code official*.

[A] 104.10 Modifications. Wherever there are practical difficulties involved in carrying out the provisions of this code, the *code official* shall have the authority to grant modifications for individual cases on application of the owner or owner's authorized representative, provided that the *code official* shall first find that special individual reason makes the strict letter of this code impractical, the modification is in compliance with the intent and purpose of this code and such modification does not lessen health, accessibility, life and fire safety, or structural requirements. The details of action granting modifications shall be recorded and entered in the files of the Department of Building Safety.

> **[A] 104.10.1 Flood hazard areas.** For *existing buildings* located in *flood hazard areas* for which *repairs*, *alterations* and *additions* constitute *substantial improvement*, the *code official* shall not grant modifications to provisions related to flood resistance unless a determination is made that:
>
> 1. The applicant has presented good and sufficient cause that the unique characteristics of the size, configuration or topography of the site render compliance with the flood-resistant construction provisions inappropriate.
> 2. Failure to grant the modification would result in exceptional hardship.
> 3. The granting of the modification will not result in increased flood heights, additional threats to public safety, extraordinary public expense nor create nuisances, cause fraud on or victimization of the public, or conflict with existing laws or ordinances.
> 4. The modification is the minimum necessary to afford relief, considering the flood hazard.
> 5. A written notice will be provided to the applicant specifying, if applicable, the difference between the design flood elevation and the elevation to which the building is to be built, stating that the cost of flood insurance will be commensurate with the increased risk resulting from the reduced floor elevation and that construction below the design flood elevation increases risks to life and property.

[A] 104.11 Alternative materials, design and methods of construction, and equipment. The provisions of this code are not intended to prevent the installation of any material or to prohibit any design or method of construction not specifically prescribed by this code, provided that any such alternative has been *approved*. An alternative material, design or method of construction shall be *approved* where the *code official* finds that the proposed design is satisfactory and complies with the intent of the provisions of this code, and that the material, method or work offered is, for the purpose intended, not less than the equivalent of that prescribed in this code in quality, strength, effectiveness, fire resistance, durability and safety. Where the alternative material, design or method of construction is not *approved*, the *code official* shall respond in writing, stating the reasons why the alternative was not *approved*.

> **[A] 104.11.1 Research reports.** Supporting data, where necessary to assist in the approval of materials or assemblies not specifically provided for in this code, shall consist of valid research reports from *approved* sources.

> **[A] 104.11.2 Tests.** Where there is insufficient evidence of compliance with the provisions of this code or evidence that a material or method does not conform to the requirements of this code, or in order to substantiate claims for alternative materials or methods, the *code official* shall have the authority to require tests as evidence of compliance to be made without expense to the jurisdiction. Test methods shall be as specified in this code or by other recognized test standards. In the absence of recognized and accepted test methods, the *code official* shall approve the testing procedures. Tests shall be performed by an *approved* agency. Reports of such tests shall be

SECTION 105
PERMITS

[A] 105.1 Required. Any owner or owner's authorized agent who intends to *repair*, add to, alter, relocate, demolish or change the occupancy of a building or to *repair*, install, add, alter, remove, convert or replace any electrical, gas, mechanical or plumbing system, the installation of which is regulated by this code, or to cause any such work to be performed, shall first make application to the *code official* and obtain the required permit.

[A] 105.1.1 Annual permit. Instead of an individual permit for each *alteration* to an already *approved* electrical, gas, mechanical, or plumbing installation, the *code official* is authorized to issue an annual permit on application therefor to any person, firm or corporation regularly employing one or more qualified trade persons in the building, structure, or on the premises owned or operated by the applicant for the permit.

[A] 105.1.2 Annual permit records. The person to whom an annual permit is issued shall keep a detailed record of *alterations* made under such annual permit. The *code official* shall have access to such records at all times, or such records shall be filed with the *code official* as designated.

[A] 105.2 Work exempt from permit. Exemptions from permit requirements of this code shall not be deemed to grant authorization for any work to be done in any manner in violation of the provisions of this code or any other laws or ordinances of this jurisdiction. Permits shall not be required for the following:

Building:
1. Sidewalks and driveways not more than 30 inches (762 mm) above grade and not over any basement or story below and that are not part of an accessible route.
2. Painting, papering, tiling, carpeting, cabinets, counter tops and similar finish work.
3. Temporary motion picture, television, and theater stage sets and scenery.
4. Shade cloth structures constructed for nursery or agricultural purposes, and not including service systems.
5. Window awnings supported by an exterior wall of Group R-3 or Group U occupancies.
6. Nonfixed and movable cases, counters and partitions not over 5 feet 9 inches (1753 mm) in height.

Electrical:
1. **Repairs and maintenance:** Minor *repair* work, including the replacement of lamps or the connection of *approved* portable electrical equipment to *approved* permanently installed receptacles.
2. **Radio and television transmitting stations:** The provisions of this code shall not apply to electrical equipment used for radio and television transmissions, but do apply to equipment and wiring for power supply, the installations of towers and antennas.
3. **Temporary testing systems:** A permit shall not be required for the installation of any temporary system required for the testing or servicing of electrical equipment or apparatus.

Gas:
1. Portable heating appliance.
2. Replacement of any minor part that does not alter approval of equipment or make such equipment unsafe.

Mechanical:
1. Portable heating appliance.
2. Portable ventilation equipment.
3. Portable cooling unit.
4. Steam, hot or chilled water piping within any heating or cooling equipment regulated by this code.
5. Replacement of any part that does not alter its approval or make it unsafe.
6. Portable evaporative cooler.
7. Self-contained refrigeration system containing 10 pounds (4.54 kg) or less of refrigerant and actuated by motors of 1 horsepower (746 W) or less.

Plumbing:
1. The stopping of leaks in drains, water, soil, waste or vent pipe; provided, however, that if any concealed trap, drainpipe, water, soil, waste or vent pipe becomes defective and it becomes necessary to remove and replace the same with new material, such work shall be considered as new work, and a permit shall be obtained and inspection made as provided in this code.
2. The clearing of stoppages or the repairing of leaks in pipes, valves or fixtures, and the removal and reinstallation of water closets, provided that such *repairs* do not involve or require the replacement or rearrangement of valves, pipes or fixtures.

[A] 105.2.1 Emergency repairs. Where equipment replacements and *repairs* must be performed in an emergency situation, the permit application shall be submitted within the next working business day to the *code official*.

[A] 105.2.2 Repairs. Application or notice to the *code official* is not required for *repairs* to structures and items listed in Section 105.2 provided that such *repairs* do not include any of the following:

1. The cutting away of any wall, partition or portion thereof.
2. The removal or cutting of any structural beam or load-bearing support.

3. The removal or change of any required means of egress or rearrangement of parts of a structure affecting the egress requirements.

4. Any *addition* to, *alteration* of, replacement or relocation of any standpipe, water supply, sewer, drainage, drain leader, gas, soil, waste, vent or similar piping, or electric wiring.

5. Mechanical or other work affecting public health or general safety.

[A] 105.2.3 Public service agencies. A permit shall not be required for the installation, *alteration* or *repair* of generation, transmission, distribution or metering, or other related equipment that is under the ownership and control of public service agencies by established right.

[A] 105.3 Application for permit. To obtain a permit, the applicant shall first file an application therefor in writing on a form furnished by the Department of Building Safety for that purpose. Such application shall:

1. Identify and describe the work in accordance with Chapter 3 to be covered by the permit for which application is made.

2. Describe the land on which the proposed work is to be done by legal description, street address or similar description that will readily identify and definitely locate the proposed building or work.

3. Indicate the use and occupancy for which the proposed work is intended.

4. Be accompanied by construction documents and other information as required in Section 106.3.

5. State the valuation of the proposed work.

6. Be signed by the applicant or the applicant's authorized agent.

7. Give such other data and information as required by the *code official*.

[A] 105.3.1 Action on application. The *code official* shall examine or cause to be examined applications for permits and amendments thereto within a reasonable time after filing. If the application or the construction documents do not conform to the requirements of pertinent laws, the *code official* shall reject such application in writing, stating the reasons therefor. If the *code official* is satisfied that the proposed work conforms to the requirements of this code and laws and ordinances applicable thereto, the *code official* shall issue a permit therefor as soon as practicable.

[A] 105.3.2 Time limitation of application. An application for a permit for any proposed work shall be deemed to have been abandoned 180 days after the date of filing, unless such application has been pursued in good faith or a permit has been issued; except that the *code official* is authorized to grant one or more extensions of time for additional periods not exceeding 90 days each. The extension shall be requested in writing and justifiable cause demonstrated.

[A] 105.4 Validity of permit. The issuance or granting of a permit shall not be construed to be a permit for, or an approval of, any violation of any of the provisions of this code or of any other ordinance of the jurisdiction. Permits presuming to give authority to violate or cancel the provisions of this code or other ordinances of the jurisdiction shall not be valid. The issuance of a permit based on construction documents and other data shall not prevent the *code official* from requiring the correction of errors in the construction documents and other data. The *code official* is authorized to prevent occupancy or use of a structure where in violation of this code or of any other ordinances of this jurisdiction.

[A] 105.5 Expiration. Every permit issued shall become invalid unless the work on the site authorized by such permit is commenced within 180 days after its issuance, or if the work authorized on the site by such permit is suspended or abandoned for a period of 180 days after the time the work is commenced. The *code official* is authorized to grant, in writing, one or more extensions of time for periods not more than 180 days each. The extension shall be requested in writing and justifiable cause demonstrated.

[A] 105.6 Suspension or revocation. The *code official* is authorized to suspend or revoke a permit issued under the provisions of this code wherever the permit is issued in error or on the basis of incorrect, inaccurate or incomplete information, or in violation of any ordinance or regulation or any of the provisions of this code.

[A] 105.7 Placement of permit. The building permit or copy shall be kept on the site of the work until the completion of the project.

SECTION 106
CONSTRUCTION DOCUMENTS

[A] 106.1 General. Submittal documents consisting of construction documents, special inspection and structural observation programs, investigation and evaluation reports, and other data shall be submitted in two or more sets, or in a digital format where allowed by the *code official,* with each application for a permit. The construction documents shall be prepared by a registered design professional where required by the statutes of the jurisdiction in which the project is to be constructed. Where special conditions exist, the *code official* is authorized to require additional construction documents to be prepared by a registered design professional.

> **Exception:** The *code official* is authorized to waive the submission of construction documents and other data not required to be prepared by a registered design professional if it is found that the nature of the work applied for is such that reviewing of construction documents is not necessary to obtain compliance with this code.

[A] 106.2 Construction documents. Construction documents shall be in accordance with Sections 106.2.1 through 106.2.6.

> **[A] 106.2.1 Construction documents.** Construction documents shall be dimensioned and drawn on suitable material. Electronic media documents are permitted to be submitted where *approved* by the *code official*. Construction documents shall be of sufficient clarity to indicate the

location, nature and extent of the work proposed and show in detail that it will conform to the provisions of this code and relevant laws, ordinances, rules and regulations, as determined by the *code official*. The *work areas* shall be shown.

[A] 106.2.2 Fire protection system(s) shop drawings. Shop drawings for the fire protection system(s) shall be submitted to indicate compliance with this code and the construction documents and shall be *approved* prior to the start of system installation. Shop drawings shall contain information as required by the referenced installation standards in Chapter 9 of the *International Building Code*.

[A] 106.2.3 Means of egress. The construction documents for *Alterations*—Level 2, *Alterations*—Level 3, *additions* and *changes of occupancy* shall show in sufficient detail the location, construction, size and character of all portions of the means of egress in compliance with the provisions of this code. The construction documents shall designate the number of occupants to be accommodated in every *work area* of every floor and in all affected rooms and spaces.

[A] 106.2.4 Exterior wall envelope. Construction documents for work affecting the *exterior wall envelope* shall describe the *exterior wall envelope* in sufficient detail to determine compliance with this code. The construction documents shall provide details of the *exterior wall envelope* as required, including windows, doors, flashing, intersections with dissimilar materials, corners, end details, control joints, intersections at roof, eaves or parapets, means of drainage, water-resistive barriers and details around openings.

The construction documents shall include manufacturer's installation instructions that provide supporting documentation that the proposed penetration and opening details described in the construction documents maintain the wind and weather resistance of the *exterior wall envelope*. The supporting documentation shall fully describe the exterior wall system that was tested, where applicable, as well as the test procedure used.

[A] 106.2.5 Exterior balconies and elevated walking surfaces. Where the scope of work involves balconies or other elevated walking surfaces have weather-exposed surfaces, and the structural framing is protected by an impervious moisture barrier, the construction documents shall include details for all elements of the impervious moisture barrier system. The construction documents shall include manufacturer's installation instructions.

[A] 106.2.6 Site plan. The construction documents submitted with the application for permit shall be accompanied by a site plan showing to scale the size and location of new construction and *existing structures* on the site, distances from lot lines, the established street grades, and the proposed finished grades; and it shall be drawn in accordance with an accurate boundary line survey. In the case of demolition, the site plan shall show construction to be demolished and the location and size of *existing structures* and construction that are to remain on the site or plot. The *code official* is authorized to waive or modify the requirement for a site plan where the application for permit is for *alteration*, *repair* or *change of occupancy*.

[A] 106.3 Examination of documents. The *code official* shall examine or cause to be examined the submittal documents and shall ascertain by such examinations whether the construction or occupancy indicated and described is in accordance with the requirements of this code and other pertinent laws or ordinances.

[A] 106.3.1 Approval of construction documents. Where the *code official* issues a permit, the construction documents shall be *approved* in writing or by stamp as "Reviewed for Code Compliance." One set of construction documents so reviewed shall be retained by the *code official*. The other set shall be returned to the applicant, shall be kept at the site of work, and shall be open to inspection by the *code official* or a duly authorized representative.

[A] 106.3.2 Previous approval. This code shall not require changes in the construction documents, construction or designated occupancy of a structure for which a lawful permit has been issued and the construction of which has been pursued in good faith within 180 days after the effective date of this code and has not been abandoned.

[A] 106.3.3 Phased approval. The *code official* is authorized to issue a permit for the construction of foundations or any other part of a building before the construction documents for the whole building or structure have been submitted, provided that adequate information and detailed statements have been filed complying with pertinent requirements of this code. The holder of such permit for the foundation or other parts of a building shall proceed at the holder's own risk with the building operation and without assurance that a permit for the entire structure will be granted.

[A] 106.3.4 Deferred submittals. Deferral of any submittal items shall have the prior approval of the *code official*. The *registered design professional in responsible charge* shall list the *deferred submittals* on the construction documents for review by the *code official*.

Submittal documents for *deferred submittal* items shall be submitted to the *registered design professional in responsible charge* who shall review them and forward them to the *code official* with a notation indicating that the *deferred submittal* documents have been reviewed and that they have been found to be in general conformance to the design of the building. The *deferred submittal* items shall not be installed until their *deferred submittal* documents have been *approved* by the *code official*.

[A] 106.4 Amended construction documents. Work shall be installed in accordance with the reviewed construction documents, and any changes made during construction that are not in compliance with the *approved* construction documents shall be resubmitted for approval as an amended set of construction documents.

[A] 106.5 Retention of construction documents. One set of *approved* construction documents shall be retained by the *code official* for a period of not less than the period required for retention of public records.

[A] 106.6 Design professional in responsible charge. Where it is required that documents be prepared by a registered design professional, the *code official* shall be authorized to require the owner or the owner's authorized agent to engage and designate on the building permit application a registered design professional who shall act as the *registered design professional in responsible charge*. If the circumstances require, the owner or the owner's authorized agent shall designate a substitute *registered design professional in responsible charge* who shall perform the duties required of the original *registered design professional in responsible charge*. The *code official* shall be notified in writing by the owner or the owner's authorized agent if the *registered design professional in responsible charge* is changed or is unable to continue to perform the duties. The *registered design professional in responsible charge* shall be responsible for reviewing and coordinating submittal documents prepared by others, including phased and *deferred submittal* items, for compatibility with the design of the building. Where structural observation is required, the inspection program shall name the individual or firms who are to perform structural observation and describe the stages of construction at which structural observation is to occur.

SECTION 107
TEMPORARY STRUCTURES AND USES

[A] 107.1 General. The *code official* is authorized to issue a permit for temporary uses. Such permits shall be limited as to time of service but shall not be permitted for more than 180 days. The *code official* is authorized to grant extensions for demonstrated cause.

[A] 107.2 Conformance. Temporary uses shall conform to the structural strength, fire safety, means of egress, accessibility, light, ventilation and sanitary requirements of this code as necessary to ensure the public health, safety and general welfare.

[A] 107.3 Temporary power. The *code official* is authorized to give permission to temporarily supply and use power in part of an electric installation before such installation has been fully completed and the final certificate of completion has been issued. The part covered by the temporary certificate shall comply with the requirements specified for temporary lighting, heat or power in NFPA 70.

[A] 107.4 Termination of approval. The *code official* is authorized to terminate such permit for a temporary use and to order the temporary use to be discontinued.

SECTION 108
FEES

[A] 108.1 Payment of fees. A permit shall not be valid until the fees prescribed by law have been paid, nor shall an amendment to a permit be released until the additional fee, if any, has been paid.

[A] 108.2 Schedule of permit fees. Where a permit is required, a fee for each permit shall be paid as required, in accordance with the schedule as established by the applicable governing authority.

[A] 108.3 Permit valuations. The applicant for a permit shall provide an estimated permit value at time of application. Permit valuations shall include total value of work, including materials and labor for which the permit is being issued, such as electrical, gas, mechanical, plumbing equipment and permanent systems. If, in the opinion of the *code official*, the valuation is underestimated on the application, the permit shall be denied unless the applicant can show detailed estimates to meet the approval of the *code official*. Final building permit valuation shall be set by the *code official*.

[A] 108.4 Work commencing before permit issuance. Any person who commences any work before obtaining the necessary permits shall be subject to a fee established by the *code official* that shall be in addition to the required permit fees.

[A] 108.5 Related fees. The payment of the fee for the construction, *alteration*, removal or demolition of work done in connection to or concurrently with the work authorized by a permit shall not relieve the applicant or holder of the permit from the payment of other fees that are prescribed by law.

[A] 108.6 Refunds. The *code official* is authorized to establish a refund policy.

SECTION 109
INSPECTIONS

[A] 109.1 General. Construction or work for which a permit is required shall be subject to inspection by the *code official*, and such construction or work shall remain visible and able to be accessed for inspection purposes until *approved*. Approval as a result of an inspection shall not be construed to be an approval of a violation of the provisions of this code or of other ordinances of the jurisdiction. Inspections presuming to give authority to violate or cancel the provisions of this code or of other ordinances of the jurisdiction shall not be valid. It shall be the duty of the permit applicant to cause the work to remain visible and able to be accessed for inspection purposes. Neither the *code official* nor the jurisdiction shall be liable for expense entailed in the removal or replacement of any material required to allow inspection.

[A] 109.2 Preliminary inspection. Before issuing a permit, the *code official* is authorized to examine or cause to be examined buildings and sites for which an application has been filed.

[A] 109.3 Required inspections. The *code official*, on notification, shall make the inspections set forth in Sections 109.3.1 through 109.3.11.

[A] 109.3.1 Footing or foundation inspection. Footing and foundation inspections shall be made after excavations for footings are complete and any required reinforcing steel is in place. For concrete foundations, any required forms shall be in place prior to inspection. Materials for the foundation shall be on the job, except where concrete is ready-mixed in accordance with ASTM C94, the concrete need not be on the job.

[A] 109.3.2 Concrete slab or under-floor inspection. Concrete slab and under-floor inspections shall be made after in-slab or under-floor reinforcing steel and building service equipment, conduit, piping accessories and other ancillary equipment items are in place but before any concrete is placed or floor sheathing installed, including the subfloor.

[A] 109.3.3 Lowest floor elevation. For *additions* and *substantial improvements* to *existing buildings* in *flood hazard areas*, on placement of the lowest floor, including basement, and prior to further vertical construction, the elevation documentation required in the *International Building Code*, or the *International Residential Code*, as applicable, shall be submitted to the *code official*.

[A] 109.3.4 Frame inspection. Framing inspections shall be made after the roof deck or sheathing, framing, fire blocking and bracing are in place and pipes, chimneys and vents to be concealed are complete and the rough electrical, plumbing, heating wires, pipes and ducts are *approved*.

[A] 109.3.5 Lath or gypsum board inspection. Lath and gypsum board inspections shall be made after lathing and gypsum board, interior and exterior, is in place but before any plastering is applied or before gypsum board joints and fasteners are taped and finished.

> **Exception:** Gypsum board that is not part of a fire-resistance-rated assembly or a shear assembly.

[A] 109.3.6 Weather-exposed balcony and walking surface waterproofing. Where the scope of work involves balconies or other elevated walking surfaces that have weather-exposed surfaces and the structural framing is protected by an impervious moisture barrier, all elements of the impervious moisture barrier system shall not be concealed until inspected and *approved*.

> **Exception:** Where special inspections are provided in accordance with Section 1705.1.1, Item 3, of the *International Building Code*.

[A] 109.3.7 Fire- and smoke-resistant penetrations. Protection of joints and penetrations in fire-resistance-rated assemblies, smoke barriers and smoke partitions shall not be concealed from view until inspected and *approved*.

[A] 109.3.8 Other inspections. In addition to the inspections specified in Sections 109.2 through 109.3.7, the *code official* is authorized to make or require other inspections of any construction work to ascertain compliance with the provisions of this code and other laws that are enforced by the Department of Building Safety.

[A] 109.3.9 Special inspections. Special inspections shall be required in accordance with the *International Building Code*.

[A] 109.3.10 Flood hazard documentation. Where a building is located in a *flood hazard area*, documentation of the elevation of the lowest floor as required in the *International Building Code* or the *International Residential Code*, as applicable, shall be submitted to the *code official* prior to the final inspection.

[A] 109.3.11 Final inspection. The final inspection shall be made after work required by the building permit is completed.

[A] 109.4 Inspection agencies. The *code official* is authorized to accept reports of *approved* inspection agencies, provided that such agencies satisfy the requirements as to qualifications and reliability.

[A] 109.5 Inspection requests. It shall be the duty of the holder of the building permit or their duly authorized agent to notify the *code official* when work is ready for inspection. It shall be the duty of the permit holder to provide access to and means for any inspections of such work that are required by this code.

[A] 109.6 Approval required. Work shall not be done beyond the point indicated in each successive inspection without first obtaining the approval of the *code official*. The *code official*, on notification, shall make the requested inspections and shall either indicate the portion of the construction that is satisfactory as completed or shall notify the permit holder or an agent of the permit holder wherein the same fails to comply with this code. Any portions that do not comply shall be corrected and such portion shall not be covered or concealed until authorized by the *code official*.

SECTION 110
CERTIFICATE OF OCCUPANCY

[A] 110.1 Change of occupancy. A structure shall not be used or occupied in whole or in part, and a *change of occupancy* of a structure or portion thereof shall not be made until the *code official* has issued a certificate of occupancy therefor as provided herein. Issuance of a certificate of occupancy shall not be construed as an approval of a violation of the provisions of this code or of other ordinances of the jurisdiction. Certificates presuming to give authority to violate or cancel the provisions of this code or other ordinances of the jurisdiction shall not be valid.

> **Exception:** Certificates of occupancy are not required for work exempt from permits in accordance with Section 105.2.

[A] 110.2 Certificate issued. After the *code official* inspects the structure and does not find violations of the provisions of this code or other laws that are enforced by the department, the *code official* shall issue a certificate of occupancy that contains the following:

1. The permit number.
2. The address of the structure.

3. The name and address of the owner or the owner's authorized agent.
4. A description of that portion of the structure for which the certificate is issued.
5. A statement that the described portion of the structure has been inspected for compliance with the requirements of this code for the occupancy and division of occupancy and the use for which the proposed occupancy is classified.
6. The name of the *code official*.
7. The edition of the code under which the permit was issued.
8. The use and occupancy in accordance with the provisions of the *International Building Code*.
9. The type of construction as defined in the *International Building Code*.
10. The design occupant load and any impact the *alteration* has on the design occupant load of the area not within the scope of the work.
11. Where an automatic sprinkler system is provided, and whether an automatic sprinkler system is required.
12. Any special stipulations and conditions of the building permit.

[A] 110.3 Temporary occupancy. The *code official* is authorized to issue a temporary certificate of occupancy before the completion of the entire work covered by the permit, provided that such portion or portions shall be occupied safely. The *code official* shall set a time period during which the temporary certificate of occupancy is valid.

[A] 110.4 Revocation. The *code official* is authorized to suspend or revoke a certificate of occupancy or completion issued under the provisions of this code, in writing, wherever the certificate is issued in error or on the basis of incorrect information supplied, or where it is determined that the building or structure or portion thereof is in violation of the provisions of this code or other ordinance of the jurisdiction.

SECTION 111
SERVICE UTILITIES

[A] 111.1 Connection of service utilities. A person shall not make connections from a utility, source of energy, fuel, power, water system or sewer system to any building or system that is regulated by this code for which a permit is required, until *approved* by the *code official*.

[A] 111.2 Temporary connection. The *code official* shall have the authority to authorize the temporary connection of the building or system to the utility, source of energy, fuel, power, water system or sewer system for the purpose of testing systems or for use under a temporary approval.

[A] 111.3 Authority to disconnect service utilities. The *code official* shall have the authority to authorize disconnection of utility service to the building, structure or system regulated by this code and the referenced codes and standards in case of emergency where necessary to eliminate an immediate hazard to life or property or where such utility connection has been made without the approval required by Section 111.1 or 111.2. The *code official* shall notify the serving utility and, wherever possible, the owner or the owner's authorized agent and the occupant of the building, structure or service system of the decision to disconnect prior to taking such action. If not notified prior to disconnecting, the owner, the owner's authorized agent or occupant of the building, structure or service system shall be notified in writing, as soon as practical thereafter.

SECTION 112
MEANS OF APPEALS

[A] 112.1 General. In order to hear and decide appeals of orders, decisions or determinations made by the *code official* relative to the application and interpretation of this code, there shall be and is hereby created a board of appeals. The board of appeals shall be appointed by the applicable governing authority and shall hold office at its pleasure. The board shall adopt rules of procedure for conducting its business and shall render all decisions and findings in writing to the appellant with a duplicate copy to the *code official*.

[A] 112.2 Limitations on authority. An application for appeal shall be based on a claim that the true intent of this code or the rules legally adopted thereunder have been incorrectly interpreted, the provisions of this code do not fully apply or an equivalent or better form of construction is proposed. The board shall not have authority to waive requirements of this code or interpret the administration of this code.

[A] 112.3 Qualifications. The board of appeals shall consist of members who are qualified by experience and training to pass on matters pertaining to building construction and are not employees of the jurisdiction.

[A] 112.4 Administration. The *code official* shall take immediate action in accordance with the decision of the board.

SECTION 113
VIOLATIONS

[A] 113.1 Unlawful acts. It shall be unlawful for any person, firm or corporation to *repair*, alter, extend, add, move, remove, demolish or change the occupancy of any building or equipment regulated by this code or cause same to be done in conflict with or in violation of any of the provisions of this code.

[A] 113.2 Notice of violation. The *code official* is authorized to serve a notice of violation or order on the person responsible for the *repair*, *alteration*, extension, *addition*, moving, removal, demolition or change in the occupancy of a building in violation of the provisions of this code or in violation of a permit or certificate issued under the provisions of this code. Such order shall direct the discontinuance of the illegal action or condition and the abatement of the violation.

[A] 113.3 Prosecution of violation. If the notice of violation is not complied with promptly, the *code official* is authorized to request the legal counsel of the jurisdiction to institute the appropriate proceeding at law or in equity to restrain, correct or abate such violation or to require the removal or termination of the unlawful occupancy of the building or structure in violation of the provisions of this code or of the order or direction made pursuant thereto.

[A] 113.4 Violation penalties. Any person who violates a provision of this code or fails to comply with any of the requirements thereof or who *repairs* or alters or changes the occupancy of a building or structure in violation of the approved construction documents or directive of the *code official* or of a permit or certificate issued under the provisions of this code shall be subject to penalties as prescribed by law.

SECTION 114
STOP WORK ORDER

[A] 114.1 Authority. Where the *code official* finds any work regulated by this code being performed in a manner contrary to the provisions of this code or in a *dangerous* or *unsafe* manner, the *code official* is authorized to issue a stop work order.

[A] 114.2 Issuance. The stop work order shall be in writing and shall be given to the owner of the property, the owner's authorized agent or the person performing the work. Upon issuance of a stop work order, the cited work shall immediately cease. The stop work order shall state the reason for the order and the conditions under which the cited work is authorized to resume.

[A] 114.3 Emergencies. Where an emergency exists, the *code official* shall not be required to give a written notice prior to stopping the work.

[A] 114.4 Failure to comply. Any person who shall continue any work after having been served with a stop work order, except such work as that person is directed to perform to remove a violation or *unsafe* condition, shall be subject to fines established by the authority having jurisdiction.

SECTION 115
UNSAFE STRUCTURES AND EQUIPMENT

[A] 115.1 Unsafe conditions. Structures or existing equipment that are or hereafter become *unsafe*, insanitary or deficient because of inadequate means of egress facilities, inadequate light and ventilation, or that constitute a fire hazard, or are otherwise dangerous to human life or the public welfare, or that involve illegal or improper occupancy or inadequate maintenance, shall be deemed an *unsafe* condition. *Unsafe* structures shall be taken down and removed or made safe as the *code official* deems necessary and as provided for in this code. A vacant structure that is not secured against unauthorized entry shall be deemed *unsafe*.

[A] 115.2 Record. The *code official* shall cause a report to be filed on an *unsafe* condition. The report shall state the occupancy of the structure and the nature of the *unsafe* condition.

[A] 115.3 Notice. If an *unsafe* condition is found, the *code official* shall serve on the owner of the structure or the owner's authorized agent a written notice that describes the condition deemed *unsafe* and specifies the required *repairs* or improvements to be made to abate the *unsafe* condition, or that requires the *unsafe* building to be demolished within a stipulated time. Such notice shall require the person thus notified to declare immediately to the *code official* acceptance or rejection of the terms of the order.

[A] 115.4 Method of service. Such notice shall be deemed properly served where a copy thereof is served in accordance with one of the following methods:

1. A copy is delivered to the owner or the owner's authorized agent personally.
2. A copy is sent by certified or registered mail addressed to the owner at the last known address with the return receipt requested.
3. A copy is delivered in any other manner as prescribed by local law.

If the certified or registered letter is returned showing that the letter was not delivered, a copy thereof shall be posted in a conspicuous place in or about the structure affected by such notice. Service of such notice in the foregoing manner on the owner's authorized agent shall constitute service of notice on the owner.

[A] 115.5 Restoration or abatement. The structure or equipment determined to be *unsafe* by the *code official* is permitted to be restored to a safe condition. The owner, the owner's authorized agent, operator or occupant of a structure, premises or equipment deemed *unsafe* by the *code official* shall abate or cause to be abated or corrected such *unsafe* conditions either by *repair*, rehabilitation, demolition or other *approved* corrective action. To the extent that *repairs*, *alterations* or *additions* are made, or a *change of occupancy* occurs during the restoration of the structure, such *repairs*, *alterations*, *additions* or *change of occupancy* shall comply with the requirements of this code.

SECTION 116
EMERGENCY MEASURES

[A] 116.1 Imminent danger. Where, in the opinion of the *code official*, there is imminent danger of failure or collapse of a building that endangers life, or where any building or part of a building has fallen and life is endangered by the occupation of the building, or where there is actual or potential danger to the building occupants or those in the proximity of any structure because of explosives, explosive fumes or vapors, or the presence of toxic fumes, gases, or materials, or operation of defective or dangerous equipment, the *code official* is hereby authorized and empowered to order and require the occupants to vacate the premises forthwith. The *code official* shall cause to be posted at each entrance to such structure a notice reading as follows: "This Structure Is Unsafe and Its Occupancy Has Been Prohibited by the Code Official." It shall be unlawful for any person to enter such structure except for the purpose of securing the structure, making the required *repairs*, removing the hazardous condition, or of demolishing the same.

[A] 116.2 Temporary safeguards. Notwithstanding other provisions of this code, whenever, in the opinion of the *code official*, there is imminent danger due to an *unsafe* condition, the *code official* shall order the necessary work to be done, including the boarding up of openings, to render such structure temporarily safe whether or not the legal procedure herein described has been instituted; and shall cause such other action to be taken as the *code official* deems necessary to meet such emergency.

[A] 116.3 Closing streets. Where necessary for public safety, the *code official* shall temporarily close structures and close or order the authority having jurisdiction to close sidewalks, streets, public ways and places adjacent to *unsafe* structures, and prohibit the same from being utilized.

[A] 116.4 Emergency repairs. For the purposes of this section, the *code official* shall employ the necessary labor and materials to perform the required work as expeditiously as possible.

[A] 116.5 Costs of emergency repairs. Costs incurred in the performance of emergency work shall be paid by the jurisdiction. The legal counsel of the jurisdiction shall institute appropriate action against the owner of the premises or the owner's authorized agent where the *unsafe* structure is or was located for the recovery of such costs.

[A] 116.6 Hearing. Any person ordered to take emergency measures shall comply with such order forthwith. Any affected person shall thereafter, on petition directed to the appeals board, be afforded a hearing as described in this code.

SECTION 117
DEMOLITION

[A] 117.1 General. The *code official* shall order the owner or owner's authorized agent of any premises on which is located any structure that in the *code official's* judgment is so old or dilapidated, or has become so out of *repair* as to be *dangerous*, *unsafe*, insanitary or otherwise unfit for human habitation of occupancy, and such that it is unreasonable to *repair* the structure, to demolish and remove such structure; or if such structure is capable of being made safe by *repairs*, to *repair* and make safe and sanitary or to demolish and remove to the owner's or the owner's authorized agent's option; or where there has been a cessation of normal construction of any structure for a period of more than two years, to demolish and remove such structure.

[A] 117.2 Notices and orders. Notices and orders shall comply with Section 113.

[A] 117.3 Failure to comply. If the owner or the owner's authorized agent of a premises fails to comply with a demolition order within the time prescribed, the *code official* shall cause the structure to be demolished and removed, either through an available public agency or by contract or arrangement with private persons, and the cost of such demolition and removal shall be charged against the real estate on which the structure is located and shall be a lien on such real estate.

[A] 117.4 Salvage materials. Where any structure has been ordered demolished and removed, the governing body or other designated officer under said contract or arrangement aforesaid shall have the right to sell the salvage and valuable materials at the highest price obtainable. The net proceeds of such sale, after deducting the expenses of such demolition and removal, shall be promptly remitted with a report of such sale or transaction, including the items of expense and the amounts deducted, for the person who is entitled thereto, subject to any order of a court. If such a surplus does not remain to be turned over, the report shall so state.

CHAPTER 2
DEFINITIONS

User note:

About this chapter: Codes, by their very nature, are technical documents. Every word, term and punctuation mark can add to or change the meaning of a technical requirement. It is necessary to maintain a consensus on the specific meaning of each term contained in the code. Chapter 2 performs this function by stating clearly what specific terms mean for the purpose of the code.

SECTION 201
GENERAL

201.1 Scope. Unless otherwise expressly stated, the following words and terms shall, for the purposes of this code, have the meanings shown in this chapter.

201.2 Interchangeability. Words used in the present tense include the future; words stated in the masculine gender include the feminine and neuter; the singular number includes the plural and the plural, the singular.

201.3 Terms defined in other codes. Where terms are not defined in this code and are defined in the other International Codes, such terms shall have the meanings ascribed to them in those codes.

201.4 Terms not defined. Where terms are not defined through the methods authorized by this chapter, such terms shall have ordinarily accepted meanings such as the context implies.

SECTION 202
GENERAL DEFINITIONS

[A] ADDITION. An extension or increase in floor area, number of stories, or height of a building or structure.

[A] ALTERATION. Any construction or renovation to an *existing structure* other than a *repair* or *addition*.

[A] APPROVED. Acceptable to the *code official*.

[A] BUILDING. Any structure utilized or intended for supporting or sheltering any occupancy.

[A] CHANGE OF OCCUPANCY. Any of the following shall be considered as a change of occupancy where the current *International Building Code* requires a greater degree of safety, accessibility, structural strength, fire protection, means of egress, ventilation or sanitation than is existing in the current building or structure:

1. Any change in the occupancy classification of a building or structure.
2. Any change in the purpose of, or a change in the level of activity within, a building or structure.
3. A change of use.

[A] CHANGE OF USE. A change in the use of a building or a portion of a building, within the same group classification, for which there is a change in application of the code requirements.

[A] CODE OFFICIAL. The officer or other designated authority charged with the administration and enforcement of this code.

[BS] DANGEROUS. Any building, structure or portion thereof that meets any of the conditions described below shall be deemed dangerous:

1. The building or structure has collapsed, has partially collapsed, has moved off its foundation or lacks the necessary support of the ground.
2. There exists a significant risk of collapse, detachment or dislodgement of any portion, member, appurtenance or ornamentation of the building or structure under permanent, routine or frequent loads; under actual loads already in effect; or under snow, wind, rain, flood, earthquake or other environmental loads when such loads are imminent.

[A] DEFERRED SUBMITTAL. Those portions of the design that are not submitted at the time of the application and that are to be submitted to the *code official* within a specified period.

[BS] DISPROPORTIONATE EARTHQUAKE DAMAGE. A condition of earthquake-related damage where both of the following occur:

1. The 0.3-second spectral acceleration at the building site as estimated by the United States Geological Survey for the earthquake in question is less than 40 percent of the mapped acceleration parameter SS.
2. The vertical elements of the lateral force-resisting system have suffered damage such that the lateral load-carrying capacity of any story in any horizontal direction has been reduced by more than 10 percent from its predamage condition.

[BE] EMERGENCY ESCAPE AND RESCUE OPENING. An operable exterior window, door or other similar device that provides for a means of escape and access for rescue in the event of an emergency.

EQUIPMENT OR FIXTURE. Any plumbing, heating, electrical, ventilating, air conditioning, refrigerating and fire protection equipment; and elevators, dumbwaiters, escalators, boilers, pressure vessels and other mechanical facilities; or installations that are related to building services. Equipment or fixture shall not include manufacturing, production or process equipment, but shall include connections from building service to process equipment.

[A] EXISTING BUILDING. A building erected prior to the date of adoption of the appropriate code, or one for which a legal building permit has been issued.

[A] EXISTING STRUCTURE. A structure erected prior to the date of adoption of the appropriate code, or one for which a legal building permit has been issued.

[BF] EXTERIOR WALL COVERING. A material or assembly of materials applied on the exterior side of exterior walls for the purpose of providing a weather-resisting barrier, insulation or for aesthetics, including but not limited to, veneers, siding, exterior insulation and finish systems, architectural trim and embellishments, such as cornices, soffits, facias, gutters and leaders.

[BF] EXTERIOR WALL ENVELOPE. A system or assembly of exterior wall components, including exterior wall finish materials, that provides protection of the building structural members, including framing and sheathing materials, and conditioned interior space from the detrimental effects of the exterior environment.

[A] FACILITY. All or any portion of buildings, structures, site improvements, elements and pedestrian or vehicular routes located on a site.

[BS] FLOOD HAZARD AREA. The greater of the following two areas:

1. The area within a flood plain subject to a 1-percent or greater chance of flooding in any year.
2. The area designated as a *flood hazard area* on a community's flood hazard map, or otherwise legally designated.

[A] HISTORIC BUILDING. Any building or structure that is one or more of the following:

1. Listed, or certified as eligible for listing, by the State Historic Preservation Officer or the Keeper of the National Register of Historic Places, in the National Register of Historic Places.
2. Designated as historic under an applicable state or local law.
3. Certified as a contributing resource within a National Register, state designated or locally designated historic district.

[BF] NONCOMBUSTIBLE MATERIAL. A material that, under the conditions anticipated, will not ignite or burn when subjected to fire or heat. Materials that pass ASTM E136 are considered *noncombustible materials*.

PRIMARY FUNCTION. A *primary function* is a major activity for which the *facility* is intended. Areas that contain a *primary function* include, but are not limited to, the customer services lobby of a bank, the dining area of a cafeteria, the meeting rooms in a conference center, as well as offices and other work areas in which the activities of the public accommodation or other private entity using the *facility* are carried out. Mechanical rooms, boiler rooms, supply storage rooms, employee lounges or locker rooms, janitorial closets, entrances, corridors and restrooms are not areas containing a *primary function*.

[A] REGISTERED DESIGN PROFESSIONAL IN RESPONSIBLE CHARGE. A registered design professional engaged by the owner or the owner's authorized agent to review and coordinate certain aspects of the project, as determined by the *code official*, for compatibility with the design of the building or structure, including submittal documents prepared by others, *deferred submittal* documents and phased submittal documents.

REHABILITATION. Any work, as described by the categories of work defined herein, undertaken in an *existing building*.

[A] RELOCATABLE BUILDING. A partially or completely assembled building constructed and designed to be reused multiple times and transported to different building sites.

[A] REPAIR. The reconstruction, replacement or renewal of any part of an *existing building* for the purpose of its maintenance or to correct damage.

[BS] REROOFING. The process of recovering or replacing an existing roof covering. See "*Roof recover*" and "*Roof replacement*."

[BS] RISK CATEGORY. A categorization of buildings and other structures for determination of flood, wind, snow, ice and earthquake loads based on the risk associated with unacceptable performance, as provided in Section 1604.5 of the *International Building Code*.

[BS] ROOF COATING. A fluid-applied adhered coating used for roof maintenance, *roof repair* or as a component of a roof covering system or roof assembly.

[BS] ROOF RECOVER. The process of installing an additional roof covering over a prepared existing roof covering without removing the existing roof covering.

[BS] ROOF REPAIR. Reconstruction or renewal of any part of an existing roof for the purpose of correcting damage or restoring the predamage condition.

[BS] ROOF REPLACEMENT. The process of removing the existing roof covering, repairing any damaged substrate and installing a new roof covering.

[BS] SEISMIC FORCES. The loads, forces and requirements prescribed herein, related to the response of the building to earthquake motions, to be used in the analysis and design of the structure and its components. Seismic forces are considered either full or reduced, as provided in Chapter 3.

[BS] SUBSTANTIAL DAMAGE. For the purpose of determining compliance with the flood provisions of this code, damage of any origin sustained by a structure whereby the cost of restoring the structure to its before-damaged condition would equal or exceed 50 percent of the market value of the structure before the damage occurred.

[BS] SUBSTANTIAL IMPROVEMENT. For the purpose of determining compliance with the flood provisions of this code, any *repair*, *alteration*, *addition* or improvement of a building or structure, the cost of which equals or exceeds 50 percent of the market value of the structure, before the improvement or *repair* is started. If the structure has sustained *substantial damage*, any *repairs* are considered *substantial improvement* regardless of the actual *repair* work performed. The term does not, however, include either of the following:

1. Any project for improvement of a building required to correct existing health, sanitary or safety code violations identified by the *code official* and that is the minimum necessary to ensure safe living conditions.

2. Any *alteration* of a historic structure, provided that the *alteration* will not preclude the structure's continued designation as a historic structure.

[BS] SUBSTANTIAL STRUCTURAL ALTERATION. An *alteration* in which the gravity load-carrying structural elements altered within a 5-year period support more than 30 percent of the total floor and roof area of the building or structure. The areas to be counted toward the 30 percent shall include mezzanines, penthouses, and in-filled courts and shafts tributary to the altered structural elements.

[BS] SUBSTANTIAL STRUCTURAL DAMAGE. A condition where any of the following apply:

1. The vertical elements of the lateral force-resisting system have suffered damage such that the lateral load-carrying capacity of any story in any horizontal direction has been reduced by more than 33 percent from its predamage condition.

2. The capacity of any vertical component carrying gravity load, or any group of such components, that has a tributary area more than 30 percent of the total area of the structure's floor(s) and roof(s) has been reduced more than 20 percent from its predamage condition, and the remaining capacity of such affected elements, with respect to all dead and live loads, is less than 75 percent of that required by the *International Building Code* for new buildings of similar structure, purpose and location.

3. The capacity of any structural component carrying snow load, or any group of such components, that supports more than 30 percent of the roof area of similar construction has been reduced more than 20 percent from its predamage condition, and the remaining capacity with respect to dead, live and snow loads is less than 75 percent of that required by the *International Building Code* for new buildings of similar structure, purpose and location.

TECHNICALLY INFEASIBLE. An *alteration* of a *facility* that has little likelihood of being accomplished because the existing structural conditions require the removal or *alteration* of a load-bearing member that is an essential part of the structural frame, or because other existing physical or site constraints prohibit modification or addition of elements, spaces or features which are in full and strict compliance with the minimum requirements for new construction and which are necessary to provide accessibility.

UNSAFE. Buildings, structures or equipment that are unsanitary, or that are deficient due to inadequate means of egress *facilities*, inadequate light and ventilation, or that constitute a fire hazard, or in which the structure or individual structural members meet the definition of *"Dangerous,"* or that are otherwise dangerous to human life or the public welfare, or that involve illegal or improper occupancy or inadequate maintenance shall be deemed *unsafe*. A vacant structure that is not secured against entry shall be deemed *unsafe*.

WORK AREA. That portion or portions of a building consisting of all reconfigured spaces as indicated on the construction documents. Work area excludes other portions of the building where incidental work entailed by the intended work must be performed and portions of the building where work not initially intended by the owner is specifically required by this code.

CHAPTER 3

PROVISIONS FOR ALL COMPLIANCE METHODS

User note:

About this chapter: Chapter 3 explains the three compliance options for alterations and additions available in the code. In addition, this chapter also lays out the methods to be used for seismic design and evaluation throughout this code. Finally, this chapter clarifies that provisions in other I-Codes® related to repairs, alterations, additions, relocation and changes of occupancy must also be addressed unless they conflict with this code. In that case, this code takes precedence.

SECTION 301
ADMINISTRATION

301.1 Applicability. The *repair, alteration, change of occupancy, addition* or relocation of all *existing buildings* shall comply with Section 301.2, 301.3 or 301.4. The provisions of Sections 302 through 309 shall apply to all *alterations, repairs, additions,* relocation of structures and *changes of occupancy* regardless of compliance method.

301.1.1 Bleachers, grandstands and folding and telescopic seating. Existing bleachers, grandstands and folding and telescopic seating shall comply with ICC 300.

301.2 Repairs. *Repairs* shall comply with the requirements of Chapter 4.

301.3 Alteration, addition or change of occupancy. The *alteration, addition* or *change of occupancy* of all *existing buildings* shall comply with one of the methods listed in Section 301.3.1, 301.3.2 or 301.3.3 as selected by the applicant. Sections 301.3.1 through 301.3.3 shall not be applied in combination with each other.

> **Exception:** Subject to the approval of the *code official*, *alterations* complying with the laws in existence at the time the building or the affected portion of the building was built shall be considered in compliance with the provisions of this code. New structural members added as part of the *alteration* shall comply with the *International Building Code*. This exception shall not apply to the following:
>
> 1. *Alterations* for accessibility required by Section 306.
> 2. *Alterations* that constitute *substantial improvement* in *flood hazard areas*, which shall comply with Sections 503.2, 701.3 or 1301.3.3.
> 3. Structural provisions of Section 304, Chapter 5 or to the structural provisions of Sections 706, 805 and 906.

301.3.1 Prescriptive compliance method. *Alterations, additions* and *changes of occupancy* complying with Chapter 5 of this code in buildings complying with the *International Fire Code* shall be considered in compliance with the provisions of this code.

301.3.2 Work area compliance method. *Alterations, additions* and *changes of occupancy* complying with the applicable requirements of Chapters 6 through 12 of this code shall be considered in compliance with the provisions of this code.

301.3.3 Performance compliance method. *Alterations, additions* and *changes of occupancy* complying with Chapter 13 of this code shall be considered in compliance with the provisions of this code.

301.4 Relocated buildings. Relocated buildings shall comply with the requirements of Chapter 14.

SECTION 302
GENERAL PROVISIONS

302.1 Dangerous conditions. The *code official* shall have the authority to require the elimination of conditions deemed *dangerous*.

302.2 Additional codes. *Alterations, repairs, additions* and *changes of occupancy* to, or relocation of, *existing buildings* and structures shall comply with the provisions for *alterations, repairs, additions* and *changes of occupancy* or relocation, respectively, in this code and the *International Energy Conservation Code, International Fire Code, International Fuel Gas Code, International Mechanical Code, International Plumbing Code, International Private Sewage Disposal Code, International Property Maintenance Code, International Residential Code* and NFPA 70. Where provisions of the other codes conflict with provisions of this code, the provisions of this code shall take precedence.

302.2.1 Additional codes in health care. In existing Group I-2 occupancies, ambulatory health care *facilities*, outpatient clinics and hyperbaric *facilities*, *alterations*, *repairs, additions* and *changes of occupancy* to, or relocation of, *existing buildings* and structures shall also comply with NFPA 99.

302.3 Existing materials. Materials already in use in a building in compliance with requirements or approvals in effect at the time of their erection or installation shall be permitted to remain in use unless determined by the code official to be *unsafe*.

302.4 New and replacement materials. Except as otherwise required or permitted by this code, materials permitted by the applicable code for new construction shall be used. Like materials shall be permitted for *repairs* and *alterations*, provided that *unsafe* conditions are not created. Hazardous materials shall not be used where the code for new construction would not permit their use in buildings of similar occupancy, purpose and location.

[BS] 302.4.1 New structural members and connections. New structural members and connections shall comply with the detailing provisions of the *International Building Code* for new buildings of similar structure, purpose and location.

> **Exception:** Where alternative design criteria are specifically permitted.

302.5 Occupancy and use. Where determining the appropriate application of the referenced sections of this code, the occupancy and use of a building shall be determined in accordance with Chapter 3 of the *International Building Code*.

SECTION 303
STORM SHELTERS

303.1 Storm shelters. This section applies to the construction of storm shelters constructed as rooms or spaces within *existing buildings* for the purpose of providing protection during storms that produce high winds, such as tornados and hurricanes. Such structures shall be designated to be hurricane shelters, tornado shelters, or combined hurricane and tornado shelters. Such structures shall be constructed in accordance with this code and ICC 500.

303.2 Addition to a Group E occupancy. Where an *addition* is added to an existing Group E occupancy located in an area where the shelter design wind speed for tornados is 250 mph (402.3 km/h) in accordance with Figure 304.2(1) of ICC 500 and the occupant load in the *addition* is 50 or more, the *addition* shall have a storm shelter constructed in accordance with ICC 500.

> **Exceptions:**
> 1. Group E day care *facilities*.
> 2. Group E occupancies accessory to places of religious worship.
> 3. *Additions* meeting the requirements for shelter design in ICC 500.

303.2.1 Required occupant capacity. The required occupant capacity of the storm shelter shall include all buildings on the site, and shall be the total occupant load of the classrooms, vocational rooms and offices in the Group E occupancy.

> **Exceptions:**
> 1. Where an *addition* is being added on an existing Group E site, and where the *addition* is not of sufficient size to accommodate the required occupant capacity of the storm shelter for all of the buildings on-site, the storm shelter shall at a minimum accommodate the required capacity for the *addition*.
> 2. Where *approved* by the *code official*, the required occupant capacity of the shelter shall be permitted to be reduced by the occupant capacity of any existing storm shelters on the site.

303.2.2 Occupancy classification. The occupancy classification for storm shelters shall be determined in accordance with Section 423.3 of the *International Building Code*.

SECTION 304
STRUCTURAL DESIGN LOADS AND EVALUATION AND DESIGN PROCEDURES

[BS] 304.1 Live loads. Where an *addition* or *alteration* does not result in increased design live load, existing gravity load-carrying structural elements shall be permitted to be evaluated and designed for live loads *approved* prior to the *addition* or *alteration*. If the *approved* live load is less than that required by Section 1607 of the *International Building Code*, the area designated for the nonconforming live load shall be posted with placards of *approved* design indicating the *approved* live load. Where the *addition* or *alteration* results in increased design live load, the live load required by Section 1607 of the *International Building Code* shall be used.

[BS] 304.2 Snow loads on adjacent buildings. Where an *alteration* or *addition* changes the potential snow drift effects on an adjacent building, the *code official* is authorized to enforce Section 7.12 of ASCE 7.

[BS] 304.3 Seismic evaluation and design procedures. Where required, seismic evaluation or design shall be based on the procedures and criteria in this section, regardless of which compliance method is used.

[BS] 304.3.1 Compliance with full seismic forces. Where compliance requires the use of full seismic forces, the criteria shall be in accordance with one of the following:

1. One-hundred percent of the values in the *International Building Code*. Where the existing seismic force-resisting system is a type that can be designated as "Ordinary," values of R, Ω_0 and C_d used for analysis in accordance with Chapter 16 of the *International Building Code* shall be those specified for structural systems classified as "Ordinary" in accordance with Table 12.2-1 of ASCE 7, unless it can be demonstrated that the structural system will provide performance equivalent to that of a "Detailed," "Intermediate" or "Special" system.
2. ASCE 41, using a Tier 3 procedure and the two-level performance objective in Table 304.3.1 for the applicable *risk category*.

PROVISIONS FOR ALL COMPLIANCE METHODS

[BS] TABLE 304.3.1
PERFORMANCE OBJECTIVES FOR USE IN ASCE 41 FOR COMPLIANCE WITH FULL SEISMIC FORCES

RISK CATEGORY (Based on IBC Table 1604.5)	STRUCTURAL PERFORMANCE LEVEL FOR USE WITH BSE-1N EARTHQUAKE HAZARD LEVEL	STRUCTURAL PERFORMANCE LEVEL FOR USE WITH BSE-2N EARTHQUAKE HAZARD LEVEL
I	Life Safety (S-3)	Collapse Prevention (S-5)
II	Life Safety (S-3)	Collapse Prevention (S-5)
III	Damage Control (S-2)	Limited Safety (S-4)
IV	Immediate Occupancy (S-1)	Life Safety (S-3)

[BS] TABLE 304.3.2
PERFORMANCE OBJECTIVES FOR USE IN ASCE 41 FOR COMPLIANCE WITH REDUCED SEISMIC FORCES

RISK CATEGORY (Based on IBC Table 1604.5)	STRUCTURAL PERFORMANCE LEVEL FOR USE WITH BSE-1E EARTHQUAKE HAZARD LEVEL	STRUCTURAL PERFORMANCE LEVEL FOR USE WITH BSE-2E EARTHQUAKE HAZARD LEVEL
I	Life Safety (S-3). See Note a	Collapse Prevention (S-5)
II	Life Safety (S-3). See Note a	Collapse Prevention (S-5)
III	Damage Control (S-2). See Note a	Limited Safety (S-4). See Note b
IV	Immediate Occupancy (S-1)	Life Safety (S-3). See Note c

a. For Risk Categories I, II and III, the Tier 1 and Tier 2 procedures need not be considered for the BSE-1E earthquake hazard level.
b. For Risk Category III, the Tier 1 screening checklists shall be based on the Collapse Prevention, except that checklist statements using the Quick Check provisions shall be based on *MS*-factors that are the average of the values for Collapse Prevention and Life Safety.
c. For Risk Category IV, the Tier 1 screening checklists shall be based on Collapse Prevention, except that checklist statements using the Quick Check provisions shall be based on *MS*-factors for Life Safety.

[BS] 304.3.2 Compliance with reduced seismic forces. Where seismic evaluation and design is permitted to use reduced seismic forces, the criteria used shall be in accordance with one of the following:

1. The *International Building Code* using 75 percent of the prescribed forces. Values of R, Ω_0 and C_d used for analysis shall be as specified in Section 304.3.1 of this code.

2. Structures or portions of structures that comply with the requirements of the applicable chapter in Appendix A as specified in Items 2.1 through 2.4 and subject to the limitations of the respective Appendix A chapters shall be deemed to comply with this section.

    2.1. The seismic evaluation and design of unreinforced masonry bearing wall buildings in *Risk Category* I or II are permitted to be based on the procedures specified in Appendix Chapter A1.

    2.2. Seismic evaluation and design of the wall anchorage system in reinforced concrete and reinforced masonry wall buildings with flexible diaphragms in *Risk Category* I or II are permitted to be based on the procedures specified in Chapter A2.

    2.3. Seismic evaluation and design of cripple walls and sill plate anchorage in residential buildings of light-frame wood construction in *Risk Category* I or II are permitted to be based on the procedures specified in Chapter A3.

    2.4. Seismic evaluation and design of soft, weak or open-front wall conditions in multiple-unit residential buildings of wood construction in *Risk Category* I or II are permitted to be based on the procedures specified in Chapter A4.

3. ASCE 41, using the performance objective in Table 304.3.2 for the applicable *risk category*.

SECTION 305
IN-SITU LOAD TESTS

[BS] 305.1 General. Where used, in-situ load tests shall be conducted in accordance with Section 1708 of the *International Building Code*.

SECTION 306
ACCESSIBILITY FOR EXISTING BUILDINGS

306.1 Scope. The provisions of Sections 306.1 through 306.7.16 apply to maintenance and *repair*, *change of occupancy*, *additions* and *alterations* to *existing buildings*, including those identified as *historic buildings*.

306.2 Design. Buildings and *facilities* shall be designed and constructed to be accessible in accordance with this code and the *alteration* and *existing building* provisions in ICC A117.1, as applicable.

306.3 Maintenance and repair. A *facility* that is constructed or altered to be accessible shall be maintained accessible during occupancy. Required accessible means of egress shall be maintained during construction, demolition, remodeling or *alterations* and *additions* to any occupied building.

Exception: Existing means of egress need not be maintained where *approved* temporary means of egress and accessible means of egress systems and *facilities* are provided.

306.3.1 Prohibited reduction in accessibility. An *alteration* that decreases or has the effect of decreasing accessibility of a building, *facility* or element, thereof, below the requirements for new construction at the time of the *alteration* is prohibited. The number of accessible elements need not exceed that required for new construction at the time of *alteration*.

306.4 Extent of application. An *alteration* of an existing *facility* shall not impose a requirement for greater accessibility than that which would be required for new construction.

306.5 Change of occupancy. *Existing buildings* that undergo a change of group or occupancy shall comply with Section 306.7.

> **Exception:** Type B dwelling or sleeping units required by Section 1108 of the *International Building Code* are not required to be provided in *existing buildings* and *facilities* undergoing a *change of occupancy* in conjunction with *alterations* where the *work area* is 50 percent or less of the aggregate area of the building.

306.6 Additions. Provisions for new construction shall apply to *additions*. An *addition* that affects the accessibility to, or contains an area of, a *primary function* shall comply with the requirements in Section 306.7.1.

306.7 Alterations. A *facility* that is altered shall comply with the applicable provisions in Chapter 11 of the *International Building Code,* ICC A117.1 and the provisions of Sections 306.7.1 through 306.7.16, unless *technically infeasible*. Where compliance with this section is *technically infeasible*, the *alteration* shall provide access to the maximum extent technically feasible.

> **306.7.1 Alterations affecting an area containing a primary function.** Where an *alteration* affects the accessibility to, or contains an area of *primary function*, the route to the *primary function* area shall be accessible. The accessible route to the *primary function* area shall include toilet *facilities* and drinking fountains serving the area of *primary function*.
>
> **Exceptions:**
> 1. The costs of providing the accessible route are not required to exceed 20 percent of the costs of the *alterations* affecting the area of *primary function*.
> 2. This provision does not apply to *alterations* limited solely to windows, hardware, operating controls, electrical outlets and signs.
> 3. This provision does not apply to *alterations* limited solely to mechanical systems, electrical systems, installation or *alteration* of fire protection systems and abatement of hazardous materials.
> 4. This provision does not apply to *alterations* undertaken for the primary purpose of increasing the accessibility of a *facility*.
> 5. This provision does not apply to altered areas limited to Type B dwelling and sleeping units.

306.7.2 Accessible means of egress. Accessible means of egress required by Chapter 10 of the *International Building Code* are not required to be added in existing *facilities*.

306.7.3 Alteration of Type A units. The *alteration* to Type A individually owned dwelling units within a Group R-2 occupancy shall be permitted to meet the provision for a Type B dwelling unit.

306.7.4 Type B units. Type B dwelling or sleeping units required by Section 1108 of the International Building Code are not required to be provided in *existing buildings* and *facilities* undergoing *alterations* where the *work area* is 50 percent or less of the aggregate area of the building.

306.7.5 Entrances. Where an *alteration* includes *alterations* to an entrance that is not accessible, and the *facility* has an accessible entrance, the altered entrance is not required to be accessible unless required by Section 306.7.1. Signs complying with Section 1112 of the *International Building Code* shall be provided.

306.7.6 Accessible route. Exterior accessible routes, including curb ramps, shall be not less than 36 inches (914 mm) minimum in width.

306.7.7 Elevators. Altered elements of existing elevators shall comply with ASME A17.1. Such elements shall also be altered in elevators programmed to respond to the same hall call control as the altered elevator.

306.7.8 Platform lifts. Platform (wheelchair) lifts installed in accordance with ASME A18.1 shall be permitted as a component of an accessible route.

306.7.9 Stairways and escalators in existing buildings. Where an escalator or stairway is added where none existed previously and major structural modifications are necessary for installation, an accessible route complying with Section 1104.4 of the *International Building Code* is required between levels served by such escalator or stairway.

306.7.10 Determination of number of units. Where Chapter 11 of the *International Building Code* requires Accessible, Type A or Type B units and where such units are being altered or added, the number of Accessible, Type A and Type B units shall be determined in accordance with Sections 306.7.10.1 through 306.7.10.3.

> **306.7.10.1 Accessible dwelling or sleeping units.** Where Group I-1, I-2, I-3, R-1, R-2 or R-4 dwelling or sleeping units are being altered or added, the requirements of Section 1108 of the *International Building Code* for Accessible units apply only to the quantity of spaces being altered or added.
>
> **306.7.10.2 Type A dwelling or sleeping units.** Where more than 20 Group R-2 dwelling or sleeping units are being altered or added, the requirements of Section 1108 of the *International Building Code* for Type A units apply only to the quantity of the spaces being altered or added.
>
> **306.7.10.3 Type B dwelling or sleeping units.** Where four or more Group I-1, I-2, R-1, R-2, R-3 or R-4 dwelling or sleeping units are being added, the requirements of Section 1108 of the *International Building Code* for Type B units apply only to the quantity of the spaces being added. Where Group I-1, I-2, R-1, R-2, R-3 or R-4 dwelling or sleeping units are being altered

and where the *work area* is greater than 50 percent of the aggregate area of the building, the requirements of Section 1108 of the *International Building Code* for Type B units apply only to the quantity of the spaces being altered.

306.7.11 Toilet rooms. Where it is *technically infeasible* to alter existing toilet rooms to be accessible, one accessible single-user toilet room or one accessible family or assisted-use toilet room constructed in accordance with Section 1110.2.1 of the *International Building Code* is permitted. This toilet room shall be located on the same floor and in the same area as the existing toilet rooms. At the inaccessible toilet rooms, directional signs indicating the location of the nearest such toilet room shall be provided. These directional signs shall include the International Symbol of Accessibility, and sign characters shall meet the visual character requirements in accordance with ICC A117.1.

306.7.12 Bathing rooms. Where it is *technically infeasible* to alter existing bathing rooms to be accessible, one accessible single-user bathing room or one accessible family or assisted-use bathing room constructed in accordance with Section 1110.2.1 of the *International Building Code* is permitted. This accessible bathing room shall be located on the same floor and in the same area as the existing bathing rooms. At the inaccessible bathing rooms, directional signs indicating the location of the nearest such bathing room shall be provided. These directional signs shall include the International Symbol of Accessibility, and sign characters shall meet the visual character requirements in accordance with ICC A117.1.

306.7.13 Additional toilet and bathing facilities. In assembly and mercantile occupancies, where additional toilet fixtures are added, not fewer than one accessible family or assisted-use toilet room shall be provided where required by Section 1110.2.1 of the *International Building Code*. In recreational *facilities*, where additional bathing rooms are being added, not fewer than one family or assisted-use bathing room shall be provided where required by Section 1110.2.1 of the *International Building Code*.

306.7.14 Dressing, fitting and locker rooms. Where it is *technically infeasible* to provide accessible dressing, fitting or locker rooms at the same location as similar types of rooms, one accessible room on the same level shall be provided. Where separate-sex *facilities* are provided, accessible rooms for each sex shall be provided. Separate-sex *facilities* are not required where only unisex rooms are provided.

306.7.15 Amusement rides. Where the structural or operational characteristics of an amusement ride are altered to the extent that the amusement ride's performance differs from that specified by the manufacturer or the original design, the amusement ride shall comply with requirements for new construction in Section 1111.4.8 of the *International Building Code*.

306.7.16 Historic structures. Where compliance with the requirements for accessible routes, entrances or toilet rooms would threaten or destroy the historic significance of the historic structure, as determined by the authority having jurisdiction, the alternative requirements of Sections 306.7.16.1 through 306.7.16.5 for that element shall be permitted.

Exceptions:

1. Accessible means of egress required by Chapter 10 of the *International Building Code* are not required to be provided in historic structures.
2. The altered element or space is not required to be on an accessible route, unless required by Sections 306.7.16.1 or 306.7.16.2.

306.7.16.1 Site arrival points. Not fewer than one exterior accessible route, including curb ramps from a site arrival point to an accessible entrance, shall be provided and shall not be less than 36 inches (914 mm) minimum in width.

306.7.16.2 Multiple-level buildings and facilities. An accessible route from an accessible entrance to public spaces on the level of the accessible entrance shall be provided.

306.7.16.3 Entrances. Where an entrance cannot be made accessible in accordance with Section 306.7.5, an accessible entrance that is unlocked while the building is occupied shall be provided; or, a locked accessible entrance with a notification system or remote monitoring shall be provided.

Signs complying with Section 1112 of the *International Building Code* shall be provided at the public entrances and the accessible entrance.

306.7.16.4 Toilet facilities. Where toilet rooms are provided, not fewer than one accessible single-user toilet room or one accessible family or assisted-use toilet room complying with Section 1110.2.1 of the *International Building Code* shall be provided.

306.7.16.5 Bathing facilities. Where bathing rooms are provided, not fewer than one accessible single-user bathing room or one accessible family or assisted-use bathing rooms complying with Section 1110.2.1 of the *International Building Code* shall be provided.

306.7.16.6 Type A units. The *alteration* to Type A individually owned dwelling units within a Group R-2 occupancy shall be permitted to meet the provision for a Type B dwelling unit.

306.7.16.7 Type B units. Type B dwelling or sleeping units required by Section 1108 of the *International Building Code* are not required to be provided in *historic buildings*.

SECTION 307
SMOKE ALARMS

307.1 Smoke alarms. Where an *alteration*, addition, change *of occupancy* or relocation of a building is made to an *existing building* or structure of a Group R and I-1 occupancy, the *existing building* shall be provided with smoke alarms in accordance with the *International Fire Code* or Section R314 of the *International Residential Code*.

Exception: Work classified as Level 1 *Alterations* in accordance with Chapter 7.

SECTION 308
CARBON MONOXIDE DETECTION

308.1 Carbon monoxide detection. Where an *addition*, *alteration*, *change of occupancy* or relocation of a building is made to Group I-1, I-2, I-4 and R occupancies and classrooms of Group E occupancies, the *existing building* shall be provided with carbon monoxide detection in accordance with the *International Fire Code* or Section R315 of the *International Residential Code*.

Exceptions:

1. Work involving the exterior surfaces of buildings, such as the replacement of roofing or siding, the addition or replacement of windows or doors, or the addition of porches or decks.

2. Installation, alteration or *repairs* of plumbing or mechanical systems, other than fuel-burning appliances.

3. Work classified as Level 1 *Alterations* in accordance with Chapter 7.

SECTION 309
ADDITIONS AND REPLACEMENTS OF EXTERIOR WALL COVERINGS AND EXTERIOR WALL ENVELOPES

309.1 General. The provisions of Section 309 apply to all *alterations*, *repairs*, *additions*, relocations of structures and *changes of occupancy* regardless of compliance method.

309.2 Additions and replacements. Where an *exterior wall covering* or *exterior wall envelope* is added or replaced, the materials and methods used shall comply with the requirements for new construction in Chapter 14 and Chapter 26 of the *International Building Code* if the added or replaced *exterior wall covering* or *exterior wall envelope* involves two or more contiguous stories and comprises more than 15 percent of the total wall area on any side of the building.

CHAPTER 4
REPAIRS

User note:

About this chapter: Chapter 4 provides requirements for repairs of existing buildings. The provisions define conditions under which repairs may be made using materials and methods like those of the original construction or the extent to which repairs must comply with requirements for new buildings.

SECTION 401
GENERAL

401.1 Scope. *Repairs* shall comply with the requirements of this chapter. *Repairs* to *historic buildings* need only comply with Chapter 12.

401.1.1 Bleachers, grandstands and folding and telescopic seating. *Repairs* to existing bleachers, grandstands and folding and telescopic seating shall comply with ICC 300.

401.2 Compliance. The work shall not make the building less complying than it was before the *repair* was undertaken.

[BS] 401.3 Flood hazard areas. In flood hazard areas, *repairs* that constitute *substantial improvement* shall require that the building comply with Section 1612 of the *International Building Code*, or Section R322 of the *International Residential Code*, as applicable.

SECTION 402
BUILDING ELEMENTS AND MATERIALS

402.1 Glazing in hazardous locations. Replacement glazing in hazardous locations shall comply with the safety glazing requirements of the *International Building Code* or *International Residential Code* as applicable.

> **Exception:** Glass block walls, louvered windows and jalousies repaired with like materials.

SECTION 403
FIRE PROTECTION

403.1 General. *Repairs* shall be done in a manner that maintains the level of fire protection provided.

SECTION 404
MEANS OF EGRESS

404.1 General. *Repairs* shall be done in a manner that maintains the level of protection provided for the means of egress.

SECTION 405
STRUCTURAL

[BS] 405.1 General. Structural *repairs* shall be in compliance with this section and Section 401.2.

[BS] 405.2 Repairs to damaged buildings. *Repairs* to damaged buildings shall comply with this section.

[BS] 405.2.1 Repairs for less than substantial structural damage. Unless otherwise required by this section, for damage less than *substantial structural damage*, the damaged elements shall be permitted to be restored to their predamage condition.

[BS] 405.2.1.1 Snow damage. Structural components whose damage was caused by or related to snow load effects shall be repaired, replaced or altered to satisfy the requirements of Section 1608 of the *International Building Code*.

[BS] 405.2.2 Disproportionate earthquake damage. A building assigned to Seismic Design Category D, E or F that has sustained *disproportionate earthquake damage* shall be subject to the requirements for buildings with substantial structural damage to vertical elements of the lateral force-resisting system.

[BS] 405.2.3 Substantial structural damage to vertical elements of the lateral force-resisting system. A building that has sustained *substantial structural damage* to the vertical elements of its lateral force-resisting system shall be evaluated in accordance with Section 405.2.3.1, and either repaired in accordance with Section 405.2.3.2 or repaired and retrofitted in accordance with Section 405.2.3.3, depending on the results of the evaluation.

> **Exceptions:**
> 1. Buildings assigned to Seismic Design Category A, B or C whose *substantial structural damage* was not caused by earthquake need not be evaluated or retrofitted for load combinations that include earthquake effects.
> 2. One- and two-family dwellings need not be evaluated or retrofitted for load combinations that include earthquake effects.

[BS] 405.2.3.1 Evaluation. The building shall be evaluated by a registered design professional, and the evaluation findings shall be submitted to the *code official*. The evaluation shall establish whether the damaged building, if repaired to its predamage state, would comply with the provisions of the *International Building Code* for load combinations that include wind or earthquake effects, except that the seismic forces shall be the reduced seismic forces.

[BS] 405.2.3.2 Extent of repair for compliant buildings. If the evaluation establishes that the building in its predamage condition complies with the provisions of Section 405.2.3.1, then the damaged elements shall be permitted to be restored to their predamage condition.

[BS] 405.2.3.3 Extent of repair for noncompliant buildings. If the evaluation does not establish that the building in its predamage condition complies with the provisions of Section 405.2.3.1, then the building shall be retrofitted to comply with the provisions of this section. The wind loads for the *repair* and *retrofit* shall be those required by the building code in effect at the time of original construction, unless the damage was caused by wind, in which case the wind loads shall be in accordance with the *International Building Code*. The seismic loads for this *retrofit* design shall be those required by the building code in effect at the time of original construction, but not less than the reduced seismic forces.

[BS] 405.2.4 Substantial structural damage to gravity load-carrying components. Gravity load-carrying components that have sustained *substantial structural damage* shall be rehabilitated to comply with the applicable provisions for dead, live and snow loads in the *International Building Code*. Undamaged gravity load-carrying components that receive dead, live or snow loads from rehabilitated components shall also be rehabilitated if required to comply with the design loads of the *rehabilitation* design.

[BS] 405.2.4.1 Lateral force-resisting elements. Regardless of the level of damage to vertical elements of the lateral force-resisting system, if *substantial structural damage* to gravity load-carrying components was caused primarily by wind or seismic effects, then the building shall be evaluated in accordance with Section 405.2.3.1 and, if noncompliant, retrofitted in accordance with Section 405.2.3.3.

Exceptions:

1. Buildings assigned to Seismic Design Category A, B or C whose *substantial structural damage* was not caused by earthquake need not be evaluated or retrofitted for load combinations that include earthquake effects.
2. One- and two-family dwellings need not be evaluated or retrofitted for load combinations that include earthquake effects.

[BS] 405.2.5 Substantial structural damage to snow load-carrying components. Where substantial structural damage to any snow load-carrying components is caused by or related to snow load effects, any components required to carry snow loads on roof framing of similar construction shall be repaired, replaced or retrofitted to satisfy the requirements of Section 1608 of the *International Building Code*.

[BS] 405.2.6 Flood hazard areas. In *flood hazard* areas, buildings that have sustained *substantial damage* shall be brought into compliance with Section 1612 of the *International Building Code*, or Section R322 of the *International Residential Code*, as applicable.

SECTION 406
ELECTRICAL

406.1 Material. Existing electrical wiring and equipment undergoing *repair* shall be allowed to be repaired or replaced with like material.

406.1.1 Receptacles. Replacement of electrical receptacles shall comply with the applicable requirements of Section 406.4(D) of NFPA 70.

406.1.2 Plug fuses. Plug fuses of the Edison-base type shall be used for replacements only where there is no evidence of over fusing or tampering per applicable requirements of Section 240.51(B) of NFPA 70.

406.1.3 Nongrounding-type receptacles. For replacement of nongrounding-type receptacles with grounding-type receptacles and for branch circuits that do not have an equipment grounding conductor in the branch circuitry, the grounding conductor of a grounding-type receptacle outlet shall be permitted to be grounded to any accessible point on the grounding electrode system or to any accessible point on the grounding electrode conductor in accordance with Section 250.130(C) of NFPA 70.

406.1.4 Health care facilities. Portions of electrical systems being repaired in Group I-2, ambulatory care *facilities* and outpatient clinics shall comply with NFPA 99 requirements for *repairs*.

406.1.5 Grounding of appliances. Frames of electric ranges, wall-mounted ovens, counter-mounted cooking units, clothes dryers and outlet or junction boxes that are part of the existing branch circuit for these appliances shall be permitted to be grounded to the grounded circuit conductor in accordance with Section 250.140 of NFPA 70.

SECTION 407
MECHANICAL

407.1 General. Existing mechanical systems undergoing *repair* shall not make the building less complying than it was before the damaged occurred.

407.2 Mechanical draft systems for manually fired appliances and fireplaces. A mechanical draft system shall be permitted to be used with manually fired appliances and fireplaces where such a system complies with all of the following requirements:

1. The mechanical draft device shall be listed and installed in accordance with the manufacturer's installation instructions.
2. A device shall be installed that produces visible and audible warning upon failure of the mechanical draft device or loss of electrical power at any time that the

mechanical draft device is turned on. This device shall be equipped with a battery backup if it receives power from the building wiring.

3. A smoke detector shall be installed in the room with the appliance or fireplace. This device shall be equipped with a battery backup if it receives power from the building wiring.

SECTION 408
PLUMBING

408.1 Materials. Plumbing materials and supplies shall not be used for *repairs* that are prohibited in the *International Plumbing Code*.

408.2 Water closet replacement. The maximum water consumption flow rates and quantities for all replaced water closets shall be 1.6 gallons (6 L) per flushing cycle.

Exception: Blowout-design water closets [3.5 gallons (13 L) per flushing cycle].

408.3 Health care facilities. Portions of medical gas systems being repaired in Group I-2, ambulatory care *facilities* and outpatient clinics shall comply with NFPA 99 requirements for *repairs*.

CHAPTER 5
PRESCRIPTIVE COMPLIANCE METHOD

User note:

About this chapter: Chapter 5 provides details for the prescriptive compliance method—one of the three main options of compliance available in this code for buildings and structures undergoing alteration, addition or change of occupancy.

SECTION 501
GENERAL

501.1 Scope. The provisions of this chapter shall control the *alteration, addition* and *change of occupancy* of *existing buildings* and structures, including *historic buildings* and structures as referenced in Section 301.3.1.

501.1.1 Compliance with other methods. *Alterations, additions* and *changes of occupancy* to *existing buildings* and structures shall comply with the provisions of this chapter or with one of the methods provided in Section 301.3.

501.2 Fire-resistance ratings. Where *approved* by the *code official*, in buildings where an automatic sprinkler system installed in accordance with Section 903.3.1.1 or 903.3.1.2 of the *International Building Code* has been added, and the building is now sprinklered throughout, the required fire-resistance ratings of building elements and materials shall be permitted to meet the requirements of the current building code. The building is required to meet the other applicable requirements of the *International Building Code*.

Plans, investigation and evaluation reports, and other data shall be submitted indicating which building elements and materials the applicant is requesting the *code official* to review and approve for determination of applying the current building code fire-resistance ratings. Any special construction features, including fire-resistance-rated assemblies and smoke-resistive assemblies, conditions of occupancy, means of egress conditions, fire code deficiencies, *approved* modifications or *approved* alternative materials, design and methods of construction, and equipment applying to the building that impact required fire-resistance ratings shall be identified in the evaluation reports submitted.

501.3 Health care facilities. In Group I-2 *facilities*, ambulatory care *facilities* and outpatient clinics, any altered or added portion of an existing electrical or medical gas systems shall be required to meet installation and equipment requirements in NFPA 99.

SECTION 502
ADDITIONS

502.1 General. *Additions* to any building or structure shall comply with the requirements of the *International Building Code* for new construction. *Alterations* to the *existing building* or structure shall be made to ensure that the *existing building* or structure together with the *addition* are not less complying with the provisions of the *International Building Code* than the *existing building* or structure was prior to the *addition*. An *existing building* together with its *additions* shall comply with the height and area provisions of Chapter 5 of the *International Building Code*.

[BS] 502.2 Disproportionate earthquake damage. A building assigned to Seismic Design Category D, E or F that has sustained *disproportionate earthquake damage* shall be subject to the requirements for buildings with *substantial structural damage* to vertical elements of the lateral force-resisting system.

[BS] 502.3 Flood hazard areas. For buildings and structures in *flood hazard* areas established in Section 1612.3 of the *International Building Code*, or Section R322 of the *International Residential Code*, as applicable, any *addition* that constitutes *substantial improvement* of the *existing structure* shall comply with the flood design requirements for new construction, and all aspects of the *existing structure* shall be brought into compliance with the requirements for new construction for flood design.

For buildings and structures in *flood hazard areas* established in Section 1612.3 of the *International Building Code*, or Section R322 of the *International Residential Code*, as applicable, any *additions* that do not constitute *substantial improvement* of the *existing structure* are not required to comply with the flood design requirements for new construction.

[BS] 502.4 Existing structural elements carrying gravity load. Any existing gravity load-carrying structural element for which an *addition* and its related *alterations* cause an increase in design dead, live or snow load, including snow drift effects, of more than 5 percent shall be replaced or altered as needed to carry the gravity loads required by the *International Building Code* for new structures. Any existing gravity load-carrying structural element whose vertical load-carrying capacity is decreased as part of the *addition* and its related *alterations* shall be considered to be an altered element subject to the requirements of Section 503.3. Any existing element that will form part of the lateral load path for any part of the *addition* shall be considered to be an existing lateral load-carrying structural element subject to the requirements of Section 502.5.

Exception: Buildings of Group R occupancy with not more than five dwelling or sleeping units used solely for residential purposes where the *existing building* and the *addition* together comply with the conventional light-frame construction methods of the *International Building*

Code or the provisions of the *International Residential Code*.

[BS] 502.5 Existing structural elements carrying lateral load. Where the *addition* is structurally independent of the *existing structure*, existing lateral load-carrying structural elements shall be permitted to remain unaltered. Where the *addition* is not structurally independent of the *existing structure*, the *existing structure* and its *addition* acting together as a single structure shall be shown to meet the requirements of Sections 1609 and 1613 of the *International Building Code* using full seismic forces.

Exceptions:

1. Any existing lateral load-carrying structural element whose demand-capacity ratio with the *addition* considered is not more than 10 percent greater than its demand-capacity ratio with the *addition* ignored shall be permitted to remain unaltered. For purposes of calculating demand-capacity ratios, the demand shall consider applicable load combinations with design lateral loads or forces in accordance with Sections 1609 and 1613 of the *International Building Code*. For purposes of this exception, comparisons of demand-capacity ratios and calculation of design lateral loads, forces and capacities shall account for the cumulative effects of *additions* and *alterations* since original construction.

2. Buildings of Group R occupancy with not more than five dwelling or sleeping units used solely for residential purposes where the *existing building* and the *addition* together comply with the conventional light-frame construction methods of the *International Building Code* or the provisions of the *International Residential Code*.

502.6 Enhanced classroom acoustics. In Group E occupancies, enhanced classroom acoustics shall be provided in all classrooms in the *addition* with a volume of 20,000 cubic feet (565 m^3) or less. Enhanced classroom acoustics shall comply with the reverberation time in Section 808 of ICC A117.1.

SECTION 503
ALTERATIONS

503.1 General. *Alterations* to any building or structure shall comply with the requirements of the *International Building Code* for new construction. *Alterations* shall be such that the *existing building* or structure is not less complying with the provisions of the *International Building Code* than the *existing building* or structure was prior to the *alteration*.

Exceptions:

1. An existing stairway shall not be required to comply with the requirements of Section 1011 of the *International Building Code* where the existing space and construction does not allow a reduction in pitch or slope.

2. Handrails otherwise required to comply with Section 1011.11 of the *International Building Code* shall not be required to comply with the requirements of Section 1014.6 of the *International Building Code* regarding full extension of the handrails where such extensions would be hazardous because of plan configuration.

3. Where provided in below-grade transportation stations, existing and new escalators shall be permitted to have a clear width of less than 32 inches (815 mm).

[BS] 503.2 Flood hazard areas. For buildings and structures in *flood hazard areas* established in Section 1612.3 of the *International Building Code*, or Section R322 of the *International Residential Code*, as applicable, any *alteration* that constitutes *substantial improvement* of the *existing structure* shall comply with the flood design requirements for new construction, and all aspects of the *existing structure* shall be brought into compliance with the requirements for new construction for flood design.

For buildings and structures in *flood hazard areas* established in Section 1612.3 of the *International Building Code*, or Section R322 of the *International Residential Code*, as applicable, any *alterations* that do not constitute *substantial improvement* of the *existing structure* are not required to comply with the flood design requirements for new construction.

[BS] 503.3 Existing structural elements carrying gravity load. Any existing gravity load-carrying structural element for which an *alteration* causes an increase in design dead, live or snow load, including snow drift effects, of more than 5 percent shall be replaced or altered as needed to carry the gravity loads required by the *International Building Code* for new structures. Any existing gravity load-carrying structural element whose gravity load-carrying capacity is decreased as part of the *alteration* shall be shown to have the capacity to resist the applicable design dead, live and snow loads including snow drift effects required by the *International Building Code* for new structures.

Exceptions:

1. Buildings of Group R occupancy with not more than five dwelling or sleeping units used solely for residential purposes where the altered building complies with the conventional light-frame construction methods of the *International Building Code* or the provisions of the *International Residential Code*.

2. Buildings in which the increased dead load is due entirely to the addition of a second layer of roof covering weighing 3 pounds per square foot (0.1437 kN/m^2) or less over an existing single layer of roof covering.

[BS] 503.4 Existing structural elements carrying lateral load. Except as permitted by Section 503.13, where the *alteration* increases design lateral loads, results in a prohibited structural irregularity as defined in ASCE 7, or decreases the capacity of any existing lateral load-carrying structural element, the structure of the altered building or structure shall meet the requirements of Sections 1609 and 1613 of the *International Building Code*. Reduced seismic forces shall be permitted.

Exceptions:

1. Any existing lateral load-carrying structural element whose demand-capacity ratio with the alteration considered is not more than 10 percent greater than its demand-capacity ratio with the alteration ignored shall be permitted to remain unaltered. For purposes of calculating demand-capacity ratios, the demand shall consider applicable load combinations with design lateral loads or forces in accordance with Sections 1609 and 1613 of the *International Building Code*. Reduced seismic forces shall be permitted. For purposes of this exception, comparisons of demand-capacity ratios and calculation of design lateral loads, forces and capacities shall account for the cumulative effects of additions and alterations since original construction.

2. Buildings in which the increase in the demand-capacity ratio is due entirely to the addition of rooftop-supported mechanical equipment individually having an operating weight less than 400 pounds (181.4 kg) and where the total additional weight of all rooftop equipment placed after initial construction of the building is less than 10 percent of the roof dead load. For purposes of this exception, "roof" shall mean the roof level above a particular story.

[BS] 503.5 Seismic Design Category F. Where the *work area* exceeds 50 percent of the building area, and where the building is assigned to Seismic Design Category F, the structure of the altered building shall meet the requirements of Sections 1609 and 1613 of the *International Building Code*. Reduced seismic forces shall be permitted.

[BS] 503.6 Bracing for unreinforced masonry parapets on reroofing. Where the intended *alteration* requires a permit for reroofing and involves removal of roofing materials from more than 25 percent of the roof area of a building assigned to Seismic Design Category D, E or F that has parapets constructed of unreinforced masonry, the work shall include installation of parapet bracing to resist out-of-plane seismic forces, unless an evaluation demonstrates compliance of such items. Reduced seismic forces shall be permitted.

[BS] 503.7 Anchorage for concrete and reinforced masonry walls. Where the *work area* exceeds 50 percent of the building area, the building is assigned to Seismic Design Category C, D, E or F and the building's structural system includes concrete or reinforced masonry walls with a flexible roof diaphragm, the *alteration* work shall include installation of wall anchors at the roof line, unless an evaluation demonstrates compliance of existing wall anchorage. Use of reduced seismic forces shall be permitted.

[BS] 503.8 Anchorage for unreinforced masonry walls in major alterations. Where the *work area* exceeds 50 percent of the building area, the building is assigned to Seismic Design Category C, D, E or F and the building's structural system includes unreinforced masonry bearing walls, the *alteration* work shall include installation of wall anchors at the floor and roof lines, unless an evaluation demonstrates compliance of existing wall anchorage. Reduced seismic forces shall be permitted.

[BS] 503.9 Bracing for unreinforced masonry parapets in major alterations. Where the *work area* exceeds 50 percent of the building area, and where the building is assigned to Seismic Design Category C, D, E or F, parapets constructed of unreinforced masonry shall have bracing installed as needed to resist out-of-plane seismic forces, unless an evaluation demonstrates compliance of such items. Reduced seismic forces shall be permitted.

[BS] 503.10 Anchorage of unreinforced masonry partitions in major alterations. Where the *work area* exceeds 50 percent of the building area, and where the building is assigned to Seismic Design Category C, D, E or F, unreinforced masonry partitions and nonstructural walls within the *work area* and adjacent to egress paths from the *work area* shall be anchored, removed or altered to resist out-of-plane seismic forces, unless an evaluation demonstrates compliance of such items. Use of reduced seismic forces shall be permitted.

[BS] 503.11 Substantial structural alteration. Where the *work area* exceeds 50 percent of the building area and where work involves a *substantial structural alteration*, the lateral load-resisting system of the altered building shall satisfy the requirements of Sections 1609 and 1613 of the *International Building Code*. Reduced seismic forces shall be permitted.

Exceptions:

1. Buildings of Group R occupancy with not more than five dwelling or sleeping units used solely for residential purposes that are altered based on the conventional light-frame construction methods of the *International Building Code* or in compliance with the provisions of the *International Residential Code*.

2. Where the intended *alteration* involves only the lowest story of a building, only the lateral load-resisting components in and below that story need comply with this section.

[BS] 503.12 Roof diaphragms resisting wind loads in high-wind regions. Where the intended *alteration* requires a permit for reroofing and involves removal of roofing materials from more than 50 percent of the roof diaphragm of a building or section of a building located where the ultimate design wind speed is greater than 130 mph (58 m/s) in accordance with Figure 1609.3(1) of the *International Building Code*, roof diaphragms, connections of the roof diaphragm to roof framing members, and roof-to-wall connections shall be evaluated for the wind loads specified in Section 1609 of the *International Building Code*, including wind uplift. If the diaphragms and connections in their current condition are not capable of resisting 75 percent of those wind loads, they shall be replaced or strengthened in accordance with the loads specified in Section 1609 of the *International Building Code*.

> **Exception:** Buildings that have been demonstrated to comply with the wind load provisions in ASCE 7-88 or later editions.

[BS] 503.13 Voluntary lateral force-resisting system alterations. Structural *alterations* that are intended exclusively to improve the lateral force-resisting system and are not required by other sections of this code shall not be required to meet the requirements of Section 1609 or 1613 of the *International Building Code*, provided that all of the following apply:

1. The capacity of existing structural systems to resist forces is not reduced.
2. New structural elements are detailed and connected to existing or new structural elements as required by the *International Building Code* for new construction.
3. New or relocated nonstructural elements are detailed and connected to existing or new structural elements as required by the *International Building Code* for new construction.
4. The *alterations* do not create a structural irregularity as defined in ASCE 7 or make an existing structural irregularity more severe.

503.14 Smoke compartments. In Group I-2 occupancies where the *alteration* is on a story used for sleeping rooms for more than 30 care recipients, the story shall be divided into not less than two compartments by smoke barrier walls in accordance with Section 407.5 of the *International Building Code* as required for new construction.

503.15 Refuge areas. Where *alterations* affect the configuration of an area utilized as a refuge area, the capacity of the refuge area shall not be reduced below the required capacity of the refuge area for horizontal exits in accordance with Section 1026.4 of the *International Building Code*.

Where the horizontal exit also forms a smoke compartment, the capacity of the refuge area for Group I-1, I-2 and I-3 occupancies and Group B ambulatory care *facilities* shall not be reduced below that required in Sections 407.5.3, 408.6.2, 420.6.1 and 422.3.2 of the *International Building Code*, as applicable.

503.16 Enhanced classroom acoustics. In Group E occupancies, where the *work area* exceeds 50 percent of the building area, enhanced classroom acoustics shall be provided in all classrooms with a volume of 20,000 cubic feet (565 m^3) or less. Enhanced classroom acoustics shall comply with the reverberation time in Section 808 of ICC A117.1.

503.17 Locking arrangements in educational occupancies. In Group E occupancies, Group B educational occupancies and Group I-4 occupancies, egress doors with locking arrangements designed to keep intruders from entering the room shall comply with Section 1010.2.8 of the *International Building Code*.

503.18 Two-way communications systems. Where the *work area* for *alterations* exceeds 50 percent of the building area and the building has elevator service, a two-way communication systems shall be provided where required by Section 1009.8 of the *International Building Code*.

SECTION 504
FIRE ESCAPES

[BE] 504.1 Where permitted. Fire escapes shall be permitted only as provided for in Sections 504.1.1 through 504.1.4.

[BE] 504.1.1 New buildings. Fire escapes shall not constitute any part of the required means of egress in new buildings.

[BE] 504.1.2 Existing fire escapes. Existing fire escapes shall continue to be accepted as a component in the means of egress in *existing buildings* only.

[BE] 504.1.3 New fire escapes. New fire escapes for *existing buildings* shall be permitted only where exterior stairways cannot be utilized because of lot lines limiting stairway size or because of sidewalks, alleys or roads at grade level. New fire escapes shall not incorporate ladders or access by windows.

[BE] 504.1.4 Limitations. Fire escapes shall comply with this section and shall not constitute more than 50 percent of the required number of exits nor more than 50 percent of the required exit capacity.

[BE] 504.2 Location. Where located on the front of the building and where projecting beyond the building line, the lowest landing shall be not less than 7 feet (2134 mm) or more than 12 feet (3658 mm) above grade, and shall be equipped with a counterbalanced stairway to the street. In alleyways and thoroughfares less than 30 feet (9144 mm) wide, the clearance under the lowest landing shall be not less than 12 feet (3658 mm).

[BE] 504.3 Construction. The fire escape shall be designed to support a live load of 100 pounds per square foot (4788 Pa) and shall be constructed of steel or other *approved noncombustible materials*. Fire escapes constructed of wood not less than nominal 2 inches (51 mm) thick are permitted on buildings of Type V construction. Walkways and railings located over or supported by combustible roofs in buildings of Type III and IV construction are permitted to be of wood not less than nominal 2 inches (51 mm) thick.

[BE] 504.4 Dimensions. Stairways shall be not less than 22 inches (559 mm) wide with risers not more than, and treads not less than, 8 inches (203 mm) and landings at the foot of stairways not less than 40 inches (1016 mm) wide by 36 inches (914 mm) long, located not more than 8 inches (203 mm) below the door.

[BE] 504.5 Opening protectives. Doors and windows within 10 feet (3048 mm) of fire escape stairways shall be protected with $^3/_4$-hour opening protectives.

> **Exception:** Opening protection shall not be required in buildings equipped throughout with an *approved* automatic sprinkler system.

SECTION 505
WINDOWS AND EMERGENCY ESCAPE OPENINGS

505.1 Replacement windows. The installation or replacement of windows shall be as required for new installations.

505.2 Window opening control devices on replacement windows. In Group R-2 or R-3 buildings containing dwelling units, and one- and two-family dwellings and townhouses regulated by the *International Residential Code*, window opening control devices or fall prevention devices complying with ASTM F2090 shall be installed where an existing window is replaced and where all of the following apply to the replacement window:

1. The window is operable.
2. One of the following applies:
   2.1. The window replacement includes replacement of the sash and frame.
   2.2. The window replacement includes the sash only where the existing frame remains.
3. One of the following applies:
   3.1. In Group R-2 or R-3 buildings containing dwelling units, the bottom of the clear opening of the window opening is at a height less than 36 inches (915 mm) above the finished floor.
   3.2. In one- and two-family dwellings and townhouses regulated by the *International Residential Code*, the bottom of the clear opening of the window opening is at a height less than 24 inches (610 mm) above the finished floor.
4. The window will permit openings that will allow passage of a 4-inch-diameter (102 mm) sphere when the window is in its largest opened position.
5. The vertical distance from the bottom of the clear opening of the window opening to the finished grade or other surface below, on the exterior of the building, is greater than 72 inches (1829 mm).

Exception: Operable windows where the bottom of the clear opening of the window opening is located more than 75 feet (22 860 mm) above the finished grade or other surface below, on the exterior of the room, space or building, and that are provided with window fall prevention devices that comply with ASTM F2006.

505.3 Replacement window emergency escape and rescue openings. Where windows are required to provide *emergency escape and rescue openings* in Group R-2 and R-3 occupancies and one- and two-family dwellings and townhouses regulated by the *International Residential Code*, replacement windows shall be exempt from the requirements of Section 1031.3 of the *International Building Code* and Section R310.2 of the *International Residential Code*, provided that the replacement window meets the following conditions:

1. The replacement window is the manufacturer's largest standard size window that will fit within the existing frame or existing rough opening. The replacement window shall be permitted to be of the same operating style as the existing window or a style that provides for an equal or greater window opening area than the existing window.
2. Where the replacement of the window is part of a *change of occupancy*, it shall comply with Section 1011.5.6.

505.3.1 Control devices. Window opening control devices or fall prevention devices complying with ASTM F2090 shall be permitted for use on windows required to provide *emergency escape and rescue openings*. After operation to release the control device allowing the window to fully open, the control device shall not reduce the net clear opening area of the window unit. *Emergency escape and rescue openings* shall be operational from the inside of the room without the use of keys or tools.

505.4 Bars, grilles, covers or screens. Bars, grilles, covers, screens or similar devices are permitted to be placed over *emergency escape and rescue openings*, bulkhead enclosure or window wells that serve such openings, provided all of the following conditions are met:

1. The minimum net clear opening size complies with the code that was in effect at the time of construction.
2. Such devices shall be releasable or removable from the inside without the use of a key, tool or force greater than that which is required for normal operation of the escape and rescue opening.
3. Where such devices are installed, they shall not reduce the net clear opening of the emergency escape and rescue openings.
4. Smoke alarms shall be installed in accordance with Section 907.2.10 of the *International Building Code*.

SECTION 506
CHANGE OF OCCUPANCY

506.1 Compliance. A *change of occupancy* shall not be made in any building unless that building is made to comply with the requirements of the *International Building Code* for the use or occupancy. Changes of occupancy in a building or portion thereof shall be such that the *existing building* is not less complying with the provisions of this code than the *existing building* or structure was prior to the change. Subject to the approval of the *code official*, changes of occupancy shall be permitted without complying with all of the requirements of this code for the new occupancy, provided that the new occupancy is less hazardous, based on life and fire risk, than the existing occupancy.

Exception: The building need not be made to comply with Chapter 16 of the *International Building Code* unless required by Section 506.5.

506.1.1 Change in the character of use. A change of occupancy with no *change of occupancy* classification shall not be made to any structure that will subject the structure to any special provisions of the applicable International Codes, without approval of the *code official*. Compliance shall be only as necessary to meet the specific provisions and is not intended to require the entire building be brought into compliance.

506.2 Certificate of occupancy. A certificate of occupancy shall be issued where it has been determined that the requirements for the new occupancy classification have been met.

506.3 Stairways. An existing stairway shall not be required to comply with the requirements of Section 1011 of the *International Building Code* where the existing space and construction does not allow a reduction in pitch or slope.

506.4 Existing emergency escape and rescue openings. Where a *change of occupancy* would require an *emergency escape and rescue opening* in accordance with Section 1031.1 of the *International Building Code*, operable windows serving as the *emergency escape and rescue opening* shall comply with the following

1. An existing operable window shall provide a minimum net clear opening of 4 square feet (0.38 m^2) with a minimum net clear opening height of 22 inches (559 mm) and a minimum net clear opening width of 20 inches (508 mm).

2. A replacement window where such window complies with both of the following:

    2.1. The replacement window meets the size requirements in Item 1.

    2.2. The replacement window is the manufacturer's largest standard size window that will fit within the existing frame or existing rough opening. The replacement window shall be permitted to be of the same operating style as the existing window or a style that provides for an equal or greater window opening area than the existing window.

506.5 Structural. Any building undergoing a *change of occupancy* shall satisfy the requirements of this section.

506.5.1 Live loads. Structural elements carrying tributary live loads from an area with a *change of occupancy* shall satisfy the requirements of Section 1607 of the *International Building Code*. Design live loads for areas of new occupancy shall be based on Section 1607 of the *International Building Code*. Design live loads for other areas shall be permitted to use previously *approved* design live loads.

> **Exception:** Structural elements whose demand-capacity ratio considering the *change of occupancy* is not more than 5 percent greater than the demand-capacity ratio based on previously *approved* live loads need not comply with this section.

506.5.2 Snow and wind loads. Where a *change of occupancy* results in a structure being assigned to a higher *risk category*, the structure shall satisfy the requirements of Sections 1608 and 1609 of the *International Building Code* for the new *risk category*.

> **Exception:** Where the area of the new occupancy is less than 10 percent of the building area, compliance with this section is not required. The cumulative effect of occupancy changes over time shall be considered.

506.5.3 Seismic loads (seismic force-resisting system). Where a *change of occupancy* results in a building being assigned to a higher *risk category*, or where the change is from a Group S or Group U occupancy to any occupancy other than Group S or Group U, the building shall satisfy the requirements of Section 1613 of the *International Building Code* for the new *risk category* using full seismic forces.

> **Exceptions:**
>
> 1. Where the area of the new occupancy is less than 10 percent of the building area, the occupancy is not changing from a Group S or Group U occupancy, and the new occupancy is not assigned to *Risk Category* IV, compliance with this section is not required. The cumulative effect of occupancy changes over time shall be considered.
>
> 2. Where a *change of use* results in a building being reclassified from *Risk Category* I or II to Risk Category III and the seismic coefficient, S_{DS}, is less than 0.33, compliance with this section is not required.
>
> 3. Unreinforced masonry bearing wall buildings assigned to *Risk Category* III and to Seismic Design Category A or B, shall be permitted to use Appendix Chapter A1 of this code.
>
> 4. Where the change is from a Group S or Group U occupancy and there is no change of risk category, use of reduced seismic forces shall be permitted.

506.5.4 Access to Risk Category IV. Any structure that provides operational access to an adjacent structure assigned to *Risk Category* IV as the result of a *change of occupancy* shall itself satisfy the requirements of Sections 1608, 1609 and 1613 of the *International Building Code*. For compliance with Section 1613, *International Building Code*-level seismic forces shall be used. Where operational access to the *Risk Category* IV structure is less than 10 feet (3048 mm) from either an interior lot line or from another structure, access protection from potential falling debris shall be provided.

506.6 Enhanced classroom acoustics. In Group E occupancies, where the *work area* exceeds 50 percent of the building area, enhanced classroom acoustics shall be provided in all classrooms with a volume of 20,000 cubic feet (565 m^3) or less. Enhanced classroom acoustics shall comply with the reverberation time in Section 808 of ICC A117.1.

SECTION 507
HISTORIC BUILDINGS

507.1 Historic buildings. The provisions of this code that require improvements relative to a building's existing condition or, in the case of *repairs*, that require improvements relative to a building's predamage condition, shall not be mandatory for *historic buildings* unless specifically required by this section.

507.2 Life safety hazards. The provisions of this code shall apply to *historic buildings* judged by the *code official* to constitute a distinct life safety hazard.

[BS] 507.3 Flood hazard areas. Within *flood hazard areas* established in accordance with Section 1612.3 of the *International Building Code*, or Section R322 of the *International Residential Code*, as applicable, where the work proposed constitutes *substantial improvement*, the building shall be brought into compliance with Section 1612 of the *International Building Code*, or Section R322 of the *International Residential Code*, as applicable.

> **Exception:** *Historic buildings* meeting any of the following criteria need not be brought into compliance:
>
> 1. Listed or preliminarily determined to be eligible for listing in the National Register of Historic Places.
> 2. Determined by the Secretary of the US Department of Interior as contributing to the historical significance of a registered historic district or a district preliminarily determined to qualify as an historic district.
> 3. Designated as historic under a state or local historic preservation program that is approved by the Department of Interior.

[BS] 507.4 Structural. *Historic buildings* shall comply with the applicable structural provisions in this chapter.

> **Exceptions:**
>
> 1. The *code official* shall be authorized to accept existing floors and existing live loads and to approve operational controls that limit the live load on any floor.
> 2. *Repair* of *substantial structural damage* is not required to comply with Sections 405.2.3, and 405.2.4. *Substantial structural damage* shall be repaired in accordance with Section 405.2.1.

CHAPTER 6
CLASSIFICATION OF WORK

User note:

About this chapter: Chapter 6 provides an overview of the Work Area Method available as an option for rehabilitation of a building. The chapter defines the different classifications of alterations and provides general requirements for alterations, change of occupancy, additions and historic buildings. Detailed requirements for all of these are given in Chapters 7 through 12.

SECTION 601
GENERAL

601.1 Scope. The provisions of this chapter shall be used in conjunction with Chapters 7 through 12 and shall apply to the *alteration*, *addition* and *change of occupancy* of *existing structures*, including historic and moved structures, as referenced in Section 301.3.2. The work performed on an *existing building* shall be classified in accordance with this chapter.

    601.1.1 Compliance with other alternatives. *Alterations*, *additions* and *changes of occupancy* to *existing structures* shall comply with the provisions of Chapters 7 through 12 or with one of the alternatives provided in Section 301.3.

601.2 Work area. The *work area*, as defined in Chapter 2, shall be identified on the construction documents.

SECTION 602
ALTERATION—LEVEL 1

602.1 Scope. Level 1 alterations include the removal and replacement or the covering of existing materials, elements, *equipment* or *fixtures* using new materials, elements, *equipment* or *fixtures* that serve the same purpose.

602.2 Application. Level 1 *alterations* shall comply with the provisions of Chapter 7.

SECTION 603
ALTERATION—LEVEL 2

603.1 Scope. Level 2 *alterations* include the addition or elimination of any door or window, the reconfiguration or extension of any system, or the installation of any additional equipment, and shall apply where the w*ork area* is equal to or less than 50 percent of the building area.

    Exception: The movement or addition of nonfixed and movable fixtures, cases, racks, counters and partitions not over 5 feet 9 inches (1753 mm) in height shall not be considered a Level 2 *alteration*.

603.2 Application. Level 2 *alterations* shall comply with the provisions of Chapter 7 for Level 1 *alterations* as well as the provisions of Chapter 8.

SECTION 604
ALTERATION—LEVEL 3

604.1 Scope. Level 3 *alterations* apply where the *work area* exceeds 50 percent of the *building area*.

604.2 Application. Level 3 *alterations* shall comply with the provisions of Chapters 7 and 8 for Level 1 and 2 *alterations*, respectively, as well as the provisions of Chapter 9.

SECTION 605
CHANGE OF OCCUPANCY

605.1 Scope. *Change of occupancy* provisions apply where the activity is classified as a *change of occupancy* as defined in Chapter 2.

605.2 Application. *Changes of occupancy* shall comply with the provisions of Chapter 10.

SECTION 606
ADDITIONS

606.1 Scope. Provisions for *additions* shall apply where work is classified as an *addition* as defined in Chapter 2.

606.2 Application. *Additions* to *existing buildings* shall comply with the provisions of Chapter 11.

SECTION 607
HISTORIC BUILDINGS

607.1 Scope. *Historic building* provisions shall apply to buildings classified as historic as defined in Chapter 2.

607.2 Application. Except as specifically provided for in Chapter 12, *historic buildings* shall comply with applicable provisions of this code for the type of work being performed.

CHAPTER 7
ALTERATIONS—LEVEL 1

User note:

About this chapter: Chapter 7 provides the technical requirements for those existing buildings that undergo Level 1 alterations as described in Section 603, which includes replacement or covering of existing materials, elements, equipment or fixtures using new materials for the same purpose. This chapter, similar to other chapters of this code, covers all building-related subjects, such as structural, mechanical, plumbing, electrical and accessibility as well as the fire and life safety issues when the alterations are classified as Level 1. The purpose of this chapter is to provide detailed requirements and provisions to identify the required improvements in the existing building elements, building spaces and building structural system. This chapter is distinguished from Chapters 8 and 9 by involving only replacement of building components with new components. In contrast, Level 2 alterations involve more space reconfiguration, and Level 3 alterations involve more extensive space reconfiguration, exceeding 50 percent of the building area.

SECTION 701
GENERAL

701.1 Scope. Level 1 *alterations* as described in Section 602 shall comply with the requirements of this chapter. Level 1 *alterations* to *historic buildings* shall comply with this chapter, except as modified in Chapter 12.

701.2 Conformance. An *existing building* or portion thereof shall not be altered such that the building becomes less safe than its existing condition.

> **Exception:** Where the current level of safety or sanitation is proposed to be reduced, the portion altered shall conform to the requirements of the *International Building Code*.

[BS] 701.3 Flood hazard areas. In *flood hazard areas*, *alterations* that constitute *substantial improvement* shall require that the building comply with Section 1612 of the *International Building Code*, or Section R322 of the *International Residential Code*, as applicable.

SECTION 702
BUILDING ELEMENTS AND MATERIALS

702.1 Interior finishes. Newly installed interior wall and ceiling finishes shall comply with Chapter 8 of the *International Building Code*.

702.2 Interior floor finish. New interior floor finish, including new carpeting used as an interior floor finish material, shall comply with Section 804 of the *International Building Code*.

702.3 Interior trim. Newly installed interior trim materials shall comply with Section 806 of the *International Building Code*.

702.4 Window opening control devices on replacement windows. In Group R-2 or R-3 buildings containing dwelling units and one- and two-family dwellings and townhouses regulated by the *International Residential Code*, window opening control devices complying with ASTM F2090 shall be installed where an existing window is replaced and where all of the following apply to the replacement window:

1. The window is operable.
2. One of the following applies:
   2.1. The window replacement includes replacement of the sash and frame.
   2.2. The window replacement includes the sash only where the existing frame remains.
3. One of the following applies:
   3.1. In Group R-2 or R-3 buildings containing dwelling units, the bottom of the clear opening of the window opening is at a height less than 36 inches (915 mm) above the finished floor.
   3.2. In one- and two-family dwellings and townhouses regulated by the *International Residential Code*, the bottom of the clear opening of the window opening is at a height less than 24 inches (610 mm) above the finished floor.
4. The window will permit openings that will allow passage of a 4-inch-diameter (102 mm) sphere when the window is in its largest opened position.
5. The vertical distance from the bottom of the clear opening of the window opening to the finished grade or other surface below, on the exterior of the building, is greater than 72 inches (1829 mm).

> **Exception:** Operable windows where the **bottom of the clear opening** of the window opening is located more than 75 feet (22 860 mm) above the finished grade or other surface below, on the exterior of the room, space or building, and that are provided with window fall prevention devices that comply with ASTM F2006.

702.5 Replacement window for emergency escape and rescue openings. Where windows are required to provide *emergency escape and rescue openings* in Group R-2 and R-3 occupancies and one- and two-family dwellings and townhouses regulated by the *International Residential Code*, replacement windows shall be exempt from the requirements of Section 1031.3 of the *International Building Code* and Section R310.2 of the *International Residential Code*, provided that the replacement window meets the following conditions:

1. The replacement window is the manufacturer's largest standard size window that will fit within the existing frame or existing rough opening.

2. Where the replacement window is part of a *change of occupancy* it shall comply with Section 1011.5.6.

702.5.1 Control devices. Window opening control devices or fall prevention devices complying with ASTM F2090 shall be permitted for use on windows required to provide *emergency escape and rescue openings*. After operation to release the control device allowing the window to fully open, the control device shall not reduce the net clear opening area of the window unit. *Emergency escape and rescue openings* shall be operational from the inside of the room without the use of keys or tools.

702.6 Bars, grilles, covers or screens. Bars, grilles, covers, screens or similar devices are permitted to be placed over emergency escape and rescue openings, bulkhead enclosure or window wells that serve such openings, provided all of the following conditions are met:

1. The minimum net clear opening size complies with the code that was in effect at the time of construction.

2. Such devices shall be releasable or removable from the inside without the use of a key, tool or force greater than that which is required for normal operation of the escape and rescue opening.

3. Where such devices are installed, they shall not reduce the net clear opening of the emergency escape and rescue openings.

4. Smoke alarms shall be installed in accordance with Section 907.2.11 of the *International Building Code*.

702.7 Materials and methods. New work shall comply with the materials and methods requirements in the *International Building Code*, *International Energy Conservation Code*, *International Mechanical Code* and *International Plumbing Code*, as applicable, that specify material standards, detail of installation and connection, joints, penetrations and continuity of any element, component or system in the building.

[FG] 702.7.1 International Fuel Gas Code. The following sections of the *International Fuel Gas Code* shall constitute the fuel gas materials and methods requirements for Level 1 *alterations*.

1. Chapter 3, entitled "General Regulations," except Sections 303.7 and 306.

2. Chapter 4, entitled "Gas Piping Installations," except Sections 401.8 and 402.3.

   2.1. Sections 401.8 and 402.3 shall apply where the work being performed increases the load on the system such that the existing pipe does not meet the size required by code. Existing systems that are modified shall not require resizing as long as the load on the system is not increased and the system length is not increased even if the altered system does not meet code minimums.

3. Chapter 5, entitled "Chimneys and Vents."

4. Chapter 6, entitled "Specific Appliances."

SECTION 703
FIRE PROTECTION

703.1 General. *Alterations* shall be done in a manner that maintains the level of fire protection provided.

SECTION 704
MEANS OF EGRESS

704.1 General. *Alterations* shall be done in a manner that maintains the level of protection provided for the means of egress.

704.1.1 Projections in nursing home corridors. In Group I-2, Condition 1 occupancies, where the corridor is at least 96 inches (2438 mm) wide, projections into the corridor width are permitted in accordance with Section 407.4.3 of the *International Building Code*.

704.2 Casework. Addition, alteration or reconfiguration of nonfixed and movable cases, counters and partitions not over 5 feet 9 inches (1753 mm) in height shall maintain the required means of egress path.

704.3 Locking arrangements in educational occupancies. In Group E occupancies, Group B educational occupancies and Group I-4 occupancies, egress doors with locking arrangements designed to keep intruders from entering the room shall comply with Section 1010.2.8 of the *International Building Code*.

SECTION 705
REROOFING

[BS] 705.1 General. Materials and methods of application used for recovering or replacing an existing roof covering shall comply with the requirements of Chapter 15 of the *International Building Code*.

Exceptions:

1. *Roof replacement* or roof recover of existing low-slope roof coverings shall not be required to meet the minimum design slope requirement of $^1/_4$ unit vertical in 12 units horizontal (2-percent slope) in Section 1507 of the *International Building Code* for roofs that provide positive roof drainage.

2. Recovering or replacing an existing roof covering shall not be required to meet the requirement for secondary (emergency overflow) drains or scuppers in Section 1502 of the *International Building Code* for roofs that provide for positive roof drainage. For the purposes of this exception, existing secondary drainage or scupper systems required in accordance with this code shall not be removed unless they are replaced by secondary drains or

scuppers designed and installed in accordance with Section 1502 of the *International Building Code*.

[BS] 705.2 Roof replacement. *Roof replacement* shall include the removal of all existing layers of roof coverings down to the roof deck.

Exception: Where the existing roof assembly includes an ice barrier membrane that is adhered to the roof deck, the existing ice barrier membrane shall be permitted to remain in place and covered with an additional layer of ice barrier membrane in accordance with Section 1507 of the *International Building Code*.

[BS] 705.2.1 Roof recover. The installation of a new roof covering over an existing roof covering shall be permitted where any of the following conditions occur:

1. The new roof covering is installed in accordance with the roof covering manufacturer's *approved* instructions.

2. Complete and separate roofing systems, such as standing-seam metal roof panel systems, that are designed to transmit the roof loads directly to the building's structural system and that do not rely on existing roofs and roof coverings for support, are installed.

3. Metal panel, metal shingle and concrete and clay tile roof coverings are installed over existing wood shake roofs in accordance with Section 705.3.

4. A new protective *roof coating* is applied over an existing protective *roof coating*, a metal roof panel, metal roof shingles, mineral-surfaced roll roofing, a built-up roof, modified bitumen roofing, thermoset and thermoplastic single-ply roofing or a spray polyurethane foam roofing system.

[BS] 705.2.1.1 Exceptions. A roof recover shall not be permitted where any of the following conditions occur:

1. The existing roof or roof covering is water soaked or has deteriorated to the point that the existing roof or roof covering is not adequate as a base for additional roofing.

2. The existing roof covering is slate, clay, cement or asbestos-cement tile.

3. The existing roof has two or more applications of any type of roof covering.

[BS] 705.3 Roof recovering. Where the application of a new roof covering over wood shingle or shake roofs creates a combustible concealed space, the entire existing surface shall be covered with gypsum board, mineral fiber, glass fiber or other *approved* materials securely fastened in place.

[BS] 705.4 Reinstallation of materials. Existing slate, clay or cement tile shall be permitted for reinstallation, except that damaged, cracked or broken slate or tile shall not be reinstalled. Existing vent flashing, metal edgings, drain outlets, collars and metal counterflashings shall not be reinstalled where rusted, damaged or deteriorated. Existing ballast that is damaged, cracked or broken shall not be reinstalled. Existing aggregate surfacing materials from built-up roofs shall not be reinstalled.

[BS] 705.5 Flashings. Flashings shall be reconstructed in accordance with *approved* manufacturer's installation instructions. Metal flashing to which bituminous materials are to be adhered shall be primed prior to installation.

SECTION 706
STRUCTURAL

[BS] 706.1 General. Where *alteration* work includes replacement of equipment that is supported by the building or where a reroofing permit is required, the provisions of this section shall apply.

[BS] 706.2 Addition or replacement of roofing or replacement of equipment. Any existing gravity load-carrying structural element for which an *alteration* causes an increase in design dead, live or snow load, including snow drift effects, of more than 5 percent shall be replaced or altered as needed to carry the gravity loads required by the *International Building Code* for new structures.

Exceptions:

1. Buildings of Group R occupancy with not more than five dwelling or sleeping units used solely for residential purposes where the altered building complies with the conventional light-frame construction methods of the *International Building Code* or the provisions of the *International Residential Code*.

2. Buildings in which the increased dead load is due entirely to the addition of a second layer of roof covering weighing 3 pounds per square foot (0.1437 kN/m^2) or less over an existing single layer of roof covering.

[BS] 706.3 Additional requirements for reroof permits. The requirements of this section shall apply to *alteration* work requiring reroof permits.

[BS] 706.3.1 Bracing for unreinforced masonry bearing wall parapets. Where a permit is issued for reroofing for more than 25 percent of the roof area of a building assigned to Seismic Design Category D, E or F that has parapets constructed of unreinforced masonry, the work shall include installation of parapet bracing unless an evaluation demonstrates compliance of such items. Reduced seismic forces shall be permitted.

[BS] 706.3.2 Roof diaphragms resisting wind loads in high-wind regions. Where roofing materials are removed from more than 50 percent of the roof diaphragm or section of a building located where the ultimate design wind speed, V_{ult}, determined in accordance with Figure 1609.3(1) of the *International Building Code*, is greater than 130 mph (58 m/s), roof diaphragms, connections of the roof diaphragm to roof framing members, and roof-to-wall connections shall be evaluated for the wind loads specified in the *International Building Code*, including wind uplift. If the diaphragms and connections in their current condition are not capable of resisting 75 percent of those wind loads, they shall be replaced or strengthened in accordance with the loads specified in the *International Building Code*.

Exception: Buildings that have been demonstrated to comply with the wind load provisions in ASCE 7-88 or later editions.

SECTION 707
ELECTRICAL

707.1 Health care facilities. In Group I-2 facilities, ambulatory care facilities and outpatient clinics, any altered portion of an existing electrical systems shall be required to meet installation and equipment requirements in NFPA 99.

SECTION 708
ENERGY CONSERVATION

708.1 Minimum requirements. Level 1 *alterations* to *existing buildings* or structures do not require the entire building or structure to comply with the energy requirements of the *International Energy Conservation Code* or *International Residential Code*. The *alterations* shall conform to the energy requirements of the *International Energy Conservation Code* or *International Residential Code* as they relate to new construction only.

CHAPTER 8
ALTERATIONS—LEVEL 2

User note:

About this chapter: Like Chapter 7, the purpose of this chapter is to provide detailed requirements and provisions to identify the required improvements in the existing building elements, building spaces and building structural system when a building is being altered. This chapter is distinguished from Chapters 7 and 9 by involving space reconfiguration that could be up to and including 50 percent of the area of the building. In contrast, Level 1 alterations (Chapter 7) do not involve space reconfiguration, and Level 3 alterations (Chapter 9) involve extensive space reconfiguration that exceeds 50 percent of the building area. Depending on the nature of alteration work, its location within the building, and whether it encompasses one or more tenants, improvements and upgrades could be required for the open floor penetrations, sprinkler system or the installation of additional means of egress such as stairs or fire escapes.

SECTION 801
GENERAL

801.1 Scope. Level 2 *alterations* as described in Section 603 shall comply with the requirements of this chapter.

Exception: Buildings in which the reconfiguration is exclusively the result of compliance with the accessibility requirements of Section 306.7.1 shall be permitted to comply with Chapter 7.

801.2 Alteration Level 1 compliance. In addition to the requirements of this chapter, all work shall comply with the requirements of Chapter 7.

801.3 System installations. Requirements related to *work area* are not applicable where the Level 2 *alterations* are limited solely to one or more of the following:

1. Mechanical systems, electrical systems, fire protection systems and abatement of hazardous materials.
2. Windows, hardware, operating controls, electrical outlets and signs.
3. *Alterations* undertaken for the primary purpose of increasing the accessibility of a *facility*.

801.4 Compliance. New construction elements, components, systems and spaces shall comply with the requirements of the *International Building Code*.

Exceptions:

1. Where windows are added they are not required to comply with the light and ventilation requirements of the *International Building Code*.
2. Newly installed electrical equipment shall comply with the requirements of Section 806.
3. The length of dead-end corridors in newly constructed spaces shall only be required to comply with the provisions of Section 804.7.
4. The minimum ceiling height of the newly created habitable and occupiable spaces and corridors shall be 7 feet (2134 mm).
5. Where provided in below-grade transportation stations, existing and new escalators shall be permitted to have a clear width of less than 32 inches (815 mm).
6. New structural members and connections shall be permitted to comply with alternative design criteria in accordance with Section 302.

SECTION 802
BUILDING ELEMENTS AND MATERIALS

802.1 Scope. The requirements of this section are limited to *work areas* in which Level 2 *alterations* are being performed and shall apply beyond the *work area* where specified.

802.2 Vertical openings. Existing vertical openings shall comply with the provisions of Sections 802.2.1, 802.2.2 and 802.2.3.

802.2.1 Existing vertical openings. Existing interior vertical openings connecting two or more floors shall be enclosed with *approved* assemblies having a fire-resistance rating of not less than 1 hour with *approved* opening protectives.

Exceptions:

1. Where vertical opening enclosure is not required by the *International Building Code* or the *International Fire Code*.
2. Interior vertical openings other than stairways may be blocked at the floor and ceiling of the *work area* by installation of not less than 2 inches (51 mm) of solid wood or equivalent construction.
3. The enclosure shall not be required where:
   3.1. Connecting the main floor and mezzanines; or
   3.2. All of the following conditions are met:
       3.2.1. The communicating area has a low-hazard occupancy or has a moderate-hazard occupancy that is protected throughout by an automatic sprinkler system.
       3.2.2. The lowest or next-to-the-lowest level is a street floor.

3.2.3. The entire area is open and unobstructed in a manner such that it is reasonable to assume that a fire in any part of the interconnected spaces will be readily obvious to all of the occupants.

3.2.4. Exit capacity is sufficient to provide egress simultaneously for all occupants of all levels by considering all areas to be a single floor area for the determination of required exit capacity.

3.2.5. Each floor level, considered separately, has not less than one-half of its individual required exit capacity provided by an exit or exits leading directly out of that level without having to traverse another communicating floor level or be exposed to the smoke or fire spreading from another communicating floor level.

4. In Group A occupancies, a minimum 30-minute enclosure shall be provided to protect all vertical openings not exceeding three stories.

5. In Group B occupancies, a minimum 30-minute enclosure shall be provided to protect all vertical openings not exceeding three stories. This enclosure, or the enclosure specified in Section 802.2.1, shall not be required in the following locations:

    5.1. Buildings not exceeding 3,000 square feet (279 m²) per floor.

    5.2. Buildings protected throughout by an *approved* automatic fire sprinkler system.

6. In Group E occupancies, the enclosure shall not be required for vertical openings not exceeding three stories where the building is protected throughout by an *approved* automatic fire sprinkler system.

7. In Group F occupancies, the enclosure shall not be required in the following locations:

    7.1. Vertical openings not exceeding three stories.

    7.2. Special-purpose occupancies where necessary for manufacturing operations and direct access is provided to not fewer than one protected stairway.

    7.3. Buildings protected throughout by an *approved* automatic sprinkler system.

8. In Group H occupancies, the enclosure shall not be required for vertical openings not exceeding three stories where necessary for manufacturing operations and every floor level has direct access to not fewer than two remote enclosed stairways or other *approved* exits.

9. In Group M occupancies, a minimum 30-minute enclosure shall be provided to protect all vertical openings not exceeding three stories. This enclosure, or the enclosure specified in Section 802.2.1, shall not be required in the following locations:

    9.1. Openings connecting only two floor levels.

    9.2. Occupancies protected throughout by an *approved* automatic sprinkler system.

10. In Group R-1 occupancies, the enclosure shall not be required for vertical openings not exceeding three stories in the following locations:

    10.1. Buildings protected throughout by an *approved* automatic sprinkler system.

    10.2. Buildings with less than 25 dwelling units or sleeping units where every sleeping room above the second floor is provided with direct access to a fire escape or other *approved* second exit by means of an *approved* exterior door or window having a sill height of not greater than 44 inches (1118 mm) and where both of the following conditions are met:

        10.2.1. Any exit access corridor exceeding 8 feet (2438 mm) in length that serves two means of egress, one of which is an unprotected vertical opening, shall have not fewer than one of the means of egress separated from the vertical opening by a 1-hour fire barrier.

        10.2.2. The building is protected throughout by an automatic fire alarm system, installed and supervised in accordance with the *International Building Code*.

11. In Group R-2 occupancies, a minimum 30-minute enclosure shall be provided to protect all vertical openings not exceeding three stories. This enclosure, or the enclosure specified in Section 802.2.1, shall not be required in the following locations:

    11.1. Vertical openings not exceeding two stories with not more than four dwelling units per floor.

11.2. Buildings protected throughout by an *approved* automatic sprinkler system.

11.3. Buildings with not more than four dwelling units per floor where every sleeping room above the second floor is provided with direct access to a fire escape or other *approved* second exit by means of an *approved* exterior door or window having a sill height of not greater than 44 inches (1118 mm) and the building is protected throughout by an automatic fire alarm system complying with Section 803.4.

12. One- and two-family dwellings.

13. Group S occupancies where connecting not more than two floor levels or where connecting not more than three floor levels and the structure is equipped throughout with an *approved* automatic sprinkler system.

14. Group S occupancies where vertical opening protection is not required for open parking garages and ramps.

802.2.2 Supplemental shaft and floor opening enclosure requirements. Where the *work area* on any floor exceeds 50 percent of that floor area, the enclosure requirements of Section 802.2 shall apply to vertical openings other than stairways throughout the floor.

Exception: Vertical openings located in tenant spaces that are entirely outside the *work area*.

802.2.3 Supplemental stairway enclosure requirements. Where the *work area* on any floor exceeds 50 percent of that floor area, stairways that are part of the means of egress serving the *work area* shall, at a minimum, be enclosed with smoketight construction on the highest *work area* floor and all floors below.

Exception: Where stairway enclosure is not required by the *International Building Code* or the *International Fire Code*.

802.3 Smoke compartments. In Group I-2 occupancies where the *work area* is on a story used for sleeping rooms for more than 30 care recipients, the story shall be divided into not less than two compartments by smoke barrier walls in accordance with Section 407.5 of the *International Building Code* as required for new construction.

802.4 Interior finish. The interior finish and trim of walls and ceilings in exits and corridors in any *work area* shall comply with the requirements of the *International Building Code*.

Exception: Existing materials that do not comply with the requirements of the *International Building Code* shall be permitted to be treated with an approved fire-retardant coating in accordance with the manufacturer's instructions to achieve the required classification. Compliance with this section shall be demonstrated by testing the fire-retardant coating on the same material and achieving the required performance. Where the same material is not available, testing on a similar material shall be permitted.

802.4.1 Supplemental interior finish requirements. Where the *work area* on any floor exceeds 50 percent of the floor area, Section 802.4 shall apply to the interior finish and trim in exits and corridors serving the *work area* throughout the floor.

Exception: Interior finish within tenant spaces that are entirely outside the *work area*.

802.5 Guards. The requirements of Sections 802.5.1 and 802.5.2 shall apply in all *work areas*.

802.5.1 Minimum requirement. Every portion of a floor, such as a balcony or a loading dock, that is more than 30 inches (762 mm) above the floor or grade below and is not provided with guards, or those in which the existing guards are judged to be in danger of collapsing, shall be provided with guards.

802.5.2 Design. Where there are no guards or where existing guards must be replaced, the guards shall be designed and installed in accordance with the *International Building Code*.

802.6 Fire-resistance ratings. Where *approved* by the *code official*, buildings where an automatic sprinkler system installed in accordance with Section 903.3.1.1 or 903.3.1.2 of the *International Building Code* has been added, and the building is now sprinklered throughout, the required fire-resistance ratings of building elements and materials shall be permitted to meet the requirements of the current building code. The building is required to meet the other applicable requirements of the *International Building Code*.

Plans, investigation and evaluation reports, and other data shall be submitted indicating which building elements and materials the applicant is requesting the *code official* to review and approve for determination of applying the current building code fire-resistance ratings. Any special construction features, including fire-resistance-rated assemblies and smoke-resistive assemblies, conditions of occupancy, means-of-egress conditions, fire code deficiencies, *approved* modifications or *approved* alternative materials, design and methods of construction, and equipment applying to the building that impact required fire-resistance ratings shall be identified in the evaluation reports submitted.

SECTION 803
FIRE PROTECTION

803.1 Scope. The requirements of this section shall be limited to *work areas* in which Level 2 *alterations* are being performed, and where specified they shall apply throughout the floor on which the *work areas* are located or otherwise beyond the *work area*.

803.1.1 Corridor ratings. Where an *approved* automatic sprinkler system is installed throughout the story, the required fire-resistance rating for any corridor located on the story shall be permitted to be reduced in accordance with the *International Building Code*. In order to be considered for a corridor rating reduction, such system shall provide coverage for the stairway landings serving the floor and the intermediate landings immediately below.

803.2 Automatic sprinkler systems. Automatic sprinkler systems shall be provided in accordance with the requirements of Sections 803.2.1 through 803.2.6. Installation requirements shall be in accordance with the *International Building Code*.

803.2.1 High-rise buildings. In high-rise buildings, *work areas* that have exits or corridors shared by more than one tenant or that have exits or corridors serving an occupant load greater than 30 shall be provided with automatic sprinkler protection in the entire *work area* where the *work area* is located on a floor that has a sufficient sprinkler water supply system from an existing standpipe or a sprinkler riser serving that floor.

803.2.1.1 Supplemental automatic sprinkler system requirements. Where the *work area* on any floor exceeds 50 percent of that floor area, Section 803.2.1 shall apply to the entire floor on which the *work area* is located.

Exception: Occupied tenant spaces that are entirely outside the *work area*.

803.2.2 Groups A, B, E, F-1, H, I-1, I-3, I-4, M, R-1, R-2, R-4, S-1 and S-2. In buildings with occupancies in Groups A, B, E, F-1, H, I-1, I-3, I-4, M, R-1, R-2, R-4, S-1 and S-2, *work areas* that have exits or corridors shared by more than one tenant or that have exits or corridors serving an occupant load greater than 30 shall be provided with automatic sprinkler protection where both of the following conditions occur:

1. The *work area* is required to be provided with automatic sprinkler protection in accordance with the *International Building Code* as applicable to new construction.
2. The *work area* exceeds 50 percent of the floor area.

Exception: If the building does not have sufficient municipal water supply for design of a fire sprinkler system available to the floor without installation of a new fire pump, *work areas* shall be protected by an automatic smoke detection system throughout all occupiable spaces other than sleeping units or individual dwelling units that activates the occupant notification system in accordance with Sections 907.4, 907.5 and 907.6 of the *International Building Code*.

803.2.2.1 Mixed uses. In *work areas* containing mixed uses, one or more of which requires automatic sprinkler protection in accordance with Section 803.2.2, such protection shall not be required throughout the *work area* provided that the uses requiring such protection are separated from those not requiring protection by fire-resistance-rated construction having a minimum 2-hour rating for Group H and a minimum 1-hour rating for all other occupancy groups.

803.2.3 Group I-2. In Group I-2 occupancies, an automatic sprinkler system installed in accordance with Section 903.3.1.1 of the *International Fire Code* shall be provided in the following

1. In Group I-2, Condition 1, throughout the *work area*.
2. In Group I-2, Condition 2, throughout the *work area* where the *work area* is 50 percent or less of the smoke compartment.
3. In Group I-2, Condition 2, throughout the smoke compartment in which the work occurs where the *work area* exceeds 50 percent of the smoke compartment.

803.2.4 Windowless stories. Work located in a windowless story, as determined in accordance with the *International Building Code*, shall be sprinklered where the *work area* is required to be sprinklered under the provisions of the *International Building Code* for newly constructed buildings and the building has a sufficient municipal water supply without installation of a new fire pump.

803.2.5 Other required automatic sprinkler systems. In buildings and areas listed in Table 903.2.11.6 of the *International Building Code*, *work areas* that have exits or corridors shared by more than one tenant or that have exits or corridors serving an occupant load greater than 30 shall be provided with an automatic sprinkler system under the following conditions

1. The *work area* is required to be provided with an automatic sprinkler system in accordance with the *International Building Code* applicable to new construction; and
2. The building has sufficient municipal water supply for design of an automatic sprinkler system available to the floor without installation of a new fire pump.

803.2.6 Supervision. Fire sprinkler systems required by this section shall be supervised by one of the following methods:

1. *Approved* central station system in accordance with NFPA 72.
2. *Approved* proprietary system in accordance with NFPA 72.
3. *Approved* remote station system of the jurisdiction in accordance with NFPA 72.
4. Where *approved* by the *code official*, *approved* local alarm service that will cause the sounding of an alarm in accordance with NFPA 72.

Exception: Supervision is not required for the following:

1. Underground key or hub gate valves in roadway boxes.
2. Halogenated extinguishing systems.
3. Carbon dioxide extinguishing systems.
4. Dry- and wet-chemical extinguishing systems.

5. Automatic sprinkler systems installed in accordance with NFPA 13R where a common supply main is used to supply both domestic and automatic sprinkler systems and a separate shutoff valve for the automatic sprinkler system is not provided.

803.3 Standpipes. Where the *work area* includes exits or corridors shared by more than one tenant and is located more than 50 feet (15 240 mm) above or below the lowest level of fire department access, a standpipe system shall be provided. Standpipes shall have an *approved* fire department connection with hose connections at each floor level above or below the lowest level of fire department access. Standpipe systems shall be installed in accordance with the *International Building Code*.

Exceptions:

1. A pump shall not be required provided that the standpipes are capable of accepting delivery by fire department apparatus of not less than 250 gallons per minute (gpm) at 65 pounds per square inch (psi) (946 L/m at 448 KPa) to the topmost floor in buildings equipped throughout with an automatic sprinkler system or not less than 500 gpm at 65 psi (1892 L/m at 448 KPa) to the topmost floor in all other buildings. Where the standpipe terminates below the topmost floor, the standpipe shall be designed to meet (gpm/psi) (L/m/KPa) requirements of this exception for possible future extension of the standpipe.
2. The interconnection of multiple standpipe risers shall not be required.

803.4 Fire alarm and detection. An *approved* fire alarm system shall be installed in accordance with Sections 803.4.1 through 803.4.2. Where automatic sprinkler protection is provided in accordance with Section 803.2 and is connected to the building fire alarm system, automatic heat detection shall not be required.

An *approved* automatic fire detection system shall be installed in accordance with the provisions of this code and NFPA 72. Devices, combinations of devices, appliances, and equipment shall be *approved*. The automatic fire detectors shall be smoke detectors, except that an *approved* alternative type of detector shall be installed in spaces such as boiler rooms, where products of combustion are present during normal operation in sufficient quantity to actuate a smoke detector.

803.4.1 Occupancy requirements. A fire alarm system shall be installed in accordance with Sections 803.4.1.1 through 803.4.1.6. Existing alarm-notification appliances shall be automatically activated throughout the building. Where the building is not equipped with a fire alarm system, alarm-notification appliances within the *work area* shall be provided and automatically activated.

Exceptions:

1. Occupancies with an existing, previously *approved* fire alarm system.

2. Where selective notification is permitted, alarm-notification appliances shall be automatically activated in the areas selected.

803.4.1.1 Group E. A fire alarm system shall be installed in *work areas* of Group E occupancies as required by the *International Fire Code* for existing Group E occupancies.

803.4.1.2 Group I-1. An automatic fire alarm system shall be installed in *work areas* of Group I-1 *facilities* as required by Chapter 11 of the *International Fire Code* for existing Group I-1 occupancies.

803.4.1.3 Group I-2. An automatic fire alarm system shall be installed throughout Group I-2 occupancies as required by Chapter 11 of the *International Fire Code*.

803.4.1.4 Group I-3. A fire alarm system shall be installed in *work areas* of Group I-3 occupancies as required by the *International Fire Code*.

803.4.1.5 Group R-1. A fire alarm system shall be installed in Group R-1 occupancies as required by the *International Fire Code* for existing Group R-1 occupancies.

803.4.1.6 Group R-2. A fire alarm system shall be installed in *work areas* of Group R-2 apartment buildings as required by the *International Fire Code* for existing Group R-2 occupancies.

803.4.2 Supplemental fire alarm system requirements. Where the *work area* on any floor exceeds 50 percent of that floor area, Section 803.4.1 shall apply throughout the floor.

Exception: Alarm-initiating and notification appliances shall not be required to be installed in tenant spaces outside of the *work area*.

SECTION 804
MEANS OF EGRESS

804.1 Scope. The requirements of this section shall be limited to *work areas* that include exits or corridors shared by more than one tenant within the *work area* in which Level 2 *alterations* are being performed, and where specified they shall apply throughout the floor on which the *work areas* are located or otherwise beyond the *work area*.

804.2 General. The means of egress shall comply with the requirements of this section.

Exceptions:

1. Where the *work area* and the means of egress serving it complies with NFPA 101.
2. Means of egress complying with the requirements of the building code under which the building was constructed shall be considered to be compliant means of egress if, in the opinion of the *code official*, they do not constitute a distinct hazard to life.

804.3 Group I-2. In Group I-2 occupancies, in areas where corridors are used for movement of care recipients in beds, the clear width of ramps and corridors shall be not less than 48 inches (1219 mm).

804.4 Number of exits. The number of exits shall be in accordance with Sections 804.4.1 through 804.4.3.

804.4.1 Minimum number. Every story utilized for human occupancy on which there is a *work area* that includes exits or corridors shared by more than one tenant within the *work area* shall be provided with the minimum number of exits based on the occupancy and the occupant load in accordance with the *International Building Code*. In addition, the exits shall comply with Sections 804.4.1.1 and 804.4.1.2.

804.4.1.1 Single-exit buildings. A single exit or access to a single exit shall be permitted from spaces, any story or any occupied roof where one of the following conditions exists:

1. The occupant load, number of dwelling units and exit access travel distance do not exceed the values in Table 804.4.1.1(1) or Table 804.4.1.1(2).

2. In Group R-1 or R-2, buildings without an *approved* automatic sprinkler system, individual single-story or multiple-story dwelling or sleeping units shall be permitted to have a single exit or access to a single exit from the dwelling or sleeping unit provided one of the following criteria are met:

    2.1. The occupant load is not greater than 10 and the exit access travel distance within the unit does not exceed 75 feet (22 860 mm).

    2.2. The building is not more than three stories in height; all third-story space is part of dwelling with an exit access doorway on the second story; and the portion of the exit access travel distance from the door to any habitable room within any such unit to the unit entrance doors does not exceed 50 feet (15 240 mm).

3. In buildings of Group R-2 occupancy of any number of stories with not more than four dwelling units per floor served by an interior exit stairway; with a smokeproof enclosure in accordance with Sections 909.20 and 1023.12 of the *International Building Code* or an exterior stairway as an exit; and where the portion of the exit access travel distance from the dwelling unit entrance door to the exit is not greater than 20 feet (6096 mm).

TABLE 804.4.1.1(1)
STORIES WITH ONE EXIT OR ACCESS TO ONE EXIT FOR R-2 OCCUPANCIES

STORY	OCCUPANCY	MAXIMUM NUMBER OF DWELLING UNITS	MAXIMUM EXIT ACCESS TRAVEL DISTANCE (feet)
Basement, first or second story above grade plane	R-2[a]	4 dwelling units	50
Third story above grade plane and higher	NP	NA	NA

For SI: 1 foot = 304.8 mm.
NP = Not Permitted.
NA = Not Applicable.

a. Group R-2, without an approved automatic sprinkler system and provided with emergency escape and rescue openings in accordance with Section 1031 of the *International Building Code*.

TABLE 804.4.1.1(2)
STORIES WITH ONE EXIT OR ACCESS TO ONE EXIT FOR OTHER OCCUPANCIES

STORY	OCCUPANCY	MAXIMUM OCCUPANT LOAD PER STORY	MAXIMUM EXIT ACCESS TRAVEL DISTANCE (feet)
First story above or below grade plane	B, F-2, S-2[a]	35	75
Second story above grade plane	B, F-2, S-2[a]	35	75
Third story above grade plane and higher	NP	NA	NA

For SI: 1 foot = 304.8 mm.
NP = Not Permitted.
NA = Not Applicable.

a. The length of exit access travel distance in a Group S-2 open parking garage shall be not more than 100 feet.

804.4.1.2 Fire escapes required. For other than Group I-2, where more than one exit is required, an existing or newly constructed fire escape complying with Section 804.4.1.2.1 shall be accepted as providing one of the required means of egress.

804.4.1.2.1 Fire escape access and details. Fire escapes shall comply with all of the following requirements:

1. Occupants shall have unobstructed access to the fire escape without having to pass through a room subject to locking.

2. Access to a new fire escape shall be through a door, except that windows shall be permitted to provide access from single dwelling units or sleeping units in Group R-1, R-2 and I-1 occupancies or to provide access from spaces having a maximum occupant load of 10 in other occupancy classifications.

    2.1. The window shall have a minimum net clear opening of 5.7

square feet (0.53 m^2) or 5 square feet (0.46 m^2) where located at grade.

2.2. The minimum net clear opening height shall be 24 inches (610 mm) and net clear opening width shall be 20 inches (508 mm).

2.3. The bottom of the clear opening shall not be greater than 44 inches (1118 mm) above the floor.

2.4. The operation of the window shall comply with the operational constraints of the *International Building Code*.

3. Newly constructed fire escapes shall be permitted only where exterior stairways cannot be utilized because of lot lines limiting the stairway size or because of the sidewalks, alleys, or roads at grade level.

4. Openings within 10 feet (3048 mm) of fire escape stairways shall be protected by fire assemblies having minimum $^3/_4$-hour fire-resistance ratings.

   Exception: Opening protection shall not be required in buildings equipped throughout with an *approved* automatic sprinkler system.

5. In all buildings of Group E occupancy, up to and including the 12th grade, buildings of Group I occupancy, rooming houses and childcare centers, ladders of any type are prohibited on fire escapes used as a required means of egress.

804.4.1.2.2 Construction. The fire escape shall be designed to support a live load of 100 pounds per square foot (4788 Pa) and shall be constructed of steel or other *approved noncombustible materials*. Fire escapes constructed of wood not less than nominal 2 inches (51 mm) thick are permitted on buildings of Type V construction. Walkways and railings located over or supported by combustible roofs in buildings of Types III and IV construction are permitted to be of wood not less than nominal 2 inches (51 mm) thick.

804.4.1.2.3 Dimensions. Stairways shall be not less than 22 inches (559 mm) wide with risers not more than, and treads not less than, 8 inches (203 mm). Landings at the foot of stairways shall be not less than 40 inches (1016 mm) wide by 36 inches (914 mm) long and located not more than 8 inches (203 mm) below the door.

804.4.2 Mezzanines. Mezzanines in the *work area* and with an occupant load of more than 50 or in which the travel distance to an exit exceeds 75 feet (22 860 mm) shall have access to not fewer than two independent means of egress.

Exception: Two independent means of egress are not required where the travel distance to an exit does not exceed 100 feet (30 480 mm) and the building is protected throughout with an automatic sprinkler system.

804.4.3 Main entrance—Group A. Buildings of Group A with an occupant load of 300 or more shall be provided with a main entrance capable of serving as the main exit with an egress capacity of not less than one-half of the total occupant load. The remaining exits shall be capable of providing one-half of the total required exit capacity.

Exception: Where a main exit is not well defined or where multiple main exits are provided, exits shall be permitted to be distributed around the perimeter of the building provided that the total width of egress is not less than 100 percent of the required width.

804.5 Egress doorways. Egress doorways in any *work area* shall comply with Sections 804.5.1 through 804.5.5.

804.5.1 Two egress doorways required. Work areas shall be provided with two egress doorways in accordance with the requirements of Sections 804.5.1.1 and 804.5.1.2.

804.5.1.1 Occupant load and travel distance. In any *work area*, all rooms and spaces having an occupant load greater than 50 or in which the travel distance to an exit exceeds 75 feet (22 860 mm) shall have not fewer than two egress doorways.

Exceptions:

1. Storage rooms having a maximum occupant load of 10.
2. Where the *work area* is served by a single exit in accordance with Section 804.4.1.1.

804.5.1.2 Group I-2. In Group I-2, Condition 2 *work areas* that include altered care suites shall comply with Sections 407.4.4 through 407.4.4.6.2 of the *International Building Code*.

804.5.2 Door swing. In the *work area* and in the egress path from any *work area* to the exit discharge, all egress doors serving an occupant load greater than 50 shall swing in the direction of exit travel.

804.5.2.1 Supplemental requirements for door swing. Where the *work area* exceeds 50 percent of the floor area, door swing shall comply with Section 804.5.2 throughout the floor.

Exception: Means of egress within or serving only a tenant space that is entirely outside the *work area*.

804.5.3 Door closing. In any *work area*, all doors opening onto an exit passageway at grade or an exit stairway shall be self-closing or automatic-closing by listed closing devices.

Exceptions:

1. Where exit enclosure is not required by the *International Building Code*.
2. Means of egress within or serving only a tenant space that is entirely outside the *work area*.

804.5.3.1 Supplemental requirements for door closing. Where the *work area* exceeds 50 percent of the floor area, doors shall comply with Section 804.5.3 throughout the exit stairway from the *work area* to, and including, the level of exit discharge.

804.5.4 Panic and fire exit hardware. In any *work area*, and in the egress path from any *work area* to the exit discharge, in buildings or portions thereof of Group A assembly occupancies with an occupant load greater than 100, all required exit doors equipped with latching devices shall be equipped with *approved* panic or fire exit hardware in accordance with Section 1010.2.9 of the *International Building Code*.

804.5.4.1 Supplemental requirements for panic hardware. Where the *work area* exceeds 50 percent of the floor area, panic hardware shall comply with Section 804.5.4 throughout the floor.

Exception: Means of egress within a tenant space that is entirely outside the *work area*.

804.5.5 Emergency power source in Group I-3. Power-operated sliding doors or power-operated locks for swinging doors shall be operable by a manual release mechanism at the door. Emergency power shall be provided for the doors and locks in accordance with Section 2702 of the *International Building Code*.

Exceptions:

1. Emergency power is not required in *facilities* with 10 or fewer locks complying with the exception to Section 408.4.1 of the *International Building Code*.
2. Emergency power is not required where remote mechanical operating releases are provided.

804.6 Openings in corridor walls. Openings in corridor walls in any *work area* shall comply with Sections 804.6.1 through 804.6.4.

Exception: Openings in corridors where such corridors are not required to be rated in accordance with the *International Building Code*.

804.6.1 Corridor doors. Corridor doors in the *work area* shall not be constructed of hollow core wood and shall not contain louvers. Dwelling unit or sleeping unit corridor doors in *work areas* in buildings of Groups R-1, R-2 and I-1 shall be not less than 1$^3/_8$-inch (35 mm) solid core wood or *approved* equivalent and shall not have any glass panels, other than *approved* wired glass or other *approved* glazing material in metal frames. Dwelling unit or sleeping unit corridor doors in *work areas* in buildings of Groups R-1, R-2 and I-1 shall be equipped with *approved* door closers. Replacement doors shall be 1$^3/_4$-inch (44 mm) solid bonded wood core or *approved* equivalent, unless the existing frame will accommodate only a 1$^3/_8$-inch (35 mm) door.

Exceptions:

1. Corridor doors within a dwelling unit or sleeping unit.
2. Existing doors meeting the requirements of *Guidelines on Fire Ratings of Archaic Materials and Assemblies* (Resource A) for a rating of 15 minutes or more shall be accepted as meeting the provisions of this requirement.
3. Existing doors in buildings protected throughout with an *approved* automatic sprinkler system shall be required only to resist smoke, be reasonably tight fitting and shall not contain louvers.
4. In group homes with not more than 15 occupants and that are protected with an *approved* automatic detection system, closing devices are not required.
5. Door assemblies having a fire protection rating of not less than 20 minutes.

804.6.2 Transoms. In all buildings of Group I-1, I-2, R-1 and R-2 occupancies, all transoms in corridor walls in *work areas* shall be either glazed with $^1/_4$-inch (6.4 mm) wired glass set in metal frames or other glazing assemblies having a fire protection rating as required for the door and permanently secured in the closed position or sealed with materials consistent with the corridor construction.

804.6.3 Other corridor openings. In any *work area*, unless protected in accordance with Section 716 of the *International Building Code*, any other sash, grille or opening in a corridor, and any window in a corridor not opening to the outside air, shall be sealed with materials consistent with the corridor construction.

804.6.3.1 Supplemental requirements for other corridor opening. Where the *work area* exceeds 50 percent of the floor area, Section 804.6.3 shall be applicable to all corridor windows, grills, sashes and other openings on the floor.

Exception: Means of egress within or serving only a tenant space that is entirely outside the *work area*.

804.6.4 Supplemental requirements for corridor openings. Where the *work area* on any floor exceeds 50 percent of the floor area, the requirements of Sections 804.6.1 through 804.6.3 shall apply throughout the floor.

804.7 Dead-end corridors. Dead-end corridors in any *work area* shall not exceed 35 feet (10 670 mm). In Group I-2 occupancies, dead-end corridors shall not exceed 30 feet (9144 mm).

Exceptions:

1. Where dead-end corridors of greater length are permitted by the *International Building Code*.

2. In other than Group A, I-2 and H occupancies, the maximum length of an existing dead-end corridor shall be 50 feet (15 240 mm) in buildings equipped throughout with an automatic fire alarm system installed in accordance with the *International Building Code*.

3. In other than Group A, I-2 and H occupancies, the maximum length of an existing dead-end corridor shall be 70 feet (21 356 mm) in buildings equipped throughout with an automatic sprinkler system installed in accordance with the *International Building Code*.

4. In other than Group A, I-2 and H occupancies, the maximum length of an existing, newly constructed, or extended dead-end corridor shall not exceed 50 feet (15 240 mm) on floors equipped with an automatic sprinkler system installed in accordance with the *International Building Code*.

804.8 Means-of-egress lighting. Means-of-egress lighting shall be in accordance with this section, as applicable.

804.8.1 Artificial lighting required. Means of egress in all *work areas* shall be provided with artificial lighting in accordance with the requirements of the *International Building Code*.

804.8.2 Supplemental requirements for means-of-egress lighting. Where the *work area* on any floor exceeds 50 percent of that floor area, means of egress throughout the floor shall comply with Section 804.8.1.

Exception: Means of egress within or serving only a tenant space that is entirely outside the *work area*.

804.9 Exit signs. Exit signs shall be in accordance with this section, as applicable.

804.9.1 Work areas. Means of egress in all *work areas* shall be provided with exit signs in accordance with the requirements of the *International Building Code*.

804.9.2 Supplemental requirements for exit signs. Where the *work area* on any floor exceeds 50 percent of that floor area, means of egress throughout the floor shall comply with Section 804.9.1.

Exception: Means of egress within a tenant space that is entirely outside the *work area*.

804.10 Handrails. The requirements of Sections 804.10.1 and 804.10.2 shall apply to handrails from the *work area* floor to, and including, the level of exit discharge.

804.10.1 Minimum requirement. Every required exit stairway that is part of the means of egress for any *work area* and that has three or more risers and is not provided with not fewer than one handrail, or in which the existing handrails are judged to be in danger of collapsing, shall be provided with handrails for the full length of the stairway on not fewer than one side. Exit stairways with a required egress width of more than 66 inches (1676 mm) shall have handrails on both sides.

804.10.2 Design. Handrails required in accordance with Section 804.10.1 shall be designed and installed in accordance with the provisions of the *International Building Code*.

804.11 Refuge areas. Where *alterations* affect the configuration of an area utilized as a refuge area, the capacity of the refuge area shall not be reduced below the required capacity of the refuge area for horizontal exits in accordance with Section 1026.4 of the *International Building Code*. Where the horizontal exit also forms a smoke compartment, the capacity of the refuge area for Group I-1, I-2 and I-3 occupancies and Group B ambulatory care *facilities* shall not be reduced below that required in Sections 407.5.3, 408.6.2, 420.6.1 and 422.3.2 of the *International Building Code*, as applicable.

804.12 Guards. The requirements of Sections 804.12.1 and 804.12.2 shall apply to guards from the *work area* floor to, and including, the level of exit discharge but shall be confined to the egress path of any *work area*.

804.12.1 Minimum requirement. Every open portion of a stairway, landing, or balcony that is more than 30 inches (762 mm) above the floor or grade below and is not provided with guards, or those portions in which existing guards are judged to be in danger of collapsing, shall be provided with guards.

804.12.2 Design. Guards required in accordance with Section 804.12.1 shall be designed and installed in accordance with the *International Building Code*.

SECTION 805
STRUCTURAL

[BS] 805.1 General. Structural elements and systems within buildings undergoing Level 2 *alterations* shall comply with this section.

[BS] 805.2 Existing structural elements carrying gravity loads. Any existing gravity load-carrying structural element for which an *alteration* causes an increase in design dead, live or snow load, including snow drift effects, of more than 5 percent shall be replaced or altered as needed to carry the gravity loads required by the *International Building Code* for new structures. Any existing gravity load-carrying structural element whose gravity load-carrying capacity is decreased as part of the *alteration* shall be shown to have the capacity to resist the applicable design dead, live and snow loads, including snow drift effects, required by the *International Building Code* for new structures.

Exceptions:

1. Buildings of Group R occupancy with not more than five dwelling or sleeping units used solely for residential purposes where the altered building complies with the conventional light-frame construction methods of the *International Building Code* or the provisions of the *International Residential Code*.

2. Buildings in which the increased dead load is attributable to the addition of a second layer of roof covering weighing 3 pounds per square foot

ALTERATIONS—LEVEL 2

(0.1437 kN/m^2) or less over an existing single layer of roof covering.

[BS] 805.3 Existing structural elements resisting lateral loads. Except as permitted by Section 805.4, where the *alteration* increases design lateral loads, or where the alteration results in prohibited structural irregularity as defined in ASCE 7, or where the *alteration* decreases the capacity of any existing lateral load-carrying structural element, the structure of the altered building or structure shall meet the requirements of Sections 1609 and 1613 of the *International Building Code*. Reduced seismic forces shall be permitted.

> **Exception:** Any existing lateral load-carrying structural element whose demand-capacity ratio with the *alteration* considered is not more than 10 percent greater than its demand-capacity ratio with the *alteration* ignored shall be permitted to remain unaltered. For purposes of calculating demand-capacity ratios, the demand shall consider applicable load combinations with design lateral loads or forces in accordance with Sections 1609 and 1613 of the *International Building Code*. Reduced seismic forces shall be permitted. For purposes of this exception, comparisons of demand-capacity ratios and calculation of design lateral loads, forces and capacities shall account for the cumulative effects of *additions* and *alterations* since original construction.

[BS] 805.4 Voluntary lateral force-resisting system alterations. Structural *alterations* that are intended exclusively to improve the lateral force-resisting system and are not required by other sections of this code shall not be required to meet the requirements of Section 1609 or Section 1613 of the *International Building Code*, provided that the following conditions are met:

1. The capacity of existing structural systems to resist forces is not reduced.
2. New structural elements are detailed and connected to existing or new structural elements as required by the *International Building Code* for new construction.
3. New or relocated nonstructural elements are detailed and connected to existing or new structural elements as required by the *International Building Code* for new construction.
4. The *alterations* do not create a structural irregularity as defined in ASCE 7 or make an existing structural irregularity more severe.

SECTION 806
ELECTRICAL

806.1 New installations. Newly installed electrical equipment and wiring relating to work done in any *work area* shall comply with all applicable requirements of NFPA 70 except as provided for in Section 806.4.

806.2 Existing installations. Existing wiring in all *work areas* in Group A-1, A-2, A-5, H and I occupancies shall be upgraded to meet the materials and methods requirements of Chapter 7.

806.3 Health care facilities. In Group I-2 *facilities*, ambulatory care *facilities* and outpatient clinics, any added portion of an existing electrical system shall be required to meet installation and equipment requirements in NFPA 99.

806.4 Residential occupancies. In Group R-2, R-3 and R-4 occupancies and buildings regulated by the *International Residential Code*, the requirements of Sections 806.4.1 through 806.4.7 shall be applicable only to *work areas* located within a dwelling unit.

> **806.4.1 Enclosed areas.** Enclosed areas, other than closets, kitchens, basements, garages, hallways, laundry areas, utility areas, storage areas and bathrooms shall have not fewer than two duplex receptacle outlets or one duplex receptacle outlet and one ceiling or wall-type lighting outlet.
>
> **806.4.2 Kitchens.** Kitchen areas shall have not fewer than two duplex receptacle outlets.
>
> **806.4.3 Laundry areas.** Laundry areas shall have not fewer than one duplex receptacle outlet located near the laundry equipment and installed on an independent circuit.
>
> **806.4.4 Ground fault circuit interruption.** Newly installed receptacle outlets shall be provided with ground fault circuit interruption as required by NFPA 70.
>
> **806.4.5 Minimum lighting outlets.** Not fewer than one lighting outlet shall be provided in every bathroom, hallway, stairway, attached garage and detached garage with electric power, and to illuminate outdoor entrances and exits.
>
> **806.4.6 Utility rooms and basements.** Not fewer than one lighting outlet shall be provided in utility rooms and basements where such spaces are used for storage or contain equipment requiring service.
>
> **806.4.7 Clearance for equipment.** Clearance for electrical service equipment shall be provided in accordance with NFPA 70.

SECTION 807
MECHANICAL

807.1 Reconfigured or converted spaces. Reconfigured spaces intended for occupancy and spaces converted to habitable or occupiable space in any *work area* shall be provided with natural or mechanical ventilation in accordance with the *International Mechanical Code*.

> **Exception:** Existing mechanical ventilation systems shall comply with the requirements of Section 807.2.

807.2 Altered existing systems. In mechanically ventilated spaces, existing mechanical ventilation systems that are altered, reconfigured or extended shall provide not less than 5 cubic feet per minute (cfm) (0.0024 m^3/s) per person of outdoor air and not less than 15 cfm (0.0071 m^3/s) of ventilation air per person, or not less than the amount of ventilation air determined by the Indoor Air Quality Procedure of ASHRAE 62.1.

807.3 Local exhaust. Newly introduced devices, equipment or operations that produce airborne particulate matter, odors, fumes, vapor, combustion products, gaseous contaminants, pathogenic and allergenic organisms, and microbial contaminants in such quantities as to affect adversely or impair health or cause discomfort to occupants shall be provided with local exhaust.

SECTION 808
PLUMBING

808.1 Health care facilities. In Group I-2 *facilities*, ambulatory care *facilities* and outpatient clinics, any added portion of an existing medical gas system shall be required to meet installation and equipment requirements in NFPA 99.

SECTION 809
ENERGY CONSERVATION

809.1 Minimum requirements. Level 2 *alterations* to *existing buildings* or structures are permitted without requiring the entire building or structure to comply with the energy requirements of the *International Energy Conservation Code* or *International Residential Code*. The *alterations* shall conform to the energy requirements of the *International Energy Conservation Code* or *International Residential Code* as they relate to new construction only.

CHAPTER 9
ALTERATIONS—LEVEL 3

User note:

About this chapter: Chapter 9 provides the technical requirements for those existing buildings that undergo Level 3 alterations. The purpose of this chapter is to provide detailed requirements and provisions to identify the required improvements in the existing building elements, building spaces and building structural system. This chapter is distinguished from Chapters 7 and 8 by involving alterations that cover 50 percent or more of the aggregate area of the building. In contrast, Level 1 alterations do not involve space reconfiguration, and Level 2 alterations involve extensive space reconfiguration that does not exceed 50 percent of the building area. Depending on the nature of alteration work, its location within the building, and whether it encompasses one or more tenants, improvements and upgrades could be required for the open floor penetrations, sprinkler system or the installation of additional means of egress such as stairs or fire escapes. At times and under certain situations, this chapter also is intended to improve the safety of certain building features beyond the work area and in other parts of the building where no alteration work might be taking place.

SECTION 901
GENERAL

901.1 Scope. Level 3 *alterations* as described in Section 604 shall comply with the requirements of this chapter.

901.2 Compliance. In addition to the provisions of this chapter, work shall comply with all of the requirements of Chapters 7 and 8. The requirements of Sections 802, 803, 804 and 805 shall apply within all *work areas* whether or not they include exits and corridors shared by more than one tenant and regardless of the occupant load.

Exception: Buildings in which the reconfiguration of space affecting exits or shared egress access is exclusively the result of compliance with the accessibility requirements of Section 306.7.1 shall not be required to comply with this chapter.

SECTION 902
SPECIAL USE AND OCCUPANCY

902.1 High-rise buildings. Any building having occupied floors more than 75 feet (22 860 mm) above the lowest level of fire department vehicle access shall comply with the requirements of Sections 902.1.1 and 902.1.2.

902.1.1 Recirculating air or exhaust systems. Where a floor is served by a recirculating air or exhaust system with a capacity greater than 15,000 cubic feet per minute (701 m^3/s), that system shall be equipped with *approved* smoke and heat detection devices installed in accordance with the *International Mechanical Code*.

902.1.2 Elevators. Where there is an elevator or elevators for public use, not fewer than one elevator serving the *work area* shall comply with this section. Existing elevators with a travel distance of 25 feet (7620 mm) or more above or below the main floor or other level of a building and intended to serve the needs of emergency personnel for fire-fighting or rescue purposes shall be provided with emergency operation in accordance with ASME A17.3.

New elevators shall be provided with Phase I emergency recall operation and Phase II emergency in-car operation in accordance with ASME A17.1/CSA B44.1.

902.2 Boiler and furnace equipment rooms. Boiler and furnace equipment rooms adjacent to or within Group I-1, I-2, I-4, R-1, R-2 and R-4 occupancies shall be enclosed by 1-hour fire-resistance-rated construction.

Exceptions:

1. Steam boiler equipment operating at pressures of 15 pounds per square inch gauge (psig) (103.4 kPa) or less is not required to be enclosed.

2. Hot water boilers operating at pressures of 170 psig (1171 kPa) or less are not required to be enclosed.

3. Furnace and boiler equipment with 400,000 British thermal units (Btu) (4.22 × 10^8 J) per hour input rating or less is not required to be enclosed.

4. Furnace rooms protected with an automatic sprinkler system are not required to be enclosed.

SECTION 903
BUILDING ELEMENTS AND MATERIALS

903.1 Existing shafts and vertical openings. Existing stairways that are part of the means of egress shall be enclosed in accordance with Section 802.2.1 from the highest *work area* floor to, and including, the level of exit discharge and all floors below.

903.2 Fire partitions in Group R-3. Fire separation in Group R-3 occupancies shall be in accordance with Section 903.2.1.

903.2.1 Separation required. Where the *work area* is in any attached dwelling unit in Group R-3 or any multiple single-family dwelling (townhouse), walls separating the dwelling units that are not continuous from the foundation to the underside of the roof sheathing shall be constructed to provide a continuous fire separation using

ALTERATIONS—LEVEL 3

construction materials consistent with the existing wall or complying with the requirements for new structures. Work shall be performed on the side of the dwelling unit wall that is part of the *work area*.

> **Exception:** Where *alterations* or *repairs* do not result in the removal of wall or ceiling finishes exposing the structure, walls are not required to be continuous through concealed floor spaces.

903.3 Interior finish. Interior finish in exits serving the *work area* shall comply with Section 802.4 between the highest floor on which there is a *work area* to the floor of exit discharge.

903.4 Enhanced classroom acoustics. In Group E occupancies, where the *work area* is a Level 3 alteration, enhanced classroom acoustics shall be provided in all classrooms with a volume of 20,000 cubic feet (565 m^3) or less. Enhanced classroom acoustics shall comply with the reverberation time in Section 808 of ICC A117.1.

SECTION 904
FIRE PROTECTION

904.1 Automatic sprinkler systems. An automatic sprinkler system shall be provided in a *work area* where required by Section 803.2 or this section.

904.1.1 High-rise buildings. An automatic sprinkler system shall be provided in *work areas* where the high-rise building has a sufficient municipal water supply for the design and installation of an automatic sprinkler system at the site.

904.1.2 Rubbish and linen chutes. Rubbish and linen chutes located in the *work area* shall be provided with automatic sprinkler system protection or an *approved* automatic fire-extinguishing system where protection of the rubbish and linen chute would be required under the provisions of the *International Building Code* for new construction.

904.1.3 Upholstered furniture or mattresses. *Work areas* shall be provided with an automatic sprinkler system in accordance with the *International Building Code* where any of the following conditions exist:

1. A Group F-1 occupancy used for the manufacture of upholstered furniture or mattresses exceeds 2,500 square feet (232 m^2).
2. A Group M occupancy used for the display and sale of upholstered furniture or mattresses exceeds 5,000 square feet (464 m^2).
3. A Group S-1 occupancy used for the storage of upholstered furniture or mattresses exceeds 2,500 square feet (232 m^2).

904.1.4 Groups A, B, E, F-1, H, I-1, I-3, I-4, M, R-1, R-2, R-4, S-1 and S-2. In buildings with occupancies in Groups A, B, E, F-1, H, I-1, I-3, I-4, M, R-1, R-2, R-4, S-1 and S-2 work areas shall be provided with automatic sprinkler protection where all of the following conditions occur:

1. The *work area* is required to be provided with automatic sprinkler protection in accordance with the *International Building Code* as applicable to new construction.
2. The building site has sufficient municipal water supply for design and installation of an automatic sprinkler system.

> **Exception:** If the building site does not have sufficient municipal water supply for design of an automatic sprinkler system, work areas shall be protected by an automatic smoke detection system throughout all occupiable spaces other than sleeping units or individual dwelling units that activates the occupant notification system in accordance with Sections 907.4, 907.5 and 907.6 of the *International Building Code.*

904.1.5 Group I-2. In Group I-2 occupancies, an automatic sprinkler system installed in accordance with Section 903.3.1.1 of the *International Fire Code* shall be provided in the following:

1. In Group I-2, Condition 1, throughout the *work area*.
2. In Group I-2, Condition 2, throughout the *work area* where the *work area* is 50 percent or less of the smoke compartment.
3. In Group I-2, Condition 2, throughout the smoke compartment in which the work occurs where the *work area* exceeds 50 percent of the smoke compartment.

904.1.6 Windowless stories. Work located in a windowless story, as determined in accordance with the *International Building Code*, shall be sprinklered where the *work area* is required to be sprinklered under the provisions of the *International Building Code* for newly constructed buildings and the building site has a sufficient municipal water supply for the design and installation of an automatic sprinkler system.

904.1.7 Other required automatic sprinkler systems. In buildings and areas listed in Table 903.2.11.6 of the *International Building Code*, *work areas* shall be provided with an automatic sprinkler system under the following conditions:

1. The *work area* is required to be provided with an automatic sprinkler system in accordance with the *International Building Code* applicable to new construction.
2. The building site has sufficient municipal water supply for design and installation of an automatic sprinkler system.

904.2 Fire alarm and detection systems. Fire alarm and detection shall be provided in accordance with Section 907 of the *International Building Code* as required for new construction.

904.2.1 Manual fire alarm systems. Where required by the *International Building Code*, a manual fire alarm system shall be provided throughout the *work area*. Alarm notification appliances shall be provided on such floors and shall be automatically activated as required by the *International Building Code*.

Exceptions:

1. Alarm-initiating and notification appliances shall not be required to be installed in tenant spaces outside of the *work area*.
2. Visual alarm notification appliances are not required, except where an existing alarm system is upgraded or replaced or where a new fire alarm system is installed.

904.2.2 Automatic fire detection. Where required by the *International Building Code* for new buildings, automatic fire detection systems shall be provided throughout the *work area*.

SECTION 905
MEANS OF EGRESS

905.1 General. The means of egress shall comply with the requirements of Section 804 except as specifically required in Sections 905.2 and 905.3.

905.2 Means-of-egress lighting. Means of egress from the highest *work area* floor to the floor of exit discharge shall be provided with artificial lighting within the exit enclosure in accordance with the requirements of the *International Building Code*.

905.3 Exit signs. Means of egress from the highest *work area* floor to the floor of exit discharge shall be provided with exit signs in accordance with the requirements of the *International Building Code*.

905.4 Two-way communications systems. In buildings with elevator service, a two-way communication system shall be provided where required by Section 1009.8 of the *International Building Code*.

SECTION 906
STRUCTURAL

[BS] 906.1 General. Where buildings are undergoing Level 3 *alterations*, the provisions of this section shall apply.

[BS] 906.2 Existing structural elements resisting lateral loads. Where work involves a *substantial structural alteration*, the lateral load-resisting system of the altered building shall be shown to satisfy the requirements of Sections 1609 and 1613 of the *International Building Code*. Reduced seismic forces shall be permitted.

Exceptions:

1. Buildings of Group R occupancy with not more than five dwelling or sleeping units used solely for residential purposes that are altered based on the conventional light-frame construction methods of the *International Building Code* or in compliance with the provisions of the *International Residential Code*.
2. Where the intended alteration involves only the lowest story of a building, only the lateral load resisting components in and below that story need comply with this section.

[BS] 906.3 Seismic Design Category F. Where the building is assigned to Seismic Design Category F, the structure of the altered building shall meet the requirements of Sections 1609 and 1613 of the *International Building Code*. Reduced seismic forces shall be permitted.

[BS] 906.4 Anchorage for concrete and masonry buildings. For any building assigned to Seismic Design Category D, E or F with a structural system that includes concrete or reinforced masonry walls with a flexible roof diaphragm, the *alteration* work shall include installation of wall anchors at the roof line of all subject buildings and at the floor lines of unreinforced masonry buildings unless an evaluation demonstrates compliance of existing wall anchorage. Reduced seismic forces shall be permitted.

[BS] 906.5 Anchorage for unreinforced masonry walls. For any building assigned to Seismic Design Category C, D, E or F with a structural system that includes unreinforced masonry bearing walls, the *alteration* work shall include installation of wall anchors at the roof line, unless an evaluation demonstrates compliance of existing wall anchorage. Reduced seismic forces shall be permitted.

[BS] 906.6 Bracing for unreinforced masonry parapets. Parapets constructed of unreinforced masonry in buildings assigned to Seismic Design Category C, D, E or F shall have bracing installed as needed to resist the reduced *International Building Code*-level seismic forces in accordance with Section 304.3, unless an evaluation demonstrates compliance of such items. Use of reduced seismic forces shall be permitted.

[BS] 906.7 Anchorage of unreinforced masonry partitions. Where the building is assigned to Seismic Design Category C, D, E or F, unreinforced masonry partitions and nonstructural walls within the *work area* and adjacent to egress paths from the *work area* shall be anchored, removed, or altered to resist out-of-plane seismic forces, unless an evaluation demonstrates compliance of such items. Use of reduced seismic forces shall be permitted.

SECTION 907
ENERGY CONSERVATION

907.1 Minimum requirements. Level 3 *alterations* to *existing buildings* or structures are permitted without requiring the entire building or structure to comply with the energy requirements of the *International Energy Conservation Code* or *International Residential Code*. The *alterations* shall conform to the energy requirements of the *International Energy Conservation Code* or *International Residential Code* as they relate to new construction only.

CHAPTER 10
CHANGE OF OCCUPANCY

User note:

About this chapter: The purpose of this chapter is to provide regulations for the circumstances where an existing building is subject to a change of occupancy or a change of occupancy classification. A change of occupancy is not to be confused with a change of occupancy classification. The International Building Code® *defines different occupancy classifications in Chapter 3 and special occupancy requirements in Chapter 4. Within specific occupancy classifications there can be many different types of actual activities that can take place. For instance, a Group A-3 occupancy classification deals with a wide variation of different types of activities, including bowling alleys and courtrooms, indoor tennis courts and dance halls. When a facility changes use from, for example, a bowling alley to a dance hall, the occupancy classification remains A-3, but the different uses could lead to drastically different code requirements. Therefore, this chapter deals with the special circumstances that are associated with a change in the use of a building within the same occupancy classification as well as a change of occupancy classification.*

SECTION 1001
GENERAL

1001.1 Scope. The provisions of this chapter shall apply where a *change of occupancy* occurs, as defined in Section 202.

1001.2 Certificate of occupancy. A *change of occupancy* or a *change of occupancy* within a space where there is a different fire protection system threshold requirement in Chapter 9 of the *International Building Code* shall not be made to any structure without the approval of the *code official*. A certificate of occupancy shall be issued where it has been determined that the requirements for the *change of occupancy* have been met.

> **1001.2.1 Change of use.** Any work undertaken in connection with a change in use that does not involve a *change of occupancy* classification or a change to another group within an occupancy classification shall conform to the applicable requirements for the work as classified in Chapter 6 and to the requirements of Sections 1002 through 1010.
>
> **Exception:** As modified in Section 1204 for *historic buildings*.
>
> **1001.2.2 Change of occupancy classification or group.** Where the occupancy classification of a building changes, the provisions of Sections 1002 through 1011 shall apply. This includes a change of occupancy classification and a change to another group within an occupancy classification.
>
>> **1001.2.2.1 Partial change of occupancy.** Where the occupancy classification or group of a portion of an *existing building* is changed, Section 1011 shall apply.

1001.3 Certificate of occupancy required. A certificate of occupancy shall be issued where a *change of occupancy* occurs that results in a different occupancy classification as determined by the *International Building Code*.

SECTION 1002
SPECIAL USE AND OCCUPANCY

1002.1 Compliance with the building code. Where an *existing building* or part of an *existing building* undergoes a *change of occupancy* to one of the special use or occupancy categories as described in Chapter 4 in the *International Building Code*, the building shall comply with all of the requirements of Chapter 4 of the *International Building Code* applicable to the special use or occupancy.

1002.2 Incidental uses. Where a portion of a building undergoes a *change of occupancy* to one of the incidental uses listed in Table 509.1 of the *International Building Code*, the incidental use shall comply with Section 509 of the *International Building Code* applicable to the incidental use.

1002.3 Change of occupancy in health care. Where a *change of occupancy* occurs to a Group I-2 or I-1 *facility*, the *work area* with the *change of occupancy* shall comply with the *International Building Code*.

> **Exception:** A change in use or occupancy in the following cases shall not be required to meet the *International Building Code*:
>
> 1. Group I-2, Condition 2 to Group I-2, Condition 1.
> 2. Group I-2 to ambulatory health care.
> 3. Group I-2 to Group I-1.
> 4. Group I-1, Condition 2 to Group I-1, Condition 1.

1002.4 Storage. In Group I-2 occupancies, equipped throughout with an automatic sprinkler in accordance with Section 903.3.1.1 of the *International Building Code*, where a room 250 square feet (23.2 m^2) or less undergoes a change in occupancy to a storage room, the room shall be separated from the remainder of the building by construction capable of resisting the passage of smoke in accordance with Section 509.4.2 of the *International Building Code*.

SECTION 1003
BUILDING ELEMENTS AND MATERIALS

1003.1 General. Building elements and materials in portions of buildings undergoing a change of occupancy classification shall comply with Section 1011.

SECTION 1004
FIRE PROTECTION

1004.1 General. Fire protection requirements of Section 1011 shall apply where a building or portions thereof undergo a *change of occupancy* classification or where there is a change of occupancy within a space where there is a different fire protection system threshold requirement in Chapter 9 of the *International Building Code*.

SECTION 1005
MEANS OF EGRESS

1005.1 General. Means of egress in portions of buildings undergoing a change of occupancy classification shall comply with Section 1011.

SECTION 1006
STRUCTURAL

[BS] 1006.1 Live loads. Structural elements carrying tributary live loads from an area with a *change of occupancy* shall satisfy the requirements of Section 1607 of the *International Building Code*. Design live loads for areas of new occupancy shall be based on Section 1607 of the *International Building Code*. Design live loads for other areas shall be permitted to use previously *approved* design live loads.

> **Exception:** Structural elements whose demand-capacity ratio considering the *change of occupancy* is not more than 5 percent greater than the demand-capacity ratio based on previously *approved* live loads.

[BS] 1006.2 Snow and wind loads. Where a *change of occupancy* results in a structure being assigned to a higher *risk category*, the structure shall satisfy the requirements of Sections 1608 and 1609 of the *International Building Code* for the new risk category.

> **Exception:** Where the area of the new occupancy is less than 10 percent of the building area. The cumulative effect of occupancy changes over time shall be considered.

[BS] 1006.3 Seismic loads. Where a *change of occupancy* results in a building being assigned to a higher *risk category*, or where the change is from a Group S or Group U occupancy to any occupancy other than Group S or Group U, the building shall satisfy the requirements of Section 1613 of the *International Building Code* for the new *risk category* using full seismic forces.

> **Exceptions:**
> 1. Where a *change of use* results in a building being reclassified from *Risk Category* I or II to *Risk Category* III and the seismic coefficient, S_{DS}, is less than 0.33, compliance with this section is not required.
> 2. Where the area of the new occupancy is less than 10 percent of the building area, the occupancy is not changing from a Group S or Group U occupancy, and the new occupancy is not assigned to *Risk Category* IV, compliance with this section is not required. The cumulative effect of occupancy changes over time shall be considered.
> 3. Unreinforced masonry bearing wall buildings assigned to *Risk Category* III and to Seismic Design Category A or B shall be permitted to use Appendix Chapter A1 of this code.
> 4. Where the change is from a Group S or Group U occupancy and there is no change of *risk category*, use of reduced seismic forces shall be permitted.

[BS] 1006.4 Access to Risk Category IV. Any structure that provides operational access to an adjacent structure assigned to *Risk Category* IV as the result of a change of occupancy shall itself satisfy the requirements of Sections 1608, 1609 and 1613 of the *International Building Code*. For compliance with Section 1613 of the *International Building Code*, the full seismic forces shall be used. Where operational access to *Risk Category* IV is less than 10 feet (3048 mm) from either an interior lot line or from another structure, access protection from potential falling debris shall be provided.

SECTION 1007
ELECTRICAL

1007.1 Special occupancies. Where the occupancy of an *existing building* or part of an *existing building* is changed to one of the following special occupancies as described in NFPA 70, the electrical wiring and equipment of the building or portion thereof that contains the proposed occupancy shall comply with the applicable requirements of NFPA 70. Health care *facilities*, including Group I-2, ambulatory health care *facilities* and outpatient clinics, shall also comply with the applicable requirements of NFPA 99:

1. Hazardous locations.
2. Commercial garages, repair and storage.
3. Aircraft hangars.
4. Gasoline dispensing and service stations.
5. Bulk storage plants.
6. Spray application, dipping and coating processes.
7. Health care *facilities*, including Group I-2, ambulatory health care *facilities* and outpatient clinics.
8. Places of assembly.
9. Theaters, audience areas of motion picture and television studios, and similar locations.
10. Motion picture and television studios and similar locations.
11. Motion picture projectors.
12. Agricultural buildings.

1007.2 Unsafe conditions. Where the occupancy of an *existing building* or part of an *existing building* is changed, all *unsafe* conditions shall be corrected without requiring that all parts of the electrical system comply with NFPA 70.

1007.3 Service upgrade. Where the occupancy of an *existing building* or part of an *existing building* is changed, electrical service shall be upgraded to meet the requirements of NFPA 70 for the new occupancy.

1007.4 Number of electrical outlets. Where the occupancy of an *existing building* or part of an *existing building* is changed, the number of electrical outlets shall comply with NFPA 70 for the new occupancy.

SECTION 1008
MECHANICAL

1008.1 Mechanical requirements. Where the occupancy of an *existing building* or part of an *existing building* is changed such that the new occupancy is subject to different kitchen exhaust requirements or to increased mechanical ventilation requirements in accordance with the *International Mechanical Code*, the new occupancy shall comply with the respective *International Mechanical Code* provisions.

SECTION 1009
PLUMBING

1009.1 Increased demand. Where the occupancy of an *existing building* or part of an *existing building* is changed such that the new occupancy is subject to increased or different plumbing fixture requirements or to increased water supply requirements in accordance with the *International Plumbing Code*, the new occupancy shall comply with the intent of the respective *International Plumbing Code* provisions.

> **Exception:** Only where the occupant load of the story is increased by more than 20 percent, plumbing fixtures for the story shall be provided in quantities specified in the *International Plumbing Code* based on the increased occupant load.

1009.2 Food-handling occupancies. If the new occupancy is a food-handling establishment, all existing sanitary waste lines above the food or drink preparation or storage areas shall be panned or otherwise protected to prevent leaking pipes or condensation on pipes from contaminating food or drink. New drainage lines shall not be installed above such areas and shall be protected in accordance with the *International Plumbing Code*.

1009.3 Interceptor required. If the new occupancy will produce grease or oil-laden wastes, interceptors shall be provided as required in the *International Plumbing Code*.

1009.4 Chemical wastes. If the new occupancy will produce chemical wastes, the following shall apply:

1. If the existing piping is not compatible with the chemical waste, the waste shall be neutralized prior to entering the drainage system or the piping shall be changed to a compatible material.

2. Chemical waste shall not discharge to a public sewer system without the approval of the sewage authority.

1009.5 Group I-2. If the occupancy group is changed to Group I-2, the plumbing system and medical gas system shall comply with the applicable requirements of the *International Plumbing Code*.

SECTION 1010
OTHER REQUIREMENTS

1010.1 Light and ventilation. Light and ventilation shall comply with the requirements of the *International Building Code* for the new occupancy.

SECTION 1011
CHANGE OF OCCUPANCY CLASSIFICATION

1011.1 General. The provisions of this section shall apply to buildings or portions thereof undergoing a change of occupancy classification. This includes a change of occupancy classification within a group as well as a change of occupancy classification from one group to a different group or where there is a *change of occupancy* within a space where there is a different fire protection system threshold requirement in Chapter 9 of the *International Building Code*. Such buildings shall also comply with Sections 1002 through 1010 of this code.

1011.2 Fire protection systems. Fire protection systems shall be provided in accordance with Sections 1011.2.1 and 1011.2.2.

> **1011.2.1 Fire sprinkler system.** Where a change in occupancy classification occurs or where there is a *change of occupancy* within a space where there is a different fire protection system threshold requirement in Chapter 9 of the *International Building Code* that requires an automatic fire sprinkler system to be provided based on the new occupancy in accordance with Chapter 9 of the *International Building Code*. The installation of the automatic sprinkler system shall be required within the area of the *change of occupancy* and areas of the building not separated horizontally and vertically from the change of occupancy by one of the following:
>
> 1. Nonrated permanent partition and horizontal assemblies.
> 2. Fire partition.
> 3. Smoke partition.
> 4. Smoke barrier.
> 5. Fire barrier.
> 6. Fire wall.
>
> **Exceptions:**
>
> 1. An automatic sprinkler system shall not be required in a one- or two-family dwelling constructed in accordance with the *International Residential Code*.

CHANGE OF OCCUPANCY

2. Automatic sprinkler system shall not be required in a townhouse constructed in accordance with the *International Residential Code*.

3. The townhouse shall be separated from adjoining units in accordance with Section R302.2 of the *International Residential Code*.

1011.2.2 Fire alarm and detection system. Where a change in occupancy classification occurs or where there is a *change of occupancy* within a space where there is a different fire protection system threshold requirement in Chapter 9 of the *International Building Code* that requires a fire alarm and detection system to be provided based on the new occupancy in accordance with Chapter 9 of the *International Building Code*, such system shall be provided throughout the area where the *change of occupancy* occurs. Existing alarm notification appliances shall be automatically activated throughout the building. Where the building is not equipped with a fire alarm system, alarm notification appliances shall be provided throughout the area where the *change of occupancy* occurs in accordance with Section 907 of the *International Building Code* as required for new construction.

1011.3 Interior finish. In areas of the building undergoing the *change of occupancy* classification, the interior finish of walls and ceilings shall comply with the requirements of the *International Building Code* for the new occupancy classification.

1011.4 Enhanced classroom acoustics. In Group E occupancies, where the *work area* is a Level 3 *alteration*, enhanced classroom acoustics shall be provided in all classrooms with a volume of 20,000 cubic feet (565 m^3) or less. Enhanced classroom acoustics shall comply with the reverberation time in Section 808 of ICC A117.1.

1011.5 Means of egress, general. Hazard categories in regard to life safety and means of egress shall be in accordance with Table 1011.5.

TABLE 1011.5
MEANS OF EGRESS HAZARD CATEGORIES

RELATIVE HAZARD	OCCUPANCY CLASSIFICATIONS
1 (Highest Hazard)	H
2	I-2; I-3; I-4
3	A; E; I-1; M; R-1; R-2; R-4, Condition 2
4	B; F-1; R-3; R-4, Condition 1; S-1
5 (Lowest Hazard)	F-2; S-2; U

1011.5.1 Means of egress for change to a higher-hazard category. Where a change of occupancy classification is made to a higher-hazard category (lower number) as shown in Table 1011.5, the means of egress shall comply with the requirements of Chapter 10 of the *International Building Code*.

Exceptions:

1. Stairways shall be enclosed in compliance with the applicable provisions of Section 903.1.

2. Existing stairways including handrails and guards complying with the requirements of Chapter 9 shall be permitted for continued use subject to approval of the *code official*.

3. Any stairway replacing an existing stairway within a space where the pitch or slope cannot be reduced because of existing construction shall not be required to comply with the maximum riser height and minimum tread depth requirements.

4. Existing corridor walls constructed on both sides of wood lath and plaster in good condition or $^1/_2$-inch-thick (12.7 mm) gypsum wallboard shall be permitted. Such walls shall either terminate at the underside of a ceiling of equivalent construction or extend to the underside of the floor or roof next above.

5. Existing corridor doorways, transoms and other corridor openings shall comply with the requirements in Sections 804.6.1, 804.6.2 and 804.6.3.

6. Existing dead-end corridors shall comply with the requirements in Section 804.7.

7. An operable window complying with Section 1011.5.6 shall be accepted as an *emergency escape and rescue opening*.

1011.5.2 Means of egress for change of use to an equal or lower-hazard category. Where a change of occupancy classification is made to an equal or lesser-hazard category (higher number) as shown in Table 1011.5, existing elements of the means of egress shall comply with the requirements of Section 905 for the new occupancy classification. Newly constructed or configured means of egress shall comply with the requirements of Chapter 10 of the *International Building Code*.

> **Exception:** Any stairway replacing an existing stairway within a space where the pitch or slope cannot be reduced because of existing construction shall not be required to comply with the maximum riser height and minimum tread depth requirements.

1011.5.3 Egress capacity. Egress capacity shall meet or exceed the occupant load as specified in the *International Building Code* for the new occupancy.

1011.5.4 Handrails. Existing stairways shall comply with the handrail requirements of Section 804.10 in the area of the *change of occupancy* classification.

1011.5.5 Guards. Existing guards shall comply with the requirements in Section 804.12 in the area of the *change of occupancy* classification.

1011.5.6 Existing emergency escape and rescue openings. Where a *change of occupancy* would require an *emergency escape and rescue opening* in accordance with Section 1031 of the *International Building Code*, operable windows serving as the *emergency escape and rescue opening* shall comply with the following:

1. An existing operable window shall provide a minimum net clear opening of 4 square feet (0.38 m^2) with a minimum net clear opening height of

22 inches (559 mm) and a minimum net clear opening width of 20 inches (508 mm).

2. A replacement window where such window complies with both of the following:

   2.1. The replacement window meets the size requirements in Item 1.

   2.2. The replacement window is the manufacturer's largest standard size window that will fit within the existing frame or existing rough opening. The replacement window shall be permitted to be of the same operating style as the existing window or a style that provides for an equal or greater window opening area than the existing window.

1011.6 Heights and areas. Hazard categories in regard to height and area shall be in accordance with Table 1011.6.

TABLE 1011.6
HEIGHTS AND AREAS HAZARD CATEGORIES

RELATIVE HAZARD	OCCUPANCY CLASSIFICATIONS
1 (Highest Hazard)	H
2	A-1; A-2; A-3; A-4; I; R-1; R-2; R-4, Condition 2
3	E; F-1; S-1; M
4 (Lowest Hazard)	B; F-2; S-2; A-5; R-3; R-4, Condition 1; U

1011.6.1 Height and area for change to a higher-hazard category. Where a change of occupancy classification is made to a higher-hazard category as shown in Table 1011.6, heights and areas of buildings and structures shall comply with the requirements of Chapter 5 of the *International Building Code* for the new occupancy classification.

Exception: For high-rise buildings constructed in compliance with a previously issued permit, the type of construction reduction specified in Section 403.2.1 of the *International Building Code* is permitted. This shall include the reduction for columns. The high-rise building is required to be equipped throughout with an automatic sprinkler system in accordance with Section 903.3.1.1 of the *International Building Code*.

1011.6.1.1 Fire wall alternative. In other than Groups H, F-1 and S-1, fire barriers and horizontal assemblies constructed in accordance with Sections 707 and 711, respectively, of the *International Building Code* shall be permitted to be used in lieu of fire walls to subdivide the building into separate buildings for the purpose of complying with the area limitations required for the new occupancy where all of the following conditions are met:

1. The buildings are protected throughout with an automatic sprinkler system in accordance with Section 903.3.1.1 of the *International Fire Code*.

2. The maximum allowable area between fire barriers, horizontal assemblies or any combination thereof shall not exceed the maximum allowable area determined in accordance with Chapter 5 of the *International Building Code* without an increase allowed for an automatic sprinkler system in accordance with Section 506 of the *International Building Code*.

3. The fire-resistance rating of the fire barriers and horizontal assemblies shall be not less than that specified for fire walls in Table 706.4 of the *International Building Code*.

Exception: Where horizontal assemblies are used to limit the maximum allowable area, the required fire-resistance rating of the horizontal assemblies shall be permitted to be reduced by 1 hour provided that the height and number of stories increases allowed for an automatic sprinkler system by Section 504 of the *International Building Code* are not used for the buildings.

1011.6.2 Height and area for change to an equal or lesser-hazard category. Where a change of occupancy classification is made to an equal or lesser-hazard category as shown in Table 1011.6, the height and area of the *existing building* shall be deemed acceptable.

1011.6.3 Fire barriers. Where a *change of occupancy* classification is made to a higher-hazard category as shown in Table 1011.6, fire barriers in separated mixed use buildings shall comply with the fire-resistance requirements of the *International Building Code*.

Exception: Where the fire barriers are required to have a 1-hour fire-resistance rating, existing wood lath and plaster in good condition or existing $^1/_2$-inch-thick (12.7 mm) gypsum wallboard shall be permitted.

1011.7 Exterior wall fire-resistance ratings. Hazard categories in regard to fire-resistance ratings of exterior walls shall be in accordance with Table 1011.7.

TABLE 1011.7
EXPOSURE OF EXTERIOR WALLS HAZARD CATEGORIES

RELATIVE HAZARD	OCCUPANCY CLASSIFICATION
1 (Highest Hazard)	H
2	F-1; M; S-1
3	A; B; E; I; R
4 (Lowest Hazard)	F-2; S-2; U

1011.7.1 Exterior wall rating for change of occupancy classification to a higher-hazard category. Where a change of occupancy classification is made to a higher hazard category as shown in Table 1011.7, exterior walls shall have fire resistance and exterior opening protectives as required by the *International Building Code*.

Exception: A 2-hour fire-resistance rating shall be allowed where the building does not exceed three stories in height and is classified as one of the following groups: A-2 and A-3 with an occupant load of less than 300, B, F, M or S.

1011.7.2 Exterior wall rating for change of occupancy classification to an equal or lesser-hazard category. Where a change of occupancy classification is made to an equal or lesser-hazard category as shown in Table 1011.7, existing exterior walls, including openings, shall be accepted.

1011.7.3 Opening protectives. Openings in exterior walls shall be protected as required by the *International Building Code*. Where openings in the exterior walls are required to be protected because of their distance from the lot line, the sum of the area of such openings shall not exceed 50 percent of the total area of the wall in each story.

Exceptions:

1. Where the *International Building Code* permits openings in excess of 50 percent.
2. Protected openings shall not be required in buildings of Group R occupancy that do not exceed three stories in height and that are located not less than 3 feet (914 mm) from the lot line.
3. Exterior opening protectives are not required where an automatic sprinkler system has been installed throughout.
4. Exterior opening protectives are not required where the *change of occupancy* group is to an equal or lower hazard classification in accordance with Table 1011.7.

1011.8 Enclosure of vertical shafts. Enclosure of vertical shafts shall be in accordance with Sections 1011.8.1 through 1011.8.4.

1011.8.1 Minimum requirements. Vertical shafts shall be designed to meet the *International Building Code* requirements for atriums or the requirements of this section.

1011.8.2 Stairways. Where a change of occupancy classification is made to a higher-hazard category as shown in Table 1011.5, interior stairways shall be enclosed as required by the *International Building Code*.

Exceptions:

1. In other than Group I occupancies, an enclosure shall not be required for openings serving only one adjacent floor and that are not connected with corridors or stairways serving other floors.
2. Unenclosed existing stairways need not be enclosed in a continuous vertical shaft if each story is separated from other stories by 1-hour fire-resistance-rated construction or *approved* wired glass set in steel frames and all exit corridors are sprinklered. The openings between the corridor and the occupant space shall have not fewer than one sprinkler head above the openings on the tenant side. The sprinkler system shall be permitted to be supplied from the domestic water-supply systems, provided that the system is of adequate pressure, capacity and sizing for the combined domestic and sprinkler requirements.
3. Existing penetrations of stairway enclosures shall be accepted if they are protected in accordance with the *International Building Code*.

1011.8.3 Other vertical shafts. Interior vertical shafts other than stairways, including but not limited to elevator hoistways and service and utility shafts, shall be enclosed as required by the *International Building Code* where there is a *change of use* to a higher-hazard category as specified in Table 1011.5.

Exceptions:

1. Existing 1-hour interior shaft enclosures shall be accepted where a higher rating is required.
2. Vertical openings, other than stairways, in buildings of other than Group I occupancy and connecting less than six stories shall not be required to be enclosed if the entire building is provided with an *approved* automatic sprinkler system.

1011.8.4 Openings. Openings into existing vertical shaft enclosures shall be protected by fire assemblies having a fire protection rating of not less than 1 hour and shall be maintained self-closing or shall be automatic-closing by actuation of a smoke detector. Other openings shall be fire protected in an *approved* manner. Existing fusible link-type automatic door-closing devices shall be permitted in all shafts except stairways if the fusible link rating does not exceed 135°F (57°C).

CHAPTER 11
ADDITIONS

User note:

About this chapter: Chapter 11 provides the requirements for additions, which correlate to the code requirements for new construction. There are, however, some exceptions that are specifically stated within this chapter. An "Addition" is defined in Chapter 2 as "an extension or increase in the floor area, number of stories or height of a building or structure." Chapter 11 contains the minimum requirements for an addition that is not separated from the existing building by a fire wall.

SECTION 1101
GENERAL

1101.1 Scope. An *addition* to a building or structure shall comply with the International Codes as adopted for new construction without requiring the *existing building* or structure to comply with any requirements of those codes or of these provisions, except as required by this chapter. Where an *addition* impacts the *existing building* or structure, that portion shall comply with this code.

1101.2 Creation or extension of nonconformity. An *addition* shall not create or extend any nonconformity in the *existing building* to which the *addition* is being made with regard to accessibility, structural strength, fire safety, means of egress or the capacity of mechanical, plumbing or electrical systems.

1101.3 Other work. Any *repair* or *alteration* work within an *existing building* to which an *addition* is being made shall comply with the applicable requirements for the work as classified in Chapter 6.

1101.4 Enhanced classroom acoustics. In Group E occupancies, enhanced classroom acoustics shall be provided in all classrooms in the *addition* with a volume of 20,000 cubic feet (565 m^3) or less. Enhanced classroom acoustics shall comply with the reverberation time in Section 808 of ICC A117.1.

SECTION 1102
HEIGHTS AND AREAS

1102.1 Height limitations. An *addition* shall not increase the height of an *existing building* beyond that permitted under the applicable provisions of Chapter 5 of the *International Building Code* for new buildings.

1102.2 Area limitations. An *addition* shall not increase the area of an *existing building* beyond that permitted under the applicable provisions of Chapter 5 of the *International Building Code* for new buildings unless fire separation as required by the *International Building Code* is provided.

> **Exception:** In-filling of floor openings and nonoccupiable appendages such as elevator and exit stairway shafts shall be permitted beyond that permitted by the *International Building Code*.

1102.3 Fire protection systems. Existing fire areas increased by the *addition* shall comply with Chapter 9 of the *International Building Code*.

SECTION 1103
STRUCTURAL

[BS] 1103.1 Additional gravity loads. Any existing gravity load-carrying structural element for which an *addition* and its related *alterations* cause an increase in design dead, live or snow load, including snow drift effects, of more than 5 percent shall be replaced or altered as needed to carry the gravity loads required by the *International Building Code* for new structures. Any existing gravity load-carrying structural element whose gravity load-carrying capacity is decreased as part of the *addition* and its related *alterations* shall be considered to be an altered element subject to the requirements of Section 805.2. Any existing element that will form part of the lateral load path for any part of the *addition* shall be considered to be an existing lateral load-carrying structural element subject to the requirements of Section 1103.3.

> **Exception:** Buildings of Group R occupancy with not more than five dwelling units or sleeping units used solely for residential purposes where the *existing building* and the *addition* together comply with the conventional light-frame construction methods of the *International Building Code* or the provisions of the *International Residential Code*.

[BS] 1103.2 Lateral force-resisting system. Where the *addition* is structurally independent of the *existing structure*, existing lateral load-carrying structural elements shall be permitted to remain unaltered. Where the *addition* is not structurally independent of the *existing structure*, the *existing structure* and its *addition* acting together as a single structure shall meet the requirements of Sections 1609 and 1613 of the *International Building Code* using full seismic forces.

> **Exceptions:**
> 1. Buildings of Group R occupancy with not more than five dwelling or sleeping units used solely for residential purposes where the *existing building* and the *addition* comply with the conventional light-frame construction methods of the *International Building Code* or the provisions of the *International Residential Code*.

2. Any existing lateral load-carrying structural element whose demand-capacity ratio with the *addition* considered is not more than 10 percent greater than its demand-capacity ratio with the *addition* ignored shall be permitted to remain unaltered. For purposes of calculating demand-capacity ratios, the demand shall consider applicable load combinations with design lateral loads or forces in accordance with Sections 1609 and 1613 of the *International Building Code*. For purposes of this exception, comparisons of demand-capacity ratios and calculation of design lateral loads, forces and capacities shall account for the cumulative effects of *additions* and *alterations* since original construction.

[BS] **1103.3 Flood hazard areas.** *Additions* and *foundations* in *flood hazard areas* shall comply with the following requirements:

1. For horizontal *additions* that are structurally interconnected to the *existing building*:

   1.1. If the *addition* and all other proposed work, when combined, constitute *substantial improvement*, the *existing building* and the *addition* shall comply with Section 1612 of the *International Building Code*, or Section R322 of the *International Residential Code*, as applicable.

   1.2. If the *addition* constitutes *substantial improvement*, the *existing building* and the *addition* shall comply with Section 1612 of the *International Building Code*, or Section R322 of the *International Residential Code*, as applicable.

2. For horizontal *additions* that are not structurally interconnected to the *existing building*:

   2.1. The *addition* shall comply with Section 1612 of the *International Building Code*, or Section R322 of the *International Residential Code*, as applicable.

   2.2. If the *addition* and all other proposed work, when combined, constitute *substantial improvement*, the *existing building* and the *addition* shall comply with Section 1612 of the *International Building Code*, or Section R322 of the *International Residential Code*, as applicable.

3. For vertical *additions* and all other proposed work that, when combined, constitute *substantial improvement*, the *existing building* shall comply with Section 1612 of the *International Building Code*, or Section R322 of the *International Residential Code*, as applicable.

4. For a raised or extended foundation, if the foundation work and all other proposed work, when combined, constitute *substantial improvement*, the *existing building* shall comply with Section 1612 of the *International Building Code*, or Section R322 of the *International Residential Code*, as applicable.

5. For a new foundation or replacement foundation, the foundation shall comply with Section 1612 of the *International Building Code*, or Section R322 of the *International Residential Code*, as applicable.

SECTION 1104
ENERGY CONSERVATION

1104.1 Minimum requirements. *Additions* to *existing buildings* shall conform to the energy requirements of the *International Energy Conservation Code* or *International Residential Code* as they relate to new construction.

CHAPTER 12
HISTORIC BUILDINGS

User note:

About this chapter: Chapter 12 provides some exceptions from code requirements when the building in question has historic value. The most important criterion for application of this chapter is that the building must be essentially accredited as being of historic significance by a state or local authority after careful review of the historical value of the building. Most, if not all, states have such authorities, as do many local jurisdictions. The agencies with such authority can be located at the state or local government level or through the local chapter of the American Institute of Architects (AIA). Other considerations include the structural condition of the building (i.e., is the building structurally sound), its proposed use, its impact on life safety and how the intent of the code, if not the letter, will be achieved.

SECTION 1201
GENERAL

1201.1 Scope. This chapter is intended to provide means for the preservation of *historic buildings*. *Historic buildings* shall comply with the provisions of this chapter relating to their *repair*, *alteration*, relocation and *change of occupancy*.

[BS] 1201.2 Report. A *historic building* undergoing *alteration* or *change of occupancy* shall be investigated and evaluated. If it is intended that the building meet the requirements of this chapter, a written report shall be prepared and filed with the *code official* by a *registered design professional* where such a report is necessary in the opinion of the *code official*. Such report shall be in accordance with Chapter 1 and shall identify each required safety feature that is in compliance with this chapter and where compliance with other chapters of these provisions would be damaging to the contributing historic features. For buildings assigned to Seismic Design Category D, E or F, a structural evaluation describing, at a minimum, the vertical and horizontal elements of the lateral force-resisting system and any strengths or weaknesses therein shall be prepared. Additionally, the report shall describe each feature that is not in compliance with these provisions and shall demonstrate how the intent of these provisions is complied with in providing an equivalent level of safety.

1201.3 Special occupancy exceptions—museums. Where a building in Group R-3 is used for Group A, B or M purposes such as museum tours, exhibits and other public assembly activities, or for museums less than 3,000 square feet (279 m^2), the *code official* is authorized to determine that the occupancy is Group B where life safety conditions can be demonstrated in accordance with Section 1201.2. Adequate means of egress in such buildings, including, but not limited to, a means of maintaining doors in an open position to permit egress, a limit on building occupancy to an occupant load permitted by the means of egress capacity, a limit on occupancy of certain areas or floors, or supervision by a person knowledgeable in the emergency exiting procedures, shall be provided.

[BS] 1201.4 Flood hazard areas. In *flood hazard areas*, if all proposed work, including *repairs*, work required because of a *change of occupancy*, and *alterations*, constitutes *substantial improvement*, then the *existing building* shall comply with Section 1612 of the *International Building Code*, or Section R322 of the *International Residential Code*, as applicable.

Exception: If a *historic building* will continue to be a *historic building* after the proposed work is completed, then the proposed work is not considered a *substantial improvement*. For the purposes of this exception, a *historic building* is any of the following:

1. Listed or preliminarily determined to be eligible for listing in the National Register of Historic Places.
2. Determined by the Secretary of the US Department of Interior to contribute to the historical significance of a registered historic district or a district preliminarily determined to qualify as a historic district.
3. Designated as historic under a state or local historic preservation program that is approved by the Department of Interior.

1201.5 Unsafe conditions. Conditions determined by the *code official* to be *unsafe* shall be remedied. Work shall not be required beyond what is required to remedy the *unsafe* conditions.

SECTION 1202
REPAIRS

1202.1 General. Repairs to any portion of a *historic building* or structure shall be permitted with original or like materials and original methods of construction, subject to the provisions of this chapter. Hazardous materials, such as asbestos and lead-based paint, shall not be used where the code for new construction would not permit their use in buildings of similar occupancy, purpose and location.

1202.2 Replacement. Replacement of existing or missing features using original materials shall be permitted. Partial replacement for *repairs* that match the original in configuration, height and size shall be permitted.

Replacement glazing in hazardous locations shall comply with the safety glazing requirements of Chapter 24 of the *International Building Code*.

Exception: Glass block walls, louvered windows and jalousies repaired with like materials.

SECTION 1203
FIRE SAFETY

1203.1 Scope. *Historic buildings* undergoing *alterations*, *changes of occupancy* or that are moved shall comply with Section 1203.

1203.2 General. Every *historic building* that does not conform to the construction requirements specified in this code for the occupancy or use and that constitutes a distinct fire hazard as defined herein shall be provided with an *approved* automatic fire-extinguishing system as determined appropriate by the *code official*. However, an automatic fire-extinguishing system shall not be used to substitute for, or act as an alternative to, the required number of exits from any *facility*.

1203.3 Means of egress. Existing door openings and corridor and stairway widths less than those specified elsewhere in this code may be *approved*, provided that, in the opinion of the *code official*, there is sufficient width and height for a person to pass through the opening or traverse the means of egress. Where *approved* by the *code official*, the front or main exit doors need not swing in the direction of the path of exit travel, provided that other *approved* means of egress having sufficient capacity to serve the total occupant load are provided.

1203.4 Transoms. In buildings with automatic sprinkler systems of Group R-1, R-2 or R-3, existing transoms in corridors and other fire-resistance-rated walls may be maintained if fixed in the closed position. A sprinkler shall be installed on each side of the transom.

1203.5 Interior finishes. The existing interior finishes shall be accepted where it is demonstrated that they are the historic finishes.

1203.6 Stairway enclosure. In buildings of three stories or less, exit enclosure construction shall limit the spread of smoke by the use of tight-fitting doors and solid elements. Such elements are not required to have a fire-resistance rating.

1203.7 One-hour fire-resistant assemblies. Where 1-hour fire-resistance-rated construction is required by these provisions, it need not be provided, regardless of construction or occupancy, where the existing wall and ceiling finish is wood or metal lath and plaster.

1203.8 Glazing in fire-resistance-rated systems. Historic glazing materials are permitted in interior walls required to have a 1-hour fire-resistance rating where the opening is provided with *approved* smoke seals and the area affected is provided with an automatic sprinkler system.

1203.9 Stairway railings. Grand stairways shall be accepted without complying with the handrail and guard requirements. Existing handrails and guards at all stairways shall be permitted to remain, provided they are not structurally *dangerous*.

1203.10 Guards. Guards shall comply with Sections 1203.10.1 and 1203.10.2.

1203.10.1 Height. Existing guards shall comply with the requirements of Section 404.

1203.10.2 Guard openings. The spacing between existing intermediate railings or openings in existing ornamental patterns shall be accepted. Missing elements or members of a guard may be replaced in a manner that will preserve the historic appearance of the building or structure.

1203.11 Exit signs. Where exit sign or egress path marking location would damage the historic character of the building, alternative exit signs are permitted with approval of the *code official*. Alternative signs shall identify the exits and egress path.

1203.12 Automatic fire-extinguishing systems. Every *historic building* that cannot be made to conform to the construction requirements specified in the *International Building Code* for the occupancy or use and that constitutes a distinct fire hazard shall be deemed to be in compliance if provided with an *approved* automatic fire-extinguishing system.

> **Exception:** Where the *code official* approves an alternative life-safety system.

SECTION 1204
CHANGE OF OCCUPANCY

1204.1 General. *Historic buildings* undergoing a *change of occupancy* shall comply with the applicable provisions of Chapter 10, except as specifically permitted in this chapter. Where Chapter 10 requires compliance with specific requirements of Chapter 7, Chapter 8 or Chapter 9 and where those requirements are subject to the exceptions in Section 1202, the same exceptions shall apply to this section.

1204.2 Building area. The allowable floor area for *historic buildings* undergoing a *change of occupancy* shall be permitted to exceed by 20 percent the allowable areas specified in Chapter 5 of the *International Building Code*.

1204.3 Location on property. Historic structures undergoing a *change of use* to a higher-hazard category in accordance with Section 1011.7 may use alternative methods to comply with the fire-resistance and exterior opening protective requirements. Such alternatives shall comply with Section 1201.2.

1204.4 Occupancy separation. Required occupancy separations of 1 hour may be omitted where the building is provided with an *approved* automatic sprinkler system throughout.

1204.5 Roof covering. Regardless of occupancy or use group, roof-covering materials not less than Class C, where tested in accordance with ASTM E108 or UL 790, shall be permitted where a fire-retardant roof covering is required.

1204.6 Means of egress. Existing door openings and corridor and stairway widths less than those that would be acceptable for nonhistoric buildings under these provisions shall be *approved*, provided that, in the opinion of the *code official*, there is sufficient width and height for a person to pass through the opening or traverse the exit and that the capacity of the exit system is adequate for the occupant load, or where other operational controls to limit occupancy are *approved* by the *code official*.

1204.7 Door swing. Where *approved* by the *code official*, existing front doors need not swing in the direction of exit travel, provided that other *approved* exits having sufficient capacity to serve the total occupant load are provided.

1204.8 Transoms. In corridor walls required by these provisions to be fire-resistance rated, existing transoms may be maintained if fixed in the closed position, and fixed wired glass set in a steel frame or other *approved* glazing shall be installed on one side of the transom.

> **Exception:** Transoms conforming to Section 1203.4 shall be accepted.

1204.9 Interior finishes. Where interior finish materials are required to comply with the fire test requirements of Section 803.1 of the *International Building Code*, existing nonconforming materials shall be permitted to be surfaced with an *approved* fire-retardant coating to achieve the required classification. Compliance with this section shall be demonstrated by testing the fire-retardant coating on the same material and achieving the required fire classification. Where the same material is not available, it shall be permitted to test on a similar material.

> **Exception:** Existing nonconforming materials need not be surfaced with an *approved* fire-retardant coating where the building is equipped throughout with an automatic sprinkler system installed in accordance with the *International Building Code* and the nonconforming materials can be substantiated as being historic in character.

1204.10 One-hour fire-resistant assemblies. Where 1-hour fire-resistance-rated construction is required by these provisions, it need not be provided, regardless of construction or occupancy, where the existing wall and ceiling finish is wood lath and plaster.

1204.11 Stairways and guards. Existing stairways shall comply with the requirements of these provisions. The *code official* shall grant alternatives for stairways and guards if alternative stairways are found to be acceptable or are judged to meet the intent of these provisions. Existing stairways shall comply with Section 1203.

> **Exception:** For buildings less than 3,000 square feet (279 m^2), existing conditions are permitted to remain at all stairways and guards.

1204.12 Exit signs. The *code official* may accept alternative exit sign locations where the location of such signs would damage the historic character of the building or structure. Such signs shall identify the exits and exit path.

[BS] 1204.13 Exit stair live load. Existing historic stairways in buildings changed to a Group R-1 or R-2 occupancy shall be accepted where it can be shown that the stairway can support a 75-pounds-per-square-foot (366 kg/m^2) live load.

1204.14 Natural light. Where it is determined by the *code official* that compliance with the natural light requirements of Section 1010.1 will lead to loss of historic character or historic materials in the building, the existing level of natural lighting shall be considered to be acceptable.

SECTION 1205
STRUCTURAL

[BS] 1205.1 General. *Historic buildings* shall comply with the applicable structural provisions for the work as classified in Chapter 4 or 5.

> **Exceptions:**
> 1. The *code official* shall be authorized to accept existing floors and existing live loads and to approve operational controls that limit the live load on any floor.
> 2. *Repair* of *substantial structural damage* is not required to comply with Sections 405.2.3 and 405.2.4. *Substantial structural damage* shall be repaired in accordance with Section 405.2.1.

[BS] 1205.2 Dangerous conditions. Conditions determined by the *code official* to be *dangerous* shall be remedied. Work shall not be required beyond what is required to remedy the *dangerous* condition.

SECTION 1206
RELOCATED BUILDINGS

1206.1 Relocated buildings. Foundations of relocated *historic buildings* and structures shall comply with the *International Building Code*. Relocated *historic buildings* shall otherwise be considered a *historic building* for the purposes of this code. Relocated *historic buildings* and structures shall be sited so that exterior wall and opening requirements comply with the *International Building Code* or with the compliance alternatives of this code.

CHAPTER 13

PERFORMANCE COMPLIANCE METHODS

User note:

About this chapter: Chapter 13 allows for existing buildings to be evaluated so as to show that alterations, while not meeting new construction requirements, will improve the current existing situation. Provisions are based on a numerical scoring system involving 19 various safety parameters and the degree of code compliance for each issue.

SECTION 1301
GENERAL

1301.1 Scope. The provisions of this chapter shall apply to the *alteration, addition* and *change of occupancy* of *existing structures*, including historic structures, as referenced in Section 301.3.3. The provisions of this chapter are intended to maintain or increase the current degree of public safety, health and general welfare in *existing buildings* while permitting, *alteration, addition* and *change of occupancy* without requiring full compliance with Chapters 6 through 12, except where compliance with the prescriptive method of Chapter 5 or the work area method of other provisions of this code is specifically required in this chapter.

1301.1.1 Compliance with other methods. *Alterations, additions* and *changes of occupancy* to *existing structures* shall comply with the provisions of this chapter or with one of the methods provided in Section 301.3.

1301.2 Applicability. *Existing buildings* in which there is work involving *additions, alterations* or *changes of occupancy* shall be made to conform to the requirements of this chapter or the provisions of Chapters 6 through 12. The provisions of Sections 1301.2.1 through 1301.2.6 shall apply to existing occupancies that will continue to be, or are proposed to be, in Groups A, B, E, F, I-2, M, R and S. These provisions shall also apply to Group U occupancies where such occupancies are undergoing a *change of occupancy* or a partial change in occupancy with separations in accordance with Section 1301.2.2. These provisions shall not apply to buildings with occupancies in Group H, I-1, I-3 or I-4.

1301.2.1 Change in occupancy. Where an *existing building* is changed to a new occupancy classification and this section is applicable, the provisions of this section for the new occupancy shall be used to determine compliance with this code.

1301.2.2 Partial change in occupancy. Where a portion of the building is changed to a new occupancy classification and that portion is separated from the remainder of the building with fire barrier or horizontal assemblies having a fire-resistance rating as required by Table 508.4 of the *International Building Code* or Section R302 of the *International Residential Code* for the separate occupancies, or with *approved* compliance alternatives, the portion changed shall be made to conform to the provisions of this section. Only the portion separated shall be required to be evaluated for compliance.

Where a portion of the building is changed to a new occupancy classification and that portion is not separated from the remainder of the building with fire barriers or horizontal assemblies having a fire-resistance rating as required by Table 508.4 of the *International Building Code* or Section R302 of the *International Residential Code* for the separate occupancies, or with *approved* compliance alternatives, the provisions of this section which apply to each occupancy shall apply to the entire building. Where there are conflicting provisions, those requirements which secure the greater public safety shall apply to the entire building or structure.

1301.2.3 Additions. *Additions* to *existing buildings* shall comply with the requirements of the *International Building Code* or the *International Residential Code* for new construction. The combined height and area of the *existing building* and the new *addition* shall not exceed the height and area allowed by Chapter 5 of the *International Building Code*. Where a fire wall that complies with Section 706 of the *International Building Code* is provided between the *addition* and the *existing building*, the *addition* shall be considered a separate building.

1301.2.4 Alterations. An *existing building* or portion thereof shall not be altered in such a manner that results in the building being less safe or sanitary than such building is currently.

Exception: Where the current level of safety or sanitation is proposed to be reduced, the portion altered shall conform to the requirements of the *International Building Code*.

1301.2.5 Escalators. Where escalators are provided in below-grade transportation stations, existing and new escalators shall be permitted to have a clear width of less than 32 inches (815 mm).

1301.2.6 Plumbing fixtures. Plumbing fixtures shall be provided in accordance with Section 1009 for a change of occupancy and Section 808 for *alterations*. Plumbing fixtures for *additions* shall be in accordance with the *International Plumbing Code*.

1301.3 Acceptance. For *repairs, alterations, additions* and *changes of occupancy* to *existing buildings* that are evaluated in accordance with this section, compliance with this section shall be accepted by the *code official*.

1301.3.1 Hazards. Where the *code official* determines that an *unsafe* condition exists as provided for in Section

115, such *unsafe* condition shall be abated in accordance with Section 115.

1301.3.2 Compliance with other codes. Buildings that are evaluated in accordance with this section shall comply with the *International Fire Code* and *International Property Maintenance Code*.

[BS] 1301.3.3 Compliance with flood hazard provisions. In *flood hazard areas*, buildings that are evaluated in accordance with this section shall comply with Section 1612 of the *International Building Code*, or Section R322 of the *International Residential Code*, as applicable, if the work covered by this section constitutes *substantial improvement*.

1301.4 Investigation and evaluation. For proposed work covered by this chapter, the building owner shall cause the *existing building* to be investigated and evaluated in accordance with the provisions of Sections 1301.4 through 1301.9.

[BS] 1301.4.1 Structural analysis. The owner shall have a structural analysis of the *existing building* made to determine adequacy of structural systems for the proposed *alteration, addition* or *change of occupancy*. The analysis shall demonstrate that the building with the work completed is capable of resisting the loads specified in Chapter 16 of the *International Building Code*.

1301.4.2 Submittal. The results of the investigation and evaluation as required in Section 1301.4, along with proposed compliance alternatives, shall be submitted to the *code official*.

1301.4.3 Determination of compliance. The *code official* shall determine whether the *existing building*, with the proposed *addition, alteration* or *change of occupancy*, complies with the provisions of this section in accordance with the evaluation process in Sections 1301.5 through 1301.9.

1301.5 Evaluation. The evaluation shall be composed of three categories: fire safety, means of egress and general safety, as defined in Sections 1301.5.1 through 1301.5.3.

1301.5.1 Fire safety. Included within the fire safety category are the structural fire resistance, automatic fire detection, fire alarm, automatic sprinkler system and fire suppression system features of the *facility*.

1301.5.2 Means of egress. Included within the means of egress category are the configuration, characteristics and support features for means of egress in the *facility*.

1301.5.3 General safety. Included within the general safety category are the fire safety parameters and the means of egress parameters.

1301.6 Evaluation process. The evaluation process specified herein shall be followed in its entirety to evaluate *existing buildings* in Groups A, B, E, F, M, R, S and U. For *existing buildings* in Group I-2, the evaluation process specified herein shall be followed and applied to each and every individual smoke compartment. Table 1301.7 shall be utilized for tabulating the results of the evaluation. References to other sections of this code or other codes indicate that compliance with those sections is required in order to gain credit in the evaluation herein outlined. In applying this section to a building with mixed occupancies, where the separation between the mixed occupancies does not qualify for any category indicated in Section 1301.6.16, the score for each occupancy shall be determined, and the lower score determined for each section of the evaluation process shall apply to the entire building or to each smoke compartment for Group I-2 occupancies.

Where the separation between the mixed occupancies qualifies for any category indicated in Section 1301.6.16, the score for each occupancy shall apply to each portion or smoke compartment of the building based on the occupancy of the space.

1301.6.1 Building height and number of stories. The value for building height and number of stories shall be the lesser value determined by the formula in Section 1301.6.1.1. Section 504 of the *International Building Code* shall be used to determine the allowable height and number of stories of the building. Subtract the actual building height from the allowable height and divide by $12^1/_2$ feet (3810 mm). Enter the height value and its sign (positive or negative) in Table 1301.7 under Safety Parameter 1301.6.1, Building Height, for fire safety, means of egress and general safety. The maximum score for a building shall be 10.

1301.6.1.1 Height formula. The following formulas shall be used in computing the building height value.

$$\text{Height value, feet} = \frac{(AH) - (EBH)}{12.5} \times CF$$

(Equation 13-1)

$$\text{Height value, stories} = (AS - EBS) \times CF$$

(Equation 13-2)

where:

- AH = Allowable height in feet (mm) from Section 504 of the *International Building Code*.
- EBH = *Existing building* height in feet (mm).
- AS = Allowable height in stories from Section 504 of the *International Building Code*.
- EBS = *Existing building* height in stories.
- CF = 1 if $(AH) - (EBH)$ is positive.
- CF = Construction-type factor shown in Table 1301.6.6(2) if $(AH) - (EBH)$ is negative.

Note: Where mixed occupancies are separated and individually evaluated as indicated in Section 1301.6, the values AH, AS, EBH and EBS shall be based on the height of the occupancy being evaluated.

1301.6.2 Building area. The value for building area shall be determined by the formula in Section 1301.6.2.2. Section 506 of the *International Building Code* and the formula in Section 1301.6.2.1 shall be used to determine the allowable area of the building. Enter the area value and its sign (positive or negative) in Table 1301.7 under Safety Parameter 1301.6.2, Building Area, for fire safety, means of egress and general safety. In determining the area value, the maximum permitted positive value for area is 50 percent of the fire safety score as listed in Table 1301.8, Mandatory Safety Scores. Group I-2 occupancies shall be scored zero.

1301.6.2.1 Allowable area formula. The following formula shall be used in computing allowable area:

$$A_a = A_t + (NS \times I_f) \qquad \text{(Equation 13-3)}$$

where:

A_a = Allowable building area per story (square feet).

A_t = Tabular allowable area factor (NS, S1, S13R, or SM value, as applicable) in accordance with Table 506.2 of the *International Building Code*.

NS = Tabular allowable area factor in accordance with Table 506.2 of the *International Building Code* for a nonsprinklered building (regardless of whether the building is sprinklered).

I_f = Area factor increase due to frontage as calculated in accordance with Section 506.3 of the *International Building Code*.

1301.6.2.2 Area formula. The following formulas shall be used in computing the area value. Equation 13-4 shall be used for a single occupancy buildings and Equation 13-5 shall be used for multiple occupancy buildings. Determine the area value for each occupancy floor area on a floor-by-floor basis. For multiple occupancy, buildings with the minimum area value of the set of values obtained for the particular occupancy shall be used as the area value for that occupancy.

For single occupancy buildings:

Area value$_i$ = (Allowable area − Actual area)/1200 square feet (Equation 13-4)

For multiple occupancy buildings:

$$\text{Area value}_i = \frac{\text{Allowable area}_i}{1200 \text{ square feet}} \left[1 - \left(\frac{\text{Actual area}_i}{\text{Allowable area}_i} + \ldots + \frac{\text{Actual area}_n}{\text{Allowable area}_n} \right) \right]$$

(Equation 13-5)

where:

i = Value for an individual separated occupancy on a floor.

n = Number of separated occupancies on a floor.

1301.6.3 Compartmentation. Evaluate the compartments created by fire barriers or horizontal assemblies which comply with Sections 1301.6.3.2 and 1301.6.3.3 and which are exclusive of the wall elements considered under Sections 1301.6.4 and 1301.6.5. Conforming compartments shall be figured as the net area and do not include shafts, chases, stairways, walls or columns. Using Table 1301.6.3, determine the appropriate compartmentation value (CV) and enter that value into Table 1301.7 under Safety Parameter 1301.6.3, Compartmentation, for fire safety, means of egress and general safety.

**TABLE 1301.6.3
COMPARTMENTATION VALUES**

OCCUPANCY	CATEGORIES[a]				
	a	b	c	d	e
A-1, A-3	0	6	10	14	18
A-2	0	4	10	14	18
A-4, B, E, S-2	0	5	10	15	20
F, M, R, S-1	0	4	10	16	22
I-2	0	2	8	10	14

a. For compartment sizes between categories, the compartmentation value shall be obtained by linear interpolation.

1301.6.3.1 Categories. The categories for compartment separations are:

1. Category a—Compartment size of 15,000 square feet (1394 m^2) or more.
2. Category b—Maximum compartment size of 10,000 square feet (929 m^2).
3. Category c—Maximum compartment size of 7,500 square feet (697 m^2).
4. Category d—Maximum compartment size of 5,000 square feet (464 m^2).
5. Category e—Maximum compartment size of 2,500 square feet (232 m^2).

1301.6.3.2 Wall construction. A wall used to create separate compartments shall be a fire barrier conforming to Section 707 of the *International Building Code* with a fire-resistance rating of not less than 2 hours. Where the building is not divided into more than one compartment, the compartment size shall be taken as the total floor area on all floors. Where there is more than one compartment within a story, each compartmented area on such story shall be provided with a horizontal exit conforming to Section 1026 of the *International Building Code*. The fire door serving as the horizontal exit between compartments shall be so installed, fitted and gasketed that such fire door will provide a substantial barrier to the passage of smoke.

1301.6.3.3 Floor/ceiling construction. A floor/ceiling assembly used to create compartments shall conform to Section 711 of the *International Building Code* and shall have a fire-resistance rating of not less than 2 hours.

1301.6.4 Tenant and dwelling unit separations. Evaluate the fire-resistance rating of floors and walls separating tenants, including dwelling units, and not evaluated under Sections 1301.6.3 and 1301.6.5. Group I-2 occupancies shall evaluate the rating of the separations between care recipient sleeping rooms.

Under the categories and occupancies in Table 1301.6.4, determine the appropriate value and enter that value in Table 1301.7 under Safety Parameter 1301.6.4, Tenant and Dwelling Unit Separation, for fire safety, means of egress and general safety. The value shall be zero for single tenant buildings and buildings without dwelling units.

TABLE 1301.6.4
SEPARATION VALUES

OCCUPANCY	CATEGORIES				
	a	b	c	d	e
A-1	0	0	0	0	1
A-2	-5	-3	0	1	3
R	-4	-2	0	2	4
A-3, A-4, B, E, F, M, S-1	-4	-3	0	2	4
I-2	0	1	2	3	4
S-2	-5	-2	0	2	4

1301.6.4.1 Categories. The categories for tenant and dwelling unit separations are:

1. Category a—No fire partitions; incomplete fire partitions; no doors; doors not self-closing or automatic-closing.

2. Category b—Fire partitions or floor assemblies with less than 1-hour fire-resistance ratings or not constructed in accordance with Section 708 or 711 of the *International Building Code*, respectively.

3. Category c—Fire partitions with 1-hour or greater fire-resistance ratings constructed in accordance with Section 708 of the *International Building Code* and floor assemblies with 1-hour but less than 2-hour fire-resistance ratings constructed in accordance with Section 711 of the *International Building Code* or with only one tenant within the floor area.

4. Category d—Fire barriers with 1-hour but less than 2-hour fire-resistance ratings constructed in accordance with Section 707 of the *International Building Code* and floor assemblies with 2-hour or greater fire-resistance ratings constructed in accordance with Section 711 of the *International Building Code*.

5. Category e—Fire barriers and floor assemblies with 2-hour or greater fire-resistance ratings and constructed in accordance with Sections 707 and 711 of the *International Building Code*, respectively.

1301.6.5 Corridor walls. Evaluate the fire-resistance rating and degree of completeness of walls which create corridors serving the floor and that are constructed in accordance with Section 1020 of the *International Building Code*. This evaluation shall not include the wall elements considered under Sections 1301.6.3 and 1301.6.4. Under the categories and groups in Table 1301.6.5, determine the appropriate value and enter that value into Table 1301.7 under Safety Parameter 1301.6.5, Corridor Walls, for fire safety, means of egress and general safety.

TABLE 1301.6.5
CORRIDOR WALL VALUES

OCCUPANCY	CATEGORIES			
	a	b	c[a]	d[a]
A-1	-10	-4	0	2
A-2	-30	-12	0	2
A-3, F, M, R, S-1	-7	-3	0	2
A-4, B, E, S-2	-5	-2	0	5
I-2	-10	0	1	2

a. Corridors not providing at least one-half the exit access travel distance for all occupants on a floor shall use Category b.

1301.6.5.1 Categories. The categories for corridor walls are:

1. Category a—No fire partitions; incomplete fire partitions; no doors; or doors not self-closing.

2. Category b—Less than 1-hour fire-resistance rating or not constructed in accordance with Section 708.4 of the *International Building Code*.

3. Category c—1-hour to less than 2-hour fire-resistance rating, with doors conforming to Section 716 of the *International Building Code* or corridors as permitted by Section 1020 of the *International Building Code* to be without a fire-resistance rating.

4. Category d—2-hour or greater fire-resistance rating, with doors conforming to Section 716 of the *International Building Code*.

1301.6.6 Vertical openings. Evaluate the fire-resistance rating of interior exit stairways or ramps, hoistways, escalator openings and other shaft enclosures within the building, and openings between two or more floors. Table 1301.6.6(1) contains the appropriate protection values. Multiply that value by the construction-type factor found in Table 1301.6.6(2). Enter the vertical opening value and its sign (positive or negative) in Table 1301.7 under Safety Parameter 1301.6.6, Vertical Openings, for fire safety, means of egress and general safety. If the structure is a one-story building or if all the unenclosed vertical openings within the building conform to the requirements of Section 712 of the *International Building Code*, enter a value of 2. The maximum positive value for this requirement (VO) shall be 2.

TABLE 1301.6.6(1)
VERTICAL OPENING PROTECTION VALUE

PROTECTION	VALUE
None (unprotected opening)	-2 times number of floors connected
Less than 1 hour	-1 times number of floors connected
1 to less than 2 hours	1
2 hours or more	2

PERFORMANCE COMPLIANCE METHODS

TABLE 1301.6.6(2)
CONSTRUCTION-TYPE FACTOR

FACTOR	TYPE OF CONSTRUCTION								
	IA	IB	IIA	IIB	IIIA	IIIB	IV	VA	VB
	1.2	1.5	2.2	3.5	2.5	3.5	2.3	3.3	7

1301.6.6.1 Vertical opening formula. The following formula shall be used in computing vertical opening value.

$$VO = PV \times CF \quad \text{(Equation 13-6)}$$

where:

VO = Vertical opening value. The calculated value shall not be greater than positive 2.0.

PV = Protection value from Table 1301.6.6(1).

CF = Construction-type factor from Table 1301.6.6(2).

1301.6.7 HVAC systems. Evaluate the ability of the HVAC system to resist the movement of smoke and fire beyond the point of origin. Under the categories in Section 1301.6.7.1, determine the appropriate value and enter that value into Table 1301.7 under Safety Parameter 1301.6.7, HVAC Systems, for fire safety, means of egress and general safety. *Facilities* in Group I-2 occupancies meeting Category a, b or c shall be considered to fail the evaluation.

1301.6.7.1 Categories. The categories for HVAC systems are:

1. Category a—Plenums not in accordance with Section 602 of the *International Mechanical Code*. -10 points.
2. Category b—Air movement in egress elements not in accordance with Section 1020.6 of the *International Building Code*. -5 points.
3. Category c—Both Categories a and b are applicable. -15 points.
4. Category d—Compliance of the HVAC system with Section 1020.6 of the *International Building Code* and Section 602 of the *International Mechanical Code*. 0 points.
5. Category e—Systems serving one story; or a central boiler/chiller system without ductwork connecting two or more stories or where systems have no ductwork. +5 points.

1301.6.8 Automatic fire detection. Evaluate the smoke detection capability based on the location and operation of automatic fire detectors in accordance with the *International Mechanical Code* and Section 907 of the *International Building Code*. Under the categories and occupancies in Table 1301.6.8, determine the appropriate value and enter that value into Table 1301.7 under Safety Parameter 1301.6.8, Automatic Fire Detection, for fire safety, means of egress and general safety. *Facilities* in Group I-2 occupancies meeting Category a, b or c shall be considered to fail the evaluation.

TABLE 1301.6.8
AUTOMATIC FIRE DETECTION VALUES

OCCUPANCY	CATEGORIES					
	a	b	c	d	e	f
A-1, A-3, F, M, R, S-1	-10	-5	0	2	6	NA
A-2	-25	-5	0	5	9	NA
A-4, B, E, S-2	-4	-2	0	4	8	NA
I-2	NP	NP	NP	4	5	2

NA = Not Applicable.
NP = Not Permitted.

1301.6.8.1 Categories. The categories for automatic fire detection are:

1. Category a—None.
2. Category b—Existing smoke detectors in HVAC systems and maintained in accordance with the *International Fire Code*.
3. Category c—Smoke detectors in HVAC systems. The detectors are installed in accordance with the requirements for new buildings in the *International Mechanical Code*.
4. Category d—Smoke detectors throughout all floor areas other than individual sleeping units, tenant spaces and dwelling units.
5. Category e—Smoke detectors installed throughout the floor area.
6. Category f—Smoke detectors in corridors only.

1301.6.9 Fire alarm systems. Evaluate the capability of the fire alarm system in accordance with Section 907 of the *International Building Code*. Under the categories and occupancies in Table 1301.6.9, determine the appropriate value and enter that value into Table 1301.7 under Safety Parameter 1301.6.9, Fire Alarm System, for fire safety, means of egress and general safety.

TABLE 1301.6.9
FIRE ALARM SYSTEM VALUES

OCCUPANCY	CATEGORIES			
	a	b[a]	c	d
A-1, A-2, A-3, A-4, B, E, R	-10	-5	0	5
F, M, S	0	5	10	15
I-2	-4	1	2	5

a. For buildings equipped throughout with an automatic sprinkler system, add 2 points for activation by a sprinkler water-flow device.

1301.6.9.1 Categories. The categories for fire alarm systems are:

1. Category a—None.
2. Category b—Fire alarm system with manual fire alarm boxes in accordance with Section 907.4 of the *International Building Code* and alarm notification appliances in accordance with Section 907.5.2 of the *International Building Code*.

3. Category c—Fire alarm system in accordance with Section 907 of the *International Building Code*.
4. Category d—Category c plus a required emergency voice/alarm communications system and a fire command station that conforms to Section 911 of the *International Building Code* and contains the emergency voice/alarm communications system controls, fire department communication system controls, and any other controls specified in Section 911 of the *International Building Code* where those systems are provided.

1301.6.10 Smoke control. Evaluate the ability of a natural or mechanical venting, exhaust or pressurization system to control the movement of smoke from a fire. Under the categories and occupancies in Table 1301.6.10, determine the appropriate value and enter that value into Table 1301.7 under Safety Parameter 1301.6.10, Smoke Control, for means of egress and general safety.

TABLE 1301.6.10
SMOKE CONTROL VALUES

OCCUPANCY	CATEGORIES					
	a	b	c	d	e	f
A-1, A-2, A-3	0	1	2	3	6	6
A-4, E	0	0	0	1	3	5
B, M, R	0	2a	3a	3a	3a	4a
F, S	0	2a	2a	3a	3a	3a
I-2	-4	0	0	0	3	0

a. This value shall be 0 if compliance with Category d or e in Section 1301.6.8.1 has not been obtained.

1301.6.10.1 Categories. The categories for smoke control are:

1. Category a—None.
2. Category b—The building is equipped throughout with an automatic sprinkler system. Openings are provided in exterior walls at the rate of 20 square feet (1.86 m^2) per 50 linear feet (15 240 mm) of exterior wall in each story and distributed around the building perimeter at intervals not exceeding 50 feet (15 240 mm). Such openings shall be readily openable from the inside without a key or separate tool and shall be provided with ready access thereto. In lieu of operable openings, clearly and permanently marked tempered glass panels shall be used.
3. Category c—One enclosed exit stairway, with ready access thereto, from each occupied floor of the building. The stairway has operable exterior windows, and the building has openings in accordance with Category b.
4. Category d—One smokeproof enclosure and the building has openings in accordance with Category b.
5. Category e—The building is equipped throughout with an automatic sprinkler system. Each floor area is provided with a mechanical air-handling system designed to accomplish smoke containment. Return and exhaust air shall be moved directly to the outside without recirculation to other floor areas of the building under fire conditions. The system shall exhaust not less than six air changes per hour from the floor area. Supply air by mechanical means to the floor area is not required. Containment of smoke shall be considered as confining smoke to the floor area involved without migration to other floor areas. Any other tested and *approved* design that will adequately accomplish smoke containment is permitted.
6. Category f—Each stairway shall be one of the following: a smokeproof enclosure in accordance with Section 1023.12 of the *International Building Code*; pressurized in accordance with Section 909.20.5 of the *International Building Code*; or shall have operable exterior windows.

1301.6.11 Means of egress capacity and number. Evaluate the means of egress capacity and the number of exits available to the building occupants. In applying this section, the means of egress are required to conform to the following sections of the *International Building Code*: 1003.7, 1004, 1005, 1006, 1007, 1016.2, 1026.1, 1028.3, 1028.5, 1030.2, 1030.3, 1030.4 and 1031. The number of exits credited is the number that is available to each occupant of the area being evaluated. Existing fire escapes shall be accepted as a component in the means of egress when conforming to Section 504.

Under the categories and occupancies in Table 1301.6.11, determine the appropriate value and enter that value into Table 1301.7 under Safety Parameter 1301.6.11, Means of Egress Capacity, for means of egress and general safety.

TABLE 1301.6.11
MEANS OF EGRESS VALUES

OCCUPANCY	CATEGORIES				
	aa	b	c	d	e
A-1, A-2, A-3, A-4, E, I-2	-10	0	2	8	10
M	-3	0	1	2	4
B, F, S	-1	0	0	0	0
R	-3	0	0	0	0

a. The values indicated are for buildings six stories or less in height. For buildings over six stories above grade plane, add an additional -10 points.

1301.6.11.1 Categories. The categories for means-of-egress capacity and number of exits are:

1. Category a—Compliance with the minimum required means-of-egress capacity or number of exits is achieved through the use of a fire escape in accordance with Section 405.

2. Category b—Capacity of the means of egress complies with Section 1005 of the *International Building Code*, and the number of exits complies with the minimum number required by Section 1006 of the *International Building Code*.

3. Category c—Capacity of the means of egress is equal to or exceeds 125 percent of the required means-of-egress capacity, the means of egress complies with the minimum required width dimensions specified in the *International Building Code*, and the number of exits complies with the minimum number required by Section 1006 of the *International Building Code*.

4. Category d—The number of exits provided exceeds the number of exits required by Section 1006 of the *International Building Code*. Exits shall be located a distance apart from each other equal to not less than that specified in Section 1007 of the *International Building Code*.

5. Category e—The area being evaluated meets both Categories c and d.

1301.6.12 Dead ends. In spaces required to be served by more than one means of egress, evaluate the length of the exit access travel path in which the building occupants are confined to a single path of travel. Under the categories and occupancies in Table 1301.6.12, determine the appropriate value and enter that value into Table 1301.7 under Safety Parameter 1301.6.12, Dead Ends, for means of egress and general safety.

TABLE 1301.6.12
DEAD-END VALUES

OCCUPANCY	CATEGORIES[a]			
	a	b	c	d
A-1, A-3, A-4, B, F, M, R, S	-2	0	2	-4
A-2, E	-2	0	2	-4
I-2	-2	0	2	-6

a. For dead-end distances between categories, the dead-end value shall be obtained by linear interpolation.

1301.6.12.1 Categories. The categories for dead ends are:

1. Category a—Dead end of 35 feet (10 670 mm) in nonsprinklered buildings or 70 feet (21 340 mm) in sprinklered buildings.

2. Category b—Dead end of 20 feet (6096 mm); or 50 feet (15 240 mm) in Group B in accordance with Section 1020.5, Exception 2, of the *International Building Code*.

3. Category c—No dead ends; or ratio of length to width (l/w) is less than 2.5:1.

4. Category d—Dead ends exceeding Category a.

1301.6.13 Maximum exit access travel distance to an exit. Evaluate the length of exit access travel to an *approved* exit. Determine the appropriate points in accordance with the following equation and enter that value into Table 1301.7 under Safety Parameter 1301.6.13, Maximum Exit Access Travel Distance for means of egress and general safety. The maximum allowable exit access travel distance shall be determined in accordance with Section 1017.1 of the *International Building Code*.

$$\text{Points} = 20 \times \frac{\text{Maximum allowable travel distance} - \text{Maximum actual travel distance}}{\text{Maximum allowable travel distance}}$$

(Equation 13-7)

1301.6.14 Elevator control. Evaluate the passenger elevator equipment and controls that are available to the fire department to reach all occupied floors. Emergency recall and in-car operation of elevators shall be provided in accordance with the *International Fire Code*. Under the categories and occupancies in Table 1301.6.14, determine the appropriate value and enter that value into Table 1301.7 under Safety Parameter 1301.6.14, Elevator Control, for fire safety, means of egress and general safety. The values shall be zero for a single-story building.

TABLE 1301.6.14
ELEVATOR CONTROL VALUES

ELEVATOR TRAVEL	CATEGORIES			
	a	b	c	d
Less than 25 feet of travel above or below the primary level of elevator access for emergency fire-fighting or rescue personnel	-2	0	0	+2
Travel of 25 feet or more above or below the primary level of elevator access for emergency fire-fighting or rescue personnel	-4	NP	0	+4

For SI: 1 foot = 304.8 mm.
NP = Not Permitted.

1301.6.14.1 Categories. The categories for elevator controls are:

1. Category a—No elevator.

2. Category b—Any elevator without Phase I emergency recall operation and Phase II emergency in-car operation.

3. Category c—All elevators with Phase I emergency recall operation and Phase II emergency in-car operation as required by the *International Fire Code*.

4. Category d—All meet Category c; or Category b where permitted to be without Phase I emergency recall operation and Phase II emergency in-car operation; and at least one elevator that complies with new construction requirements serves all occupied floors.

1301.6.15 Means of egress emergency lighting. Evaluate the presence of and reliability of means of egress emergency lighting. Under the categories and occupancies in Table 1301.6.15, determine the appropriate value and enter that value into Table 1301.7 under Safety Parameter 1301.6.15, Means of Egress Emergency Lighting, for means of egress and general safety.

TABLE 1301.6.15
MEANS OF EGRESS EMERGENCY LIGHTING VALUES

NUMBER OF EXITS REQUIRED BY SECTION 1006 OF THE *INTERNATIONAL BUILDING CODE*	CATEGORIES		
	a	b	c
Two or more exits	NP	0	4
Minimum of one exit	0	1	1

NP = Not Permitted.

1301.6.15.1 Categories. The categories for means of egress emergency lighting are:

1. Category a—Means-of-egress lighting and exit signs not provided with emergency power in accordance with Section 2702 of the *International Building Code*.

2. Category b—Means of egress lighting and exit signs provided with emergency power in accordance with Section 2702 of the *International Building Code*.

3. Category c—Emergency power provided to means of egress lighting and exit signs, which provides protection in the event of power failure to the site or building.

1301.6.16 Mixed occupancies. Where a building has two or more occupancies that are not in the same occupancy classification, the separation between the mixed occupancies shall be evaluated in accordance with this section. Where there is no separation between the mixed occupancies or the separation between mixed occupancies does not qualify for any of the categories indicated in Section 1301.6.16.1, the building shall be evaluated as indicated in Section 1301.6, and the value for mixed occupancies shall be zero. Under the categories and occupancies in Table 1301.6.16, determine the appropriate value and enter that value into Table 1301.7 under Safety Parameter 1301.6.16, Mixed Occupancies, for fire safety and general safety. For buildings without mixed occupancies, the value shall be zero. *Facilities* in Group I-2 occupancies meeting Category a shall be considered to fail the evaluation.

TABLE 1301.6.16
MIXED OCCUPANCY VALUES[a]

OCCUPANCY	CATEGORIES		
	a	b	c
A-1, A-2, R	-10	0	10
A-3, A-4, B, E, F, M, S	-5	0	5
I-2	NP	0	5

NP = Not Permitted.
a. For fire-resistance ratings between categories, the value shall be obtained by linear interpolation.

1301.6.16.1 Categories. The categories for mixed occupancies are:

1. Category a—Occupancies separated by minimum 1-hour fire barriers or minimum 1-hour horizontal assemblies, or both.

2. Category b—Separations between occupancies in accordance with Section 508.4 of the *International Building Code*.

3. Category c—Separations between occupancies having a fire-resistance rating of not less than twice that required by Section 508.4 of the *International Building Code*.

1301.6.17 Automatic sprinklers. Evaluate the ability to suppress or control a fire based on the installation of an automatic sprinkler system in accordance with Section 903.3.1 of the *International Building Code*. "Required sprinklers" shall be based on the requirements of the *International Building Code*. Under the categories and occupancies in Table 1301.6.17, determine the appropriate value and enter that value into Table 1301.7 under Safety Parameter 1301.6.17, Automatic Sprinklers, for fire safety, means of egress divided by 2, and general safety. High-rise buildings defined in Chapter 2 of the *International Building Code* that undergo a *change of occupancy* to Group R shall be equipped throughout with an automatic sprinkler system in accordance with Section 403 of the *International Building Code* and Chapter 9 of the *International Building Code*. *Facilities* in Group I-2 occupancies meeting Category a, b, c or f shall be considered to fail the evaluation.

TABLE 1301.6.17
SPRINKLER SYSTEM VALUES

OCCUPANCY	CATEGORIES					
	a[a]	b[a]	c	d	e	f
A-1, A-3, F, M, R, S-1	-6	-3	0	2	4	6
A-2	-4	-2	0	1	2	4
A-4, B, E, S-2	-12	-6	0	3	6	12
I-2	NP	NP	NP	8	10	NP

NP = Not Permitted.
a. These options cannot be taken if Category a in Section 1301.6.18 is used.

1301.6.17.1 Categories. The categories for automatic sprinkler system protection are:

1. Category a— An *approved* automatic sprinkler system is required throughout; an *approved* automatic sprinkler system is not provided.

2. Category b—An *approved* automatic sprinkler system is required in a portion of a building; an *approved* automatic sprinkler system is not provided; the sprinkler system design is not adequate for the hazard protected in accordance with Chapter 9 of the *International Building Code*.

3. Category c—An *approved* automatic sprinkler system is not required; none are provided.

4. Category d—An *approved* automatic sprinkler system is required in a portion of a building; an *approved* automatic sprinkler system is provided in a portion of a building in accordance with Chapter 9 of the *International Building Code*.

5. Category e—An *approved* automatic sprinkler system is required throughout; an *approved* automatic sprinkler system is provided throughout in accordance with Chapter 9 of the *International Building Code*.

6. Category f—An *approved* automatic sprinkler system is not required throughout; an *approved* automatic sprinkler system is provided throughout in accordance with Chapter 9 of the *International Building Code*.

1301.6.18 Standpipes. Evaluate the ability to initiate attack on a fire by making a supply of water readily available through the installation of standpipes in accordance with Section 905 of the *International Building Code*. "Required Standpipes" shall be based on the requirements of the *International Building Code*. Under the categories and occupancies in Table 1301.6.18, determine the appropriate value and enter that value into Table 1301.7 under Safety Parameter 1301.6.18, Standpipes, for fire safety, means of egress and general safety.

TABLE 1301.6.18
STANDPIPE SYSTEM VALUES

OCCUPANCY	CATEGORIES			
	a[a]	b	c	d
A-1, A-3, F, M, R, S-1	-6	0	4	6
A-2	-4	0	2	4
A-4, B, E, S-2	-12	0	6	12
I-2	-2	0	1	2

a. This option cannot be taken if Category a or Category b in Section 1301.6.17 is used.

1301.6.18.1 Standpipe categories. The categories for standpipe systems are:

1. Category a—Standpipes are required; standpipe is not provided or the standpipe system design is not in compliance with Section 905.3 of the *International Building Code*.

2. Category b—Standpipes are not required; none are provided.

3. Category c—Standpipes are required; standpipes are provided in accordance with Section 905 of the *International Building Code*.

4. Category d—Standpipes are not required; standpipes are provided in accordance with Section 905 of the *International Building Code*.

1301.6.19 Incidental uses. Evaluate the protection of incidental uses in accordance with Section 509.4.2 of the *International Building Code*. Do not include those where this code requires automatic sprinkler systems throughout the building including covered and open mall buildings, high-rise buildings, public garages and unlimited area buildings. Assign the lowest score from Table 1301.6.19 for the building or floor area being evaluated and enter that value into Table 1301.7 under Safety Parameter 1301.6.19, Incidental Uses, for fire safety, means of egress and general safety. If there are no specific occupancy areas in the building or floor area being evaluated, the value shall be zero.

TABLE 1301.6.19
INCIDENTAL USE AREA VALUES

PROTECTION REQUIRED BY TABLE 509.1 OF THE *INTERNATIONAL BUILDING CODE*	PROTECTION PROVIDED						
	None	1 hour	AS	AS with CRS	1 hour and AS	2 hours	2 hours and AS
2 hours and AS	-4	-3	-2	-2	-1	-2	0
2 hours, or 1 hour and AS	-3	-2	-1	-1	0	0	0
1 hour and AS	-3	-2	-1	-1	0	-1	0
1 hour	-1	0	-1	-1	0	0	0
1 hour, or AS with CRS	-1	0	-1	-1	0	0	0
AS with CRS	-1	-1	-1	-1	0	-1	0
1 hour or AS	-1	0	0	0	0	0	0

AS = Automatic Sprinkler System.
CRS = Construction capable of resisting the passage of smoke (see Section 509.4.2 of the *International Building Code*).

1301.6.20 Smoke compartmentation. Evaluate the smoke compartments for compliance with Section 407.5 of the *International Building Code*. Under the categories and occupancies in Table 1301.6.20, determine the appropriate smoke compartmentation value (SCV) and enter that value into Table 1301.7 under Safety Parameter 1301.6.20, Smoke Compartmentation, for fire safety, means of egress and general safety. *Facilities* in Group I-2 occupancies meeting Category b or c shall be considered to fail the evaluation.

TABLE 1301.6.20
SMOKE COMPARTMENTATION VALUES

OCCUPANCY	CATEGORIES[a]		
	a	b	c
A, B, E, F, M, R and S	0	0	0
I-2	0	-10	NP

NP = Not Permitted.
a. For areas between categories, the smoke compartmentation value shall be obtained by linear interpolation.

1301.6.20.1 Categories. Categories for smoke compartment size are:

1. Category a—Smoke compartment complies with Section 407.5 of the *International Building Code*.

2. Category b—Smoke compartment are provided but do not comply with Section 407.5 of the *International Building Code*.

3. Category c—Smoke compartments are not provided.

1301.6.21 Care recipient ability, concentration, smoke compartment location and ratio to attendant. In I-2 occupancies, the ability of care recipients, their concentration and ratio to attendants shall be evaluated and applied in accordance with this section. Evaluate each smoke compartment using the categories in Sections 1301.6.21.1, 1301.6.21.2 and 1301.6.21.3 and enter the value in Table 1301.7. To determine the safety factor, multiply the three values together; if the product is less than 6, compliance has failed.

1301.6.21.1 Care recipient ability for self-preservation. Evaluate the ability of the care recipients for self-preservation in each smoke compartment in an emergency. Under the categories and occupancies in Table 1301.6.21.1, determine the appropriate value and enter that value in Table 1301.7 under Safety Parameter 1301.6.21.1, Care Recipient Ability for Self-preservation, for means of egress and general safety.

TABLE 1301.6.21.1
CARE RECIPIENT ABILITY VALUES

OCCUPANCY	CATEGORIES		
	a	b	c
I-2	3	2	1

1301.6.21.1.1 Categories. The categories for care recipient ability for self-preservation are:

1. Category a—(mobile) Care recipients are capable of self-preservation without assistance.
2. Category b—(not mobile) Care recipients rely on assistance for evacuation or relocation.
3. Category c—(not movable) Care recipients cannot be evacuated or relocated.

1301.6.21.2 Care recipient concentration. Evaluate the concentration of care recipients in each smoke compartment under Section 1301.6.21.2. Under the categories and occupancies in Table 1301.6.21.2 determine the appropriate value and enter that value in Table 1301.7 under Safety Parameter 1301.6.21.2, Care Recipient Concentration, for means of egress and general safety.

TABLE 1301.6.21.2
CARE RECIPIENT CONCENTRATION VALUES

OCCUPANCY	CATEGORIES		
	a	b	c
I-2	3	2	1

1301.6.21.2.1 Categories: The categories for care recipient concentration are:

1. Category a—smoke compartment has 1 to 10 care recipients.
2. Category b—smoke compartment has more than 10 to 40 care recipients.
3. Category c—smoke compartment has more than 40 care recipients.

1301.6.21.3 Attendant-to-care recipients ratio. Evaluate the attendant-to-care recipients ratio for each compartment under Section 1301.6.21.3. Under the categories and occupancies in Table 1301.6.21.3 determine the appropriate value and enter that value in Table 1301.7 under Safety Parameter 1301.6.21.3, Attendant-to-Care Recipients Ratio, for means of egress and general safety.

TABLE 1301.6.21.3
ATTENDANT-TO-CARE RECIPIENTS RATIO VALUES

OCCUPANCY	CATEGORIES		
	a	b	c
I-2	3	2	1

1301.6.21.3.1 Categories. The categories for attendant-to-care recipient concentrations are:

1. Category a—attendant-to-care recipients concentration is 1:5 or no care recipients.
2. Category b—attendant-to-care recipients concentration is 1:6 to 1:10.
3. Category c—attendant-to-care recipients concentration is greater than 1:10.

1301.7 Building score. After determining the appropriate data from Section 1301.6, enter those data in Table 1301.7 and total the building score.

1301.8 Safety scores. The values in Table 1301.8 are the required mandatory safety scores for the evaluation process listed in Section 1301.6.

TABLE 1301.8
MANDATORY SAFETY SCORES[a]

OCCUPANCY	FIRE SAFETY(MFS)	MEANS OF EGRESS (MME)	GENERAL SAFETY (MGS)
A-1	20	31	31
A-2	21	32	32
A-3	22	33	33
A-4, E	29	40	40
B	30	40	40
F	24	34	34
I-2	19	34	34
M	23	40	40
R	21	38	38
S-1	19	29	29
S-2	29	39	39

a. MFS = Mandatory Fire Safety.
MME = Mandatory Means of Egress.
MGS = Mandatory General Safety.

PERFORMANCE COMPLIANCE METHODS

TABLE 1301.7
SUMMARY SHEET—BUILDING CODE

Existing occupancy: _____	Proposed occupancy: _____
Year building was constructed: _____	Number of stories: _____ Height in feet: _____
Type of construction: _____	Area per floor: _____
Percentage of open perimeter increase: ____ %	
Completely suppressed: Yes____ No____	Corridor wall rating: _____
	Type: _____
Compartmentation: Yes____ No____	Required door closers: Yes____ No____
Fire-resistance rating of vertical opening enclosures: _____	
Type of HVAC system: _____, serving number of floors: _____	
Automatic fire detection: Yes____ No____	Type and location: _____
Fire alarm system: Yes____ No____	Type: _____
Smoke control: Yes____ No____	Type: _____
Adequate exit routes: Yes____ No____	Dead ends: Yes____ No____
Maximum exit access travel distance: _____	Elevator controls: Yes____ No____
Means of egress emergency lighting: Yes____ No____	Mixed occupancies: Yes____ No____
Standpipes: Yes____ No____	Care recipients ability for self-preservation: _____
Incidental use: Yes____ No____	Care recipients concentration: _____
Smoke compartmentation less than 22,500 sq. feet (2092 m²): Yes____ No____	Attendant-to-care recipients ratio: _____

SAFETY PARAMETERS	FIRE SAFETY (FS)	MEANS OF EGRESS (ME)	GENERAL SAFETY (GS)
1301.6.1 Building height			
1301.6.2 Building area			
1301.6.3 Compartmentation			
1301.6.4 Tenant and dwelling unit separations			
1301.6.5 Corridor walls			
1301.6.6 Vertical openings			
1301.6.7 HVAC systems			
1301.6.8 Automatic fire detection			
1301.6.9 Fire alarm system			
1301.6.10 Smoke control	* * * *		
1301.6.11 Means of egress	* * * *		
1301.6.12 Dead ends	* * * *		
1301.6.13 Maximum exit access travel distance	* * * *		
1301.6.14 Elevator control			
1301.6.15 Means of egress emergency lighting	* * * *		
1301.6.16 Mixed occupancies		* * * *	
1301.6.17 Automatic sprinklers		÷ 2 =	
1301.6.18 Standpipes			
1301.6.19 Incidental use			
1301.6.20 Smoke compartmentation			
1301.6.21.1 Care recipients ability for self-preservation[a]	* * * *		
1301.6.21.2 Care recipients concentration[a]	* * * *		
1301.6.21.3 Attendant-to-care recipients ratio[a]	* * * *		
Building score–total value			

* * * *No applicable value to be inserted.

a. Only applicable to Group I-2 occupancies.

1301.9 Evaluation of building safety. The mandatory safety score in Table 1301.8 shall be subtracted from the building score in Table 1301.7 for each category in accordance with the evaluation formulas in Table 1301.9. Where the final score for any category equals zero or more, the building is in compliance with the requirements of this section for that category. Where the final score for any category is less than zero, the building is not in compliance with the requirements of this section.

1301.9.1 Mixed occupancies. For mixed occupancies, the following provisions shall apply:

1. Where the separation between mixed occupancies does not qualify for any category indicated in Section 1301.6.16, the mandatory safety scores for the occupancy with the lowest general safety score in Table 1301.8 shall be utilized (see Section 1301.6).

2. Where the separation between mixed occupancies qualifies for any category indicated in Section 1301.6.16, the mandatory safety scores for each occupancy shall be placed against the evaluation scores for the appropriate occupancy. An evaluation is not required for areas of the building with separated occupancies in accordance with Table 508.4 of the *International Building Code* in which there are no *alterations* or *change of occupancy*.

**TABLE 1301.9
EVALUATION FORMULAS**[a]

FORMULA	TABLE 1301.7	TABLE 1301.8		SCORE	PASS	FAIL
FS − MFS ≥ 0	_____(FS) −	_____(MFS)	=	_____	_____	_____
ME − MME ≥ 0	_____(ME) −	_____(MME)	=	_____	_____	_____
GS − MGS ≥ 0	_____(GS) −	_____(MGS)	=	_____	_____	_____

a. FS = Fire Safety.
   ME = Means of Egress.
   GS = General Safety.
   MFS = Mandatory Fire Safety.
   MME = Mandatory Means of Egress.
   MGS = Mandatory General Safety.

CHAPTER 14

RELOCATED OR MOVED BUILDINGS

User note:

About this chapter: Chapter 14 is applicable to any building that is moved or relocated. The relocation of a building will automatically cause an inspection and evaluation process that enables the jurisdiction to determine the level of compliance with the International Fire Code® *and the* International Property Maintenance Code®. *These two codes, by their scope, are applicable to existing buildings. This is the case regardless of any repair, remodeling, alteration work or change of occupancy occurring (see the* International Fire Code *and* International Property Maintenance Code*).*

SECTION 1401
GENERAL

1401.1 Scope. This chapter provides requirements for relocated or moved structures, including *relocatable buildings* as defined in Chapter 2.

1401.1.1 Bleachers, grandstands and folding and telescopic seating. Relocated or moved bleachers, grandstands and folding and telescopic seating shall comply with ICC 300.

1401.2 Conformance. The building shall be safe for human occupancy as determined by the *International Fire Code* and the *International Property Maintenance Code*. Any *repair, alteration* or *change of occupancy* undertaken within the moved structure shall comply with the requirements of this code applicable to the work being performed. Any field-fabricated elements shall comply with the requirements of the *International Building Code* or the *International Residential Code*, as applicable.

SECTION 1402
REQUIREMENTS

1402.1 Location on the lot. The building shall be located on the lot in accordance with the requirements of the *International Building Code* or the *International Residential Code*, as applicable.

[BS] 1402.2 Foundation. The foundation system of relocated buildings shall comply with the *International Building Code* or the *International Residential Code*, as applicable.

[BS] 1402.2.1 Connection to the foundation. The connection of the relocated building to the foundation shall comply with the *International Building Code* or the *International Residential Code*, as applicable.

[BS] 1402.3 Wind loads. Buildings shall comply with *International Building Code* or *International Residential Code* wind provisions, as applicable.

Exceptions:

1. Detached one- and two-family dwellings and Group U occupancies where wind loads at the new location are not higher than those at the previous location.
2. Structural elements whose stress is not increased by more than 10 percent.

[BS] 1402.4 Seismic loads. Buildings shall comply with *International Building Code* or *International Residential Code* seismic provisions at the new location, as applicable.

Exceptions:

1. Structures in Seismic Design Categories A and B and detached one- and two-family dwellings in Seismic Design Categories A, B and C where the seismic loads at the new location are not higher than those at the previous location.
2. Structural elements whose stress is not increased by more than 10 percent.

[BS] 1402.5 Snow loads. Structures shall comply with *International Building Code* or *International Residential Code* snow loads, as applicable, where snow loads at the new location are higher than those at the previous location.

Exception: Structural elements whose stress is not increased by more than 5 percent.

[BS] 1402.6 Flood hazard areas. If relocated or moved into a *flood hazard area*, structures shall comply with Section 1612 of the *International Building Code*, or Section R322 of the *International Residential Code*, as applicable.

[BS] 1402.7 Required inspection and repairs. The *code official* shall be authorized to inspect, or to require *approved* professionals to inspect at the expense of the owner, the various structural parts of a relocated building to verify that structural components and connections have not sustained structural damage. Any *repairs* required by the *code official* as a result of such inspection shall be made prior to the final approval.

CHAPTER 15

CONSTRUCTION SAFEGUARDS

User note:

About this chapter: Chapter 15 looks to the construction process. Parameters are provided for demolition and for protecting adjacent property during demolition and construction. Issues such as how to provide egress and adequate water supply while the building is growing, the timing of standpipe and sprinkler installation, and protection of pedestrians are addressed. Note that this chapter is consistent with Chapter 33 of the International Building Code *and Chapter 33 of the* International Fire Code.

SECTION 1501 GENERAL

[BG] 1501.1 Scope. The provisions of this chapter shall govern safety during construction and the protection of adjacent public and private properties.

[BG] 1501.2 Storage and placement. Construction equipment and materials shall be stored and placed so as not to endanger the public, the workers or adjoining property for the duration of the construction project.

[BS] 1501.2.1 Structural and construction loads. Structural roof components shall be capable of supporting the roof-covering system and the material and equipment loads that will be encountered during installation of the system.

[BG] 1501.3 Alterations, repairs and additions. Required exits, existing structural elements, fire protection devices and sanitary safeguards shall be maintained at all times during *alterations*, *repairs* or *additions* to any building or structure.

Exceptions:

1. Where such required elements or devices are being altered or repaired, adequate substitute provisions shall be made.
2. Maintenance of such elements and devices is not required where the *existing building* is not occupied.

[BG] 1501.4 Manner of removal. Waste materials shall be removed in a manner that prevents injury or damage to persons, adjoining properties and public rights-of-way.

[BG] 1501.5 Fire safety during construction. Fire safety during construction shall comply with the applicable requirements of the *International Building Code* and the applicable provisions of Chapter 33 of the *International Fire Code*.

[BS] 1501.6 Protection of pedestrians. Pedestrians shall be protected during construction and demolition activities as required by Sections 1501.6.1 through 1501.6.7 and Table 1501.6. Signs shall be provided to direct pedestrian traffic.

**[BS] TABLE 1501.6
PROTECTION OF PEDESTRIANS**

HEIGHT OF CONSTRUCTION	DISTANCE OF CONSTRUCTION TO LOT LINE	TYPE OF PROTECTION REQUIRED
8 feet or less	Less than 5 feet	Construction railings
	5 feet or more	None
More than 8 feet	Less than 5 feet	Barrier and covered walkway
	5 feet or more, but not more than one-fourth the height of construction	Barrier and covered walkway
	5 feet or more, but between one-fourth and one-half the height of construction	Barrier
	5 feet or more, but exceeding one-half the height of construction	None

For SI: 1 foot = 304.8 mm.

[BS] 1501.6.1 Walkways. A walkway shall be provided for pedestrian travel in front of every construction and demolition site unless the applicable governing authority authorizes the sidewalk to be fenced or closed. A walkway shall be provided for pedestrian travel that leads from a building entrance or exit of an occupied structure to a public way. Walkways shall be of sufficient width to accommodate the pedestrian traffic, but shall be not less than 4 feet (1219 mm) in width. Walkways shall be provided with a durable walking surface and shall be accessible in accordance with Chapter 11 of the *International Building Code*. Walkways shall be designed to support all imposed loads and the design live load shall be not less than 150 pounds per square foot (psf) (7.2 kN/m^2).

[BS] 1501.6.2 Directional barricades. Pedestrian traffic shall be protected by a directional barricade where the walkway extends into the street. The directional barricade shall be of sufficient size and construction to direct vehicular traffic away from the pedestrian path.

[BS] 1501.6.3 Construction railings. Construction railings shall be not less than 42 inches (1067 mm) in height and shall be sufficient to direct pedestrians around construction areas.

[BS] 1501.6.4 Barriers. Barriers shall be not less than 8 feet (2438 mm) in height and shall be placed on the side of the walkway nearest the construction. Barriers shall extend the entire length of the construction site. Openings in such barriers shall be protected by doors that are normally kept closed.

[BS] 1501.6.4.1 Barrier design. Barriers shall be designed to resist loads required in Chapter 16 of the *International Building Code* unless constructed as follows:

1. Barriers shall be provided with 2-inch by 4-inch (51 mm by 102 mm) top and bottom plates.
2. The barrier material shall be boards not less than $^3/_4$ inch (19.1 mm) in thickness or wood structural use panels not less than $^1/_4$ inch (6.4 mm) in thickness.
3. Wood structural use panels shall be bonded with an adhesive identical to that for exterior wood structural use panels.
4. Wood structural use panels $^1/_4$ inch (6.4 mm) or $^{15}/_{16}$ inch (23.8 mm) in thickness shall have studs spaced not more than 2 feet (610 mm) on center.
5. Wood structural use panels $^3/_8$ inch (9.5 mm) or $^1/_2$ inch (12.7 mm) in thickness shall have studs spaced not more than 4 feet (1219 mm) on center, provided that a 2-inch by 4-inch (51 mm by 102 mm) stiffener is placed horizontally at mid-height where the stud spacing is greater than 2 feet (610 mm) on center.
6. Wood structural use panels $^5/_8$ inch (15.9 mm) or thicker shall not span over 8 feet (2438 mm).

[BS] 1501.6.5 Covered walkways. Covered walkways shall have a clear height of not less than 8 feet (2438 mm) as measured from the floor surface to the canopy overhead. Adequate lighting shall be provided at all times. Covered walkways shall be designed to support all imposed loads. The design live load shall be not less than 150 psf (7.2 kN/m^2) for the entire structure.

Exception: Roofs and supporting structures of covered walkways for new, light-frame construction not exceeding two stories above grade plane are permitted to be designed for a live load of 75 psf (3.6 kN/m^2) or the loads imposed on them, whichever is greater. In lieu of such designs, the roof and supporting structure of a covered walkway are permitted to be constructed as follows:

1. Footings shall be continuous 2-inch by 6-inch (51 mm by 152 mm) members.
2. Posts not less than 4 inches by 6 inches (102 mm by 152 mm) shall be provided on both sides of the roof and spaced not more than 12 feet (3658 mm) on center.
3. Stringers not less than 4 inches by 12 inches (102 mm by 305 mm) shall be placed on edge on the posts.
4. Joists resting on the stringers shall be not less than 2 inches by 8 inches (51 mm by 203 mm) and shall be spaced not more than 2 feet (610 mm) on center.
5. The deck shall be planks not less than 2 inches (51 mm) thick or wood structural panels with an exterior exposure durability classification not less than $^{23}/_{32}$ inch (18.3 mm) thick nailed to the joists.
6. Each post shall be knee-braced to joists and stringers by members not less than 2 inches by 4 inches (51 mm by 102 mm); 4 feet (1219 mm) in length.
7. A curb that is not less than 2 inches by 4 inches (51 mm by 102 mm) shall be set on edge along the outside edge of the deck.

[BS] 1501.6.6 Repair, maintenance and removal. Pedestrian protection required by Section 1501.6 shall be maintained in place and kept in good order for the entire length of time pedestrians are subject to being endangered. The owner or the owner's authorized agent, on completion of the construction activity, shall immediately remove walkways, debris and other obstructions and leave such public property in as good a condition as it was before such work was commenced.

[BS] 1501.6.7 Adjacent to excavations. Every excavation on a site located 5 feet (1524 mm) or less from the street lot line shall be enclosed with a barrier not less than 6 feet (1829 mm) in height. Where located more than 5 feet (1524 mm) from the street lot line, a barrier shall be erected where required by the *code official*. Barriers shall be of adequate strength to resist wind pressure as specified in Chapter 16 of the *International Building Code*.

[BG] 1501.7 Facilities required. Sanitary facilities shall be provided during construction or demolition activities in accordance with the *International Plumbing Code*.

SECTION 1502
PROTECTION OF ADJOINING PROPERTY

[BS] 1502.1 Protection required. Adjoining public and private property shall be protected from damage during construction and demolition work. Protection must be provided for footings, foundations, party walls, chimneys, skylights and roofs. Provisions shall be made to control water runoff and erosion during construction or demolition activities. The person making or causing an excavation to be made shall provide written notice to the owners of adjoining buildings advising them that the excavation is to be made and that the adjoining buildings should be protected. Said notification shall be delivered not less than 10 days prior to the scheduled starting date of the excavation.

[BS] 1502.2 Excavation retention systems. Where a retention system is used to provide support of an excavation for protection of adjacent structures, the system shall conform to the requirements in Section 1502.2.1 through 1502.2.3.

> **[BS] 1502.2.1 Excavation retention system design.** Excavation retention systems shall be designed by a *registered design professional* to provide vertical and lateral support.
>
> **[BS] 1502.2.2 Excavation retention system monitoring.** The retention system design shall include requirements for monitoring of the system and adjacent structures for horizontal and vertical movement.
>
> **[BS] 1502.2.3 Retention system removal.** Elements of the system shall only be removed or decommissioned where adequate replacement support is provided by backfill or by the new structure. Removal or decommissioning shall be performed in such a manner that protects the adjacent property.

SECTION 1503
TEMPORARY USE OF STREETS, ALLEYS AND PUBLIC PROPERTY

[BG] 1503.1 Storage and handling of materials. The temporary use of streets or public property for the storage or handling of materials or equipment required for construction or demolition, and the protection provided to the public shall comply with the provisions of the applicable governing authority and this chapter.

[BG] 1503.2 Obstructions. Construction materials and equipment shall not be placed or stored so as to obstruct access to fire hydrants, standpipes, fire or police alarm boxes, catch basins or manholes, nor shall such material or equipment be located within 20 feet (6096 mm) of a street intersection, or placed so as to obstruct normal observations of traffic signals or to hinder the use of public transit loading platforms.

[BG] 1503.3 Utility fixtures. Building materials, fences, sheds or any obstruction of any kind shall not be placed so as to obstruct free approach to any fire hydrant, fire department connection, utility pole, manhole, fire alarm box or catch basin, or so as to interfere with the passage of water in the gutter. Protection against damage shall be provided to such utility fixtures during the progress of the work, but sight of them shall not be obstructed.

SECTION 1504
FIRE EXTINGUISHERS

[F] 1504.1 Where required. Structures under construction, *alteration* or demolition shall be provided with not fewer than one *approved* portable fire extinguisher in accordance with Section 906 of the *International Fire Code* and sized for not less than ordinary hazard as follows:

1. At each stairway on all floor levels where combustible materials have accumulated.
2. In every storage and construction shed.
3. Additional portable fire extinguishers shall be provided where special hazards exist, such as the storage and use of flammable and combustible liquids.

[F] 1504.2 Fire hazards. The provisions of this code and of the *International Fire Code* shall be strictly observed to safeguard against all fire hazards attendant upon construction operations.

SECTION 1505
MEANS OF EGRESS

[BE] 1505.1 Stairways required. Where building construction exceeds 40 feet (12 192 mm) in height above the lowest level of fire department vehicle access, a temporary or permanent stairway shall be provided. As construction progresses, such stairway shall be extended to within one floor of the highest point of construction having secured decking or flooring.

[F] 1505.2 Maintenance of means of egress. Means of egress and required accessible means of egress shall be maintained at all times during construction, demolition, remodeling or *alterations* and *additions* to any building.

> **Exception:** Existing means of egress need not be maintained where *approved* temporary means of egress and accessible means of egress systems and facilities are provided.

SECTION 1506
STANDPIPES

[F] 1506.1 Where required. In buildings required to have standpipes by Section 905.3.1 of the *International Building Code*, not less than one standpipe shall be provided for use during construction. Such standpipes shall be installed prior to construction exceeding 40 feet (12 192 mm) in height above the lowest level of fire department vehicle access. Such standpipes shall be provided with fire department hose connections at locations adjacent to *stairways*, complying with Section 1505.1. As construction progresses, such standpipes shall be extended to within one floor of the highest point of construction having secured decking or flooring.

[F] 1506.2 Buildings being demolished. Where a building or portion of a building is being demolished and a standpipe is existing within such a building, such standpipe shall be maintained in an operable condition so as to be available for use by the fire department. Such standpipe shall be demolished with the building but shall not be demolished more than one floor below the floor being demolished.

[F] 1506.3 Detailed requirements. Standpipes shall be installed in accordance with the provisions of Chapter 9 of the *International Building Code*.

> **Exception:** Standpipes shall be either temporary or permanent in nature, and with or without a water supply, provided that such standpipes conform to the requirements of Section 905 of the *International Building Code* as to capacity, outlets and materials.

SECTION 1507
AUTOMATIC SPRINKLER SYSTEM

[F] 1507.1 Completion before occupancy. In buildings where an automatic sprinkler system is required by this code or the *International Building Code*, it shall be unlawful to occupy any portions of a building or structure until the automatic sprinkler system installation has been tested and *approved*, except as provided in Section 110.3.

[F] 1507.2 Operation of valves. Operation of sprinkler control valves shall be permitted only by properly authorized personnel and shall be accompanied by notification of duly designated parties. When the sprinkler protection is being regularly turned off and on to facilitate connection of newly completed segments, the sprinkler control valves shall be checked at the end of each work period to ascertain that protection is in service.

SECTION 1508
ACCESSIBILITY

[BE] 1508.1 Construction sites. Structures, sites and equipment directly associated with the actual process of construction, including, but not limited to, scaffolding, bridging, material hoists, material storage or construction trailers, are not required to be accessible.

SECTION 1509
WATER SUPPLY FOR FIRE PROTECTION

[F] 1509.1 When required. An *approved* water supply for fire protection, either temporary or permanent, shall be made available as soon as combustible building material arrives on the site, on commencement of vertical combustible construction, and on installation of a standpipe system in buildings under construction, in accordance with Sections 1509.1 through 1509.5.

> **Exception:** The fire code official is authorized to reduce the fire-flow requirements for isolated buildings or a group of buildings in rural areas or small communities where the development of full fire-flow requirements is impractical.

[F] 1509.2 Combustible building materials. When combustible building materials of the building under construction are delivered to a site, a minimum fire flow of 500 gallons per minute (1893 L/m) shall be provided. The fire hydrant used to provide this fire flow supply shall be within 500 feet (152 m) of the combustible building materials as measured along an *approved* fire apparatus access lane. Where the site configuration is such that one fire hydrant cannot be located within 500 feet (152 m) of all combustible building materials, additional fire hydrants shall be required to provide coverage in accordance with this section.

[F] 1509.3 Vertical construction of Types III, IV and V construction. Prior to commencement of vertical construction of Type III, IV or V buildings that utilize any combustible building materials, the fire flow required by Sections 1509.3.1 through 1509.3.3 shall be provided, accompanied by fire hydrants in sufficient quantity to deliver the required fire flow and proper coverage.

[F] 1509.3.1 Fire separation up to 30 feet. Where a building of Type III, IV or V construction has a fire separation distance of less than 30 feet (9144 mm) from property lot lines, and an adjacent property has an *existing structure* or otherwise can be built on, the water supply shall provide either a minimum of 500 gallons per minute (1893 L/m), or the entire fire flow required for the building when constructed, whichever is greater.

[F] 1509.3.2 Fire separation of 30 feet up to 60 feet. Where a building of Type III, IV or V construction has a fire separation distance of 30 feet (9144 mm) up to 60 feet (18 288 mm) from property lot lines, and an adjacent property has an *existing structure* or otherwise can be constructed upon, the water supply shall provide a minimum of 500 gallons per minute (1893 L/m), or 50 percent of the fire flow required for the building when constructed, whichever is greater.

[F] 1509.3.3 Fire separation of 60 feet or greater. Where a building of Type III, IV or V construction has a fire separation distance of 60 feet (18 288 mm) or greater from a property lot line, a water supply of 500 gallons per minute (1893 L/m) shall be provided.

[F] 1509.4 Vertical construction, Types I and II construction. If combustible construction materials are delivered to the construction site, water supply in accordance with Section 1509.2 shall be provided. Additional water supply for fire flow is not required prior to commencing vertical construction of Type I and II buildings.

[F] 1509.5 Standpipe supply. Regardless of the presence of combustible building materials, the construction type or the fire separation distance, where a standpipe is required in accordance with Section 1506, a water supply providing a minimum flow of 500 gallons per minute (1893 L/m) shall be provided. The fire hydrant used for this water supply shall be located within 100 feet (30 480 mm) of the fire department connection supplying the standpipe.

CHAPTER 16
REFERENCED STANDARDS

User note:

About this chapter: *This code contains numerous references to standards that are used to regulate materials and methods of construction. Chapter 16 contains a comprehensive list of all standards that are referenced in the code, including the appendices. The standards are part of the code to the extent of the reference to the standard. Compliance with the referenced standard is necessary for compliance with this code. By providing specifically adopted standards, the construction and installation requirements necessary for compliance with the code can be readily determined. The basis for code compliance is, therefore, established and available on an equal basis to the building code official, contractor, designer and owner.*

This chapter lists the standards that are referenced in various sections of this document. The standards are listed herein by the promulgating agency of the standard, the standard identification, the effective date and title, and the section or sections of this document that reference the standard. The application of the referenced standards shall be as specified in Section 102.4.

ASCE/SEI

American Society of Civil Engineers
Structural Engineering Institute
1801 Alexander Bell Drive
Reston, VA 20191-4400

7—1988: Minimum Design Loads and Associated Criteria for Buildings and Other Structures
503.12 , 706.3.2

7—1993: Minimum Design Loads and Associated Criteria for Buildings and Other Structures
503.12 , 706.3.2

7—1995: Minimum Design Loads and Associated Criteria for Buildings and Other Structures
503.12 , 706.3.2

7—1998: Minimum Design Loads and Associated Criteria for Buildings and Other Structures
503.12 , 706.3.2

7—2002: Minimum Design Loads and Associated Criteria for Buildings and Other Structures
503.12 , 706.3.2

7—2005: Minimum Design Loads and Associated Criteria for Buildings and Other Structures
503.12 , 706.3.2

7—2010: Minimum Design Loads and Associated Criteria for Buildings and Other Structures
503.12 , 706.3.2

7—2016: Minimum Design Loads and Associated Criteria for Buildings and Other Structures with Supplement 1
304.2, 304.3.1, 503.4, 503.12, 503.13, 706.3.2, 805.3, 805.4

41—2017: Seismic Evaluation and Retrofit of Existing Buildings
304.3.1, Table 304.3.1, 304.3.2, Table 304.3.2

ASHRAE

ASHRAE
1791 Tullie Circle NE
Atlanta, GA 30329

62.1—2019: Ventilation for Acceptable Indoor Air Quality
807.2

ASME

American Society of Mechanical Engineers
Two Park Avenue
New York, NY 10016

A17.1—2019/CSA B44—19: Safety Code for Elevators and Escalators
306.7.7, 902.1.2

A17.3—2020: Safety Code for Existing Elevators and Escalators
902.1.2

A18.1—2020: Safety Standard for Platform Lifts and Stairway Chair Lifts
306.7.8

REFERENCED STANDARDS

ASTM

ASTM International
100 Barr Harbor Drive, P.O. Box C700
West Conshohocken, PA 19428-2959

C94/C94M—17A: Specification for Ready-mixed Concrete
    109.3.1

E108—17: Standard Test Methods for Fire Tests of Roof Coverings
    1204.5

E136—16A: Test Method for Behavior of Materials in a Vertical Tube Furnace at 750°C
    202

F2006—17: Standard Safety Specification for Window Fall Prevention Devices for Non-Emergency Escape (Egress) and Rescue (Ingress) Windows
    505.2, 702.4

F2090—17: Standard Specification for Window Fall Prevention Devices with Emergency (Egress) Release Mechanisms
    505.2, 505.3.1, 702.4, 702.5.1

ICC

International Code Council, Inc.
500 New Jersey Avenue NW 6th Floor
Washington, DC 20001

IBC—21: International Building Code®
    101.4.1, 104.2.1, 106.2.2, 109.3.3, 109.3.6, 109.3.9, 109.3.10, 110.2, 202, 301.3, 302.4.1, 302.5, 303.1, 303.2.2, 304.1, 304.3.1, 304.3.2, 305.1, 306.5, 306.7, 306.7.2, 306.7.4, 306.7.5, 306.7.9, 306.7.10, 306.7.10.1, 306.7.10.2, 306.7.10.3, 306.7.11, 306.7.12, 306.7.13, 306.7.15, 306.7.16, 306.7.16.3, 306.7.16.4, 306.7.16.5, 306.7.16.7, 309.2, 401.3, 402.1, 405.2.1.1, 405.2.3.1, 405.2.3.3, 405.2.4, 405.2.5, 405.2.6, 501.2, 502.1, 502.3, 502.4, 502.5, 503.1, 503.2, 503.3, 503.4, 503.5, 503.11, 503.12, 503.13, 503.14, 503.15, 503.17, 503.18, 505.3, 505.4, 506.1, 506.3, 506.4, 506.5.1, 506.5.2, 506.5.3, 506.5.4, 507.3, 701.2, 701.3, 702.1, 702.2, 702.3, 702.5, 702.6, 702.7, 704.1.1, 704.3, 705.1, 705.2, 706.2, 706.3.2, 802.2.1, 802.2.3, 802.3, 802.4, 802.5.2, 802.6, 802.6, 803.1.1, 803.2, 803.2.2, 803.2.3, 803.2.4, 803.2.5, 803.3, 804.1, 804.4.1, 804.4.1.1, Table 804.4.1.1(1), 804.4.1.2.1, 804.5.1.2, 804.5.3, 804.5.4, 804.5.5, 804.6, 804.7, 804.8.1, 804.9.1, 804.10.2, 804.11, 804.12.2, 805.2, 805.3, 805.4, 904.1.2, 904.1.3, 904.1.4, 904.1.6, 904.1.7, 904.2, 904.2.1, 904.2.2, 905.2, 905.3, 905.4, 906.2, 906.3, 906.6, 1001.2, 1001.3, 1002.1, 1002.2, 1002.3, 1002.4, 1004.1, 1006.1, 1006.2, 1006.3, 1006.4, 1010.1, 1011.1, 1011.2.1, 1011.2.2, 1011.3, 1011.5.1, 1011.5.2, 1011.5.3, 1011.5.6, 1011.6.1, 1011.6.1.1, 1011.6.3, 1011.7.1, 1011.7.2, 1011.7.3, 1011.8.1, 1011.8.2, 1011.8.3, 1102.1, 1102.2, 1102.3, 1103.1, 1103.2, 1103.3, 1201.4, 1202.2, 1203.12, 1204.2, 1204.9, 1206.1, 1301.2.2, 1301.2.3, 1301.2.4, 1301.3.3, 1301.4.1, 1301.6.1, 1301.6.1.1, 1301.6.2, 1301.6.2.1, 1301.6.3.2, 1301.6.3.3, 1301.6.4.1, 1301.6.5, 1301.6.5.1, 1301.6.6, 1301.6.7.1, 1301.6.8, 1301.6.9, 1301.6.9.1, 1301.6.10, 1301.6.10.1, 1301.6.11, 1301.6.11.1, 1301.6.12.1, 1301.6.13, Table 1301.6.15, 1301.6.15.1, 1301.6.16.1, 1301.6.17, 1301.6.17.1, 1301.6.18, 1301.6.18.1, 1301.6.19, Table 1301.6.19, 1301.6.20, 1301.6.20.1, 1301.9.1, 1401.2, 1402.1, 1402.2, 1402.2.1, 1402.3, 1402.4, 1402.5, 1402.6, 1501.5, 1501.6, 1501.6.4.1, 1501.6.7, 1506.1, 1506.3, 1507.1

ICC 300—17: ICC Standard on Bleachers, Folding and Telescopic Seating and Grandstands
    301.1.1

ICC 500—20: Standard for the Design and Construction of Storm Shelters
    303.1, 303.2

ICC A117.1—17: Accessible and Usable Buildings and Facilities
    306.3, 306.7, 306.7.11, 306.7.12

IECC—21: International Energy Conservation Code®
    302.2, 702.7, 708.1, 809.1, 907.1, 1104.1

IFC—21: International Fire Code®
    101.2.1, 101.4.2, 301.3.1, 302.2, 307.1, 308.1, 802.2.1, 802.2.3, 803.2.3, 803.4.1.1, 803.4.1.2, 803.4.1.3, 803.4.1.4, 803.4.1.5, 803.4.1.6, 904.1.5, 1011.6.1.1, 1301.3.2, 1301.6.8.1, 1301.6.14, 1301.6.14.1, 1401.2, 1501.5, 1504.1, 1504.2

IFGC—21: International Fuel Gas Code®
    302.2, 702.7.1

ICC—continued

IMC—21: International Mechanical Code®
    302.2, 702.7, 807.1, 902.1.1, 1008.1, 1301.6.7.1, 1301.6.8, 1301.6.8.1

IPC—21: International Plumbing Code®
    302.2, 408.1, 702.7, 1009.1, 1009.2, 1009.3, 1009.5, 1501.7

IPMC—21: International Property Maintenance Code®
    101.4.2, 302.2, 1301.3.2, 1401.2

IRC—21: International Residential Code®
    101.2, 101.4.1, 104.2.1, 109.3.3, 109.3.10, 302.2, 307.1, 308.1, 401.3, 402.1, 405.2.6, 502.3, 502.4, 502.5, 503.2, 503.3, 503.11, 505.2, 505.3, 507.3, 701.3, 702.4, 702.5, 706.2, 708.1, 805.2, 806.4, 809.1, 906.2, 907.1, 1011.2.1, 1103.1, 1103.2, 1103.3, 1104.1, 1201.4, 1301.2.2, 1301.2.3, 1301.3.3, 1401.2, 1402.1, 1402.2, 1402.2.1, 1402.3, 1402.4, 1402.5, 1402.6

NFPA

National Fire Protection Association
1 Batterymarch Park
Quincy, MA 02169-7471

NFPA 13R—19: Standard for the Installation of Sprinkler Systems in Residential Occupancies up to and Including Four Stories in Height
    803.2.4

NFPA 70—20: National Electrical Code
    107.3, 302.2, 406.1.1, 406.1.2, 406.1.3, 406.1.5, 806.1, 806.4.4, 1007.1, 1007.2, 1007.3, 1007.4

NFPA 72—19: National Fire Alarm and Signaling Code
    803.2.6, 803.4

NFPA 99—21: Health Care Facilities Code
    302.2.1, 406.1.4, 408.3, 501.3, 707.1, 806.3, 808.1, 1007.1

NFPA 101—21: Life Safety Code
    804.2

UL

UL LLC
333 Pfingsten Road
Northbrook, IL 60062

790—04: Standard Test Methods for Fire Tests of Roof Coverings—with Revisions through October 2018
    1204.5

Appendix A: GUIDELINES FOR THE SEISMIC RETROFIT OF EXISTING BUILDINGS

CHAPTER A1

SEISMIC STRENGTHENING PROVISIONS FOR UNREINFORCED MASONRY BEARING WALL BUILDINGS

User note:

About this appendix: *Appendix A provides guidelines for upgrading the seismic-resistance capacity of different types of existing buildings. It is organized into separate chapters that deal with buildings of different types, including unreinforced masonry buildings, reinforced concrete and reinforced masonry wall buildings, and light-frame wood buildings.*

SECTION A101 PURPOSE

[BS] A101.1 Purpose. The purpose of this chapter is to promote public safety and welfare by reducing the risk of death or injury from the effects of earthquakes on existing unreinforced masonry bearing wall buildings.

The provisions of this chapter are intended as minimum standards for structural seismic resistance, and are established primarily to reduce the risk of life loss or injury. Compliance with these provisions will not necessarily prevent loss of life or injury, or prevent earthquake damage to retrofitted buildings.

SECTION A102 SCOPE

[BS] A102.1 General. The provisions of this chapter shall apply to all *existing buildings* not more than six stories in height above the base of the structure and having not fewer than one unreinforced masonry bearing wall. The elements regulated by this chapter shall be determined in accordance with Table A102.1. Except as provided herein, other structural provisions of the building code shall apply. This chapter does not apply to the *alteration* of existing electrical, plumbing, mechanical or fire safety systems.

[BS] A102.2 Essential and hazardous facilities. The provisions of this chapter shall not apply to the strengthening of buildings in *Risk Category* III or IV. Such buildings shall be strengthened to meet the requirements of the *International Building Code* for new buildings of the same *risk category* or other such criteria *approved* by the *code official*.

SECTION A103 DEFINITIONS

[BS] A103.1 Definitions. For the purpose of this chapter, the applicable definitions in the building code shall also apply.

[BS] BED JOINT. The horizontal layer of mortar on which a masonry unit is laid.

[BS] COLLAR JOINT. The vertical space between adjacent wythes. A collar joint may contain mortar or grout.

[BS] CROSSWALL. A new or existing wall that meets the requirements of Section A111.3. A crosswall is not a shear wall.

[BS] CROSSWALL SHEAR CAPACITY. The unit shear value times the length of the crosswall, $v_c L_c$.

[BS] DETAILED BUILDING SYSTEM ELEMENTS. The localized elements and the interconnections of these elements that define the design of the building.

[BS] TABLE A102.1
ELEMENTS REGULATED BY THIS CHAPTER

BUILDING ELEMENTS	S_{D1}			
	$\geq 0.067_g < 0.133_g$	$\geq 0.133_g < 0.20_g$	$\geq 0.20_g < 0.30_g$	$\geq 0.30_g$
Parapets	X	X	X	X
Walls, anchorage	X	X	X	X
Walls, *h/t* ratios		X	X	X
Walls, in-plane shear		X	X	X
Diaphragms[a]			X	X
Diaphragms, shear transfer[b]		X	X	X
Diaphragms, demand-capacity ratios[b]			X	X

a. Applies only to buildings designed according to the general procedures of Section A110.
b. Applies only to buildings designed according to the special procedures of Section A111.

[BS] DIAPHRAGM EDGE. The intersection of the horizontal diaphragm and a shear wall.

[BS] DIAPHRAGM SHEAR CAPACITY. The unit shear value times the depth of the diaphragm, $v_u D$.

[BS] FLEXIBLE DIAPHRAGM. A diaphragm of wood or untopped metal deck construction in which the horizontal deformation along its length is at least two times the average story drift.

HEAD JOINT. The vertical mortar joint placed between masonry units within the wythe.

[BS] NORMAL WALL. A wall perpendicular to the direction of seismic forces.

[BS] OPEN FRONT. An exterior building wall line on one side only without vertical elements of the seismic force-resisting system in one or more stories.

[BS] POINTING. The process of removal of deteriorated mortar from between masonry units and placement of new mortar. Also known as repointing or tuckpointing for purposes of this chapter.

[BS] REPOINTING. See *"Pointing."*

[BS] RIGID DIAPHRAGM. A diaphragm of concrete construction or concrete-filled metal deck construction.

[BS] TUCKPOINTING. See *"Pointing."*

[BS] UNREINFORCED MASONRY (URM). Includes burned clay, concrete or sand-lime brick; hollow clay or concrete block; plain concrete; and hollow clay tile. These materials shall comply with the requirements of Section A106 as applicable.

[BS] UNREINFORCED MASONRY BEARING WALL. A URM wall that provides the vertical support for the reaction of floor or roof-framing members for which the total superimposed vertical load exceeds 100 pounds per linear foot (1459 N/m) of wall length.

[BS] UNREINFORCED MASONRY WALL. A masonry wall that relies on the tensile strength of masonry units, mortar and grout in resisting design loads, and in which the area of reinforcement is less than the minimum amounts as defined for reinforced masonry walls.

[BS] YIELD STORY DRIFT. The lateral displacement of one level relative to the level above or below at which yield stress is first developed in a frame member.

SECTION A104
SYMBOLS AND NOTATIONS

[BS] A104.1 Symbols and notations. For the purpose of this chapter, the following notations supplement the applicable symbols and notations in the building code.

a_n = Diameter of core multiplied by its length or the area of the side of a square prism.

A = Cross-sectional area of unreinforced masonry pier or wall, square inches (10^{-6} m^2).

A_b = Total area of the bed joints above and below the test specimen for each in-place shear test, square inches (10^{-6} m^2).

A_n = Area of net mortared or grouted section of a wall or wall pier.

D = In-plane width dimension of pier, inches (10^{-3} m), or depth of diaphragm, feet (m).

DCR = Demand-capacity ratio specified in Section A111.4.2.

f'_m = Lower bound masonry compressive strength.

f_{sp} = Tensile-splitting strength of masonry.

F_{wx} = Force applied to a wall at level x, pounds (N).

H = Least clear height of opening on either side of a pier, inches (10^{-3} m).

h/t = Height-to-thickness ratio of URM wall. Height, h, is measured between wall anchorage levels and/or slab-on-grade.

L = Span of diaphragm between shear walls, or span between shear wall and open front, feet (m).

L_c = Length of crosswall, feet (m).

L_i = Effective diaphragm span for an open-front building specified in Section A111.8, feet (m).

P = Applied force as determined by standard test method of ASTM C496 or ASTM E519, pounds (N).

P_D = Superimposed dead load at the location under consideration, pounds (N). For determination of the rocking shear capacity, dead load at the top of the pier under consideration shall be used.

P_{D+L} = Stress resulting from the dead plus actual live load in place at the time of testing, pounds per square inch (kPa).

P_{test} = Splitting tensile test load determined by standard test method ASTM C496, pounds (N).

P_w = Weight of wall, pounds (N).

R = Response modification factor for Ordinary plain masonry shear walls in Bearing Wall System from Table 12.2-1 of ASCE 7, where $R = 1.5$.

S_{DS} = Design spectral acceleration at short period, in g units.

S_{D1} = Design spectral acceleration at 1-second period, in g units.

v_a = The shear strength of any URM pier, $v_m A/1.5$ pounds (N).

v_c = Unit shear strength for a crosswall sheathed with any of the materials given in Table A108.1(1) or Table A108.1(2), pounds per foot (N/m).

v_{mL} = Shear strength of unreinforced masonry, pounds per square inch (kPa).

V_{aa} = The shear strength of any URM pier or wall, pounds (N).

V_{ca} = Total shear capacity of crosswalls in the direction of analysis immediately above the diaphragm level being investigated, $v_c L_c$, pounds (N).

V_{cb} = Total shear capacity of crosswalls in the direction of analysis immediately below the diaphragm level being investigated, $v_c L_c$, pounds (N).

V_p = Shear force assigned to a pier on the basis of its relative shear rigidity, pounds (N).

V_r = Pier rocking shear capacity of any URM wall or wall pier, pounds (N).

v_{test} = Load at incipient cracking for each in-place shear test performed in accordance with Section A106.2.3.6, pounds (N).

v_{tl} = Lower bound mortar shear strength, pounds per square inch (kPa).

v_{to} = Mortar shear test values as specified in Section A106.2.3.6, pounds per square inch (kPa).

v_u = Unit shear capacity value for a diaphragm sheathed with any of the materials given in Table A108.1(1) or A108.1(2), pounds per foot (N/m).

V_{wx} = Total shear force resisted by a shear wall at the level under consideration, pounds (N).

W = Total seismic dead load as defined in the building code, pounds (N).

W_d = Total dead load tributary to a diaphragm level, pounds (N).

W_w = Total dead load of a URM wall above the level under consideration or above an open-front building, pounds (N).

W_{wx} = Dead load of a URM wall assigned to level x halfway above and below the level under consideration, pounds (N).

$\Sigma v_u D$ = Sum of diaphragm shear capacities of both ends of the diaphragm, pounds (N).

$\Sigma\Sigma v_u D$ = For diaphragms coupled with crosswalls, $v_u D$ includes the sum of shear capacities of both ends of diaphragms coupled at and above the level under consideration, pounds (N).

ΣW_d = Total dead load of all the diaphragms at and above the level under consideration, pounds (N).

SECTION A105
GENERAL REQUIREMENTS

[BS] A105.1 General. The seismic force-resisting system specified in this chapter shall comply with the *International Building Code* and referenced standards, except as modified herein.

[BS] A105.2 Alterations and repairs. *Alterations* and *repairs* required to meet the provisions of this chapter shall comply with applicable structural requirements of the building code unless specifically provided for in this chapter.

[BS] A105.3 Requirements for plans. The following construction information shall be included in the plans required by this chapter:

1. Dimensioned floor and roof plans showing existing walls and the size and spacing of floor and roof-framing members and sheathing materials. The plans shall indicate all existing URM walls, and new crosswalls and shear walls, and their materials of construction. The location of these walls and their openings shall be fully dimensioned and drawn to scale on the plans.

2. Dimensioned URM wall elevations showing openings, piers, wall classes as defined in Section A106.2.3.9, thickness, heights, wall shear test locations, cracks or damaged portions requiring *repairs*, the general condition of the mortar joints, and if and where pointing is required. Where the exterior face is veneer, the type of veneer, its thickness and its bonding and/or ties to the structural wall masonry shall be noted.

3. The type of interior wall and ceiling materials, and framing.

4. The extent and type of existing wall anchorage to floors and roof where used in the design.

5. The extent and type of parapet corrections that were previously performed, if any.

6. *Repair* details, if any, of cracked or damaged unreinforced masonry walls required to resist forces specified in this chapter.

7. All other plans, sections and details necessary to delineate required retrofit construction.

8. The design procedure used shall be stated on both the plans and the permit application.

9. Details of the anchor prequalification program required by Section A107.5.3, if used, including location and results of all tests.

10. Quality assurance requirements of special inspection for all new construction materials and for retrofit construction including: anchor tests, pointing or repointing of mortar joints, installation of adhesive or mechanical anchors, and other elements as deemed necessary to ensure compliance with this chapter.

[BS] A105.4 Structural observation, testing and inspection. Structural observation, in accordance with Section 1704.6 of the *International Building Code*, shall be required for all structures in which seismic retrofit is being performed in accordance with this chapter. Structural observation shall include visual observation of work for compliance with the *approved* construction documents and confirmation of existing conditions assumed during design.

Structural testing and inspection for new and existing construction materials shall be in accordance with the building code, except as modified by this chapter.

Special inspection as described in Section A105.3, Item 10, shall be provided equivalent to Level 3 as prescribed in TMS 402, Table 3.1(2).

SECTION A106
MATERIALS REQUIREMENTS

[BS] A106.1 Condition of existing materials. Existing materials used as part of the required vertical load-carrying or seismic force-resisting system shall be evaluated by onsite investigation and: determined to be in good condition (free of degraded mortar, degraded masonry units or significant cracking); or shall be repaired, enhanced, retrofitted or removed and replaced with new materials. Mortar joint deterioration shall be patched by pointing or repointing of the eroded joint in accordance with Section A106.2.3.10. Existing significant cracks in solid unit unreinforced and solid grouted hollow unit masonry shall be repaired.

[BS] A106.2 Existing unreinforced masonry.

[BS] A106.2.1 General. Unreinforced masonry walls used to support vertical loads or seismic forces parallel and perpendicular to the wall plane shall be tested as specified in this section. Masonry that does not meet the minimum requirements established by this chapter shall be repaired, enhanced, removed and replaced with new materials, or alternatively, shall have its structural functions replaced with new materials and shall be anchored to supporting elements.

[BS] A106.2.2 Lay-up of walls. Unreinforced masonry walls shall be laid in a running bond pattern.

[BS] A106.2.2.1 Header in multiple-wythe solid brick. The facing and backing wythes of multiple-wythe walls shall be bonded so that not less than 10 percent of the exposed face area is composed of solid headers extending not less than 4 inches (102 mm) into the backing wythes. The clear distance between adjacent header courses shall not exceed 24 inches (610 mm) vertically or horizontally. Where backing consists of two or more wythes, the headers shall extend not less than 4 inches (102 mm) into the most distant wythe, or the backing wythes shall be bonded together with separate headers for which the area and spacing conform to the foregoing. Wythes of walls not meeting these requirements shall be considered to be veneer, and shall not be included in the effective thickness used in calculating the height-to-thickness ratio and the shear capacity strength of the wall.

Exception: Where SD1 is 0.3 g or less, veneer wythes anchored and made composite with backup masonry are permitted to be used for calculation of the effective thickness.

[BS] A106.2.2.2 Lay-up patterns. Lay-up patterns other than those specified in Section A106.2.2.1 are allowed if their performance can be justified.

[BS] A106.2.3 Testing of masonry.

[BS] A106.2.3.1 Concrete masonry units and structural clay load-bearing tile. Grouted or ungrouted hollow concrete masonry units shall be tested in accordance with ASTM C140. Grouted or ungrouted structural clay load-bearing tile shall be tested in accordance with ASTM C67.

[BS] A106.2.3.2 In-place mortar joint shear tests. Mortar joint shear test values, v_{to}, shall be obtained by one of the following:

1. ASTM C1531.
2. For masonry walls that have high shear strength mortar, or where in-place testing is not practical because of crushing or other failure mode of the masonry, alternative procedures for testing shall be used in accordance with Section A106.2.3.2.

[BS] A106.2.3.3 Alternative procedures for testing masonry. The splitting tensile strength of existing masonry, f_{sp}, or the prism strength of existing masonry, f'_m, is permitted to be determined in accordance with ASTM C496 and calculated by the following equation:

$$f_{sp} = \frac{0.494P}{a_n} \qquad \text{(Equation A1-1)}$$

[BS] A106.2.3.4 Location of tests. The shear tests shall be taken at locations representative of the mortar conditions throughout the building. Test locations shall be determined at the building site by the *registered design professional* in charge. Results of all tests and their locations shall be recorded.

[BS] A106.2.3.5 Number of tests. The minimum number of tests per masonry class shall be determined as follows:

1. At each of both the first and top stories, not less than two tests per wall or line of wall elements providing a common line of resistance to seismic forces.
2. At each of all other stories, not less than one test per wall or line of wall elements providing a common line of resistance to seismic forces.
3. In any case, not less than one test per 1,500 square feet (139.4 m^2) of wall surface and not less than a total of eight tests.

[BS] A106.2.3.6 Minimum quality of mortar.

1. Mortar shear test values, v_{to}, in pounds per square inch (kPa), shall be obtained for each in-place shear test in accordance with the following equation:

$$v_{to} = (V_{test}/A_b) - P_{D+L} \qquad \text{(Equation A1-2)}$$

where:

V_{test} = Load at first observed movement.

A_b = Total area of the bed joints above and below the test specimen.

P_{D+L} = Stress resulting from actual dead plus live loads in place at the time of testing.

2. Individual unreinforced masonry walls with more than 50 percent of mortar test values, v_{to}, less than 30 pounds per square inch (207 kPa) shall be pointed prior to and retested.

3. The lower bound mortar shear strength, v_{tL}, is defined as the mean minus one standard deviation of the mortar shear test values, v_{to}.

4. Unreinforced masonry with mortar shear strength, v_{tL}, less than 30 pounds per square inch (207 kPa) shall be pointed and retested or shall have its structural function replaced, and shall be anchored to supporting elements in accordance with Sections A106.2.1 and A113.8. When existing mortar in any wythe is pointed to increase its shear strength and is retested, the condition of the mortar in the adjacent bed joints of the inner wythe or wythes and the opposite outer wythe shall be examined for extent of deterioration. The shear strength of any wall class shall be not greater than that of the weakest wythe of that class.

[BS] A106.2.3.7 Minimum quality of masonry. Where the alternative procedures of Section A106.2.3.2 are used to determine masonry quality, the following minimums apply:

1. The minimum average value of splitting tensile strength, f_{sp}, as calculated by Equation A1-1 shall be 50 pounds per square inch (344.7 kPa).

2. Individual unreinforced masonry walls with average splitting tensile strength of less than 50 pounds per square inch (344.7 kPa) shall be pointed and retested.

3. The lower-bound mortar strength f_{spL} is defined as the mean minus one standard deviation P_{D+L} of the splitting tensile test values f_{sp}.

[BS] A106.2.3.8 Collar joints. The collar joints shall be inspected at the test locations during each in-place shear test, and estimates of the percentage of surfaces of the adjacent wythe that are covered with mortar shall be reported along with the results of the in-place shear tests.

[BS] A106.2.3.9 Unreinforced masonry classes. Existing unreinforced masonry shall be categorized into one or more classes based on shear strength, quality of construction, state of *repair*, deterioration and weathering. A class shall be characterized by the masonry shear strength determined in accordance with Section A108.2. Classes are defined for whole walls, not for small areas of masonry within a wall. Discretion in the definition of classes of masonry is permitted to avoid unnecessary testing.

[BS] A106.2.3.10 Pointing. Deteriorated mortar joints in unreinforced masonry walls shall be pointed in accordance with the following requirements:

1. **Joint preparation.** Deteriorated mortar shall be cut out by means of a toothing chisel or nonimpact power tool until sound mortar is reached, to a depth not less than $\frac{3}{4}$ inch (19.1 mm) or twice the thickness of the joint, whichever is less, but not greater than 2 inches (50 mm). Care shall be taken not to damage the masonry edges. After cutting is complete, all loose material shall be removed with a brush, or air or water stream.

2. **Mortar preparation.** The mortar mix shall be proportioned as required by the construction specifications and manufacturer's *approved* instructions.

3. **Packing.** The joint into which the mortar is to be packed shall be dampened but without free-standing water. The mortar shall be tightly packed into the joint in layers not exceeding $\frac{1}{4}$ inch (6.4 mm) deep until it is filled; then it shall be tooled to a smooth surface to match the original profile.

Nothing shall prevent pointing of any masonry wall joints before testing is performed in accordance with Section A106.2.3, except as required in Section A107.2.

SECTION A107
QUALITY CONTROL

[BS] A107.1 Pointing. Preparation and mortar pointing shall be performed with special inspection.

Exception: At the discretion of the *code official*, incidental pointing may be performed without special inspection.

[BS] A107.2 Masonry shear tests. In-place masonry shear tests shall comply with Section A106.2.3.1. Testing of masonry for determination of splitting tensile strength shall comply with Section A106.2.3.3.

[BS] A107.3 Existing wall anchors. Existing wall anchors used as all or part of the required tension anchors shall be tested in pullout according to Section A107.5.1. Not fewer than four anchors tested per floor shall be tested in pullout, with not fewer than two tests at walls with joists framing into the wall and two tests at walls with joists parallel to the wall, but not less than 10 percent of the total number of existing tension anchors at each level.

[BS] A107.4 New wall anchors. New wall anchors embedded in URM walls shall be subject to special inspection prior to placement of the anchor and grout or adhesive in the drilled hole. Five percent of all anchors that do not extend through the wall shall be subject to a direct-tension test, and an additional 20 percent shall be tested using a calibrated torque wrench. Testing shall be performed in accordance with Section A107.5.

New wall anchors embedded in URM walls resisting tension forces or a combination of tension and shear forces shall be subject to special inspection, prior to placement of the anchor and grout or adhesive in the drilled hole. Five percent of all anchors resisting tension forces shall be subject to a direct-tension test, and an additional 20 percent shall be tested using a calibrated torque wrench. Testing shall be performed in accordance with Section A107.5.

Exception: New bolts that extend through the wall with steel plates on the far side of the wall need not be tested.

[BS] A107.5 Tests of anchors in unreinforced masonry walls. Tests of anchors in unreinforced masonry walls shall be in accordance with Sections A107.5.1 through A107.5.3. Results of all tests shall be reported to the authority having jurisdiction. The report shall include the test results of maximum load for each test; pass-fail results; corresponding anchor size and type; orientation of loading; details of the anchor installation, testing apparatus and embedment; wall thickness; and joist orientation and proximity to the tested anchor.

[BS] A107.5.1 Direct tension testing of existing anchors and new anchors. The test apparatus shall be supported by the masonry wall. The test procedure for prequalification of tension and shear anchors shall comply with ASTM E488. Existing wall anchors shall be given a preload of 300 pounds (1335 N) before establishing a datum for recording elongation. The tension test load reported shall be recorded at $1/8$ inch (3.2 mm) relative movement between the existing anchor and the adjacent masonry surface. New embedded tension anchors shall be subject to a direct tension load of not less than 2.5 times the design load but not less than 1,500 pounds (6672 N) for 5 minutes.

Exception: Where obstructions occur, the distance between the anchor and the test apparatus support shall be not less than one-half the wall thickness for existing anchors and 75 percent of the embedment length for new embedded anchors.

[BS] A107.5.2 Torque testing of new anchors. Anchors embedded in unreinforced masonry walls shall be tested using a torque-calibrated wrench to the following minimum torques:

- $1/2$-inch-diameter (12.7 mm) bolts: 40 foot pounds (54.2 N-m).
- $5/8$-inch-diameter (15.9 mm) bolts: 50 foot pounds (67.8 N-m).
- $3/4$-inch-diameter (19.1 mm) bolts: 60 foot pounds (81.3 N-m).

[BS] A107.5.3 Prequalification test for bolts and other types of anchors. ASTM E488 or the test procedure in Section A107.5.1 is permitted to be used to determine tension or shear strength values for anchors greater than those permitted by Table A108.1(2). Anchors shall be installed in the same manner and using the same materials as will be used in the actual construction. Not fewer than five tests for each bolt size and type shall be performed for each class of masonry in which they are proposed to be used. The tension and shear strength values for such anchors shall be the lesser of the average ultimate load divided by 5.0 or the average load at which $1/8$ inch (3.2 mm) elongation occurs for each size and type of anchor and class of masonry.

SECTION A108
DESIGN STRENGTHS

[BS] A108.1 Strength values.

1. Strength values for existing materials are given in Table A108.1(1) and for new materials in Table A108.1(2).
2. The strength reduction factor, Φ, shall be taken equal to 1.0.
3. The use of materials not specified herein shall be based on substantiating research data or engineering judgment, as *approved* by the *code official*.

[BS] A108.2 Masonry shear strength. The unreinforced masonry shear strength, v_{mL}, shall be determined for each masonry class from one of the following equations:

1. When testing is performed in accordance with Section A106.2.3.1, the unreinforced masonry shear strength, v_m, shall be determined by Equation A1-3.

$$v_{mL} = \frac{0.75\left(0.75 v_{tL} \frac{P_D}{A_n}\right)}{1.5} \quad \text{(Equation A1-3)}$$

The mortar shear strength values, v_{tL}, shall be determined in accordance with Section A106.2.3.6.

2. When alternate testing is performed in accordance with Section A106.2.3.3, unreinforced masonry shear, v_{mL}, shall be determined by Equation A1-4.

$$v_{mL} = \frac{0.75\left(f_{sp} + \frac{P_D}{A_n}\right)}{1.5} \quad \text{(Equation A1-4)}$$

[BS] A108.3 Masonry compression. Where any increase in wall dead plus live load compression stress occurs, the maximum compression stress in unreinforced masonry, Q_G/A_n, shall not exceed 300 pounds per square inch (2070 kPa).

[BS] A108.4 Masonry tension. Unreinforced masonry shall be assumed to have no tensile capacity.

[BS] A108.5 Wall tension anchors. The tension strength of wall anchors shall be the average of the tension test values for anchors having the same wall thickness and framing orientation.

[BS] A108.6 Foundations. For existing foundations, new total dead loads are permitted to be increased over the existing dead load by 25 percent. New total dead load plus live load plus seismic forces may be increased over the existing dead load plus live load by 50 percent. Higher values may be justified only in conjunction with a geotechnical investigation.

SECTION A109
ANALYSIS AND DESIGN PROCEDURE

[BS] A109.1 General. The elements of buildings hereby required to be analyzed are specified in Table A102.1.

[BS] TABLE A108.1(1)
STRENGTH VALUES FOR EXISTING MATERIALS

EXISTING MATERIALS OR CONFIGURATION OF MATERIALS[a]		STRENGTH VALUES
		x 14.594 for N/m
Horizontal diaphragms	Roofs with straight sheathing and roofing applied directly to the sheathing.	300 lbs. per ft. for seismic shear
	Roofs with diagonal sheathing and roofing applied directly to the sheathing.	750 lbs. per ft. for seismic shear
	Floors with straight tongue-and-groove sheathing.	300 lbs. per ft. for seismic shear
	Floors with straight sheathing and finished wood flooring with board edges offset or perpendicular.	1,500 lbs. per ft. for seismic shear
	Floors with diagonal sheathing and finished wood flooring.	1,800 lbs. per ft. for seismic shear
	Metal deck welded with minimal welding.[c]	1,800 lbs. per ft. for seismic shear
	Metal deck welded for seismic resistance.[d]	3,000 lbs. per ft. for seismic shear
Crosswalls[b]	Plaster on wood or metal lath.	600 lbs. per ft. for seismic shear
	Plaster on gypsum lath.	550 lbs. per ft. for seismic shear
	Gypsum wallboard, unblocked edges.	200 lbs. per ft. for seismic shear
	Gypsum wallboard, blocked edges.	400 lbs. per ft. for seismic shear
Existing footing, wood framing, structural steel, reinforcing steel	Plain concrete footings.	f'_c = 1,500 psi unless otherwise shown by tests
	Douglas fir wood.	Same as D.F. No. 1
	Reinforcing steel.	F_y = 40,000 psi maximum
	Structural steel.	F_y = 33,000 psi maximum

For SI: 1 inch = 25.4 mm, 1 square inch = 645.16 mm^2, 1 pound = 4.4 N, 1 pound per square inch = 6894.75 N/m^2, 1 pound per foot = 14.43 N/m.
a. Material must be sound and in good condition.
b. Shear values of these materials may be combined, except the total combined value should not exceed 900 pounds per foot.
c. Minimum 22-gage steel deck with welds to supports satisfying the standards of the Steel Deck Institute.
d. Minimum 22-gage steel deck with $3/4$-inch diameter plug welds at an average spacing not exceeding 8 inches and with sidelap welds appropriate for the deck span.

[BS] A109.2 Selection of procedure. Buildings with rigid diaphragms shall be analyzed by the general procedure of Section A110. Buildings with flexible diaphragms shall be analyzed by the general procedure or, where applicable, are permitted to be analyzed by the special procedure of Section A111.

SECTION A110
GENERAL PROCEDURE

[BS] A110.1 Minimum design lateral forces. Buildings shall be analyzed to resist minimum lateral forces assumed to act nonconcurrently in the direction of each of the main axes of the structure in accordance with the following:

$$V = \frac{0.75 S_{DS} W}{R} \quad \text{(Equation A1-5)}$$

[BS] A110.2 Seismic forces on elements of structures. Parts and portions of a structure not covered in Section A110.3 shall be analyzed and designed per the current building code, using force levels defined in Section A110.1.

Exceptions:

1. Unreinforced masonry walls for which height-tothickness ratios do not exceed ratios set forth in Table A110.2 need not be analyzed for out-of-plane loading. Unreinforced masonry walls that exceed the allowable h/t ratios of Table A110.2 shall be braced according to Section A113.5.

2. Parapets complying with Section A113.6 need not be analyzed for out-of-plane loading.

3. Where walls are to be anchored to flexible floor and roof diaphragms, the anchorage shall be in accordance with Section A113.1.

[BS] A110.3 In-plane loading of URM shear walls and frames. Vertical seismic force-resisting elements shall be analyzed in accordance with Section A112.

[BS] A110.4 Redundancy and overstrength factors. Any redundancy or overstrength factors contained in the building code may be taken as unity. The vertical component of seismic force (E_v) may be taken as zero.

[BS] TABLE A108.1(2)
STRENGTH VALUES OF NEW MATERIALS USED IN CONJUNCTION WITH EXISTING CONSTRUCTION

NEW MATERIALS OR CONFIGURATION OF MATERIALS		STRENGTH VALUES
Horizontal diaphragms	Plywood sheathing applied directly over existing straight sheathing with ends of plywood sheets bearing on joists or rafters and edges of plywood located on center of individual sheathing boards.	675 lbs. per ft.
Crosswalls	Plywood sheathing applied directly over wood studs; no value should be given to plywood applied over existing plaster or wood sheathing.	1.2 times the value specified in the current building code.
	Drywall or plaster applied directly over wood studs.	The value specified in the current building code.
	Drywall or plaster applied to sheathing over existing wood studs.	50 percent of the value specified in the current building code.
Tension anchors[f]	Anchors extending entirely through unreinforced masonry wall secured with bearing plates on far side of a wall 30 square inches of area.[b, c]	5,400 lbs. per anchor for three-wythe minimum walls. 2,700 lbs. for two-wythe walls.
Shear bolts[e, f]	Anchors embedded not less than 8 inches into unreinforced masonry walls; anchors should be centered in $2\frac{1}{2}$-inch-diameter holes with dry-pack or nonshrink grout around the circumference of the anchor.	The value for plain masonry specified for solid masonry TMS 402; and no value larger than those given for $\frac{3}{4}$-inch bolts should be used.
Combined tension and shear anchors[f]	Through-anchors—anchors meeting the requirements for shear and for tension anchors.[b, c]	Tension—same as for tension anchors. Shear—same as for shear anchors.
	Embedded anchors—anchors extending to the exterior face of the wall with a $2\frac{1}{2}$-inch round plate under the head and drilled at an angle of $22\frac{1}{2}$ degrees to the horizontal; installed as specified for shear anchors.[a, b, c]	Tension—3,600 lbs. per anchor. Shear—same as for shear anchors.
Infilled walls	Reinforced masonry infilled openings in existing unreinforced masonry walls; provide keys or dowels to match reinforcing.	Same as values specified for unreinforced masonry walls.
Reinforced masonry[d]	Masonry piers and walls reinforced per the current building code.	The value specified in the current building code for strength design.
Reinforced concrete[d]	Concrete footings, walls and piers reinforced as specified in the current building code.	The value specified in the current building code for strength design.

For SI: 1 inch = 25.4 mm, 1 square inch = 645.16 mm^2, 1 pound = 4.4 N, 1 degree = 0.017 rad, 1 pound per foot = 14.43 N/m, 1 foot = 304.8 mm.

a. Embedded anchors to be tested as specified in Section A107.4.
b. Anchors shall be $\frac{1}{2}$ inch minimum in diameter.
c. Drilling for anchors shall be done with an electric rotary drill; impact tools should not be used for drilling holes or tightening anchors and shear bolt nuts.
d. Load factors or capacity reduction factors shall not be used.
e. Other bolt sizes, values and installation methods may be used, provided that a testing program is conducted in accordance with Section A107.5.3. The strength value shall be determined by multiplying the calculated allowable value, determined in accordance with Section A107.5.3, by 3.0, and the usable value shall be limited to not greater than 1.5 times the value given in the table. Bolt spacing shall not exceed 6 feet on center and shall be not less than 12 inches on center.
f. An alternative adhesive anchor bolt system is permitted to be used providing: its properties and installation conform to an ICC Evaluation Service Report; and the report states that the system's use is in unreinforced masonry as an acceptable alternative to Sections A107.4 and A113.1 or TMS 402, Section 2.1.4. The report's allowable values shall be multiplied by a factor of three to obtain strength values and the strength reduction factor, Φ, shall be taken equal to 1.0.

SECTION A111
SPECIAL PROCEDURE

[BS] A111.1 Limits for the application of this procedure. The special procedures of this section shall be applied only to buildings having the following characteristics:

1. Flexible diaphragms at all levels above the base of the structure.

2. Vertical elements of the seismic force-resisting system consisting predominantly of masonry or a combination of masonry and concrete shear walls.

3. Except for single-story buildings with an open front on one side only, not fewer than two lines of vertical elements of the seismic force-resisting system parallel to each axis of the building (see Section A111.8 for open front buildings).

[BS] A111.2 Seismic forces on elements of structures. With the exception of the provisions in Sections A111.4 through A111.7, elements of structures shall comply with Sections A110.2 through A110.4.

[BS] A111.3 Crosswalls. Crosswalls shall meet the requirements of this section.

[BS] A111.3.1 Crosswall definition. A crosswall is a wood-framed wall sheathed with any of the materials described in Table A108.1(1) or Table A108.1(2) or other

[BS] TABLE A110.2
ALLOWABLE VALUE OF HEIGHT-TO-THICKNESS RATIO OF UNREINFORCED MASONRY WALLS

WALL TYPES	$0.13_g \leq S_{D1} < 0.25_g$	$0.25_g \leq S_{D1} < 0.4_g$	$S_{D1} \geq 0.4_g$ BUILDINGS WITH CROSSWALLS[a]	$S_{D1} \geq 0.4_g$ ALL OTHER BUILDINGS
Walls of one-story buildings	20	16	16[b, c]	13
First-story wall of multiple-story building	20	18	16	15
Walls in top story of multiple-story building	14	14	14[b, c]	9
All other walls	20	16	16	13

For SI: 1 pound per square inch = 6894.75 N/m².

a. Applies to the special procedures of Section A111 only. See Section A111.7 for other restrictions.
b. This value of height-to-thickness ratio shall be used where mortar shear tests establish a tested mortar shear strength, v_t, of not less than 100 pounds per square inch. This value shall also be used where the tested mortar shear strength is not less than 60 pounds per square inch, and where a visual examination of the collar joint indicates not less than 50-percent mortar coverage.
c. Where a visual examination of the collar joint indicates not less than 50-percent mortar coverage, and the tested mortar shear strength, v_t, is greater than 30 pounds per square inch but less than 60 pounds per square inch, the allowable height-to-thickness ratio may be determined by linear interpolation between the larger and smaller ratios in direct proportion to the tested mortar shear strength.

system as defined in Section A111.3.5. Crosswalls shall be spaced not more than 40 feet (12 192 mm) on center measured perpendicular to the direction of consideration, and shall be placed in each story of the building. Crosswalls shall extend the full story height between diaphragms.

Exceptions:

1. Crosswalls need not be provided at all levels where used in accordance with Section A111.4.2, Item 4.
2. Existing crosswalls need not be continuous below a wood diaphragm at or within 4 feet (1219 mm) of grade, provided that:
   2.1. Shear connections and anchorage requirements of Section A111.5 are satisfied at all edges of the diaphragm.
   2.2. Crosswalls with total shear capacity of $0.5S_{D1}\Sigma W_d$ interconnect the diaphragm to the foundation.
   2.3. The demand-capacity ratio of the diaphragm between the crosswalls that are continuous to their foundations does not exceed 2.5, calculated as follows:

$$DCR = \frac{(2.1S_{D1}W_d + V_{ca})}{2v_uD}$$

(Equation A1-6)

[BS] A111.3.2 Crosswall shear capacity. Within any 40 feet (12 192 mm) measured along the span of the diaphragm, the sum of the crosswall shear capacities shall be not less than 30 percent of the diaphragm shear capacity of the strongest diaphragm at or above the level under consideration.

[BS] A111.3.3 Existing crosswalls. Existing crosswalls shall have a maximum height-to-length ratio between openings of 1.5 to 1. Existing crosswall connections to diaphragms need not be investigated as long as the crosswall extends to the framing of the diaphragms above and below.

[BS] A111.3.4 New crosswalls. New crosswall connections to the diaphragm shall develop the crosswall shear capacity. New crosswalls shall have the capacity to resist an overturning moment equal to the crosswall shear capacity times the story height. Crosswall overturning moments need not be cumulative over more than two stories.

[BS] A111.3.5 Other crosswall systems. Other systems, such as moment-resisting frames, may be used as crosswalls provided that the yield story drift does not exceed 1 inch (25 mm) in any story.

[BS] A111.4 Wood diaphragms.

[BS] A111.4.1 Acceptable diaphragm span. A diaphragm is acceptable if the point (L, DCR) on Figure A111.4.1 falls within Region 1, 2 or 3.

[BS] A111.4.2 Demand-capacity ratios. Demand-capacity ratios shall be calculated for the diaphragm at any level according to the following formulas:

1. For a diaphragm without qualifying crosswalls at levels immediately above or below:

$$DCR = 2.1S_{D1}W_d/\Sigma v_uD \quad \text{(Equation A1-7)}$$

2. For a diaphragm in a single-story building with qualifying crosswalls, or for a roof diaphragm coupled by crosswalls to the diaphragm directly below:

$$DCR = 2.1S_{D1}W_d/\Sigma v_uD + V_{cb}$$

(Equation A1-8)

3. For diaphragms in a multiple-story building with qualifying crosswalls in all levels:

$$DCR = 2.1S_{D1}\Sigma W_d/(\Sigma\Sigma v_uD + V_{cb})$$

(Equation A1-9)

DCR shall be calculated at each level for the set of diaphragms at and above the level under consideration. In addition, the roof diaphragm shall meet the requirements of Equation A1-10.

APPENDIX A—GUIDELINES FOR THE SEISMIC RETROFIT OF EXISTING BUILDINGS

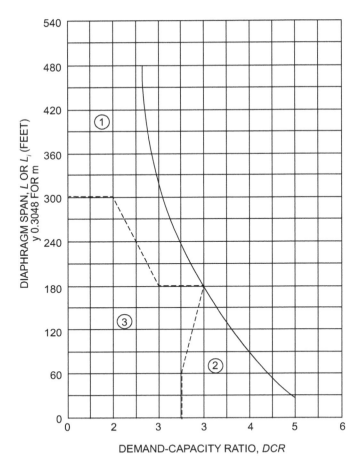

1. Region of demand-capacity ratios where crosswalls may be used to increase h/t ratios.
2. Region of demand-capacity ratios where h/t ratios of "buildings with crosswalls" may be used, whether or not crosswalls are present.
3. Region of demand-capacity ratios where h/t ratios of "all other buildings" shall be used, whether or not crosswalls are present.

For SI: 1 foot = 304.8 mm.

[BS] FIGURE A111.4.1
ACCEPTABLE DIAPHRAGM SPAN

4. For a roof diaphragm and the diaphragm directly below, if coupled by crosswalls:

$$DCR = 2.1S_{D1}\Sigma W_d / \Sigma\Sigma v_u D \quad \textbf{(Equation A1-10)}$$

[BS] A111.4.3 Chords. An analysis for diaphragm flexure need not be made, and chords need not be provided.

[BS] A111.4.4 Collectors. An analysis of diaphragm collector forces shall be made for the transfer of diaphragm edge shears into vertical elements of the lateral force-resisting system. Collector forces may be resisted by new or existing elements.

[BS] A111.4.5 Diaphragm openings.

1. Diaphragm forces at corners of openings shall be investigated and shall be developed into the diaphragm by new or existing materials.
2. In addition to the demand-capacity ratios of Section A111.4.2, the demand-capacity ratio of the portion of the diaphragm adjacent to an opening shall be calculated using the opening dimension as the span.
3. Where an opening occurs in the end quarter of the diaphragm span, the calculation of $v_u D$ for the demand-capacity ratio shall be based on the net depth of the diaphragm.

[BS] A111.5 Diaphragm shear transfer. Diaphragms shall be connected to shear walls and new vertical seismic force-resisting elements with connections capable of developing the diaphragm-loading tributary to the shear wall or new seismic force-resisting elements given by the lesser of the following formulas:

$$V = 1.2S_{D1}C_p W_d \quad \textbf{(Equation A1-11)}$$

using the C_p values in Table A111.5, or

$$V = v_u D \quad \textbf{(Equation A1-12)}$$

[BS] TABLE A111.5
HORIZONTAL FORCE FACTOR, C_p

CONFIGURATION OF MATERIALS	C_p
Roofs with straight or diagonal sheathing and roofing applied directly to the sheathing, or floors with straight tongue-and-groove sheathing.	0.50
Diaphragms with double or multiple layers of boards with edges offset, and blocked plywood systems.	0.75
Diaphragms of metal deck without topping:	
Minimal welding or mechanical attachment.	0.6
Welded or mechanically attached for seismic resistance.	0.68

[BS] A111.6 Shear walls (In-plane loading).

[BS] A111.6.1 Wall story force. The wall story force distributed to a shear wall at any diaphragm level shall be the lesser value calculated as:

$$F_{wx} = 0.8 S_{D1}(W_{wx} + W_d/2) \quad \text{(Equation A1-13)}$$

but need not exceed

$$F_{wx} = 0.8 S_{D1} W_{wx} + v_u D \quad \text{(Equation A1-14)}$$

[BS] A111.6.2 Wall story shear. The wall story shear shall be the sum of the wall story forces at and above the level of consideration.

$$V_{wx} = \Sigma F_{wx} \quad \text{(Equation A1-15)}$$

[BS] A111.6.3 Shear wall analysis. Shear walls shall comply with Section A112.

[BS] A111.6.4 New seismic force-resisting elements. New seismic force-resisting elements such as moment frames, braced frames or shear walls shall be designed as required by the building code, except that the seismic forces shall be as specified in Section A111.6.1, and the story drift ratio shall be limited to 0.015, except as further limited by Section A112.4.2 for moment frames.

[BS] A111.7 Out-of-plane forces—unreinforced masonry walls.

[BS] A111.7.1 Allowable unreinforced masonry wall height-to-thickness ratios. The provisions of Section A110.2 are applicable, except the allowable height-to-thickness ratios given in Table A110.2 shall be determined from Figure A111.4.1 as follows:

1. In Region 1, height-to-thickness ratios for buildings with crosswalls may be used if qualifying crosswalls are present in all stories.
2. In Region 2, height-to-thickness ratios for buildings with crosswalls may be used whether or not qualifying crosswalls are present.
3. In Region 3, height-to-thickness ratios for "all other buildings" shall be used whether or not qualifying crosswalls are present.

[BS] A111.7.2 Walls with diaphragms in different regions. Where diaphragms above and below the wall under consideration have demand-capacity ratios in different regions of Figure A111.4.1, the lesser height-to-thickness ratio shall be used.

[BS] A111.8 Open-front design procedure. A single-story building with an open front on one side and crosswalls parallel to the open front may be designed by the following procedure:

1. Effective diaphragm span, L_i, for use in Figure A111.4.1 shall be determined in accordance with the following formula:

$$L_i = 2[(W_w/W_d)L + L] \quad \text{(Equation A1-16)}$$

2. Diaphragm demand-capacity ratio shall be calculated as:

$$DCR = 2.1 S_{D1}(W_d + W_w)/[(v_u D) + V_{cb}] \quad \text{(Equation A1-17)}$$

SECTION A112
ANALYSIS AND DESIGN

[BS] A112.1 General. The following requirements are applicable to both the general procedure and the special procedure for analyzing vertical elements of the lateral force-resisting system.

[BS] A112.2 In-plane shear of unreinforced masonry walls.

[BS] A112.2.1 Flexural rigidity. Flexural components of deflection need not be considered in determining the rigidity of an unreinforced masonry wall.

[BS] A112.2.2 Shear walls with openings. Wall piers shall be analyzed according to the following procedure, which is diagrammed in Figure A112.2.2.

1. For any pier,
   1.1. The pier shear capacity shall be calculated as:

   $$v_a = v_m A_n \quad \text{(Equation A1-18)}$$

   where:

   A_n = area of net mortared or grouted section of a wall or wall pier.

   1.2. The pier rocking shear capacity shall be calculated as:

   $$V_r = 0.9 P_D D/H \quad \text{(Equation A1-19)}$$

2. The wall piers at any level are acceptable if they comply with one of the following modes of behavior:

   2.1. Rocking controlled mode. Where the pier rocking shear capacity is less than the pier shear capacity, in other words, $V_r < v_a$, for each pier in a level, forces in the wall at that level, V_{wx}, shall be distributed to each pier in proportion to $P_D D/H$.

   For the wall at that level:

   $$0.7 V_{wx} < \Sigma V_r \quad \text{(Equation A1-20)}$$

   2.2. Shear controlled mode. Where the pier shear capacity is less than the pier rocking capacity, in other words, $v_a < V_r$ in one or more pier(s) in a level, forces in the wall at the level, V_{wx}, shall be distributed to each pier in proportion to D/H.

For each pier at that level:

$$V_p < v_a \qquad \text{(Equation A1-21)}$$

and

$$V_p < V_r \qquad \text{(Equation A1-22)}$$

If $V_p < v_a$ for each pier and $V_p > V_r$ for one or more piers, such piers shall be omitted from the analysis, and the procedure shall be repeated for the remaining piers, unless the wall is strengthened and reanalyzed.

3. Masonry pier tension stress. Unreinforced masonry wall piers need not be analyzed for tension stress.

[BS] A112.2.3 Shear walls without openings. Shear walls without openings shall be analyzed the same as for walls with openings, except that V_r shall be calculated as follows:

$$V_r = 0.9(P_D + 0.5P_w)D/H \qquad \text{(Equation A1-23)}$$

[BS] A112.3 Plywood-sheathed shear walls. Plywood-sheathed shear walls may be used to resist lateral forces for URM buildings with flexible diaphragms analyzed according to provisions of Section A111. Plywood-sheathed shear walls shall not be used to share lateral forces with other materials along the same line of resistance.

[BS] A112.4 Combinations of vertical elements.

[BS] A112.4.1 Seismic force distribution. Seismic forces shall be distributed among the vertical-resisting elements in proportion to their relative rigidities, except that moment-resisting frames shall comply with Section A112.4.2.

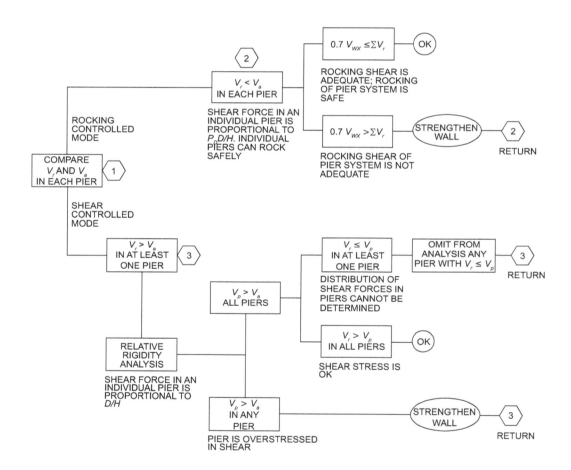

V_a = Allowable shear strength of a pier.
V_p = Shear force assigned to a pier on the basis of a relative shear rigidity analysis.
V_r = Rocking shear capacity of pier.
V_{wx} = Total shear force resisted by the wall.
ΣV_r = Rocking shear capacity of all piers in the wall.

**[BS] FIGURE A112.2.2
ANALYSIS OF URM WALL IN-PLANE SHEAR FORCES**

[BS] A112.4.2 Moment-resisting frames. Moment resisting frames shall not be used with an unreinforced masonry wall in a single line of resistance unless the wall has piers that have adequate shear capacity to sustain rocking in accordance with Section A112.2.2. The frames shall be designed in accordance with the building code to resist 100 percent of the seismic forces tributary to that line of resistance, as determined from Section A111.2. The story drift ratio shall be limited to 0.0075.

SECTION A113
DETAILED BUILDING SYSTEM DESIGN REQUIREMENTS

[BS] A113.1 Wall anchorage.

[BS] A113.1.1 Anchor locations. Unreinforced masonry walls shall be anchored at the roof and floor levels as required in Section A110.2. Ceilings of plaster or similar materials, where not attached directly to roof or floor framing and where abutting masonry walls, shall either be anchored to the walls at a maximum spacing of 6 feet (1829 mm) or be removed.

[BS] A113.1.2 Anchor requirements. Anchors shall consist of bolts installed through the wall as specified in Table A108.1(2), or an *approved* equivalent at a maximum anchor spacing of 6 feet (1829 mm). Wall anchors shall be secured to the framing members parallel or perpendicular to the wall to develop the required forces.

[BS] A113.1.3 Minimum wall anchorage. Anchorage of masonry walls to each floor or roof shall resist a minimum force determined as $0.9S_{DS}$ times the tributary weight or 200 pounds per linear foot (2920 N/m), whichever is greater, acting normal to the wall at the level of the floor or roof. Existing wall anchors, if used, must be tested and meet the requirements of Section A107.5.1 or be upgraded.

[BS] A113.1.4 Anchors at corners. At the roof and floor levels, both shear and tension anchors shall be provided within 2 feet (610 mm) horizontally from the inside of the corners of the walls.

[BS] A113.2 Diaphragm shear transfer. Anchors transmitting shear forces shall have a maximum spacing of 6 feet (1829 mm) and shall have nuts installed over malleable iron or plate washers where bearing on wood, and heavy-cut washers where bearing on steel.

[BS] A113.3 Collectors. Collector elements shall be provided that are capable of transferring the seismic forces originating in other portions of the building to the element providing the resistance to those forces.

[BS] A113.4 Ties and continuity. Ties and continuity shall conform to the requirements of the building code.

[BS] A113.5 Wall bracing.

[BS] A113.5.1 General. Where a wall height-to-thickness ratio exceeds the specified limits, the wall may be laterally supported by vertical bracing members per Section A113.5.2 or by reducing the wall height by bracing per Section A113.5.3.

[BS] A113.5.2 Vertical bracing members. Vertical bracing members shall be attached to floor and roof construction for their design loads independently of required wall anchors. Horizontal spacing of vertical bracing members shall not exceed one-half of the unsupported height of the wall or 10 feet (3048 mm). Deflection of such bracing members at design loads shall not exceed one-tenth of the wall thickness.

[BS] A113.5.3 Intermediate wall bracing. The wall height may be reduced by bracing elements connected to the floor or roof. Horizontal spacing of the bracing elements and wall anchors shall be as required by design, but shall not exceed 6 feet (1829 mm) on center. Bracing elements shall be detailed to minimize the horizontal displacement of the wall by the vertical displacement of the floor or roof.

[BS] A113.6 Parapets. Parapets and exterior wall appendages not conforming to this chapter shall be removed, or stabilized or braced to ensure that the parapets and appendages remain in their original positions.

The maximum height of an unbraced unreinforced masonry parapet above the lower of either the level of tension anchors or the roof sheathing shall not exceed the height-to-thickness ratio shown in Table A113.6. If the required parapet height exceeds this maximum height, a bracing system designed for the forces determined in accordance with the building code shall support the top of the parapet. Parapet corrective work must be performed in conjunction with the installation of tension roof anchors.

The height of a URM parapet above any wall anchor shall be not less than 12 inches (305 mm).

Exception: If a reinforced concrete beam is provided at the top of the wall, the height above the wall anchor is permitted to be not less than 6 inches (152 mm).

[BS] A113.7 Veneer.

1. Veneer shall be anchored with *approved* anchor ties conforming to the required design capacity specified in the building code and shall be placed at a maximum spacing of 24 inches (610 mm) with a maximum supported area of 4 square feet (0.372 m^2).

    Exception: Existing anchor ties for attaching brick veneer to brick backing shall be acceptable, provided that the ties are in good condition and

**[BS] TABLE A113.6
MAXIMUM ALLOWABLE HEIGHT-TO-THICKNESS RATIO FOR PARAPETS**

	S_{D1}		
	$0.13g \leq S_{D1} \leq 0.25g$	$0.25g \leq S_{D1} < 0.4g$	$S_{D1} \geq 0.4g$
Maximum allowable height-to-thickness ratios	2.5	2.5	1.5

conform to the following minimum size and material requirements.

Existing veneer anchor ties shall be considered adequate if they are of corrugated galvanized iron strips not less than 1 inch (25 mm) in width, 8 inches (203 mm) in length and $\frac{1}{16}$ inch (1.6 mm) in thickness, or the equivalent.

2. The location and condition of existing veneer anchor ties shall be verified as follows:

   2.1. An *approved* testing laboratory shall verify the location and spacing of the ties and shall submit a report to the *code official* for approval as part of the structural analysis.

   2.2. The veneer in a selected area shall be removed to expose a representative sample of ties (not less than four) for inspection by the *code official*.

[BS] A113.8 Nonstructural masonry walls. Unreinforced masonry walls that do not carry design vertical or lateral loads and that are not required by the design to be part of the lateral force-resisting system shall be adequately anchored to new or existing supporting elements. The anchors and elements shall be designed for the out-of-plane forces specified in the building code. The height- or length-to-thickness ratio between such supporting elements for such walls shall not exceed nine.

[BS] A113.9 Truss and beam supports. Where trusses and beams other than rafters or joists are supported on masonry, independent secondary columns shall be installed to support vertical loads of the roof or floor members.

Exception: Secondary supports are not required where S_{D1} is less than 0.3 g.

[BS] A113.10 Adjacent buildings. Where elements of adjacent buildings do not have a separation of 5 inches (127 mm) or greater, the allowable height-to-thickness ratios for "all other buildings" per Table A110.2 shall be used in the direction of consideration.

SECTION A114
WALLS OF UNBURNED CLAY, ADOBE OR STONE MASONRY

[BS] A114.1 General. Walls of unburned clay, adobe or stone masonry construction shall conform to the following:

1. Walls of unburned clay, adobe or stone masonry shall not exceed a height- or length-to-thickness ratio specified in Table A114.1.

2. Adobe shall be allowed a maximum value of 9 pounds per square inch (62.1 kPa) for shear unless higher values are justified by test.

3. Mortar for repointing may be of the same soil composition and stabilization as the brick, in lieu of cement mortar.

[BS] TABLE A114.1
MAXIMUM HEIGHT-TO-THICKNESS RATIO FOR ADOBE OR STONE WALLS

	S_{D1}		
	$0.13_g \leq S_{D1} < 0.25_g$	$0.25_g \leq S_{D1} < 0.4_g$	$S_{D1} \geq 0.4_g$
One-story buildings	12	10	8
Two-story buildings			
First story	14	11	9
Second story	12	10	8

CHAPTER A2

EARTHQUAKE HAZARD REDUCTION IN EXISTING REINFORCED CONCRETE AND REINFORCED MASONRY WALL BUILDINGS WITH FLEXIBLE DIAPHRAGMS

SECTION A201 PURPOSE

[BS] A201.1 Purpose. The purpose of this chapter is to promote public safety and welfare by reducing the risk of death or injury as a result of the effects of earthquakes on reinforced concrete and reinforced masonry wall buildings with flexible diaphragms. Based on past earthquakes, these buildings have been categorized as being potentially hazardous and prone to significant damage, including possible collapse in a moderate to major earthquake. The provisions of this chapter are minimum standards for structural seismic resistance established primarily to reduce the risk of life loss or injury on both subject and adjacent properties. These provisions will not necessarily prevent loss of life or injury, or prevent earthquake damage to an *existing building* that complies with these standards.

SECTION A202 SCOPE

[BS] A202.1 Scope. The provisions of this chapter shall apply to wall anchorage systems that resist out-of-plane forces and to collectors in existing reinforced concrete or reinforced masonry buildings with flexible diaphragms. Wall anchorage systems that were designed and constructed in accordance with the 1997 *Uniform Building Code* or the 2000 or subsequent editions of the *International Building Code* shall be deemed to comply with these provisions.

SECTION A203 DEFINITIONS

[BS] A203.1 Definitions. For the purpose of this chapter, the applicable definitions in the *International Building Code* and the following shall apply:

[BS] CONTINUITY CONNECTOR. A component, typically a plate, rod, strap or hold-down, that ensures load path continuity along the full length of a crosstie or strut.

[BS] CROSSTIE. A member or group of members continuous across the main diaphragm that connects opposite wall lines and transfers out-of-plane wall anchorage forces into the diaphragm.

[BS] FLEXIBLE DIAPHRAGM. A roof or floor sheathed with plywood, wood decking (1-by or 2-by) or metal deck without a concrete topping slab.

[BS] STRUT. A member or group of members continuous across a subdiaphragm that transfers out-of-plane wall anchorage forces into the subdiaphragm.

[BS] WALL ANCHORAGE SYSTEM. The components comprising a complete load path for out-of-plane wall forces from the wall to the main diaphragm, typically including anchors embedded in or fastened to the wall; rods, straps, plates, hold-downs or other hardware; subdiaphragms and their chords; crossties; struts; and continuity connectors.

[BS] WALL SEGMENT. Any length of concrete wall with continuous horizontal reinforcing and not interrupted or intersected by a pilaster or vertical construction joint, or any length of reinforced masonry wall with continuous horizontal reinforcing and not interrupted or intersected by a pilaster or vertical control joint.

SECTION A204 SYMBOLS AND NOTATIONS

[BS] A204.1 General. For the purpose of this chapter, the applicable symbols and notations in the *International Building Code* shall apply.

SECTION A205 GENERAL REQUIREMENTS

[BS] A205.1 General. The seismic-resisting elements specified in this chapter shall comply with applicable provisions of Section 1613 of the *International Building Code* and Chapter 12 of ASCE 7, except as modified herein.

[BS] A205.2 Requirements for plans. The plans shall accurately reflect the results of the engineering investigation and design and shall show all pertinent dimensions and sizes for plan review and construction. The following shall be provided:

1. Floor plans and roof plans shall show existing framing construction, diaphragm construction, proposed wall anchors, crossties and collectors. Existing nailing, anchors, crossties and collectors shall be shown on the plans if they are considered part of the lateral force-resisting systems.

2. Typical wall panel details and sections with panel thickness, height, pilasters and location of anchors shall be provided.

3. Details shall include existing and new anchors and the method of developing anchor forces into the

diaphragm framing, existing and new crossties, and existing and new or improved support of roof and floor girders at pilasters or walls.

4. The basis for design and the building code used for the design shall be stated on the plans.

[BS] A205.3 Structural observation. Structural observation, in accordance with Section 1704.6 of the *International Building Code* is required, regardless of seismic design category, height or other conditions. Structural observation shall include visual observation of work for conformance to the *approved* construction documents and confirmation of existing conditions assumed during design.

A205.3.1 Additional special inspection. In addition to the requirements of Section 1705.13 of the *International Building Code*, special inspection shall be required for:

1. Installation of anchors into existing concrete or masonry walls to form part of a wall anchorage system.
2. Fastening of new or existing steel deck forming part of a wall anchorage system.
3. Installation of continuity connectors along the length of crossties, to ensure compliance with Section A206.2. This inspection may be periodic special inspection.

A205.3.2 Testing to establish adequacy of existing wall anchors. Testing shall show that the existing anchors can sustain a test load of 1.5 times the design tension load without noticeable deformation or damage to the anchor, to the masonry or concrete element, or to any part of the existing load path between the anchor and new retrofit components. Three anchors of each existing detail type shall be tested, and all three shall satisfy the requirement. Prior to testing, the design professional shall submit a test plan for *code official* approval identifying the expected locations of the existing anchors in question, the locations of the proposed tests, and the test procedure and criteria. After testing, the design professional shall submit a report of the satisfactory testing showing the test results, the design strengths derived from them, and the size and spacing as confirmed by investigation.

A205.4 Testing and Inspection. Structural testing and inspection for new construction materials, submittals, reports and certificates of compliance shall be in accordance with Sections 1704 and 1705 of the *International Building Code*. Work done to comply with this chapter shall not be eligible for Exception 1 to Section 1704.2 of the *International Building Code* or Exception 2 to Section 1705.13 of the *International Building Code*.

SECTION A206
ANALYSIS AND DESIGN

[BS] A206.1 Reinforced concrete and reinforced masonry wall anchorage. Concrete and masonry walls shall be anchored to all floors and roofs that provide lateral support for the wall in accordance with Section 12.11.2 of ASCE 7. The anchorage shall provide a direct connection capable of resisting 75 percent of the forces specified in Section 12.11.2.1 of ASCE 7.

Exceptions:

1. Existing walls need not be evaluated or retrofitted for bending between anchors.
2. Work required by this chapter need not consider shrinkage, thermal changes or differential settlement.

A206.1.1 Seismicity parameters, site class and geologic hazards. For any site designated as Site Class E, the value of F_a shall be taken as 1.2. Site-specific procedures are not required for compliance with this chapter. Mitigation of existing geologic site hazards such as liquefiable soil, fault rupture or landslide is not required for compliance with this chapter.

[BS] A206.2 Additional requirements for wall anchorage systems. The wall anchorage system shall comply with the requirements of this section and Section 12.11.2.2 of ASCE 7.

The maximum spacing between wall anchors shall be 8 feet (2438 mm), and each wall segment shall have at least two wall anchors.

The wall anchorage system, excluding subdiaphragms and existing roof or floor framing members, shall be designed and installed to limit the relative movement between the wall and the diaphragm to no more than $^1/_8$ inch before engagement of the anchors. Wall anchors shall be provided to resist out-of-plane forces, independent of existing shear anchors.

Where new members are added as crossties, they shall be spaced no more than 24 feet (7315 mm) apart. Where existing girders are used as crossties, their actual spacing shall be deemed adequate even where the spacing exceeds 24 feet (7315 mm), as long as the girders are provided with continuity connectors as required.

Wall anchorage shall not be provided by fastening the edge of plywood sheathing to steel ledgers. Wall anchorage shall not be provided solely by fastening the edge of steel decking to steel ledgers unless analysis demonstrates acceptable capacity. The existing connections shall be subject to field verification and the new connections shall be subject to special inspection.

New wall anchors shall be provided to resist the full wall anchorage design force independent of existing shear or tension anchors.

Exception: Existing cast-in-place anchors shall be permitted as part of the wall anchorage system if the tie element can be readily attached to the anchors, and if the anchors are capable of resisting the total vertical and lateral shear load (including dead load) while being acted on by the maximum wall anchorage tension force caused by an earthquake. Acceptable tension values for the existing anchors shall be established by testing in accordance with Section A205.4.

[BS] A206.3 Development of anchor forces into the diaphragm. Development of the required anchorage forces into roof and floor diaphragms shall comply with the requirements of this section and Section 12.11.2.2 of ASCE 7.

Lengths of development of anchor loads in wood diaphragms shall be based on existing field nailing of the sheathing unless existing edge nailing is positively identified on the original construction plans or at the site.

[BS] A206.4 Anchorage at pilasters. Where pilasters are present, the wall anchorage system shall comply with the requirements of this section and Section 12.11.2.2.7 of ASCE 7. The pilasters or the walls immediately adjacent to the pilasters shall be anchored directly to the roof framing such that the existing vertical anchor bolts at the top of the pilasters are bypassed without permitting tension or shear failure at the top of the pilasters.

> **Exception:** If existing vertical anchor bolts at the top of the pilasters are used for the anchorage, additional exterior confinement shall be provided as required to resist the total anchorage force.

[BS] A206.5 Combination of anchor types. New anchors used in combination on a single framing member shall be of compatible behavior and stiffness.

[BS] A206.6 Anchorage at interior walls. Existing interior reinforced concrete or reinforced masonry walls that extend to the floor above or to the roof diaphragm shall be anchored for out-of-plane forces per Sections A206.1 and A206.3. Walls extending through the roof diaphragm shall be anchored for out-of-plane forces on both sides, and continuity ties shall be spliced across or continuous through the interior wall to provide diaphragm continuity.

[BS] A206.7 Collectors. Collectors designed in accordance with this section shall be provided at reentrant corners and at interior shear walls. Existing or new collectors shall have the capacity to develop into the diaphragm a force equal to the lesser of the rocking or shear capacity of the reentrant wall or the tributary shear based on 75 percent of the diaphragm design forces specified in Section 12.10 of ASCE 7. The capacity of the collector need not exceed the capacity of the diaphragm to deliver loads to the collector. A connection shall be provided from the collector to the reentrant wall to transfer the full collector internal force . If a truss or beam other than a rafter or purlin is supported by the reentrant wall or by a column integral with the reentrant wall, then an independent secondary column is required to support the roof or floor members whenever rocking or shear capacity of the reentrant wall is less than the tributary shear.

[BS] A206.8 Mezzanines. Existing mezzanines relying on reinforced concrete or reinforced masonry walls for vertical or lateral support shall be anchored to the walls for the tributary mezzanine load. Walls depending on the mezzanine for lateral support shall be anchored per Sections A206.1, A206.2 and A206.3.

> **Exception:** Existing mezzanines that have independent lateral and vertical support need not be anchored to the walls.

SECTION A207
MATERIALS OF CONSTRUCTION

[BS] A207.1 Materials. Materials permitted by the building code, including their appropriate strength or allowable stresses, shall be used to meet the requirements of this chapter.

CHAPTER A3

PRESCRIPTIVE PROVISIONS FOR SEISMIC STRENGTHENING OF CRIPPLE WALLS AND SILL PLATE ANCHORAGE OF LIGHT, WOOD-FRAME RESIDENTIAL BUILDINGS

SECTION A301
GENERAL

[BS] A301.1 Purpose. The provisions of this chapter are intended to promote public safety and welfare by reducing the risk of earthquake-induced damage to existing wood-frame residential buildings. The requirements contained in this chapter are prescriptive minimum standards intended to improve the seismic performance of residential buildings; however, they will not necessarily prevent earthquake damage.

This chapter sets standards for strengthening that may be *approved* by the *code official* without requiring plans or calculations prepared by a registered design professional. The provisions of this chapter are not intended to prevent the use of any material or method of construction not prescribed herein. The *code official* may require that construction documents for strengthening using alternative materials or methods be prepared by a registered design professional.

[BS] A301.2 Scope. The provisions of this chapter apply to residential buildings of light-frame wood construction containing one or more of the structural weaknesses specified in Section A303.

> **Exception:** The provisions of this chapter do not apply to the buildings, or elements thereof, listed as follows. These buildings or elements require analysis by a registered design professional in accordance with Section A301.3 to determine appropriate strengthening:
>
> 1. Group R-1.
> 2. Group R with more than four dwelling units.
> 3. Buildings with a lateral force-resisting system using poles or columns embedded in the ground.
> 4. Cripple walls that exceed 4 feet (1219 mm) in height.
> 5. Buildings exceeding three stories in height and any three-story building with cripple wall studs exceeding 14 inches (356 mm) in height.
> 6. Buildings where the *code official* determines that conditions exist that are beyond the scope of the prescriptive requirements of this chapter.
> 7. Buildings or portions thereof constructed on concrete slabs on grade.

[BS] A301.3 Alternative design procedures. The details and prescriptive provisions herein are not intended to be the only acceptable strengthening methods permitted. Alternative details and methods shall be permitted to be used where *approved* by the *code official*. Approval of alternatives shall be based on a demonstration that the method or material used is at least equivalent in terms of strength, deflection and capacity to that provided by the prescriptive methods and materials.

Where analysis by a registered design professional is required, such analysis shall be in accordance with all requirements of the building code, except that the seismic forces may be taken as 75 percent of those specified in the *International Building Code*.

SECTION A302
DEFINITIONS

[BS] A302.1 Definitions. For the purpose of this chapter, in addition to the applicable definitions in the building code, certain additional terms are defined as follows:

[BS] ADHESIVE ANCHOR. An assembly consisting of a threaded rod, washer, nut, and chemical adhesive approved by the *code official* for installation in existing concrete or masonry.

[BS] CRIPPLE WALL. A wood-frame stud wall extending from the top of the foundation to the underside of the lowest floor framing.

[BS] EXPANSION ANCHOR. An *approved* post-installed anchor, inserted into a predrilled hole in existing concrete or masonry, that transfers loads to or from the concrete or masonry by direct bearing or friction or both.

[BS] PERIMETER FOUNDATION. A foundation system that is located under the exterior walls of a building.

[BS] SNUGTIGHT. As tight as an individual can torque a nut on a bolt by hand, using a wrench with a 10-inch-long (254 mm) handle, and the point at which the full surface of the plate washer is contacting the wood member and slightly indenting the wood surface.

[BS] WOOD STRUCTURAL PANEL. A panel manufactured from veneers, wood strands or wafers or a combination of veneer and wood strands or wafers bonded together with waterproof synthetic resins or other suitable bonding systems. Examples of wood structural panels are:

Composite panels. A wood structural panel that is comprised of wood veneer and reconstituted wood-based material and bonded together with waterproof adhesive.

Oriented strand board (OSB). A mat-formed wood structural panel comprised of thin rectangular wood strands arranged in cross-aligned layers with surface layers normally arranged in the long panel direction and bonded with waterproof adhesive.

Plywood. A wood structural panel comprised of plies of wood veneer arranged in cross-aligned layers. The plies are bonded with waterproof adhesive that cures on application of heat and pressure.

SECTION A303
STRUCTURAL WEAKNESSES

[BS] A303.1 General. For the purposes of this chapter, any of the following conditions shall be deemed a structural weakness:

1. Sill plates or floor framing that are supported directly on the ground without a foundation system that conforms to the building code.
2. A perimeter foundation system that is constructed only of wood posts supported on isolated pad footings.
3. Perimeter foundation systems that are not continuous.

    Exceptions:
    1. Existing single-story exterior walls not exceeding 10 feet (3048 mm) in length, forming an extension of floor area beyond the line of an existing continuous perimeter foundation.
    2. Porches, storage rooms and similar spaces not containing fuel-burning appliances.

4. A perimeter foundation system that is constructed of unreinforced masonry or stone.
5. Sill plates that are not connected to the foundation or that are connected with less than what is required by the building code.

    Exception: Where *approved* by the *code official*, connections of a sill plate to the foundation made with other than sill bolts shall be accepted if the capacity of the connection is equivalent to that required by the building code.

6. Cripple walls that are not braced in accordance with the requirements of Section A304.4 and Table A304.3.1, or cripple walls not braced with diagonal sheathing or wood structural panels in accordance with the building code.

SECTION A304
STRENGTHENING REQUIREMENTS

[BS] A304.1 General.

[BS] A304.1.1 Scope. The structural weaknesses noted in Section A303 shall be strengthened in accordance with the requirements of this section. Strengthening work may include both new construction and *alteration* of existing construction. Except as provided herein, all strengthening work and materials shall comply with the applicable provisions of the *International Building Code*.

[BS] A304.1.2 Condition of existing wood materials. Existing wood materials that will be a part of the strengthening work (such as sills, studs and sheathing) shall be in a sound condition and free from defects that substantially reduce the capacity of the member. Any wood material found to contain fungus infection shall be removed and replaced with new material. Any wood material found to be infested with insects or to have been infested with insects shall be strengthened or replaced with new materials to provide a net dimension of sound wood equal to or greater than its undamaged original dimension.

[BS] A304.1.3 Floor joists not parallel to foundations. Floor joists framed perpendicular or at an angle to perimeter foundations shall be restrained either by an existing nominal 2-inch-wide (51 mm) continuous rim joist or by a nominal 2-inch-wide (51 mm) full-depth block between alternate joists in one- and two-story buildings, and between each joist in three-story buildings. Existing blocking for multiple-story buildings must occur at each joist space above a braced cripple wall panel.

Existing connections at the top and bottom edges of an existing rim joist or blocking need not be verified in one story buildings. In multiple-story buildings, the existing top edge connection need not be verified; however, the bottom edge connection to either the foundation sill plate or the top plate of a cripple wall shall be verified. The minimum existing bottom edge connection shall consist of 8d toenails spaced 6 inches (152 mm) apart for a continuous rim joist, or three 8d toenails per block. Where this minimum bottom edge-connection is not present or cannot be verified, a supplemental connection installed as shown in Figure A304.1.3 or A304.1.4(2) shall be provided.

Where an existing continuous rim joist or the minimum existing blocking does not occur, new $^3/_4$-inch (19.1 mm) or $^{23}/_{32}$-inch (18 mm) wood structural panel blocking installed tightly between floor joists and nailed as shown in Figure A304.1.4(3) shall be provided at the inside face of the cripple wall. In lieu of wood structural panel blocking, tight fitting, full-depth 2-inch (51 mm) blocking may be used. New blocking may be omitted where it will interfere with vents or plumbing that penetrates the wall.

APPENDIX A—GUIDELINES FOR THE SEISMIC RETROFIT OF EXISTING BUILDINGS

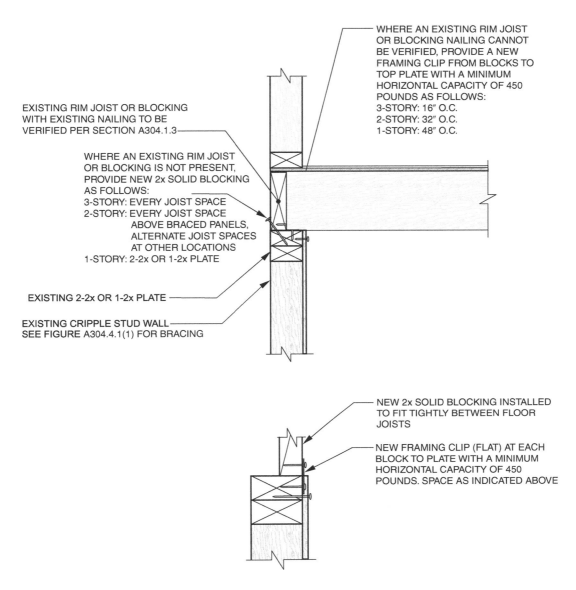

For SI: 1 inch = 25.4 mm, 1 pound = 4.4 N.
NOTE: See manufacturing instructions for nail sizes associated with metal framing clips.

[BS] FIGURE A304.1.3
TYPICAL FLOOR TO CRIPPLE WALL CONNECTION (FLOOR JOISTS NOT PARALLEL TO FOUNDATIONS)

[BS] A304.1.4 Floor joists parallel to foundations. Where existing floor joists are parallel to the perimeter foundations, the end joist shall be located over the foundation and, except for required ventilation openings, shall be continuous and in continuous contact with the foundation sill plate or the top plate of the cripple wall. Existing connections at the top and bottom edges of the end joist need not be verified in one-story buildings. In multiple-story buildings, the existing top edge connection of the end joist need not be verified; however, the bottom edge connection to either the foundation sill plate or the top plate of a cripple wall shall be verified. The minimum bottom edge connection shall be 8d toenails spaced 6 inches (152 mm) apart. If this minimum bottom edge connection is not present or cannot be verified, a supplemental connection installed as shown in Figure A304.1.4(1), A304.1.4(2) or A304.1.4(3) shall be provided.

APPENDIX A—GUIDELINES FOR THE SEISMIC RETROFIT OF EXISTING BUILDINGS

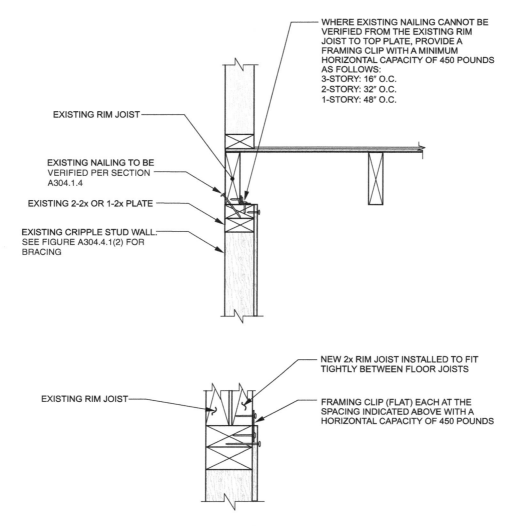

For SI: 1 inch = 25.4 mm, 1 pound = 4.4 N.
NOTE: See manufacturing instructions for nail sizes associated with metal framing clips.

[BS] FIGURE A304.1.4(1)
TYPICAL FLOOR TO CRIPPLE WALL CONNECTION (FLOOR JOISTS PARALLEL TO FOUNDATIONS)

APPENDIX A—GUIDELINES FOR THE SEISMIC RETROFIT OF EXISTING BUILDINGS

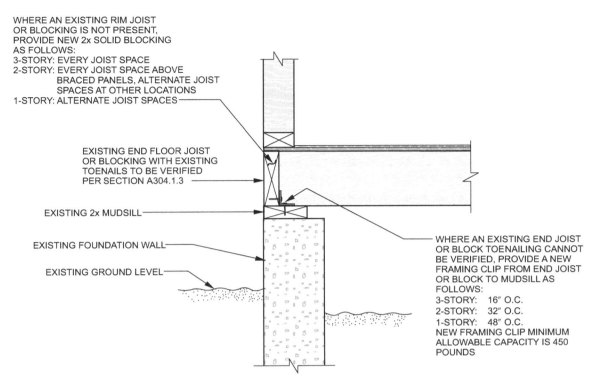

FLOOR JOISTS NOT PARALLEL TO FOUNDATIONS

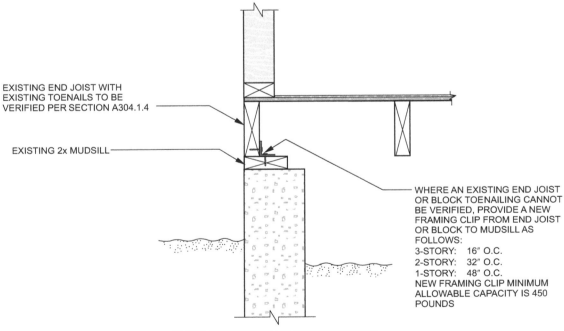

FLOOR JOISTS PARALLEL TO FOUNDATIONS

For SI: 1 inch = 25.4 mm.

NOTES:
1. See Section A304.3 for sill plate anchorage.
2. See manufacturing instructions for nail sizes associated with metal framing clips.

[BS] FIGURE A304.1.4(2)
TYPICAL FLOOR TO MUDSILL CONNECTIONS

APPENDIX A—GUIDELINES FOR THE SEISMIC RETROFIT OF EXISTING BUILDINGS

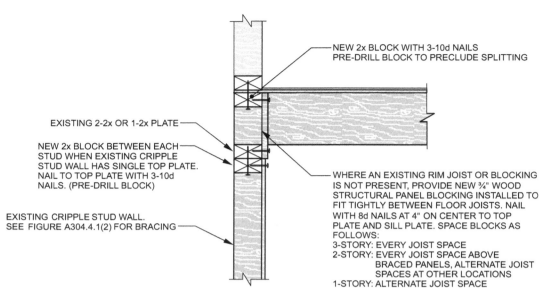

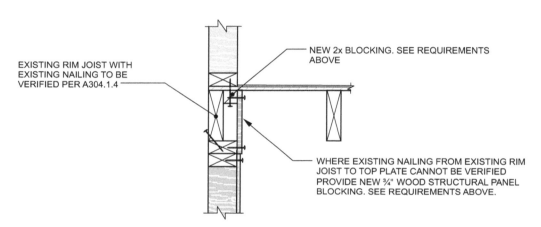

For SI: 1 inch = 25.4 mm, 1 pound = 4.4 N.
NOTE: See Section A304.4 for cripple wall bracing.

[BS] FIGURE A304.1.4(3)
ALTERNATIVE FLOOR FRAMING TO CRIPPLE WALL CONNECTION

[BS] A304.2 Foundations.

[BS] A304.2.1 New perimeter foundations. New perimeter foundations shall be provided for structures with the structural weaknesses noted in Items 1 and 2 of Section A303. Soil investigations or geotechnical studies are not required for this work unless the building is located in a special study zone as designated by the *code official* or other authority having jurisdiction.

[BS] A304.2.2 Evaluation of existing foundations. Partial perimeter foundations or unreinforced masonry foundations shall be evaluated by a registered design professional for the force levels specified in Section A301.3. Test reports or other substantiating data to determine existing foundation material strengths shall be submitted to the *code official*. Where *approved* by the *code official*, these existing foundation systems shall be strengthened in accordance with the recommendations included with the evaluation in lieu of being replaced.

Exception: In lieu of testing existing foundations to determine material strengths, and where *approved* by the *code official*, a new nonperimeter foundation system designed for the forces specified in Section A301.3 shall be used to resist lateral forces from perimeter walls. A registered design professional shall confirm the ability of the existing diaphragm to transfer seismic forces to the new nonperimeter foundations.

[BS] A304.2.3 Details for new perimeter foundations. All new perimeter foundations shall be continuous and constructed according to Figure A304.2.3(1) and Table A304.2.3(1) or Figure A304.2.3(2) and Table A304.2.3(2). New construction materials shall comply with the requirements of building code. Where *approved* by the *code official*, the existing clearance between existing floor joists or girders and existing grade below the floor need not comply with the building code.

Exception: Where designed by a registered design professional and *approved* by the *code official*, partial perimeter foundations shall be used in lieu of a continuous perimeter foundation.

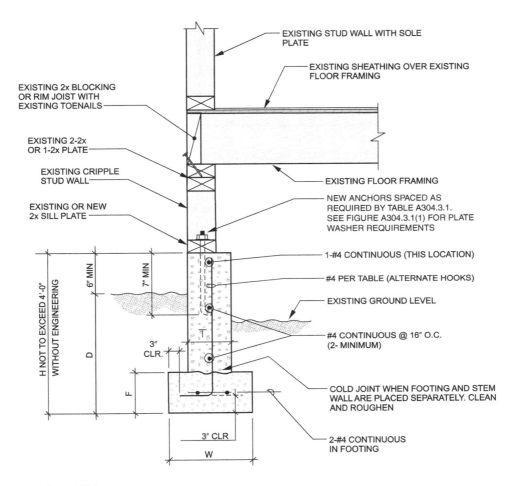

For SI: 1 inch = 25.4 mm, 1 foot = 304.8 mm.

**[BS] FIGURE A304.2.3(1)
NEW REINFORCED CONCRETE FOUNDATION SYSTEM**

APPENDIX A—GUIDELINES FOR THE SEISMIC RETROFIT OF EXISTING BUILDINGS

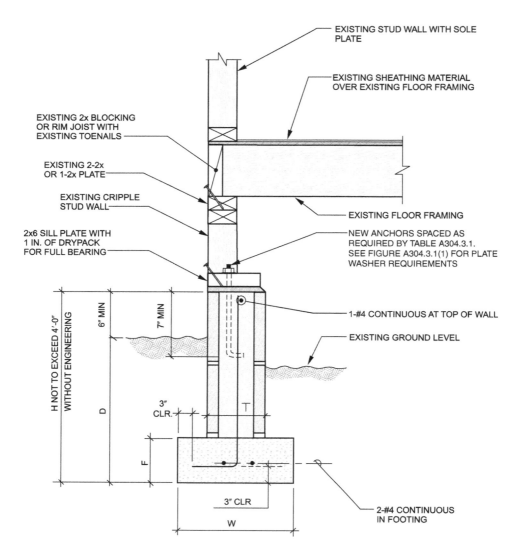

For SI: 1 inch = 25.4 mm, 1 foot = 304.8 mm.

**[BS] FIGURE A304.2.3(2)
NEW MASONRY CONCRETE FOUNDATION**

**[BS] TABLE A304.2.3(1)
NEW REINFORCED CONCRETE FOUNDATION SYSTEM**

NUMBER OF STORIES	MINIMUM FOUNDATION DIMENSIONS					MINIMUM FOUNDATION REINFORCING	
						VERTICAL REINFORCING	
	W	F	D[a, b, c]	T	H	Single-pour wall and footing	Footing placed separate from wall
1	12 inches	6 inches	12 inches	6 inches	≤ 24 inches	#4 @ 48 inches on center	#4 @ 32 inches on center
2	15 inches	7 inches	18 inches	8 inches	≥ 36 inches	#4 @ 48 inches on center	#4 @ 32 inches on center
3	18 inches	8 inches	24 inches	10 inches	≥ 36 inches	#4 @ 48 inches on center	#4 @ 18 inches on center

For SI: 1 inch = 25.4 mm.
a. Where frost conditions occur, the minimum depth shall extend below the frost line.
b. The ground surface along the interior side of the foundation may be excavated to the elevation of the top of the footing.
c. Where the soil is designated as expansive, the foundation depth and reinforcement shall be approved by the code official.

[BS] TABLE A304.2.3(2)
NEW MASONRY CONCRETE FOUNDATION

NUMBER OF STORIES	MINIMUM FOUNDATION DIMENSIONS					MINIMUM FOUNDATION REINFORCING	
	W	F	D[a, b, c]	T	H	VERTICAL REINFORCING	HORIZONTAL REINFORCING
1	12 inches	6 inches	12 inches	6 inches	≤ 24 inches	#4 @ 24 inches on center	#4 continuous at top of stem wall
2	15 inches	7 inches	18 inches	8 inches	≥ 24 inches	#4 @ 24 inches on center	#4 @ 16 inches on center
3	18 inches	8 inches	24 inches	10 inches	≥ 36 inches	#4 @ 24 inches on center	#4 @ 16 inches on center

For SI: 1 inch = 25.4 mm.

a. Where frost conditions occur, the minimum depth shall extend below the frost line.
b. The ground surface along the interior side of the foundation may be excavated to the elevation of the top of the footing.
c. Where the soil is designated as expansive, the foundation depth and reinforcement shall be approved by the code official.

[BS] A304.2.4 New concrete foundations. New concrete foundations shall have a minimum compressive strength of 2,500 pounds per square inch (17.24 MPa) at 28 days.

[BS] A304.2.5 New hollow-unit masonry foundations. New hollow-unit masonry foundations shall be solidly grouted. The grout shall have minimum compressive strength of 2,000 pounds per square inch (13.79 MPa). Mortar shall be Type M or S.

[BS] A304.2.6 New sill plates. Where new sill plates are used in conjunction with new foundations, they shall be minimum two times nominal thickness and shall be preservative-treated wood or naturally durable wood permitted by the building code for similar applications, and shall be marked or branded by an *approved* agency. Fasteners in contact with preservative-treated wood shall be hot-dip galvanized or other material permitted by the building code for similar applications. Anchors, that attach a preservative-treated sill plate to the foundation, shall be permitted to be of mechanically deposited zinc-coated steel with coating weights in accordance with ASTM B695, Class 55 minimum. Metal framing anchors in contact with preservative-treated wood shall be galvanized in accordance with ASTM A653 with a G185 coating.

[BS] A304.3 Foundation sill plate anchorage.

[BS] A304.3.1 Existing perimeter foundations. Where the building has an existing continuous perimeter foundation, all perimeter wall sill plates shall be anchored to the foundation with adhesive anchors or expansion anchors in accordance with Table A304.3.1.

Anchors shall be installed in accordance with Figure A304.3.1(1), with the plate washer installed between the nut and the sill plate. The nut shall be tightened to a snug-tight condition after curing is complete for adhesive anchors and after expansion wedge engagement for expansion anchors. Anchors shall be installed in accordance with manufacturer's recommendations. Expansion anchors shall not be used where the installation causes surface cracking of the foundation wall at the locations of the anchor.

[BS] TABLE A304.3.1
SILL PLATE ANCHORAGE AND CRIPPLE WALL BRACING

NUMBER OF STORIES ABOVE CRIPPLE WALLS	MINIMUM SILL PLATE CONNECTION AND MAXIMUM SPACING[a, b, c]	AMOUNT OF BRACING FOR EACH WALL LINE[d, e, f]	
		A Combination of Exterior Walls Finished with Portland Cement Plaster and Roofing Using Clay Tile or Concrete Tile Weighing More than 6 psf (287 N/m²)	All Other Conditions
One story	1/2 inch spaced 6 feet, 0 inch center-to-center with washer plate	Each end and not less than 50 percent of the wall length	Each end and not less than 40 percent of the wall length
Two stories	1/2 inch spaced 4 feet, 0 inch center-to-center with washer plate; or 5/8 inch spaced 6 feet, 0 inch center-to-center with washer plate	Each end and not less than 70 percent of the wall length	Each end and not less than 50 percent of the wall length
Three stories	5/8 inch spaced 4 feet, 0 inch center-to-center with washer plate	100 percent of the wall length[g]	Each end and not less than 80 percent of the wall length[g]

For SI: 1 inch = 25.4 mm, 1 foot = 304.8 mm, 1 pound per square foot = 47.88 N/m².

a. Sill plate anchors shall be adhesive anchors or expansion anchors in accordance with Section A304.3.1.
b. All washer plates shall be 3 inches by 3 inches by 0.229 inch minimum. The hole in the plate washer is permitted to be diagonally slotted with a width of up to 3/16 inch larger than the bolt diameter and a slot length not to exceed 1 3/4 inches, provided that a standard cut washer is placed between the plate washer and the nut.
c. This table shall also be permitted for the spacing of the alternative connections specified in Section A304.3.1.
d. See Figure A304.4.2 for braced panel layout.
e. Braced panels at ends of walls shall be located as near to the end as possible.
f. All panels along a wall shall be nearly equal in length and shall be nearly equal in spacing along the length of the wall.
g. The minimum required underfloor ventilation openings are permitted in accordance with Section A304.4.4.

Where existing conditions prevent anchor installations through the top of the sill plate, this connection shall be made in accordance with Figure A304.3.1(2), A304.3.1(3) or A304.3.1(4). Alternative anchorage methods having a minimum shear capacity of 900 pounds (4003 N) per connection parallel to the wall shall be permitted. The spacing of these alternative connections shall comply with the maximum spacing requirements of Table A304.3.1 for $1/2$-inch (12.7 mm) bolts.

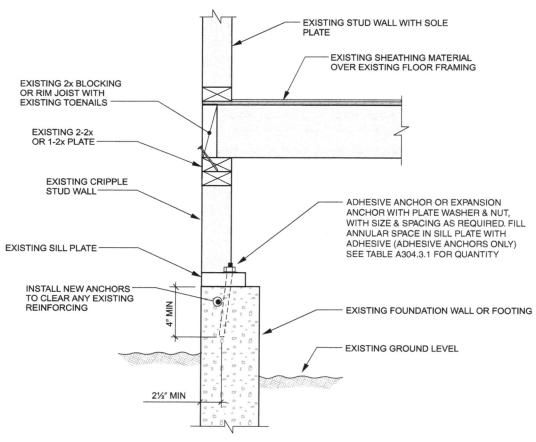

For SI: 1 inch = 25.4 mm.
a. Plate washers shall comply with the following:
 $1/2$-inch anchor or bolt—3 inches × 3 inches × 0.229 inch minimum.
 $5/8$-inch anchor or bolt—3 inches × 3 inches × 0.229 inch minimum.
 A diagonal slot in the plate washer is permitted in accordance with Table A304.3.1, Note b.
b. See Figure A304.4.1(1) or A304.4.1(2) for cripple wall bracing.

**[BS] FIGURE A304.3.1(1)
SILL PLATE BOLTING TO EXISTING FOUNDATION[a, b]**

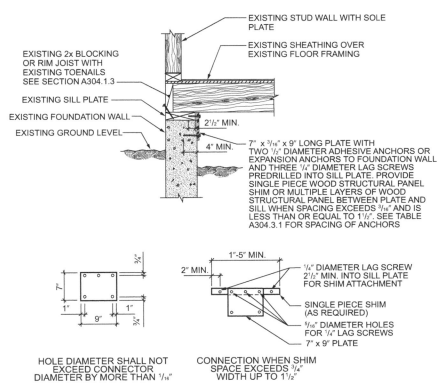

For SI: 1 inch = 25.4 mm.

NOTES:
a. If shim space exceeds 1 1/2 inches, alternative details will be required.
b. Where required, single piece shim shall be naturally durable wood or preservative-treated wood. If preservative-treated wood is used, it shall be isolated from the foundation system with a moisture barrier.

[BS] FIGURE A304.3.1(2)
ALTERNATIVE SILL PLATE ANCHORING IN EXISTING FOUNDATION—WITHOUT
CRIPPLE WALLS AND FLOOR FRAMING NOT PARALLEL TO FOUNDATIONS[a, b]

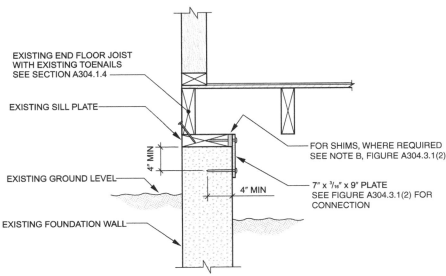

For SI: 1 inch = 25.4 mm.

[BS] FIGURE A304.3.1(3)
ALTERNATIVE SILL PLATE ANCHOR TO EXISTING FOUNDATION WITHOUT
CRIPPLE WALL AND FLOOR FRAMING PARALLEL TO FOUNDATIONS

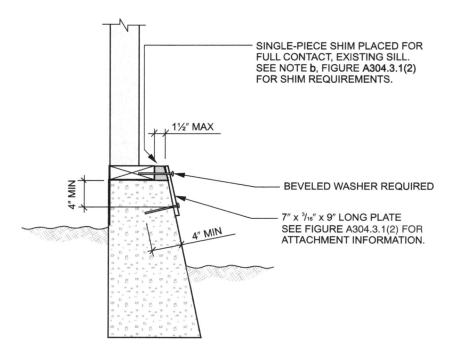

[BS] FIGURE A304.3.1(4)
SILL PLATE ANCHORING TO EXISTING FOUNDATION—ALTERNATIVE CONNECTION FOR BATTERED FOOTING

[BS] A304.3.2 Placement of anchors. Anchors shall be placed within 12 inches (305 mm), but not less than 9 inches (229 mm), from the ends of sill plates and shall be placed in the center of the stud space closest to the required spacing. New sill plates may be installed in pieces where necessary because of existing conditions. For lengths of sill plates 12 feet (3658 mm) or greater, anchors shall be spaced along the sill plate as specified in Table A304.3.1. For other lengths of sill plate, anchor placement shall be in accordance with Table A304.3.2.

Exception: Where physical obstructions such as fireplaces, plumbing or heating ducts interfere with the placement of an anchor, the anchor shall be placed as close to the obstruction as possible, but not less than 9 inches (229 mm) from the end of the plate. Center-to-center spacing of the anchors shall be reduced as necessary to provide the minimum total number of anchors required based on the full length of the wall. Center-to-center spacing shall be not less than 12 inches (305 mm).

[BS] A304.3.3 New perimeter foundations. Sill plates for new perimeter foundations shall be anchored in accordance with Table A304.3.1 and Figure A304.2.3(1) and Table A304.2.3(1) or Figure A304.2.3(2) and Table A304.2.3(2).

[BS] TABLE A304.3.2
SILL PLATE ANCHORAGE FOR VARIOUS LENGTHS OF SILL PLATE[a, b]

NUMBER OF STORIES	LENGTHS OF SILL PLATE		
	Less than 12 feet to 6 feet	Less than 6 feet to 30 inches	Less than 30 inches[c]
One story	Three connections	Two connections	One connection
Two stories	Four connections for $1/2$-inch anchors or bolts or three connections for $5/8$-inch anchors or bolts	Two connections	One connection
Three stories	Four connections	Two connections	One connection

For SI: 1 inch = 25.4 mm, 1 foot = 304.8 mm.
a. Connections shall be either adhesive anchors or expansion anchors.
b. See Section A304.3.2 for minimum end distances.
c. Connections shall be placed as near to the center of the length of plate as possible.

APPENDIX A—GUIDELINES FOR THE SEISMIC RETROFIT OF EXISTING BUILDINGS

[BS] A304.4 Cripple wall bracing.

[BS] A304.4.1 General. Exterior cripple walls not exceeding 4 feet (1219 mm) in height shall be permitted to be specified by the prescriptive bracing method in Section A304.4. Cripple walls over 4 feet (1219 mm) in height require analysis by a registered design professional in accordance with Section A301.3.

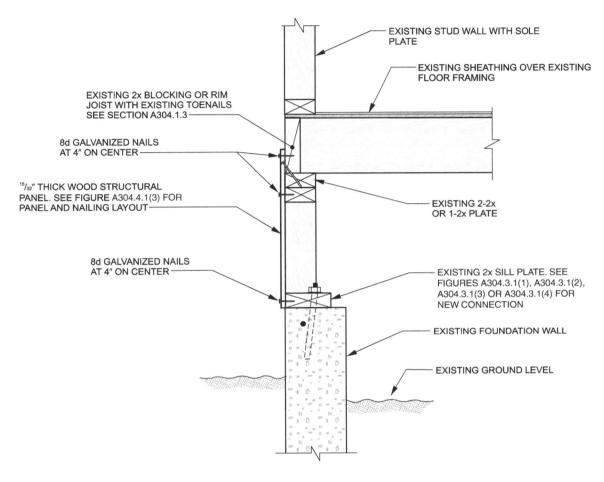

For SI: 1 inch = 25.4 mm.
NOTE: See Figure A304.3.1(1) for sill plate anchoring.

**[BS] FIGURE A304.4.1(1)
CRIPPLE WALL BRACING WITH NEW WOOD STRUCTURAL PANEL ON EXTERIOR FACE OF CRIPPLE STUDS**

APPENDIX A—GUIDELINES FOR THE SEISMIC RETROFIT OF EXISTING BUILDINGS

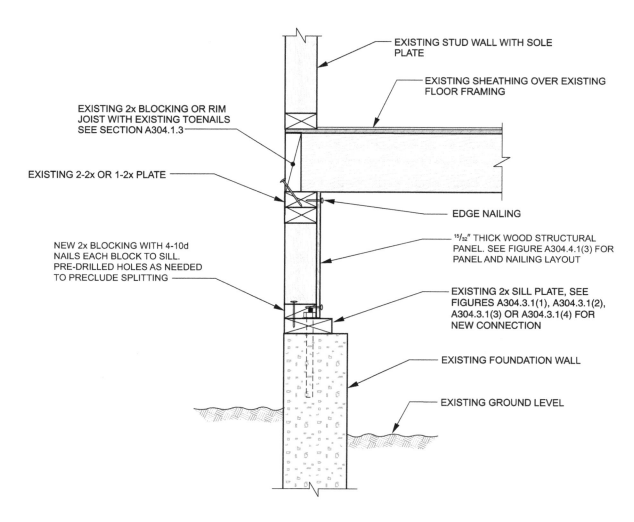

For SI: 1 inch = 25.4 mm.

[BS] FIGURE A304.4.1(2)
CRIPPLE WALL BRACING WITH WOOD STRUCTURAL PANEL ON INTERIOR FACE OF CRIPPLE STUDS

APPENDIX A—GUIDELINES FOR THE SEISMIC RETROFIT OF EXISTING BUILDINGS

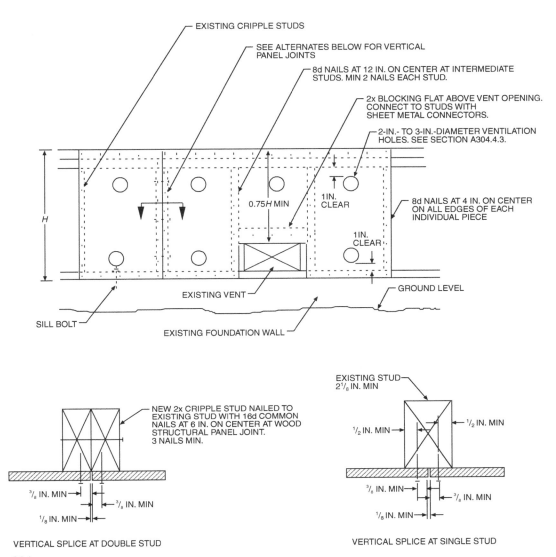

For SI: 1 inch = 25.4 mm.

**[BS] FIGURE A304.4.1(3)
PARTIAL CRIPPLE STUD WALL ELEVATION**

[BS] A304.4.1.1 Sheathing installation requirements. Wood structural panel sheathing shall be not less than $^{15}/_{32}$-inch (12 mm) thick and shall be installed in accordance with Figure A304.4.1(1) or A304.4.1(2). Individual pieces of wood structural panels shall be nailed with 8d common nails spaced 4 inches (102 mm) on center at all edges and 12 inches (305 mm) on center at each intermediate support with not less than two nails for each stud. Nails shall be driven so that their heads are flush with the surface of the sheathing and shall penetrate the supporting member not less than $1^1/_2$ inches (38 mm). When a nail fractures the surface, it shall be left in place and not counted as part of the required nailing. A new 8d nail shall be located within 2 inches (51 mm) of the discounted nail and be hand-driven flush with the sheathing surface. Where the installation involves horizontal joints, those joints shall occur over nominal 2-inch by 4-inch (51 mm by 102 mm) blocking installed with the nominal 4-inch (102 mm) dimension against the face of the plywood.

Vertical joints at adjoining pieces of wood structural panels shall be centered on studs such that there is a minimum $^1/_8$ inch (3.2 mm) between the panels. Where required edge distances cannot be maintained because of the width of the existing stud, a new stud shall be added adjacent to the existing studs and connected in accordance with Figure A304.4.1(3).

[BS] A304.4.2 Distribution and amount of bracing. See Table A304.3.1 and Figure A304.4.2 for the distribution and amount of bracing required for each wall line. Each braced panel length must be not less than two times the height of the cripple stud. Where the minimum amount of bracing prescribed in Table A304.3.1 cannot be installed

APPENDIX A—GUIDELINES FOR THE SEISMIC RETROFIT OF EXISTING BUILDINGS

along any walls, the bracing must be designed in accordance with Section A301.3.

Exception: Where physical obstructions such as fireplaces, plumbing or heating ducts interfere with the placement of cripple wall bracing, the bracing shall then be placed as close to the obstruction as possible. The total amount of bracing required shall not be reduced because of obstructions.

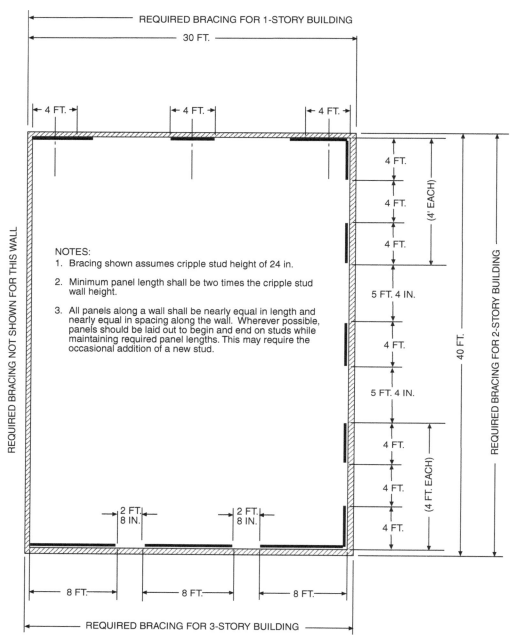

Bracing determination:
    1-story building—each end and not less than 40% of wall length.[a]
        Transverse wall—30 ft. × 0.40 = 12 ft. minimum panel length = 4 ft. 0 in.
    2-story building—each end and not less than 50% of wall length.[a]
        Longitudinal wall—40 ft. × 0.50 = 20 ft. 0 in. minimum of bracing.
    3-story building—each end and not less than 80% of wall length.[a]
        Transverse wall—30 ft. × 0.80 = 24 ft. 0 in. minimum of bracing.
  a. See Table 304.3.1 for buildings with both plaster walls and roofing exceeding 6 psf.

For SI: 1 inch = 25.4 mm, 1 foot = 304.8 mm, 1 pound per square foot = 42.88 N/m^2.

[BS] FIGURE A304.4.2
FLOOR PLAN-CRIPPLE WALL BRACING LAYOUT

[BS] A304.4.3 Stud space ventilation. Where bracing materials are installed on the interior face of studs forming an enclosed space between the new bracing and the existing exterior finish, each braced stud space must be ventilated. Adequate ventilation and access for future inspection shall be provided by drilling one 2-inch to 3-inch-diameter (51 mm to 76 mm) round hole through the sheathing, nearly centered between each stud at the top and bottom of the cripple wall. Such holes should be spaced not less than 1 inch (25 mm) clear from the sill or top plates. In stud spaces containing sill bolts, the hole shall be located on the centerline of the sill bolt but not closer than 1 inch (25 mm) clear from the nailing edge of the sheathing. Where existing blocking occurs within the stud space, additional ventilation holes shall be placed above and below the blocking, or the existing block shall be removed and a new nominal 2-inch by 4-inch (51 mm by 102 mm) block shall be installed with the nominal 4-inch (102 mm) dimension against the face of the plywood. For stud heights less than 18 inches (457 mm), only one ventilation hole need be provided.

[BS] A304.4.4 Existing underfloor ventilation. Existing underfloor ventilation shall not be reduced without providing equivalent new ventilation as close to the existing ventilation as possible. Braced panels may include underfloor ventilation openings where the height of the opening, measured from the top of the foundation wall to the top of the opening, does not exceed 25 percent of the height of the cripple stud wall; however, the length of the panel shall be increased a distance equal to the length of the opening or one stud space minimum. Where an opening exceeds 25 percent of the cripple wall height, braced panels shall not be located where the opening occurs. See Figure A304.4.1(3).

> **Exception:** For homes with a post and pier foundation system where a new continuous perimeter foundation system is being installed, new ventilation shall be provided in accordance with the building code.

[BS] A304.5 Quality control. All work shall be subject to inspection by the *code official* including, but not limited to:

1. Placement and installation of new adhesive or expansion anchors installed in existing foundations. Special inspection is not required for adhesive anchors installed in existing foundations regulated by the prescriptive provisions of this chapter.
2. Installation and nailing of new cripple wall bracing.
3. Any work shall be subject to special inspection where required by the *code official* in accordance with the building code.

[BS] A304.5.1 Nails. All nails specified in this chapter shall be common wire nails of the following diameters and lengths:

1. 8d nails = 0.131 inch (3.3 mm) by $2^1/_2$ inches (64 mm).
2. 10d nails = 0.148 inch (3.8 mm) by 3 inches (76 mm).
3. 12d nails = 0.148 inch (3.8 mm) by $3^1/_4$ inches (83 mm).
4. 16d nails = 0.162 inch (4.1 mm) by $3^1/_2$ inches (89 mm).

Nails used to attach metal framing connectors directly to wood members shall be as specified by the connector manufacturer in an *approved* report.

CHAPTER A4

EARTHQUAKE RISK REDUCTION IN WOOD-FRAME RESIDENTIAL BUILDINGS WITH SOFT, WEAK OR OPEN FRONT WALLS

SECTION A401
GENERAL

[BS] A401.1 Purpose. The purpose of this chapter is to promote public welfare and safety by reducing the risk of death or injury as a result of the effects of earthquakes on existing wood-frame, multiple-unit residential buildings. The ground motions of past earthquakes have caused the loss of human life, personal injury and property damage in these types of buildings. This chapter creates minimum standards to strengthen the more vulnerable portions of these structures. Where fully followed, these minimum standards will improve the performance of these buildings but will not necessarily prevent all earthquake-related damage.

[BS] A401.2 Scope. The provisions of this chapter apply to existing buildings of wood construction that contain residential occupancies and are assigned to *Risk Category* II, and where the structure has a soft, weak or open-front wall line, and there exists one or more stories above.

SECTION A402
DEFINITIONS

[BS] A402.1 Definitions. Notwithstanding the applicable definitions, symbols and notations in the building code, the following definitions shall apply for the purposes of this chapter:

[BS] ASPECT RATIO. The span-width ratio for horizontal diaphragms and the height-length ratio for shear walls.

[BS] NONCONFORMING STRUCTURAL MATERIALS. Wall bracing materials other than wood structural panels or diagonal sheathing.

[BS] OPEN-FRONT WALL LINE. An exterior wall line, without vertical elements of the lateral force-resisting system, that requires tributary seismic forces to be resisted by diaphragm rotation or excessive cantilever beyond parallel lines of shear walls. Diaphragms that cantilever more than 25 percent of the distance between lines of lateral force-resisting elements from which the diaphragm cantilevers shall be considered to be excessive. Exterior exit balconies of 6 feet (1829 mm) or less in width shall not be considered excessive cantilevers.

[BS] RETROFIT. An improvement of the lateral force-resisting system by *alteration* of existing structural elements or *addition* of new structural elements.

[BS] SOFT WALL LINE. A wall line whose lateral stiffness is less than that required by story drift limitations or deformation compatibility requirements of this chapter. In lieu of analysis, a soft wall line may be defined as a wall line in a story where the story stiffness is less than 70 percent of the story above for the direction under consideration.

[BS] STORY. A story as defined by the building code, including any basement or underfloor space of a building with cripple walls exceeding 4 feet (1219 mm) in height.

[BS] WALL LINE. Any length of wall along a principal axis of the building used to provide resistance to lateral loads. Parallel wall lines separated by less than 4 feet (1219 mm) shall be considered to be one wall line for the distribution of loads.

[BS] WEAK WALL LINE. A wall line in a story where the story strength is less than 80 percent of the story above in the direction under consideration.

SECTION A403
ANALYSIS AND DESIGN

[BS] A403.1 General. Modifications required by the provisions in this chapter shall be designed in accordance with the *International Building Code* provisions for new construction, except as modified by this chapter.

> **Exception:** Buildings for which the prescriptive measures provided in Section A404 apply and are used.

Alteration of the existing lateral force-resisting system or vertical load-carrying system shall not reduce the strength or stiffness of the *existing structure*, unless the altered structure would remain in conformance to the building code and this chapter.

[BS] A403.2 Scope of analysis. This chapter requires the *alteration, repair*, replacement or addition of structural elements and their connections to meet the strength and stiffness requirements herein. The lateral load-path analysis shall include the resisting elements and connections from the wood diaphragm immediately above any soft, weak or open-front wall lines to the foundation soil interface or to the uppermost story of a podium structure comprised of steel, masonry, or concrete structural systems that supports the upper wood-framed structure. Stories above the uppermost story with a soft, weak or open-front wall line shall be considered in the analysis but need not be modified. The lateral load-path analysis for added structural elements shall include evaluation of the allowable soil-bearing and lateral pressures in accordance with the building code. Where any portion of a building within the scope of this chapter is constructed on or into a slope steeper than one unit vertical in three units horizontal (33-percent slope), the lateral force-resisting system at and below the base level diaphragm shall be analyzed for the effects of concentrated lateral forces at the base caused by this hillside condition.

[BS] A403.3 Design base shear and design parameters. The design base shear in a given direction shall be permitted to be 75 percent of the value required for similar new construction in accordance with the building code. The value of R used in the design of the strengthening of any story shall not exceed the lowest value of R used in the same direction at any story above. The system overstrength factor, Ω_0, and the deflection amplification factor, C_d, shall be not less than the largest respective value corresponding to the R factor being used in the direction under consideration.

Exceptions:

1. For structures assigned to Seismic Design Category B, values of R, Ω_0 and C_d shall be permitted to be based on the seismic force-resisting system being used to achieve the required strengthening.

2. For structures assigned to Seismic Design Category C or D, values of R, Ω_0 and C_d shall be permitted to be based on the seismic force-resisting system being used to achieve the required strengthening, provided that when the strengthening is complete, the strengthened structure will not have an extreme weak story irregularity defined as Type 5b in ASCE 7 Table 12.3-2.

3. For structures assigned to Seismic Design Category E, values of R, Ω_0 and C_d shall be permitted to be based on the seismic force-resisting system being used to achieve the required strengthening, provided that when the strengthening is complete, the strengthened structure will not have an extreme soft story, a weak story, or an extreme weak story irregularity defined, respectively, as Types 1b, 5a and 5b in ASCE 7 Table 12.3-2.

4. For retrofit systems involving different seismic force-resisting systems in the same direction within the same story, resisting elements are permitted to be designed using the least value of R for the different structural systems found in each independent line of resistance if all of the following conditions are met:

    4.1. The building is assigned to *Risk Category* I or II.

    4.2. The building height is no more than four stories above grade plane.

    4.3. The seismic force-resisting systems of the retrofitted building comprise only wood structural panel shear walls, steel moment-resisting frames, steel cantilever columns and steel-braced frames. Values for C and Ω_0 shall be consistent with the R value used.

5. With reference to ASCE 7 Table 12.2-1, ordinary, intermediate and special steel systems, and all light-frame systems shall be permitted without limitation where those systems are used only for retrofit to comply with the requirements of this chapter.

[BS] A403.3.1 Expected story strength. Despite any other requirement of Section A403.3 or A403.4, the total expected strength of retrofit elements added to any story need not exceed 1.7 times the expected strength of the story immediately above in a two-story building, or 1.3 times the expected strength of the story immediately above in a three-story or taller building, as long as the retrofit elements are located symmetrically about the center of mass of the story above, or so as to minimize torsion in the retrofitted story. Calculation of expected story strength and identification of irregularities in Section A403.3 shall be based on the expected strength of all wall lines, even if sheathed with nonconforming materials. The strength of a wall line above the retrofitted story shall be permitted to be reduced to account for inadequate load path or overturning resistance.

[BS] A403.3.2 Seismicity parameters, site class and geologic hazards. For any site designated as Site Class E, the value of F shall be taken as 1.2. Site-specific procedures are not required for compliance with this chapter. Mitigation of existing geologic site hazards such as liquefiable soil, fault rupture or landslide is not required for compliance with this chapter.

[BS] A403.4 Story drift limitations. The calculated story drift for each retrofitted story shall not exceed the allowable deformation compatible with all vertical load-resisting elements and 0.025 times the story height. The calculated story drift shall not be reduced by the effects of horizontal diaphragm stiffness but shall be increased where these effects produce rotation. Drift calculations shall be in accordance with the building code.

[BS] A403.4.1 Pole structures. The effects of rotation and soil stiffness shall be included in the calculated story drift where lateral loads are resisted by vertical elements whose required depth of embedment is determined by pole formulas. The coefficient of subgrade reaction used in deflection calculations shall be based on a geotechnical investigation conducted in accordance with the building code.

[BS] A403.5 Deformation compatibility and P Δ effects. The requirements of the building code shall apply, except as modified herein. Structural framing elements and their connections not required by design to be part of the lateral force-resisting system shall be designed and detailed to be adequate to maintain support of expected gravity loads when subjected to the expected deformations caused by seismic forces. Increased demand caused by P Δ effects and story sidesway stability shall be considered in retrofit stories that rely on the strength and stiffness of cantilever columns for lateral resistance.

[BS] A403.6 Ties and continuity. All parts of the structure included in the scope of Section A403.2 shall be interconnected as required by the building code.

[BS] A403.7 Collector elements. Collector elements shall be provided to transfer the seismic forces between the elements within the scope of Section A403.2.

[BS] A403.8 Floor diaphragms. Floor diaphragms within the scope of Section A403.2 shall be shown to have adequate strength at the following locations:

1. For straight lumber sheathed diaphragms without integral hardwood flooring throughout the diaphragm: The code official is authorized to waive the requirement where it is shown that the condition occurs in areas small enough not to affect overall building performance.

2. For all other diaphragms adequate strength shall be shown to be provided at locations where forces are transferred between the diaphragm and each new or strengthened vertical element of the seismic force-resisting system. Collector elements shall be provided where needed to distribute the transferred force over a greater length of diaphragm.

Exception: Where the existing vertical elements of the seismic force-resisting system are shown to comply with this chapter, diaphragms need not be evaluated.

[BS] A403.9 Wood-framed shear walls. Wood-framed shear walls shall have strength and stiffness sufficient to resist the seismic loads and shall conform to the requirements of this section. Where new sheathing is applied to existing studs to create new wood-framed shear walls, the new wall elements shall be considered bearing wall systems for purposes of determining seismic design parameters.

[BS] A403.9.1 Gypsum or cement plaster products. Gypsum or cement plaster products shall not be used to provide the strength required by Section A403.3 or the stiffness required by Section A403.4.

[BS] A403.9.2 Wood structural panels.

[BS] A403.9.2.1 Drift limit. Wood structural panel shear walls shall meet the story drift limitation of Section A403.4. Conformance to the story drift limitation shall be determined by *approved* testing or calculation. Individual shear panels shall be permitted to exceed the maximum aspect ratio, provided that the allowable story drift and allowable shear capacities are not exceeded.

[BS] A403.9.2.2 Openings. Shear walls are permitted to be designed for continuity around openings in accordance with the building code. Blocking and steel strapping shall be provided at corners of the openings to transfer forces from discontinuous boundary elements into adjoining panel elements. Alternatively, perforated shear wall provisions of the building code are permitted to be used.

[BS] A403.9.3 Hold-down connectors.

[BS] A403.9.3.1 Expansion anchors in tension. Expansion anchors that provide tension strength by friction resistance shall not be used to connect hold-down devices to existing concrete or masonry elements.

[BS] A403.9.3.2 Required depth of embedment. The required depth of embedment or edge distance for the anchor used in the hold-down connector shall be provided in the concrete or masonry below any plain concrete slab unless satisfactory evidence is submitted to the *code official* that shows that the concrete slab and footings are of monolithic construction.

A403.10 Steel retrofit systems. Steel retrofit systems shall have strength and stiffness sufficient to resist the seismic loads and shall conform to the requirements of this section.

A403.10.1 Special moment frames. Steel special moment frames shall comply with all applicable provisions of AISC 341, except that Section E3.4a addressing strong-column/weak-beams of AISC 341, is not required for columns that carry no gravity load.

A403.10.2 Inverted moment frame systems. Cantilevered column systems shall be permitted to be designed as inverted special, intermediate or ordinary moment frames, with corresponding moment frame seismic design coefficients, where the system satisfies the following conditions:

1. The columns carry no gravity load.
2. The columns are configured in pairs or larger groups connected by a continuous reinforced concrete foundation or grade beam.
3. The foundation or grade beam shall be designed to resist the expected plastic moment at the base of each column, computed as $R_y F_y Z$ in accordance with AISC 341.
4. The flexibility of the foundation or grade beam, considering cracked section properties of the reinforced concrete, shall be included in computing the deformation of the steel frame system.
5. The column height shall be taken as twice the actual height when checking lateral torsional buckling.

SECTION A404
PRESCRIPTIVE MEASURES FOR WEAK STORY

[BS] A404.1 Limitation. These prescriptive measures shall apply only to two-story buildings and only where deemed appropriate by the *code official*. These prescriptive measures rely on rotation of the second floor diaphragm to distribute the seismic load between the side and rear walls around a ground floor open area. In the absence of an existing floor diaphragm of wood structural panel or diagonal sheathing at the top of the first story, a new wood structural panel diaphragm of minimum thickness of $^3/_4$ inch (19.1 mm) and with 10d common nails at 6 inches (152 mm) on center shall be applied.

[BS] A404.1.1 Additional conditions. To qualify for these prescriptive measures, the following additional conditions need to be satisfied by the retrofitted structure:

1. Diaphragm aspect ratio L/W is less than 0.67, where W is the diaphragm dimension parallel to the soft, weak or open-front wall line and L is the distance in the orthogonal direction between that

wall line and the rear wall of the ground floor open area.

2. Minimum length of side shear walls = 20 feet (6096 mm).
3. Minimum length of rear shear wall = three-fourths of the total rear wall length.
4. Plan or vertical irregularities shall not be other than a soft, weak or open-front wall line.
5. Roofing weight less than or equal to 5 pounds per square foot (240 N/m^2).
6. Aspect ratio of the full second floor diaphragm meets the requirements of the building code for new construction.

[BS] A404.2 Minimum required retrofit.

[BS] A404.2.1 Anchor size and spacing. The anchor size and spacing shall be not less than $^3/_4$ inch (19.1 mm) in diameter at 32 inches (813 mm) on center. Where existing anchors are inadequate, supplemental or alternative *approved* connectors (such as new steel plates bolted to the side of the foundation and nailed to the sill) shall be used.

[BS] A404.2.2 Connection to floor above. Shear wall top plates shall be connected to blocking or rim joist at upper floor with not less than 18-gage galvanized steel angle clips $4^1/_2$ inches (114 mm) long with 12-8d nails spaced not farther than 16 inches (406 mm) on center, or by equivalent shear transfer methods.

[BS] A404.2.3 Shear wall sheathing. The shear wall sheathing shall be not less than $^{15}/_{32}$-inch (11.9 mm), 5-ply Structural I with 10d nails at 4 inches (102 mm) on center at edges and 12 inches (305 mm) on center at field; blocked all edges with 3 by 4 board or larger. Where existing sill plates are less than 3-by thick, place flat 2-by on top of sill between studs, with flat 18-gage galvanized steel clips $4^1/_2$ inches (114 mm) long with 12-8d nails or $^3/_8$-inch-diameter (9.5 mm) lags through blocking for shear transfer to sill plate. Stagger nailing from wall sheathing between existing sill and new blocking. Anchor new blocking to foundation as specified in this section.

[BS] A404.2.4 Shear wall hold-downs. Shear walls shall be provided with hold-down anchors at each end. Two hold-down anchors are required at intersecting corners. Hold-downs shall be *approved* connectors with a minimum $^5/_8$-inch-diameter (15.9 mm) threaded rod or other *approved* anchor with a minimum allowable load of 4,000 pounds (17.8 kN). Anchor embedment in concrete shall be not less than 5 inches (127 mm). Tie-rod systems shall be not less than $^5/_8$ inch (15.9 mm) in diameter unless using high-strength cable. High-strength cable elongation shall not exceed $^5/_8$ inch (15.9 mm) under a 4,000 pound (17.8 kN) axial load.

SECTION A405
MATERIALS OF CONSTRUCTION

[BS] A405.1 New materials. New materials shall meet the requirements of the *International Building Code*, except where allowed by this chapter.

[BS] A405.2 Allowable foundation and lateral pressures. The use of default values from the building code for continuous and isolated concrete spread footings shall be permitted. For soil that supports embedded vertical elements, Section A403.4.1 shall apply.

[BS] A405.3 Existing materials. The physical condition, strengths and stiffnesses of existing building materials shall be taken into account in any analysis required by this chapter. The verification of existing materials conditions and their conformance to these requirements shall be made by physical observation, material testing or record drawings as determined by the registered design professional subject to the approval of the *code official*.

[BS] A405.3.1 Wood-structural-panel shear walls.

[BS] A405.3.1.1 Existing nails. Where the required calculations rely on design values for common nails or surfaced dry lumber, their use in construction shall be verified by exposure.

[BS] A405.3.1.2 Existing plywood. Where verification of the existing plywood is by use of record drawings alone, plywood shall be assumed to be of three plies.

[BS] A405.3.2 Existing wood framing. Wood framing is permitted to use the design stresses specified in the building code under which the building was constructed or other stress criteria *approved* by the *code official*.

[BS] A405.3.3 Existing structural steel. All existing structural steel shall be permitted to be assumed to comply with ASTM A36. Existing pipe or tube columns shall be assumed to be of minimum wall thickness unless verified by testing or exposure.

[BS] A405.3.4 Existing concrete. All existing concrete footings shall be permitted to be assumed to be plain concrete with a compressive strength of 2,000 pounds per square inch (13.8 MPa). Existing concrete compressive strength taken greater than 2,000 pounds per square inch (13.8 MPa) shall be verified by testing, record drawings or department records.

[BS] A405.3.5 Existing sill plate anchorage. The analysis of existing cast-in-place anchors shall be permitted to assume proper anchor embedment for purposes of evaluating shear resistance to lateral loads.

SECTION A406
CONSTRUCTION DOCUMENTS

[BS] A406.1 General. The plans shall show all information necessary for plan review and for construction and shall accurately reflect the design. The plans shall contain a note that states that this retrofit was designed in compliance with the criteria of this chapter.

[BS] A406.2 Existing construction. The plans shall show existing diaphragm and shear wall sheathing and framing materials; fastener type and spacing; diaphragm and shear wall connections; continuity ties; collector elements; and the portion of the existing materials that needs verification during construction. If the cap allowed by Section A403.3.1 is used to limit the scope of retrofit, the foregoing information shall be shown for each retrofitted story and at least one story above the uppermost retrofitted story. If the cap allowed by Section A403.3.1 is not used, the foregoing information need only be shown for each retrofitted story and for the floor at the top of that story.

[BS] A406.3 New construction.

  [BS] A406.3.1 Foundation plan elements. The foundation plan shall include the size, type, location and spacing of all anchor bolts with the required depth of embedment, edge and end distance; the location and size of all shear walls and all columns for braced frames or moment frames; referenced details for the connection of shear walls, braced frames or moment-resisting frames to their footing; and referenced sections for any grade beams and footings.

  [BS] A406.3.2 Framing plan elements. The framing plan shall include the length, location and material of shear walls; the location and material of frames; references or details for the column-to-beam connectors, beam-to-wall connections and shear transfers at floor and roof diaphragms; and the required nailing and length for wall top plate splices.

  [BS] A406.3.3 Shear wall schedule, notes and details. Shear walls shall have a referenced schedule on the plans that includes the correct shear wall capacity in pounds per foot (N/m); the required fastener type, length, gage and head size; and a complete specification for the sheathing material and its thickness. The schedule shall also show the required location of 3-inch (76 mm) nominal or two 2-inch (51 mm) nominal edge members; the spacing of shear transfer elements such as framing anchors or added sill plate nails; the required hold-down with its bolt, screw or nail sizes; and the dimensions, lumber grade and species of the attached framing member.

  Notes shall show required edge distance for fasteners of structural wood panels and framing members; required flush nailing at the plywood surface; limits of mechanical penetrations; and the sill plate material assumed in the design. The limits of mechanical penetrations shall be detailed showing the maximum notching and drilled hole sizes.

  [BS] A406.3.4 General notes. General notes shall show the requirements for material testing, special inspection and structural observation.

SECTION A407
QUALITY CONTROL

[BS] A407.1 Structural observation. Structural observation, in accordance with Section 1704.6 of the *International Building Code* is required, regardless of seismic design category, height or other conditions. Structural observation shall include visual observation of work for conformance to the *approved* construction documents and confirmation of existing conditions assumed during design.

A407.2 Contractor responsibility. Contractor responsibility shall be in accordance with Section 1704.4 of the International Building Code.

A407.3 Testing and inspection. Structural testing and inspection for new construction materials, submittals, reports and certificates of compliance shall be in accordance with Sections 1704 and 1705 of the *International Building Code*. Work done to comply with this chapter shall not be eligible for Exceptions 1, 2, or 3 of Section 1704.2 of the *International Building Code* or for the exception to Section 1705.13.2 of the *International Building Code*.

CHAPTER A5
REFERENCED STANDARDS

SECTION A501
REFERENCED STANDARDS

A501.1 General. See Table A501.1 for standards that are referenced in various sections of this appendix. Standards are listed by the standard identification with the effective date, standard title, and the section or sections of this appendix that references the standard.

TABLE A501.1
REFERENCED STANDARDS

STANDARD ACRONYM	STANDARD NAME	SECTIONS HEREIN REFERENCED
AISC 341-16	Seismic Provisions for Structural Steel Buildings	A403.10.1, A403.10.2
ASCE/SEI 7 —16	Minimum Design Loads for Buildings and Other Structures with Supplement No. 1	A104.1, A205.1, A206.1, A206.2, A206.3, A206.4, A206.7, A403.3
ASTM A36/A36M— 14	Specification for Carbon Structural Steel	A405.3.3
ASTM A653/A653M —15	Standard Specification for Steel Sheet, Zinc Coated (Galvanized) or Zinc-Iron Alloy-Coated (Galvannealed) by Hot-Dip Process	A304.2.6
ASTM B695 —04(2009)	Standard Specification for Coating of Zinc Mechanically Deposited on Iron and Steel	A304.2.6
ASTM C67- 14	Test Methods of Sampling and Testing Brick and Structural Clay Tile	A106.2.3.1
ASTM C140/C140M —15	Standard Test Methods for Sampling and Testing Concrete Masonry Units and Related Units	A106.2.3.1
ASTM C496 —96/C496M —11	Standard Test Method for Splitting Tensile Strength of Cylindrical Concrete Specimens	A104.1, A106.2.3.3
ASTM C1531—15	Standard Test Methods for In Situ Measurement of Masonry Mortar Joint Shear Strength Index	A106.2.3.2
ASTM E488/E488M —15	Standard Test Methods for Strength of Anchors in Concrete and Masonry Elements	A107.5.1, A107.5.3
ASTM E519/E519M —2010	Standard Test Method for Diagonal Tension (Shear) in Masonry Assemblages	A104.1

(continued)

TABLE A501.1—continued
REFERENCED STANDARDS

STANDARD ACRONYM	STANDARD NAME	SECTIONS HEREIN REFERENCED
IBC—00	International Building Code	A202.1
IBC—03	International Building Code	A202.1
IBC—06	International Building Code	A202.1
IBC—09	International Building Code	A202.1
IBC—12	International Building Code	A202.1
IBC—15	International Building Code	A202.1
IBC—18	International Building Code	A202.1
IBC—21	International Building Code	A102.2, A105.1, A105.4, A202.1, A203.1, A204.1, A205.1, A205.3, A205.3.1, A205.4, A301.3, A304.1.1, A403.1, A405.1, A407.1, A407.2, A407.3
UBC—97	Uniform Building Code	A202.1

APPENDIX B

SUPPLEMENTARY ACCESSIBILITY REQUIREMENTS FOR EXISTING BUILDINGS AND FACILITIES

The provisions contained in this appendix are not mandatory unless specifically referenced in the adopting ordinance.

User note:

About this appendix: Chapter 11 of the International Building Code® contains provisions that set forth requirements for accessibility to buildings and their associated sites and facilities for people with physical disabilities. Sections 306 and 1508 in the code address accessibility provisions and alternatives permitted in existing buildings. Appendix B was added to address accessibility in construction for items that are not typically enforceable through the traditional building code enforcement process.

SECTION B101
QUALIFIED HISTORIC BUILDINGS AND FACILITIES

[BE] B101.1 General. Qualified *historic buildings* and *facilities* shall comply with Sections B101.2 through B101.5.

[BE] B101.2 Qualified historic buildings and facilities. These procedures shall apply to buildings and *facilities* designated as historic structures that undergo *alterations* or a *change of occupancy*.

[BE] B101.3 Qualified historic buildings and facilities subject to Section 106 of the National Historic Preservation Act. Where an *alteration* or *change of occupancy* is undertaken to a qualified *historic building* or *facility* that is subject to Section 106 of the National Historic Preservation Act, the federal agency with jurisdiction over the undertaking shall follow the Section 106 process. Where the state historic preservation officer or Advisory Council on Historic Preservation determines that compliance with the requirements for accessible routes, ramps, entrances, or toilet *facilities* would threaten or destroy the historic significance of the building or *facility*, the alternative requirements of Section 306.7.16 for that element are permitted.

[BE] B101.4 Qualified historic buildings and facilities not subject to Section 106 of the National Historic Preservation Act. Where an *alteration* or *change of occupancy* is undertaken to a qualified *historic building* or *facility* that is not subject to Section 106 of the National Historic Preservation Act, and the entity undertaking the *alterations* believes that compliance with the requirements for accessible routes, ramps, entrances or toilet *facilities* would threaten or destroy the historic significance of the building or *facility*, the entity shall consult with the state historic preservation officer. Where the state historic preservation officer determines that compliance with the accessibility requirements for accessible routes, ramps, entrances or toilet *facilities* would threaten or destroy the historical significance of the building or *facility*, the alternative requirements of Section 306.7.16 for that element are permitted.

[BE] B101.4.1 Consultation with interested persons. Interested persons shall be invited to participate in the consultation process, including state or local accessibility officials, individuals with disabilities, and organizations representing individuals with disabilities.

[BE] B101.4.2 Certified local government historic preservation programs. Where the state historic preservation officer has delegated the consultation responsibility for purposes of this section to a local government historic preservation program that has been certified in accordance with Section 101 of the National Historic Preservation Act of 1966 [(16 U.S.C. 470a(c)] and implementing regulations (36 CFR 61.5), the responsibility shall be permitted to be carried out by the appropriate local government body or official.

[BE] B101.5 Displays. In qualified *historic buildings* and *facilities* where alternative requirements of Section 306.7.16 are permitted, displays and written information shall be located where they can be seen by a seated person. Exhibits and signs displayed horizontally shall be 44 inches (1120 mm) maximum above the floor.

SECTION B102
FIXED TRANSPORTATION FACILITIES AND STATIONS

[BE] B102.1 General. Existing fixed transportation *facilities* and stations shall comply with Section B102.2.

[BE] B102.2 Existing facilities—key stations. Rapid rail, light rail, commuter rail, intercity rail, high-speed rail and other fixed guideway systems, altered stations, and intercity rail and key stations, as defined under criteria established by the Department of Transportation in Subpart C of 49 CFR Part 37, shall comply with Sections B102.2.1 through B102.2.3.

[BE] B102.2.1 Accessible route. One accessible route, or more, from an accessible entrance to those areas necessary for use of the transportation system shall be provided. The accessible route shall include the features specified in Section E109.2 of the *International Building Code*, except that escalators shall comply with Section 3004.2.2 of the *International Building Code*. Where technical unfeasibility in existing stations requires the accessible route to lead from the public way to a paid area of the transit system, an accessible fare collection machine complying with Section E109.2.3 of the *Interna-*

tional Building Code shall be provided along such accessible route.

[BE] B102.2.2 Platform and vehicle floor coordination. Station platforms shall be positioned to coordinate with vehicles in accordance with applicable provisions of 36 CFR Part 1192. Low-level platforms shall be 8 inches (250 mm) minimum above top of rail.

> **Exception:** Where vehicles are boarded from sidewalks or street-level, low-level platforms shall be permitted to be less than 8 inches (250 mm).

[BE] B102.2.3 Direct connections. New direct connections to commercial, retail or residential *facilities* shall, to the maximum extent feasible, have an accessible route complying with Section 306.7.1 from the point of connection to boarding platforms and transportation system elements used by the public. Any elements provided to facilitate future direct connections shall be on an accessible route connecting boarding platforms and transportation system elements used by the public.

SECTION B103
DWELLING UNITS AND SLEEPING UNITS

[BE] B103.1 Communication features. Where dwelling units and sleeping units are altered or added, the requirements of Section E104.2 of the *International Building Code* shall apply only to the units being altered or added until the number of units with accessible communication features complies with the minimum number required for new construction.

SECTION B104
REFERENCED STANDARDS

[BE] B104.1 General. See Table B104.1 for standards that are referenced in various sections of this appendix. Standards are listed by the standard identification with the effective date, standard title, and the section or sections of this appendix that reference the standard.

[BE] TABLE B104.1
REFERENCED STANDARDS

STANDARD ACRONYM	STANDARD NAME	SECTIONS HEREIN REFERENCED
Y3.H626 2P	*National Historic Preservation J101.2, 43/933 Act of 1966 as amended J101.3, 3rd Edition*	B101.3, B101.4, B101.4.2
IBC—21	*International Building Code®*	B102.2.1, B103.1
36 CFR Part 1192	*Americans with Disabilities Act Guidelines for Transportation Vehicles—Rapid Rail Vehicles and Systems*	B102.2.2
49 CFR Part 37 Subpart C	*Alteration of Transportation Facilities by Public Entities Department of Transportation*	B102.2

APPENDIX C: GUIDELINES FOR THE WIND RETROFIT OF EXISTING BUILDINGS

CHAPTER C1

GABLE END RETROFIT FOR HIGH-WIND AREAS

The provisions contained in this appendix are not mandatory unless specifically referenced in the adopting ordinance.

User note:

About this appendix: *Appendix C is intended to provide guidance for retrofitting existing structures to strengthen their resistance to wind forces. This appendix is similar in scope to Appendix A, which addresses seismic retrofits for existing buildings, except that the subject matter is related to wind retrofits. These retrofits are voluntary measures that serve to better protect the public and reduce damage from high-wind events for existing buildings.*

The purpose of this appendix is to provide prescriptive alternatives for addressing retrofit of buildings in high-wind areas. Currently there are two chapters that deal with the retrofit of gable ends and the fastening of roof decks, Appendix Chapters C1 and C2, respectively.

SECTION C101
GENERAL

[BS] C101.1 Purpose. This chapter provides prescriptive methods for partial structural retrofit of an *existing building* to increase its resistance to out-of-plane wind loads. It is intended for voluntary use and for reference by mitigation programs. The provisions of this chapter do not necessarily satisfy requirements for new construction. Unless specifically cited, the provisions of this chapter do not necessarily satisfy requirements for structural improvements triggered by *addition, alteration, repair, change of occupancy*, building relocation or other circumstances.

[BS] C101.2 Eligible buildings and gable end walls. The provisions of this chapter are applicable only to buildings that meet the following eligibility requirements:

1. The building is not more than three stories tall, from adjacent grade to the bottom plate of each gable end wall being retrofitted with this chapter.
2. The building is classified as Occupancy Group R3 or is within the scope of the *International Residential Code*.
3. The structure includes one or more wood-framed gable end walls, either conventionally framed or metal-plate-connected.

In addition, the provisions of this chapter are applicable only to gable end walls that meet the following eligibility requirements:

4. Each gable end wall has or shall be provided with studs or vertical webs spaced 24 inches (610 mm) on center maximum.
5. Each gable end wall has a maximum height of 16 feet (4877 mm).

[BS] C101.3 Compliance Eligible gable end walls in eligible buildings may be retrofitted in accordance with this chapter. Other modifications required for compliance with this chapter shall be designed and constructed in accordance with the *International Building Code* or *International Residential Code* provisions for new construction, except as specifically provided for by this chapter.

SECTION C102
DEFINITIONS

[BS] C102.1 Definitions. The following words and terms shall, for the purposes of this chapter, have the meanings shown herein.

[BS] ANCHOR BLOCK. A piece of lumber secured to horizontal braces and filling the gap between existing framing members for the purpose of restraining horizontal braces from movement perpendicular to the framing members.

[BS] COMPRESSION BLOCK. A piece of lumber used to restrain in the compression mode (force directed toward the interior of the attic) an existing or retrofit stud. It is attached to a horizontal brace and bears directly against the existing or retrofit stud.

[BS] CONVENTIONALLY FRAMED GABLE END. A gable end framed with studs whose faces are perpendicular to the gable end wall.

[BS] GABLE END FRAME. A factory or site-fabricated frame, installed as a complete assembly that incorporates vertical webs with their faces parallel to the plane of the frame.

[BS] HORIZONTAL BRACE. A piece of lumber used to restrain both compression and tension loads applied by a retrofit stud. It is typically installed horizontally on the top of attic floor framing members (truss bottom chords or ceiling joists) or on the bottom of pitched roof framing members (truss top chord or rafters).

[BS] HURRICANE TIES. Manufactured metal connectors designed to provide uplift and lateral restraint for roof framing members.

[BS] NAIL PLATE. A manufactured metal plate made of galvanized steel with factory-punched holes for fasteners. A nail plate may have the geometry of a strap.

[BS] RETROFIT. The voluntary process of strengthening or improving buildings or structures, or individual components of buildings or structures for the purpose of making existing conditions better serve the purpose for which they were originally intended or the purpose that current building codes intend.

[BS] RETROFIT STUD. A lumber member used to structurally supplement an existing gable end wall stud or gable end frame web.

[BS] STUD-TO-PLATE CONNECTOR. A manufactured metal connector designed to connect studs to plates.

SECTION C103
MATERIALS OF CONSTRUCTION

[BS] C103.1 Existing materials. Existing wood materials that will be part of the retrofitting work (such as trusses, rafters, ceiling joists, top plates and wall studs) shall be in sound condition and free from defects or damage that substantially reduces the load-carrying capacity of the member. Any wood materials found to be damaged or deteriorated shall be strengthened or replaced with new materials to provide a net dimension of sound wood equivalent to its undamaged original dimensions.

[BS] C103.2 New materials. All new materials shall comply with the standards for those materials as specified in the *International Building Code* or the *International Residential Code*.

[BS] C103.3 Material specifications for retrofits. Materials for retrofitting gable end walls shall comply with Table C103.3.

[BS] C103.4 Twists in straps. Straps shall be permitted to be twisted or bent where they transition between framing members or connection points. Straps shall be bent only once at a given location though it is permissible that they be bent or twisted at multiple locations along their length.

[BS] C103.5 Fasteners. Fasteners shall meet the requirements of Table C103.5, Sections C103.5.1 and C103.5.2, and shall be permitted to be screws or nails meeting the minimum length requirement shown in the figures and specified in the tables of this appendix. Fastener spacing shall meet the requirements of Section C103.5.3.

[BS] C103.5.1 Screws. Unless otherwise indicated in the appendix, screw sizes and lengths shall be in accordance with Table C103.5. Permissible screws include deck screws and wood screws. Screws shall have not less than 1 inch (25 mm) of thread. Fine threaded screws or drywall screws shall not be permitted. Select the largest possible diameter screw such that the shank adjacent to the head fits through the hole in the strap.

[BS] C103.5.2 Nails. Unless otherwise indicated in this appendix, nail sizes and lengths shall be in accordance with Table C103.5.

[BS] C103.5.3 General fastener spacing. Fastener spacing for shear connections of lumber-to-lumber shall meet the requirements shown in Figure C103.5.3 and the following conditions.

[BS] TABLE C103.3
MATERIAL SPECIFICATIONS FOR RETROFITS[a]

COMPONENT	MINIMUM SIZE OR THICKNESS	MINIMUM MATERIAL GRADE	MINIMUM CAPACITY
Anchor blocks, compression blocks and horizontal braces	2 × 4 nominal lumber	#2 Spruce-Pine-Fir or better	NA
Nail plates	20 gage thickness 8d minimum nail holes	Galvanized sheet steel	NA
Retrofit studs	2 × 4 nominal lumber	#2 Spruce-Pine-Fir or better	NA
Gusset angle	14 gage thickness	Galvanized sheet steel	350 pounds uplift and lateral load
Stud-to-plate connector	20 gage thickness	Galvanized sheet steel	500 pounds uplift
Metal plate connectors, straps and anchors	20 gage thickness	Galvanized sheet steel	NA

For SI: 1 pound = 4.4 N.
NA = Not Applicable.

a. Metal plate connectors, nail plates, stud-to-plate connectors, straps and anchors shall be products approved for connecting wood-to-wood or wood-to-concrete as appropriate.

[BS] TABLE C103.5
NAIL AND SCREW REQUIREMENTS

FASTENER TYPE	MINIMUM SHANK DIAMETER	MINIMUM HEAD DIAMETER	MINIMUM FASTENER LENGTH
#8 screws	NA	0.28 inches	$1\frac{1}{4}$ inches
8d common nails	0.131 inches	0.28 inches	$2\frac{1}{2}$ inches
10d common nails	0.148 inches	0.28 inches	3 inches

For SI: 1 inch = 25.4 mm.
NA = Not Applicable.

APPENDIX C—GUIDELINES FOR THE WIND RETROFIT OF EXISTING BUILDINGS

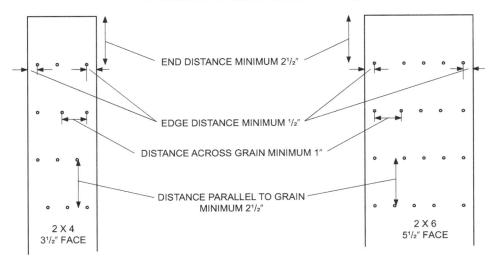

For SI: 1 inch = 25.4 mm.

[BS] FIGURE C103.5.3
FASTENER SPACINGS FOR LUMBER-TO-LUMBER CONNECTIONS OPERATING IN SHEAR PARALLEL TO GRAIN

[BS] C103.5.3.1 General fastener spacing. Fastener spacing shall meet the following conditions except as provided for in Section C103.5.3.

The distance between fasteners and the edge of lumber that is less than $3^1/_2$ inches deep (89 mm) in the direction of the fastener length shall be not less than $^3/_4$ inch (19.1 mm).

1. The distance between fasteners and the edge of lumber that is more than 2 inches (51 mm) thick in the direction of the fastener length shall be not less than $^1/_2$ inch (12.7 mm).
2. The distance between a fastener and the end of lumber shall be not less than $2^1/_2$ inches (64 mm).
3. The distance between fasteners parallel to the grain (center-to-center) shall be not less than $2^1/_2$ inches (64 mm).
4. The distance between fasteners perpendicular to the grain (center-to-center) in lumber that is less than $3^1/_2$ inches (89 mm) deep in the direction of the fastener length shall be 1 inch (25 mm).
5. The distance between fasteners perpendicular to the grain (center-to-center) in lumber that is more than 2 inches (51 mm) thick in the direction of the fastener length shall be $^1/_2$ inch (12.7 mm).

[BS] C103.5.3.2 Wood-to-wood connections of two members each 2 inches or less in thickness. Wood-to-wood connections fastener spacing shall meet the following conditions.

1. The distance between fasteners parallel to grain (center-to-center) shall be not less than $2^1/_2$ inches (64 mm).
2. The distance between fasteners across grain (center-to-center) shall be not less than 1 inch (25 mm).
3. For wood-to-wood connections of lumber at right angles, fasteners shall be spaced not less than $2^1/_2$ inches (64 mm) parallel to the grain and 1 inch (25 mm) perpendicular to the grain in any direction.

[BS] C103.5.3.3 Metal connectors for wood-to-wood connections. Metal connectors for wood-to-wood connections shall meet the following conditions.

1. Fastener spacing to edge or ends of lumber shall be as dictated by the prefabricated holes in the connectors and the connectors shall be installed in a configuration that is similar to that shown by the connector manufacturer.
2. Fasteners in $1^1/_4$-inch-wide (32 mm) metal straps that are installed on the narrow face of lumber shall be a minimum $^1/_4$ inch (6.4 mm) from either edge of the lumber. Consistent with Section C103.5.3.1, fasteners shall be permitted to be spaced according to the fastener holes fabricated into the strap.
3. Fasteners in metal nail plates shall be spaced not less than $^1/_2$ inch (12.7 mm) perpendicular to grain and not less than $1^1/_2$ inches (38 mm) parallel to grain.

SECTION C104
RETROFITTING GABLE END WALLS TO ENHANCE WIND RESISTANCE

[BS] C104.1 General. These prescriptive methods of retrofitting are intended to increase the resistance of existing gable end construction for out-of-plane wind loads resulting from high-wind events. The ceiling diaphragm shall be comprised of minimum $^1/_2$-inch-thick (12.7 mm) gypsum board, minimum nominal $^3/_8$-inch-thick (9.5 mm) wood structural panels, or plaster. An overview isometric drawing of one type of gable end retrofit to improve wind resistance is shown in Figure C104.1.

[BS] C104.2 Horizontal braces. Horizontal braces shall be installed perpendicular to the roof and ceiling framing members at the location of each existing gable end stud greater than 3 feet (91 cm) in length. Unless it is adjacent to an omitted horizontal brace location, horizontal braces shall be minimum 2-inch by 4-inch (38 mm by 89 mm) dimensional lumber as defined in Section C103.3. A single horizontal brace is required at the top and bottom of each gable end stud for Retrofit Configuration A, B, or C. Two horizontal braces are required at the top and bottom of each gable end stud for Retrofit Configuration D. Maximum heights of gable end wall studs and associated retrofit studs for each Retrofit Configuration shall not exceed the values listed in Table C104.2. Horizontal braces shall be oriented with their wide faces across the roof or ceiling framing members, be fastened to not fewer than three framing members, and extend not less than 6 feet (183 cm) measured perpendicularly from the gable end plus $2^1/_2$ inches (64 mm) beyond the last top chord or bottom chord member (rafter or ceiling joist) from the gable end as shown in Figures C104.2(1), C104.2(2), C104.2(3) and C104.2(4).

[BS] C104.2.1 Existing gable end studs. If the spacing of existing vertical gable end studs is greater than 24 inches (64 mm), a new stud and corresponding horizontal braces shall be installed such that the maximum spacing between existing and added studs shall be not greater than 24 inches (64 mm). Additional gable end wall studs shall not be required at locations where their length would be 3 feet (914 mm) or less. Each end of each required new stud shall be attached to the existing roofing framing members (truss top chord or rafter and truss bottom chord or ceiling joist) using not fewer than two 3-inch (76 mm) toenail fasteners (#8 wood screws or 10d nails) and a metal connector with minimum uplift capacity of 175 pounds (778 N), or nail plates with not fewer than four $1^1/_4$-inch-long (32 mm) fasteners (No. 8 wood screws or 8d nails).

[BS] C104.2.2 Main method of installation. Each horizontal brace shall be fastened to each existing roof or ceiling member that it crosses using three 3-inch-long (76 mm) fasteners (No. 8 wood screws or 10d nails) as indicated in Figure C104.2(1) and Figure C104.2(3) for trusses and Figure C104.2(2) and Figure C104.2(4) for conventionally framed gable end walls. Alternative methods for providing horizontal bracing of the gable end studs as provided in Sections C104.2.3 through C104.2.9 shall be permitted.

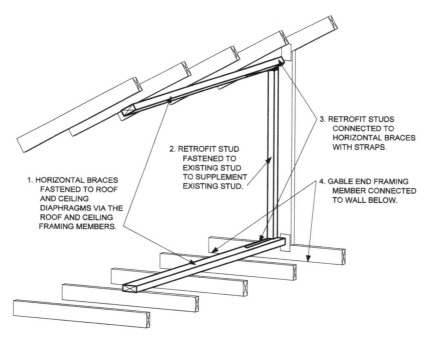

THIS FIGURE SHOWS A TRUSS GABLE END.
THE METHODOLOGY FOR A CONVENTIONALLY FRAMED GABLE END IS SIMILAR.
THE NUMBERS INDICATE A TYPICAL SEQUENCE OF INSTALLATION.
IN ORDER TO SHOW STRAPS COMPRESSION BLOCKS ARE NOT SHOWN.

**[BS] FIGURE C104.1
BASIC GABLE END RETROFIT METHODOLOGY**

APPENDIX C—GUIDELINES FOR THE WIND RETROFIT OF EXISTING BUILDINGS

[BS] TABLE C104.2
STUD LENGTH LIMITATIONS BASED ON EXPOSURE AND DESIGN WIND SPEED

EXPOSURE CATEGORY	MAXIMUM 3-SEC GUST BASIC WIND SPEED[a]	MAXIMUM HEIGHT OF GABLE END RETROFIT STUD[b]			
C	140	8'-0"	11'-3"	14'-9"	16'-0"
C	150	7'-6"	10'-6"	13'-6"	16'-0"
C	165	7'-0"	10'-0"	12'-3"	16'-0"
C	180	7'-0"	10'-0"	12'-3"	16'-0"
C	190	6'-6"	8'-9"	11'-0"	16'-0"
B	140	8'-0"	12'-3"	16'-0"	NR[c]
B	150	8'-0"	11'-3"	14'-9"	16'-0"
B	165	8'-0"	11'-3"	14'-9"	16'-0"
B	180	7'-6"	10'-6"	13'-6"	16'-0"
B	190	7'-0"	10'-0"	12'-3"	16'-0"
	Retrofit Configuration	A	B	C	D

For SI: 1 inch = 25.4 mm, 1 foot = 304.8 mm.
NR = Not Required.
a. Interpolation between given wind speeds is not permitted.
b. Existing gable end studs less than or equal to 3 feet 0 inches in height shall not require retrofitting.
c. Configuration C is acceptable to 16 feet 0 inches maximum height.

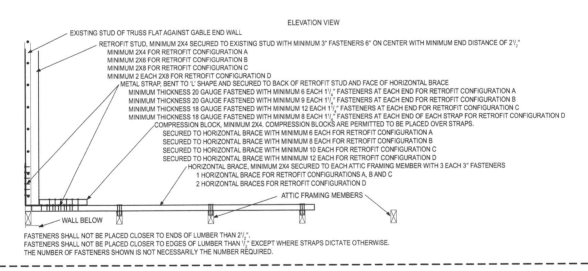

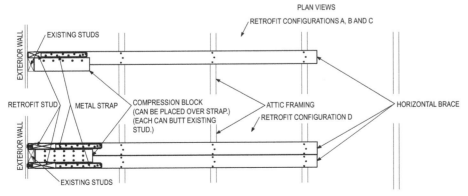

For SI: 1 inch = 25.4 mm.

[BS] FIGURE C104.2(1)
TRUSS FRAMED GABLE END

APPENDIX C—GUIDELINES FOR THE WIND RETROFIT OF EXISTING BUILDINGS

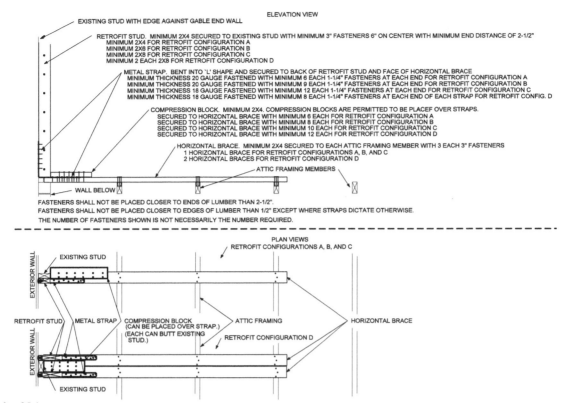

[BS] FIGURE C104.2(2)
CONVENTIONALLY FRAMED GABLE END L-BENT STRAP

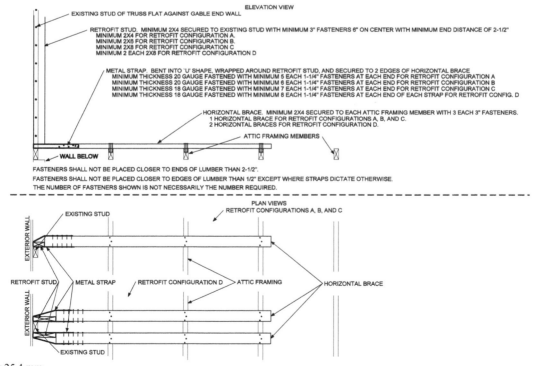

[BS] FIGURE C104.2(3)
TRUSS FRAMED GABLE END U-BENT STRAP

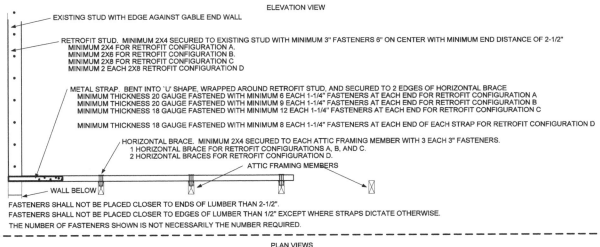

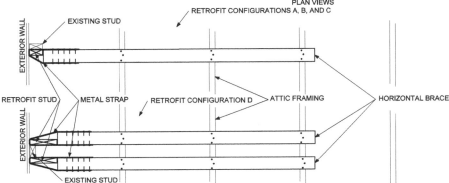

For SI: 1 inch = 25.4 mm.

**[BS] FIGURE C104.2(4)
CONVENTIONALLY FRAMED GABLE END U-BENT STRAP**

[BS] C104.2.3 Omitted horizontal brace. Where conditions exist that prevent installation in accordance with Section C104.2.2, horizontal braces shall be permitted to be omitted for height limitations corresponding to Retrofit Configurations A and B as defined in Table C104.2 provided that installation is as indicated in Figure C104.2.3 and provided that all of the following conditions are met. This method is not permitted for Retrofit Configurations C or D.

1. There shall be not fewer than two horizontal braces on each side of an omitted horizontal brace or not fewer than one horizontal brace if it is the end horizontal brace. Omitted horizontal braces must be separated by not fewer than two horizontal braces even if that location is composed of two retrofit studs and two horizontal braces.

2. Horizontal braces adjacent to the omitted horizontal brace shall be 2-inch by 6-inch (38 mm by 140 mm) lumber, shall butt against the existing studs, and shall be fastened to each existing roof or ceiling member crossed using three 3-inch-long (76 mm) fasteners (No. 8 wood screws or 10d nails). For Retrofit Configuration B, four fasteners shall be required on not fewer than one of the connections between the horizontal brace and the existing roof and ceiling framing members. Fasteners shall be spaced a not less than $3/4$ inch (19.1 mm) from the edges of the horizontal braces and not less than $1 3/4$ inches (44 mm) from adjacent fasteners.

3. Where the existing studs on each side of an omitted horizontal brace have their wide face perpendicular to the gable end wall, the retrofit studs at those locations and the retrofit stud at the omitted horizontal brace locations shall extend not less than $3 3/4$ inches (95 mm) beyond the interior edge of the existing studs for both Retrofit Configurations A and B. The edges of the three retrofit studs facing towards the interior of the attic shall be aligned such that they are the same distance from the gable end wall.

4. Retrofit studs shall be fastened to existing studs in accordance with Section C104.3.

5. Retrofit studs adjacent to the omitted horizontal brace shall be fastened to the horizontal brace using straps in accordance with Table C104.4.1 consistent with the size of the retrofit stud. The method applicable to Table C104.4.2 is not permitted.

APPENDIX C—GUIDELINES FOR THE WIND RETROFIT OF EXISTING BUILDINGS

6. A strong back made of minimum of 2-inch by 8-inch (38 mm by 184 mm) nominal lumber shall be placed parallel to the gable end and shall be located on and span between horizontal braces on the two sides of the omitted horizontal brace and shall extend beyond each horizontal brace by not less than $2^1/_2$ inches (64 mm). The strong back shall be butted to the three retrofit studs. The strong back shall be attached to each of the horizontal braces on which it rests with five 3-inch-long (76 mm) fasteners (#8 screws or 8d nails). The fasteners shall have a minimum $^3/_4$-inch (19.1 mm) edge distance and a minimum $2^1/_2$-inch (64 mm) spacing between fasteners. Additional compression blocks shall not be required at locations where a strong back butts against a retrofit stud.

7. The retrofit stud at the location of the omitted horizontal braces shall be fastened to the strong back using a connector with minimum uplift capacity of 800 pounds (3559 N) and installed such that this capacity is oriented in the direction perpendicular to the gable end wall.

8. The use of shortened horizontal braces using the alternative method of Section C104.2.5 is not permitted for horizontal braces adjacent to the omitted horizontal braces.

9. Horizontal braces shall be permitted to be interrupted in accordance with Section C104.2.8.

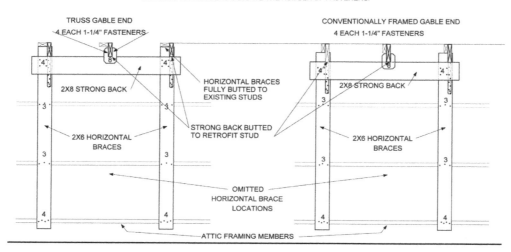

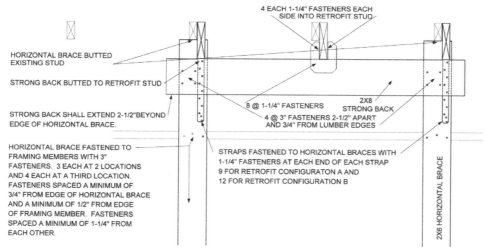

For SI: 1 inch = 25.4 mm.

**[BS] FIGURE C104.2.3
OMITTED HORIZONTAL BRACE**

[BS] C104.2.4 Omitted horizontal brace and retrofit stud. Where conditions exist that prevent installation in accordance with Section C104.2.2 or C104.2.3, then retrofit studs and horizontal braces shall be permitted to be omitted from those locations by installation of ladder assemblies for Retrofit Configurations A and B as defined in Table C104.2 provided that all of the following conditions are met. This method is not permitted for Retrofit Configurations C or D.

1. Not more than two ladder assemblies are permitted on a single gable end.
2. There shall be not fewer than two retrofit studs and horizontal brace assemblies on either side of the locations where the retrofit studs and horizontal bracing members are omitted (two ladder braces shall not bear on a single retrofit stud).
3. Where the existing studs on each side of an omitted horizontal brace have their wide face parallel to the gable end wall the retrofit studs at those locations and the retrofit stud at the omitted horizontal brace locations shall be 2-inch by 6-inch (38 mm by 180 mm) nominal lumber for Retrofit Configuration A and 2-inch by 8-inch (38 mm by 184 mm) lumber for Retrofit Configuration B.
4. Horizontal braces adjacent to the omitted horizontal brace shall be 2-inch by 6-inch (38 mm by 180 mm) nominal lumber and be fastened to each existing roof or ceiling member crossed using three 3-inchlong (76 mm) fasteners (#8 wood screws or 10d nails) as indicated in Figures C104.2(1) and C104.2(3) for gable end frames and Figures C104.2(2) and C104.2(4) for conventionally framed gable end walls. For Retrofit Configuration B, four fasteners shall be required on one of the connections between the horizontal brace and the existing roof and ceiling framing members.
5. Ladder rungs shall be provided across the location of the omitted retrofit studs as indicated in Figure C104.2.4(1) for gable end frames and Figure C104.2.4(2) for conventionally framed gable end walls.
6. Ladder rungs shall be minimum 2-inch by 4-inch (38 mm by 89 mm) lumber oriented with their wide face horizontal and spaced not greater than 16 inches (406 mm) on center vertically.
7. Where ladder rungs cross wall framing members they shall be connected to the wall framing members with a metal connector with a minimum capacity of 175 pounds (778 N) in the direction perpendicular to the gable end wall.
8. Notching of the ladder rungs shall not be permitted unless the net depth of the framing member is not less than $3^1/_2$ inches (89 mm).

[BS] C104.2.5 Short horizontal brace. Where conditions exist that prevent installation in accordance with Section C104.2.2, C104.2.3 or C104.2.4, the horizontal braces shall be permitted to be shortened provided that installation is as indicated in Figure C104.2.5 and all of the following conditions are met.

1. The horizontal brace shall be installed across not fewer than two framing spaces, extend not less than 4 feet (1220 mm) from the gable end wall plus $2^1/_2$ inches (64 mm) beyond the farthest roof or ceiling framing member from the gable end, and be fastened to each existing framing member with three 3-inch-long (76 mm) fasteners (#8 wood screws or 10d nails).
2. An anchor block shall be fastened to the side of the horizontal brace in the second framing space from the gable end wall as shown in Figure C104.2.5. The anchor block lumber shall have a minimum edge thickness of $1^1/_2$ inches (38 mm) and the depth shall be at a minimum the depth of the existing roof or ceiling framing member. Six 3-inch-long (76 mm) fasteners (#8 wood screws or 10d nails) shall be used to fasten the anchor block to the side of the horizontal brace.
3. The anchor block shall extend into the space between the roof or ceiling framing members not less than one-half the depth of the existing-framing members at the location where the anchor block is installed. The anchor block shall be installed tightly between the existing framing members such that the gap at either end shall not exceed $^1/_8$ inch (3.2 mm).
4. The use of omitted horizontal braces using the method of Section C104.2.3 adjacent to a short horizontal brace as defined in this section is not permitted.

[BS] C104.2.6 Installation of horizontal braces onto webs of trusses. Where existing conditions preclude installation of horizontal braces on truss top or bottom chords they shall be permitted to be installed on truss webs provided that all of the following conditions are met.

1. Horizontal braces shall be installed as close to the top or bottom chords as practical without altering the truss or any of its components and not more than three times the depth of the truss member to which it would ordinarily be attached.
2. A racking block, comprised of an anchor block meeting the definition of "Anchor block" in Section C102 or comprised of minimum $^{15}/_{32}$-inch (12 mm) plywood or $^7/_{16}$-inch (11.1 mm) oriented strand board (OSB), shall be fastened to the horizontal brace in the second framing space from the gable end wall. The racking block shall extend toward the roof or ceiling diaphragm so that the edge of the racking block closest to the diaphragm is within one-half the depth of the existing framing member from the diaphragm surface. The racking block shall be attached to horizontal braces using six fasteners (No. 8 wood screws or 10d nails) of sufficient length to provide $1^1/_2$ inches (38 mm) of penetration into the horizontal brace.

APPENDIX C—GUIDELINES FOR THE WIND RETROFIT OF EXISTING BUILDINGS

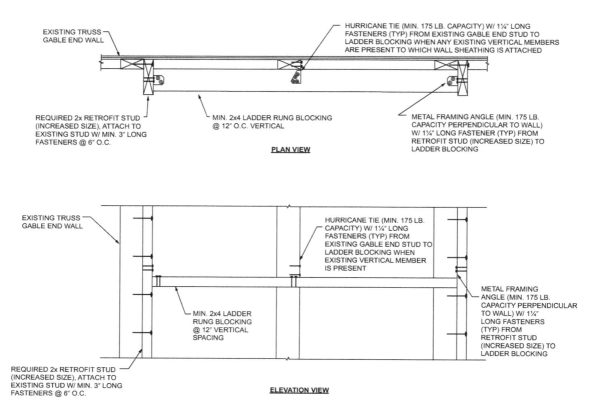

For SI: 1 inch = 25.4 mm, 1 pound = 4.4 N.

[BS] FIGURE C104.2.4(1)
LADDER BRACING FOR OMITTED RETROFIT STUD (GABLE END FRAME)

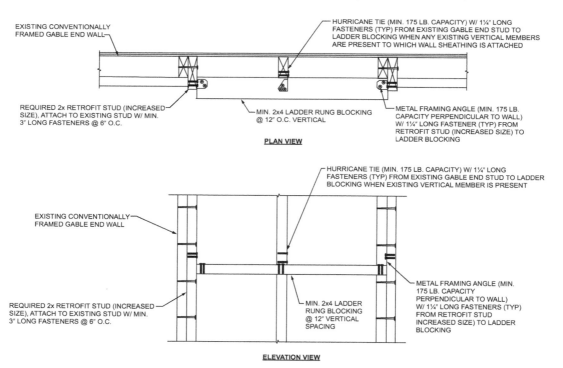

For SI: 1 inch = 25.4 mm, 1 pound = 4.4 N.

[BS] FIGURE C104.2.4(2)
LADDER BRACING FOR OMITTED RETROFIT STUD (CONVENTIONALLY FRAMED GABLE END)

APPENDIX C—GUIDELINES FOR THE WIND RETROFIT OF EXISTING BUILDINGS

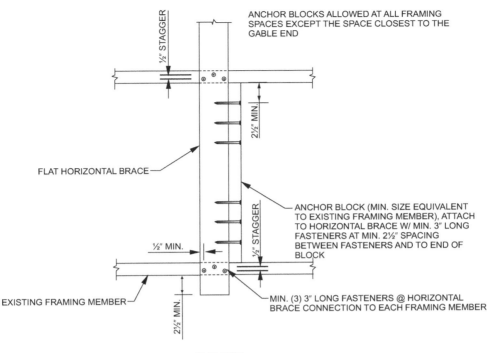

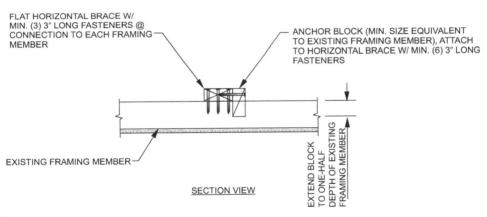

For SI: 1 inch = 25.4 mm.

**[BS] FIGURE C104.2.5
ANCHOR BLOCK INSTALLATION**

3. Racking blocks shall be permitted to be fastened to any face or edge of horizontal braces between each web or truss vertical posts to which a horizontal brace is attached. Racking blocks shall be permitted to be on alternate sides of horizontal braces. Racking blocks shall be installed tightly between the lumber of truss members or truss plates such that the gap at either end shall be not greater than $1/8$ inch (3.2 mm).

[BS] C104.2.7 Alternative method of installation of horizontal braces at truss ridges. Where conditions exist that limit or restrict installation of horizontal braces near the peak of the roof, ridge ties shall be added to provide support for the required horizontal brace. The top of additional ridge tie members shall be installed not greater than 16 inches (406 mm) below the existing ridge line or 4 inches (102 mm) below impediments. A minimum 2-inch by 4-inch (38 mm by 89 mm) nominal member shall be used for each ridge tie, and fastening shall consist of two 3-inch-long (76 mm) wood screws, four 3-inch-long (76 mm) 10d nails or two $3^{1}/_{2}$-inch-long (89 mm) 16d nails driven through and clinched at each top chord or web member intersected by the ridge tie as illustrated in Figure C104.2.7.

APPENDIX C—GUIDELINES FOR THE WIND RETROFIT OF EXISTING BUILDINGS

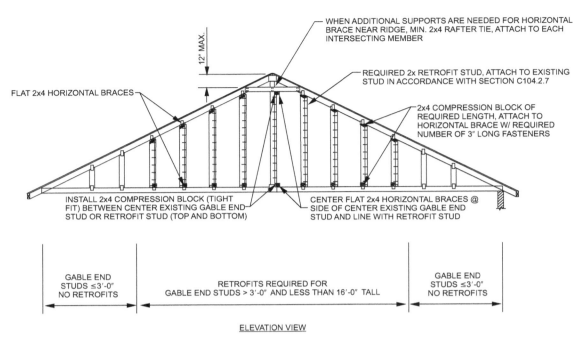

**[BS] FIGURE C104.2.7
DETAIL OF RETROFIT TIE INSTALLATION**

For SI: 1 inch = 25.4 mm, 1 foot = 304.8 mm.

[BS] C104.2.8 Interrupted horizontal braces. Where conditions exist that prevent the installation of a continuous horizontal brace then horizontal braces shall be permitted to be interrupted using the methods shown in Figures C104.2.8(1), C104.2.8(2), and C104.2.8(3). For interruptions that occur in the attic framing space closest to the gable end, nine 3-inch (76 mm) fasteners shall be used to connect each section of the interrupted horizontal braces. For interruptions that occur in the second attic space from the gable end, six 3-inch (76 mm) fasteners shall be used to connect each section of the interrupted horizontal braces. For interruptions that occur in the attic framing space farthest from the gable end, three 3-inch (76 mm) fasteners shall be used to connect each section of the interrupted horizontal braces. Horizontal braces shall be continued far enough to allow connections to three existing roof framing members as shown in Figure C104.2.8(1), C104.2.8(2) or C104.2.8(3). Fasteners shall be spaced in accordance with Section C103.5.3. Horizontal braces shall be the same width and depth as required for an uninterrupted member.

[BS] C104.2.9 Piggyback gable end frames. Piggyback gable end frames (gable end frames built in two sections one above the other) shall be permitted to be retrofitted if either of the following cases is true:

1. The existing studs in both the upper gable end frames and the lower gable end frames to which wall sheathing, panel siding, or other wall covering are attached are sufficiently in line that retrofit studs can be installed and connections made between the two with retrofit stud(s).

2. Existing studs in the upper frame are not sufficiently in line with the studs in the frame below and the existing studs in the upper frame are 3 feet (91 cm) or shorter.

For Condition 1 both the lower stud and the upper stud shall be retrofitted using the methods of Section C104.2. For Condition 2 the retrofit stud shall be connected to the lower studs using the methods of Section C104.2 and be continuous from the bottom horizontal brace to the top horizontal brace. Connection is not required between the retrofit stud and the upper stud. In both conditions the bottom chord of the piggyback truss section shall be fastened to each retrofit stud using a connector with minimum axial capacity of 175 pounds (778 N).

[BS] C104.3 Retrofit studs. Retrofit studs shall be installed in accordance with Section C104.3.1 using one of the five methods of Sections C104.3.2, C104.3.3, C104.3.4, C104.3.5 or C104.3.6. Figure C104.3 shows these methods of installation. For the Retrofit Configuration obtained from Table C104.2, the size of retrofit studs shall be as indicated in Table C104.4.1 or Table C104.4.2. Retrofit studs shall extend from the top of the lower horizontal brace to the bottom of the upper horizontal brace except that a maximum gap of $1/8$ inch (3.2 mm) is permitted at the bottom and $1/2$ inch (12.7 mm) at the top. Where wall sheathing, panel siding or other wall covering is fastened to a conventionally framed gable end, retrofit studs shall be applied in accordance with Section C104.2.1.

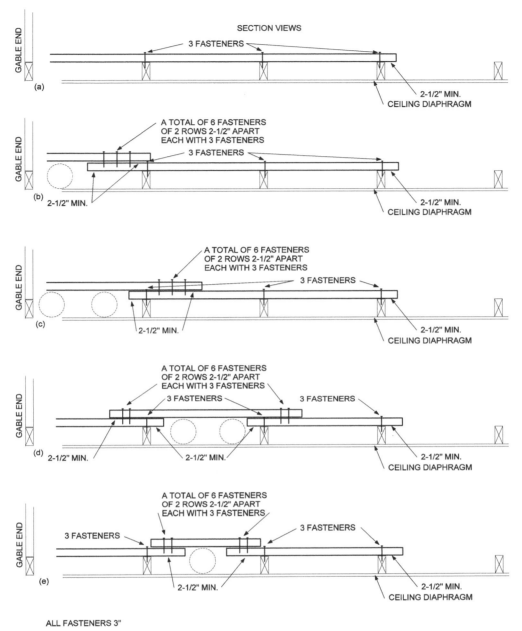

For SI: 1 inch = 25.4 mm.

[BS] FIGURE C104.2.8(1)
SPLICED HORIZONTAL BRACES

[BS] C104.3.1 Fastening. Where nail plates are not used, retrofit studs shall be attached to existing studs using 3-inch (76 mm) fasteners at not greater than 6 inches (152 mm) on center but not closer than $2^{1}/_{2}$ inches (64 mm) on center with fasteners not closer to ends of members than $2^{1}/_{2}$ inches (64 mm).

[BS] C104.3.2 Method #1: Face-to-edge or face-to-face method. Retrofit studs shall be installed immediately adjacent to existing gable end wall studs as indicated in Figure C104.3(a). The retrofit studs shall overlap the edge or side of the existing stud by not less than $1^{1}/_{4}$ inches (32 mm). Fasteners shall be installed as specified in Section C104.3.1.

[BS] C104.3.3 Method #2: Face-to-face offset method. Retrofit studs shall be installed against the face of existing studs as indicated in Figure C104.3(b) such that the faces overlap not less than $1^{1}/_{2}$ inches (38 mm) and the edge distance to fasteners is not less than $^{3}/_{4}$ inch (19.1 mm). Fasteners shall be installed as specified in Section C104.3.1.

APPENDIX C—GUIDELINES FOR THE WIND RETROFIT OF EXISTING BUILDINGS

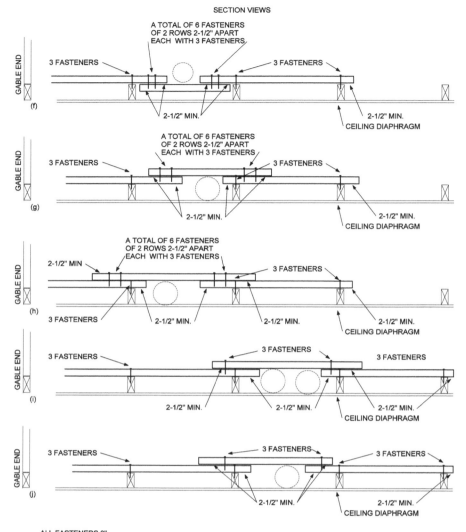

For SI: 1 inch = 25.4 mm.

[BS] FIGURE C104.2.8(2)
SPLICED HORIZONTAL BRACES

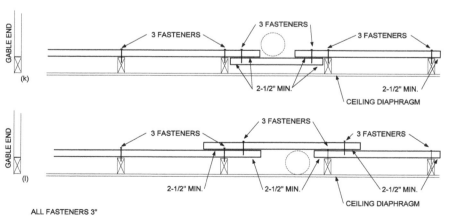

For SI: 1 inch = 25.4 mm.

[BS] FIGURE C104.2.8(3)
SPLICED HORIZONTAL BRACES

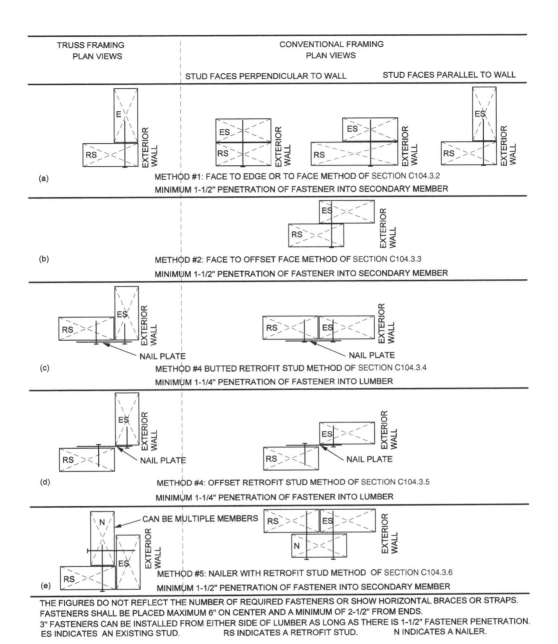

[BS] FIGURE C104.3
METHOD OF INSTALLING RETROFIT STUDS

[BS] C104.3.4 Method #3: Butted retrofit stud method. Provided that all of the following fastening conditions are met, retrofit studs shall be permitted to be butted by their edge to existing studs with the addition of nail plates as indicated in Figure C104.3(c) and Figure C104.3.4.

1. The narrow edge of retrofit studs shall be installed against the narrow or the wide face of existing studs.
2. Not fewer than two nail plates shall be used.
3. Fasteners used to secure nail plates to studs shall be a minimum $1^1/_4$ inches (32 mm) long (#8 wood screws or 8d nails).
4. Fasteners placed in nail plates shall have a minimum end distance of $2^1/_2$ inches (64 mm) for both studs and a maximum end distance of 6 inches (152 mm) from the ends of the shorter stud.
5. Fasteners shall have a minimum $^1/_2$-inch (12.7 mm) edge distance. Fasteners shall be placed not greater than $1^1/_2$ inches (38 mm) from the abutting vertical edges of existing studs and retrofit studs.

APPENDIX C—GUIDELINES FOR THE WIND RETROFIT OF EXISTING BUILDINGS

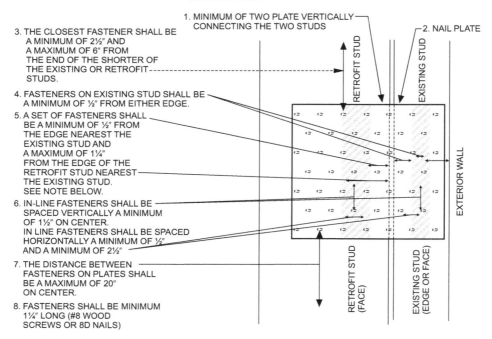

**[BS] FIGURE C104.3.4
NAIL PLATE FASTENING**

6. There shall be at least three fasteners through nail plates into all existing and retrofit studs to which the nail plate is attached.

7. Nail plates with three fasteners onto a single existing or retrofit stud shall be spaced not greater than 15 inches (38 cm) on center.

8. Nail plates with more than three fasteners onto a single existing or retrofit stud shall be spaced not greater than 20 inches (51 cm) on center.

9. Fasteners used to secure nail plates shall be spaced vertically not less than $1^1/_2$ inches (38 mm) on center. Staggered fasteners used to secure nail plates shall be spaced horizontally not less than $^1/_2$ inch (12.7 mm).

[BS] C104.3.5 Method #4: Offset retrofit stud method. Retrofit studs may be offset from existing studs by use of nail plates as shown in Figure C104.3(d) such that the vertical corner of a retrofit stud shall align with the vertical corner of an existing stud as indicated in Figure C104.3(d) and Figure C104.3.4, and the fastening conditions of Section C104.3.4 are met.

[BS] C104.3.6 Method #5: Nailer with retrofit stud method. Retrofit studs and existing studs shall be permitted to be connected using noncontinuous 2-inch by 4-inch (38 mm by 89 mm) nailers as indicated in Figure C104.3(e) provided that the following conditions are met.

1. Both the existing stud and the retrofit stud shall be butted to nailers and both shall be fastened to the nailer with 3-inch-long (76 mm) fasteners (#8 wood screws or 8d nails). Fasteners connecting each stud to the nailer shall be a spaced 6 inches (152 mm) o.c.

2. Fasteners into nailers from any direction shall be offset vertically by not less than $2^1/_2$ inches (64 mm).

3. Fasteners into nailers shall be not less than $2^1/_2$ inches (64 mm) but not more than 6 inches (152 mm) from the end of the shorter of the existing stud and retrofit stud to which they are fastened.

[BS] C104.3.7 Reduced depth of retrofit studs. Retrofit studs may be reduced in depth by notching, tapering or other methods at any number of locations along their length provided that all of the following conditions are met:

1. Retrofit studs to be reduced in depth shall be sized such that the remaining minimum depth of member at the location of the notch (including

APPENDIX C—GUIDELINES FOR THE WIND RETROFIT OF EXISTING BUILDINGS

cross-cut kerfs) shall be not less than that required by Table C104.4.1 or Table C104.4.2.

2. Reduced in-depth retrofit stud shall not be spliced within 12 inches (30 cm) of the location of notches. Splice members shall not be notched.

3. The vertical extent of notches shall not exceed 12 inches (30 cm) as measured at the depth of location of reduced depth.

4. A reduced in-depth retrofit stud member shall be fastened to the side of the existing gable end wall studs in accordance with Section C104.3.1. Two additional 3-inch (76 mm) fasteners (#8 wood screws or 10d nails) shall be installed on each side of notches in addition to those required by Section C104.3.1.

[BS] C104.3.8 Retrofit stud splices. Retrofit studs greater than 8 feet (244 cm) in height may be field spliced in accordance with Figure C104.3.8.

[BS] C104.4 Connection between horizontal braces and retrofit studs. Connections between horizontal braces and retrofit studs shall comply with Section C104.4.1 or C104.4.2. Each retrofit stud shall be connected to the top and bottom horizontal brace members with a minimum 20-gage $1^1/_4$-inch-wide (32 mm) flat or coil metal strap with prepunched holes for fasteners. Straps shall be fastened with $1^1/_4$-inch-long (32 mm) fasteners (#8 wood screws or 8d nails) with the number of fasteners as indicated in Table C104.4.1 and Table C104.4.2. Fasteners shall be not closer to the end of lumber than $2^1/_2$ inches (64 mm).

[BS] C104.4.1 L-bent strap method. Retrofit studs shall be connected to horizontal braces or to strong backs in accordance with Figure C104.2(1), C104.2(2) or C104.2.3, and shall comply with the following conditions.

1. A strap shall be applied to the edges of a retrofit stud nearest the gable end wall and to the face of horizontal braces using at each end of the strap the number of fasteners specified in Table C104.4.1. Straps shall be long enough so that each strap extends sufficient distance onto the vertical face of the retrofit stud that the fastener closest to the ends of the studs is not less than $2^1/_2$ inches (64 mm) from the end of the stud. Straps shall be permitted to be twisted to accommodate the transition between the tops of retrofit studs and horizontal bracings following roof pitches.

2. Compression blocks shall be installed on the horizontal braces directly against either the existing vertical gable end wall stud or the retrofit stud. Figure C104.2(1) (trusses) and Figure C104.2(2) (conventionally framed) show the installation of the compression block against the existing vertical gable end wall stud with the strap from the retrofit stud running beside the compression block. Compression blocks shall be permitted to be placed over straps. Compression blocks shall be fastened to the horizontal braces with not fewer than the minimum number of 3-inch-long (76 mm) fasteners (#8 wood screws or 10d nails) specified in Table C104.4.1. End and edge distances for fasteners shall be in accordance with Section C103.5.3.

[BS] C104.4.2 U-bent strap method. Retrofit studs shall be connected to horizontal braces in accordance with Figure C104.2(3) or C104.2(4), shall be limited to Retrofit Configurations A and B as defined in Table C104.2, and shall comply with the following conditions.

1. Straps of sufficient length to meet the requirements for the number of fasteners in accordance with Table C104.4.2 and meet the end distance requirements of Section C103.5.3 shall be shaped around retrofit studs and fastened to the edges of horizontal braces. Straps shall wrap the back edge of the retrofit stud snugly with a maximum gap of $^1/_4$ inch (6.4 mm). Rounded bends of straps shall be permitted. One fastener shall be installed that connects each strap to the side of the associated retrofit stud.

2. The horizontal brace shall butt snugly against the retrofit stud with a maximum gap of $^1/_4$ inch (6.4 mm).

3. Straps shall be permitted to be twisted to accommodate the transition between the tops of retrofit studs and horizontal braces that follow the roof pitch.

[BS] C104.5 Connection of gable end wall to wall below. The bottom chords or bottom members of wood-framed gable end walls shall be attached to the wall below using one of the methods prescribed in Section C104.5.1 or C104.5.2. The particular method chosen shall correspond to the framing system and type of wall construction encountered.

[BS] TABLE C104.4.1
ELEMENT SIZING AND SPACING FOR L-BENT RETROFIT METHOD

RETROFIT ELEMENTS	RETROFIT CONFIGURATION			
	A	B	C	D
Minimum size and number of Horizontal Braces	2 × 4	2 × 4	2 × 4	2 each 2 × 4
Minimum size and number of Retrofit Studs	2 × 4	2 × 6	2 × 8	2 each 2 × 8
Minimum number of fasteners connecting each end of straps to Retrofit Studs or to Horizontal Braces #8 screws or 10d nails $1^1/_4$" long	6	9	12	8 on each strap
Minimum number of fasteners to connect Compression Blocks to Horizontal Braces #8 screws or 10d nails 3" long	6	8	10	12

For SI: 1 inch = 25.4 mm, 1 foot = 304.8 mm.

APPENDIX C—GUIDELINES FOR THE WIND RETROFIT OF EXISTING BUILDINGS

[BS] TABLE C104.4.2
ELEMENT SIZING AND SPACING FOR U-BENT RETROFIT METHOD

RETROFIT ELEMENTS	RETROFIT CONFIGURATION			
	A	B	C	D
Minimum size and number of Horizontal Braces	2 × 4	2 × 4	2 × 4	2 each 2 × 4
Minimum size and number of Retrofit Studs	2 × 4	2 × 6	2 × 8	2 each 2 × 8
Minimum number of fasteners connecting Straps to each edge of Horizontal Braces #8 screws or 10d nails $1\frac{1}{4}''$ long	6	7	7	6 on each side of strap

For SI: 1 inch = 25.4 mm, 1 foot = 304.8 mm.

[BS] C104.5.1 Gable end frame. The bottom chords of the gable end frame shall be attached to the wall below using gusset angles. Not fewer than two fasteners shall be installed into the bottom chord. The gusset angles shall be installed throughout the portion of the gable end where the gable end wall height is greater than 3 feet (91 cm) at the spacing specified in Table C104.5.1. Connection to the wall below shall be by one of the following methods:

1. For a wood-frame wall below, not fewer than two fasteners shall be installed. The fasteners shall be of the same diameter and style specified by the gusset angle manufacturer and sufficient length to extend through the double top plate of the wall below.

2. For a concrete or masonry wall below without a sill plate, the type and number of fasteners into the wall shall be consistent with the gusset angle manufacturer's specifications for fasteners installed in concrete or masonry.

3. For a concrete or masonry wall below with a 2x sill plate, the fasteners into the wall below shall be of the diameter and style specified by the gusset angle manufacturer for concrete or masonry connections; but, long enough to pass through the wood sill plate and provide the required embedment into the concrete or masonry below. Alternatively, the gusset angle can be anchored to the sill plate using four each $1\frac{1}{2}$-inch-long (38 mm) fasteners of the same type as specified by the gusset angle manufacturer for wood connections, provided that the sill plate is anchored to the wall on each side of the gusset angle by a $\frac{1}{4}$-inch-diameter (6.4 mm) masonry screw with $2\frac{3}{4}$ inches (70 mm) of embedment into the concrete or masonry wall. A $\frac{1}{4}$-inch (6.4 mm) washer shall be placed under the heads of the masonry screws.

[BS] C104.5.2 Conventionally framed gable end wall. Each stud in a conventionally framed gable end wall, throughout the length of the gable end wall where the wall height is greater than 3 feet (914 mm), shall be attached to the bottom or sill plate using a stud to plate connector with minimum uplift capacity of 175 pounds (778 N). The bottom or sill plate shall then be connected to the wall below using one of the following methods:

1. For a wood frame wall below, the sill or bottom plate shall be connected to the top plate of the wall below using $\frac{1}{4}$-inch-diameter (6.4 mm) lag bolt fasteners of sufficient length to penetrate the bottom plate of the upper gable end wall and extend through the bottom top plate of the wall below. A washer sized for the diameter of the lag bolt shall be placed under the head of each lag bolt. The fasteners shall be installed at the spacing indicated in Table C104.5.2.

2. For a concrete or masonry wall below, the sill or bottom plate shall be connected to the concrete or masonry wall below using $\frac{1}{4}$-inch-diameter (6.4 mm) concrete or masonry screws of sufficient length to provide $2\frac{3}{4}$ inches (70 mm) of embedment into the top of the concrete or masonry wall. A washer sized for the diameter of the lag bolt shall be placed under the head of each lag bolt. The fasteners shall be installed at the spacing indicated in Table C104.5.2.

[BS] TABLE C104.5.1
SPACING OF GUSSET ANGLES

EXPOSURE CATEGORY	BASIC WIND SPEED (mph)	SPACING OF GUSSET ANGLES (inches)
C	140	38
C	150	32
C	165	28
C	180	24
C	190	20
B	140	48
B	150	40
B	165	36
B	180	30
B	190	26

For SI: 1 inch = 25.4 mm, 1 mile per hour = 0.447 m/s.

APPENDIX C—GUIDELINES FOR THE WIND RETROFIT OF EXISTING BUILDINGS

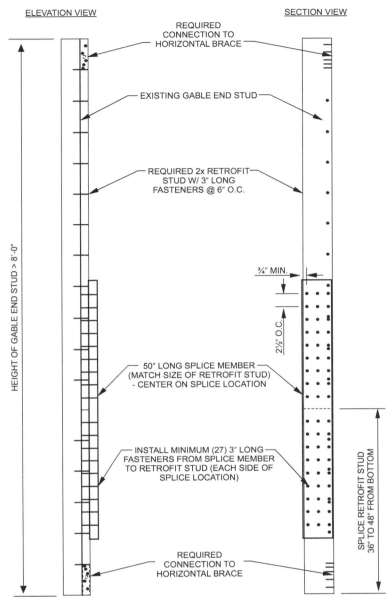

For SI: 1 inch = 25.4 mm, 1 foot = 304.8 mm.

[BS] FIGURE C104.3.8
RETROFIT STUD SPLICES

APPENDIX C—GUIDELINES FOR THE WIND RETROFIT OF EXISTING BUILDINGS

[BS] TABLE C104.5.2
SPACING OF LAG OR MASONRY SCREWS USED TO CONNECT SILL PLATE OF GABLE END WALL TO TOP OF THE WALL BELOW

EXPOSURE CATEGORY	BASIC WIND SPEED (mph)	SPACING OF LAG OR MASONRY SCREWS (inches)
C	140	19
C	150	16
C	165	14
C	180	14
C	190	10
B	140	24
B	150	20
B	165	18
B	180	15
B	190	13

For SI: 1 inch = 25.4 mm, 1 mile per hour = 0.447 m/s.

CHAPTER C2

ROOF DECK FASTENING FOR HIGH-WIND AREAS

SECTION C201
GENERAL

[BS] C201.1 Purpose. This chapter provides prescriptive methods for partial structural retrofit of an *existing building* to increase its resistance to wind loads. It is intended for voluntary use where the ultimate design wind speed, V_{ult}, determined in accordance with Figure 1609.3(1) of the *International Building Code* exceeds 130 mph (58 m/s) and for reference by mitigation programs. The provisions of this chapter do not necessarily satisfy requirements for new construction. Unless specifically cited, the provisions of this chapter do not necessarily satisfy requirements for structural improvements triggered by *addition, alteration, repair, change of occupancy*, building relocation or other circumstances.

[BS] C201.2 Eligible conditions. The provisions of this chapter are applicable only to buildings that meet either of the following eligibility requirements:

1. Buildings assigned to *Risk Category* I or II in accordance with Table 1604.5 of the *International Building Code*.
2. Buildings within the scope of the *International Residential Code*.

SECTION C202
ROOF DECK ATTACHMENT FOR WOOD ROOFS

[BS] C202.1 Roof decking attachment for one- and two-family dwellings. For one- and two-family dwellings, fastening shall be in accordance with Section C202.1.1 or C202.1.2 as appropriate for the existing construction. The diameter of 8d nails shall be not less than 0.131 inch (3 mm) and the length shall be not less than $2^1/_4$ inches (57 mm) to qualify for the provisions of this section for existing nails regardless of head shape or head diameter.

[BS] C202.1.1 Sawn lumber or wood plank roofs. Roof decking consisting of sawn lumber or wood planks up to 12 inches (30 cm) wide and secured with not fewer than two nails (minimum size 8d) to each roof framing member it crosses shall be deemed to be sufficiently connected. Sawn lumber or wood plank decking secured with smaller fasteners than 8d nails or with fewer than two nails (minimum size 8d) to each framing member it crosses shall be deemed sufficiently connected if fasteners are added such that two clipped head, round head or ring shank nails (minimum size 8d) are in place on each framing member the nail crosses.

[BS] C202.1.2 Wood structural panel roofs For roof decking consisting of wood structural panels, fasteners and spacings required in Table C202.1.2 shall be deemed to comply with the requirements of Section 706.3.

Supplemental fasteners as required by Table C202.1.2 shall be 8d ring shank nails with round heads and the following minimum dimensions:

1. 0.113-inch-nominal (3 mm) shank diameter.
2. Ring diameter not less than 0.012 inch (0.3 mm) greater than shank diameter.
3. 16 to 20 rings per inch.
4. A minimum 0.280-inch (7 mm) full round head diameter.
5. Ring shank to extend not less than $1^1/_2$ inches (38 mm) from the tip of the nail.
6. Minimum $2^1/_4$-inch (57 mm) nail length.

APPENDIX C—GUIDELINES FOR THE WIND RETROFIT OF EXISTING BUILDINGS

[BS] TABLE C202.1.2
SUPPLEMENT FASTENERS AT PANEL EDGES AND INTERMEDIATE FRAMING

EXISTING FASTENERS	EXISTING FASTENER SPACING (EDGE OR INTERMEDIATE SUPPORTS)	MAXIMUM SUPPLEMENTAL FASTENER SPACING FOR 130 MPH < V_{ult} ≤ 140 MPH	MAXIMUM SUPPLEMENTAL FASTENER SPACING FOR INTERIOR ZONE[c] LOCATIONS FOR MPH V_{ult} > 140 MPH AND EDGE ZONES NOT COVERED BY THE COLUMN TO THE RIGHT	EDGE ZONE[d] FOR V_{ult} > 160 MPH AND EXPOSURE C, OR V_{ult} > 180 MPH AND EXPOSURE B
Staples or 6d	Any	6″ o.c.[b]	6″ o.c.[b]	4″ o.c.[b] at panel edges and 4″ o.c.[b] at intermediate supports
8d clipped head or round head smooth shank	6″ o.c. or less	None necessary	None necessary along edges of panels but 6″ o.c.[b] at intermediate supports of panel	4″ o.c.[a] at panel edges and 4″ o.c.[a] at intermediate supports
8d clipped head or round head ring shank	6″ o.c. or less	None necessary	None necessary	4″ o.c.[a] at panel edges and 4″ o.c.[a] at intermediate supports
8d clipped head or round head smooth shank	Greater than 6″ o.c.	6″ o.c.[a]	6″ o.c.[a] along panel edges and 6″ o.c.[b] at intermediate supports of panel	4″ o.c.[a] at panel edges and 4″ o.c.[a] at intermediate supports
8d clipped head or round head ring shank	Greater than 6″ o.c.	6″ o.c.[a]	6″ o.c.[a]	4″ o.c.[a] at panel edges and 4″ o.c.[a] at intermediate supports

For SI: 1 inch = 25.4 mm; 1 foot = 304.8 mm; 1 mile per hour = 0.447 m/s.
a. Maximum spacing determined based on existing fasteners and supplemental fasteners.
b. Maximum spacing determined based on supplemental fasteners only.
c. Interior zone = sheathing that is not located within 4 feet of the perimeter edge of the roof or within 4 feet of each side of a ridge.
d. Edge zone = sheathing that is located within 4 feet of the perimeter edge of the roof and within 4 feet of each side of a ridge.

CHAPTER C3

REFERENCED STANDARDS

SECTION C301
REFERENCED STANDARDS

[BS] C301.1 General. See Table C301.1 for standards that are referenced in various sections of this appendix. Standards are listed by the standard identification with the effective date, standard title, and the section or sections of this appendix that reference the standard.

**[BS] TABLE C301.1
REFERENCED STANDARDS**

STANDARD ACRONYM	STANDARD NAME	SECTIONS HEREIN REFERENCED
IBC—21	*International Building Code®*	C101.3, C103.2, C201.1, C201.2
IRC—21	*International Residential Code®*	C101.2, C101.3, C103.2, C201.2

APPENDIX D

BOARD OF APPEALS

The provisions contained in this appendix are not mandatory unless specifically referenced in the adopting ordinance.

User notes:

About this appendix: Appendix D provides criteria for Board of Appeals members. Also provided are procedures by which the Board of Appeals should conduct its business.

Code development reminder: Code change proposals to this appendix will be considered by the Administrative Code Development Committee during the 2022 (Group B) Code Development Cycle.

SECTION D101
GENERAL

[A] D101.1 Scope. A board of appeals shall be established within the jurisdiction for the purpose of hearing applications for modification of the requirements of this code pursuant to the provisions of Section 112. The board shall be established and operated in accordance with this section, and shall be authorized to hear evidence from appellants and the *code official* pertaining to the application and intent of this code for the purpose of issuing orders pursuant to these provisions.

[A] D101.2 Application for appeal. Any person shall have the right to appeal a decision of the *code official* to the board. An application for appeal shall be based on a claim that the intent of this code or the rules legally adopted hereunder have been incorrectly interpreted, the provisions of this code do not fully apply or an equally good or better form of construction is proposed. The application shall be filed on a form obtained from the *code official* within 20 days after the notice was served.

[A] D101.2.1 Limitation of authority. The board shall not have authority to waive requirements of this code or interpret the administration of this code.

[A] D101.2.2 Stays of enforcement. Appeals of notice and orders, other than Imminent Danger notices, shall stay the enforcement of the notice and order until the appeal is heard by the board.

[A] D101.3 Membership of board. The board shall consist of five voting members appointed by the chief appointing authority of the jurisdiction. Each member shall serve for **[INSERT NUMBER OF YEARS]** years or until a successor has been appointed. The board member's terms shall be staggered at intervals, so as to provide continuity. The *code official* shall be an ex officio member of said board but shall not vote on any matter before the board.

[A] D101.3.1 Qualifications. The board shall consist of five individuals, who are qualified by experience and training to pass on matters pertaining to building construction and are not employees of the jurisdiction.

[A] D101.3.2 Alternate members. The chief appointing authority is authorized to appoint two alternate members who shall be called by the board chairperson to hear appeals during the absence or disqualification of a member. Alternate members shall possess the qualifications required for board membership, and shall be appointed for the same term or until a successor has been appointed.

[A] D101.3.3 Vacancies. Vacancies shall be filled for an unexpired term in the same manner in which original appointments are required to be made.

[A] D101.3.4 Chairperson. The board shall annually select one of its members to serve as chairperson.

[A] D101.3.5 Secretary. The chief appointing authority shall designate a qualified clerk to serve as secretary to the board. The secretary shall file a detailed record of all proceedings, which shall set forth the reasons for the board's decision, the vote of each member, the absence of a member and any failure of a member to vote.

[A] D101.3.6 Conflict of interest. A member with any personal, professional or financial interest in a matter before the board shall declare such interest and refrain from participating in discussions, deliberations and voting on such matters.

[A] D101.3.7 Compensation of members. Compensation of members shall be determined by law.

[A] D101.3.8 Removal from the board. A member shall be removed from the board prior to the end of their terms only for cause. Any member with continued absence from regular meeting of the board may be removed at the discretion of the chief appointing authority.

[A] D101.4 Rules and procedures. The board shall establish policies and procedures necessary to carry out its duties consistent with the provisions of this code and applicable state law. The procedures shall not require compliance with strict rules of evidence, but shall mandate that only relevant information be presented.

[A] D101.5 Notice of meeting. The board shall meet upon notice from the chairperson, within 10 days of the filing of an appeal or at stated periodic intervals.

[A] D101.5.1 Open hearing. All hearings before the board shall be open to the public. The appellant, the appellant's representative, the *code official* and any person whose interests are affected shall be given an opportunity to be heard.

[A] D101.5.2 Quorum. Three members of the board shall constitute a quorum.

[A] D101.5.3 Postponed hearing. When five members are not present to hear an appeal, either the appellant or the appellant's representative shall have the right to request a postponement of the hearing.

[A] D101.6 Legal counsel. The jurisdiction shall furnish legal counsel to the board to provide members with general legal advice concerning matters before them for consideration. Members shall be represented by legal counsel at the jurisdiction's expense in all matters arising from service within the scope of their duties.

[A] D101.7 Board decision. The board shall only modify or reverse the decision of the *code official* by a concurring vote of three or more members.

[A] D101.7.1 Resolution. The decision of the board shall be by resolution. Every decision shall be promptly filed in writing in the office of the *code official* within three days and shall be open to the public for inspection. A certified copy shall be furnished to the appellant or the appellant's representative and to the *code official*.

[A] D101.7.2 Administration. The *code official* shall take immediate action in accordance with the decision of the board.

[A] D101.8 Court review. Any person, whether or not a previous party of the appeal, shall have the right to apply to the appropriate court for a writ of certiorari to correct errors of law. Application for review shall be made in the manner and time required by law following the filing of the decision in the office of the chief administrative officer.

RESOURCE A

GUIDELINES ON FIRE RATINGS OF ARCHAIC MATERIALS AND ASSEMBLIES

User note:

About this resource: *In the process of repair and alteration of existing buildings, based on the nature and the extent of the work, this code might require certain upgrades in the fire-resistance rating of building elements, at which time it becomes critical for the designers and the code officials to be able to determine the fire-resistance rating of the existing building elements as part of the overall evaluation for the assessment of the need for improvements. This resource document provides a guideline for such an evaluation for fire-resistance rating of archaic materials that is not typically found in the modern model building codes.*

Introduction

The *International Existing Building Code®* (IEBC®) is a comprehensive code with the goal of addressing all aspects of work taking place in existing buildings and providing user-friendly methods and tools for regulation and improvement of such buildings. This resource document is included within the cover of the IEBC with that goal in mind and as a step towards accomplishing that goal.

In the process of *repair* and *alteration* of existing buildings, based on the nature and the extent of the work, the IEBC might require certain upgrades in the fire-resistance rating of building elements, at which time it becomes critical for the designers and the code officials to be able to determine the fire-resistance rating of the *existing building* elements as part of the overall evaluation for the assessment of the need for improvements. This resource document provides a guideline for such an evaluation for fire-resistance ratings of archaic materials that are not typically found in the modern model building codes.

Resource A is only a guideline and is not intended to be a document for specific adoption as it is not written in the format or language of ICC's International Codes and is not subject to the code development process.

PURPOSE

The *Guidelines on Fire Ratings of Archaic Materials and Assemblies* focuses upon the fire-related performance of archaic construction. "Archaic" encompasses construction typical of an earlier time, generally prior to 1950. "Fire-related performance" includes fire resistance, flame spread, smoke production and degree of combustibility.

The purpose of this guideline is to update the information which was available at the time of original construction, for use by architects, engineers and code officials when evaluating the fire safety of a rehabilitation project. In addition, information relevant to the evaluation of general classes of materials and types of construction is presented for those cases when documentation of the fire performance of a particular archaic material or assembly cannot be found.

It has been assumed that the building materials and their fastening, joining and incorporation into the building structure are sound mechanically. Therefore, some determination must be made that the original manufacture, the original construction practice, and the rigors of aging and use have not weakened the building. This assessment can often be difficult because process and quality control was not good in many industries, and variations among locally available raw materials and manufacturing techniques often resulted in a product which varied widely in its strength and durability. The properties of iron and steel, for example, varied widely, depending on the mill and the process used.

There is nothing inherently inferior about archaic materials or construction techniques. The pressures that promote fundamental change are most often economic or technological matters not necessarily related to concerns for safety. The high cost of labor made wood lath and plaster uneconomical. The high cost of land and the congestion of the cities provided the impetus for high-rise construction. Improved technology made it possible. The difficulty with archaic materials is not a question of suitability, but familiarity.

Code requirements for the fire performance of key building elements (e.g., walls, floor/ceiling assemblies, doors, shaft enclosures) are stated in performance terms: hours of fire resistance. It matters not whether these elements were built in 1908 or 1980, only that they provide the required degree of fire resistance. The level of performance will be defined by the local community, primarily through the enactment of a building or rehabilitation code. This guideline is only a tool to help evaluate the various building elements, regardless of what the level of performance is required to be.

The problem with archaic materials is simply that documentation of their fire performance is not readily available. The application of engineering judgment is more difficult because building officials may not be familiar with the materials or construction method involved. As a result, either a full-scale fire test is required or the archaic construction in question removed and replaced. Both alternatives are time consuming and wasteful.

RESOURCE A—GUIDELINES ON FIRE RATINGS OF ARCHAIC MATERIALS AND ASSEMBLIES

This guideline and the accompanying appendix are designed to help fill this information void. By providing the necessary documentation, there will be a firm basis for the continued acceptance of archaic materials and assemblies.

1
FIRE-RELATED PERFORMANCE OF ARCHAIC MATERIALS AND ASSEMBLIES

1.1 FIRE PERFORMANCE MEASURES

This guideline does not specify the level of performance required for the various building components. These requirements are controlled by the building occupancy and use and are set forth in the local building or rehabilitation code.

The fire resistance of a given building element is established by subjecting a sample of the assembly to a "standard" fire test which follows a "standard" time-temperature curve. This test method has changed little since the 1920s. The test results tabulated in the Appendix have been adjusted to reflect current test methods.

The current model building codes cite other fire-related properties not always tested for in earlier years: flame spread, smoke production, and degree of combustibility. However, they can generally be assumed to fall within well defined values because the principal combustible component of archaic materials is cellulose. Smoke production is more important today because of the increased use of plastics. However, the early flame spread tests, developed in the early 1940s, also included a test for smoke production.

"Plastics," one of the most important classes of contemporary materials, were not found in the review of archaic materials. If plastics are to be used in a rehabilitated building, they should be evaluated by contemporary standards. Information and documentation of their fire-related properties and performance is widely available.

Flame spread, smoke production and degree of combustibility are discussed in detail below. Test results for eight common species of lumber, published in an Underwriter's Laboratories' report (104), are noted in the following table:

TUNNEL TEST RESULTS FOR EIGHT SPECIES OF LUMBER

SPECIES OF LUMBER	FLAME SPREAD	FUEL CONTRIBUTED	SMOKE DEVELOPED
Western White Pine	75	50–60	50
Northern White Pine	120–215	120–140	60–65
Ponderosa Pine	80–215	120–135	100–110
Yellow Pine	180–190	130–145	275–305
Red Gum	140–155	125–175	40–60
Yellow Birch	105–110	100–105	45–65
Douglas Fir	65–100	50–80	10–100

Flame Spread

The flame spread of interior finishes is most often measured by the ASTM E84 "tunnel test." This test measures how far and how fast the flames spread across the surface of the test sample. The resulting flame spread rating (FSR) is expressed as a number on a continuous scale where cement-asbestos board is 0 and red oak is 100. (Materials with a flame spread greater than red oak have an FSR greater than 100.) The scale is divided into distinct groups or classes. The most commonly used flame spread classifications are: Class I or A*, with a 0–25 FSR; Class II or B, with a 26–75 FSR; and Class III or C, with a 76–200 FSR. The *NFPA Life Safety Code* also has a Class D (201–500 FSR) and Class E (over 500 FSR) interior finish.

These classifications are typically used in modern building codes to restrict the rate of fire spread. Only the first three classifications are normally permitted, though not all classes of materials can be used in all places throughout a building. For example, the interior finish of building materials used in exits or in corridors leading to exits is more strictly regulated than materials used within private dwelling units.

In general, inorganic archaic materials (e.g., bricks or tile) can be expected to be in Class I. Materials of whole wood are mostly Class II. Whole wood is defined as wood used in the same form as sawn from the tree. This is in contrast to the contemporary reconstituted wood products such as plywood, fiberboard, hardboard, or particle board. If the organic archaic material is not whole wood, the flame spread classification could be well over 200 and thus would be particularly unsuited for use in exits and other critical locations in a building. Some plywoods and various wood fiberboards have flame spreads over 200. Although they can be treated with fire retardants to reduce their flame spread, it would be advisable to assume that all such products have a flame spread over 200 unless there is information to the contrary.

Smoke Production

The evaluation of smoke density is part of the ASTM E84 tunnel test. For the eight species of lumber shown in the table above, the highest levels are 275–305 for Yellow Pine, but most of the others are less smoky than red oak which has an index of 100. The advent of plastics caused substantial increases in the smoke density values measured by the tunnel test. The ensuing limitation of the smoke production for wall and ceiling materials by the model building codes has been a reaction to the introduction of plastic materials. In general, cellulosic materials fall in the 50–300 range of smoke density which is below the general limitation of 450 adopted by many codes.

Degree of Combustibility

The model building codes tend to define "noncombustibility" on the basis of having passed ASTM E136 or if the material is totally inorganic. The acceptance of gypsum wallboard as noncombustible is based on limiting paper thickness to not over $1/8$ inch and a 0–50 flame spread rating by ASTM E84.

---

* Some codes are Roman numerals, others use letters.

At times there were provisions to define a Class I or A material (0–25 FSR) as noncombustible, but this is not currently recognized by most model building codes.

If there is any doubt whether or not an archaic material is noncombustible, it would be appropriate to send out samples for evaluation. If an archaic material is determined to be noncombustible according to ASTM E136, it can be expected that it will not contribute fuel to the fire.

1.2 COMBUSTIBLE CONSTRUCTION TYPES

One of the earliest forms of timber construction used exterior load-bearing masonry walls with columns and/or wooden walls supporting wooden beams and floors in the interior of the building. This form of construction, often called "mill" or "heavy timber" construction, has approximately 1 hour fire resistance. The exterior walls will generally contain the fire within the building.

With the development of dimensional lumber, there was a switch from heavy timber to "balloon frame" construction. The balloon frame uses load-bearing exterior wooden walls which have long timbers often extending from foundation to roof. When longer lumber became scarce, another form of construction, "platform" framing, replaced the balloon framing. The difference between the two systems is significant because platform framing is automatically fire-blocked at every floor while balloon framing commonly has concealed spaces that extend unblocked from basement to attic. The architect, engineer, and *code official* must be alert to the details of construction and the ease with which fire can spread in concealed spaces.

2 BUILDING EVALUATION

A given rehabilitation project will most likely go through several stages. The preliminary evaluation process involves the designer in surveying the prospective building. The fire resistance of *existing building* materials and construction systems is identified; potential problems are noted for closer study. The final evaluation phase includes: developing design solutions to upgrade the fire resistance of building elements, if necessary; preparing working drawings and specifications; and the securing of the necessary code approvals.

2.1 PRELIMINARY EVALUATION

A preliminary evaluation should begin with a building survey to determine the existing materials, the general arrangement of the structure and the use of the occupied spaces, and the details of construction. The designer needs to know "what is there" before a decision can be reached about what to keep and what to remove during the rehabilitation process. This preliminary evaluation should be as detailed as necessary to make initial plans. The fire-related properties need to be determined from the applicable building or rehabilitation code, and the materials and assemblies existing in the building then need to be evaluated for these properties. Two work sheets are shown below to facilitate the preliminary evaluation.

Two possible sources of information helpful in the preliminary evaluation are the original building plans and the building code in effect at the time of original construction. Plans may be on file with the local building department or in the offices of the original designers (e.g., architect, engineer) or their successors. If plans are available, the investigator should verify that the building was actually constructed as called for in the plans, as well as incorporate any later alterations or changes to the building. Earlier editions of the local building code should be on file with the building official. The code in effect at the time of construction will contain fire performance criteria. While this is no guarantee that the required performance was actually provided, it does give the investigator some guidance as to the level of performance which may be expected. Under some code administration and enforcement systems, the code in effect at the time of construction also defines the level of performance that must be provided at the time of rehabilitation.

Figure 1 illustrates one method for organizing preliminary field notes. Space is provided for the materials, dimensions, and condition of the principal building elements. Each floor of the structure should be visited and the appropriate information obtained. In practice, there will often be identical materials and construction on every floor, but the exception may be of vital importance. A schematic diagram should be prepared of each floor showing the layout of exits and hallways and indicating where each element described in the field notes fits into the structure as a whole. The exact arrangement of interior walls within apartments is of secondary importance from a fire safety point of view and need not be shown on the drawings unless these walls are required by code to have a fire-resistance rating.

The location of stairways and elevators should be clearly marked on the drawings. All exterior means of escape (e.g., fire escapes) should be identified.[1]

The following notes explain the entries in Figure 1.

<u>Exterior Bearing Walls</u>: Many old buildings utilize heavily constructed walls to support the floor/ceiling assemblies at the exterior of the building. There may be columns and/or interior bearing walls within the structure, but the exterior walls are an important factor in assessing the fire safety of a building.

The field investigator should note how the floor/ceiling assemblies are supported at the exterior of the building. If columns are incorporated in the exterior walls, the walls may be considered nonbearing.

<u>Interior Bearing Walls</u>: It may be difficult to determine whether or not an interior wall is load bearing, but the field investigator should attempt to make this determination. At a later stage of the rehabilitation process, this question will

---

1. Problems providing adequate exiting are discussed at length in the *Egress Guideline for Residential Rehabilitation*.

need to be determined exactly. Therefore, the field notes should be as accurate as possible.

Exterior Nonbearing Walls: The fire resistance of the exterior walls is important for two reasons. These walls (both bearing and nonbearing) are depended upon to: a) contain a fire within the building of origin; or b) keep an exterior fire *outside* the building. It is therefore important to indicate on the drawings where any openings are located as well as the materials and construction of all doors or shutters. The drawings should indicate the presence of wired glass, its thickness and framing, and identify the materials used for windows and door frames. The protection of openings adjacent to exterior means of escape (e.g., exterior stairways, fire escapes) is particularly important. The ground floor drawing should locate the building on the property and indicate the precise distances to adjacent buildings.

Interior Nonbearing Walls (Partitions): A partition is a "wall that extends from floor to ceiling and subdivides space within any story of a building." (48) Figure 1 has two categories (A & B) for Interior Nonbearing Walls (Partitions) which can be used for different walls, such as hallway walls as compared to inter-apartment walls. Under some circumstances there may be only one type of wall construction; in others, three or more types of wall construction may occur.

The field investigator should be alert for differences in function as well as in materials and construction details. In general, the details within apartments are not as important as the major exit paths and exit stairways. The preliminary field investigation should attempt to determine the thickness of all walls. A term introduced below called "thickness design" will depend on an accurate ($\pm\ ^1/_4$ inch) determination. Even though this initial field survey is called "preliminary," the data generated should be as accurate and complete as possible.

The field investigator should note the exact location from which observations are recorded. For instance, if a hole is found through a wall enclosing an exit stairway which allows a cataloguing of the construction details, the field investigation notes should reflect the location of the "find." At the preliminary stage it is not necessary to core every wall; the interior details of construction can usually be determined at some location.

Structural Frame: There may or may not be a complete skeletal frame, but usually there are columns, beams, trusses, or other like elements. The dimensions and spacing of the structural elements should be measured and indicated on the drawings. For instance, if there are 10-inch square columns located on a 30-foot square grid throughout the building, this should be noted. The structural material and cover or protective materials should be identified wherever possible. The thickness of the cover materials should be determined to an accuracy of $\pm\ ^1/_4$ inch. As discussed above, the preliminary field survey usually relies on accidental openings in the cover materials rather than a systematic coring technique.

Floor/Ceiling Structural Systems: The span between supports should be measured. If possible, a sketch of the cross-section of the system should be made. If there is no location where accidental damage has opened the floor/ceiling construction to visual inspection, it is necessary to make such an opening. An evaluation of the fire resistance of a floor/ceiling assembly requires detailed knowledge of the materials and their arrangement. Special attention should be paid to the cover on structural steel elements and the condition of suspended ceilings and similar membranes.

**FIGURE 1
PRELIMINARY EVALUATION FIELD NOTES**

BUILDING ELEMENT		MATERIALS	THICKNESS	CONDITION	NOTES
Exterior Bearing Walls					
Interior Bearing Walls					
Exterior Nonbearing Walls					
Interior Nonbearing Walls or Partitions:	A				
	B				
Structural Frame: Columns					
Beams					
Other					
Floor/Ceiling Structural System Spanning					
Roofs					
Doors (including frame and hardware): a) Enclosed vertical exitway					
b) Enclosed horizontal exitway					
c) Other					

Roofs: The preliminary field survey of the roof system is initially concerned with watertightness. However, once it is apparent that the roof is sound for ordinary use and can be retained in the rehabilitated building, it becomes necessary to evaluate the fire performance. The field investigator must measure the thickness and identify the types of materials which have been used. Be aware that there may be several layers of roof materials.

Doors: Doors to stairways and hallways represent some of the most important fire elements to be considered within a building. The uses of the spaces separated largely controls the level of fire performance necessary. Walls and doors enclosing stairways or elevator shafts would normally require a higher level of performance than between the bedroom and bath. The various uses are differentiated in Figure 1.

Careful measurements of the thickness of door panels must be made, and the type of core material within each door must be determined. It should be noted whether doors have self-closing devices; the general operation of the doors should be checked. The latch should engage and the door should fit tightly in the frame. The hinges should be in good condition. If glass is used in the doors, it should be identified as either plain glass or wired glass mounted in either a wood or steel frame.

Materials: The field investigator should be able to identify ordinary building materials. In situations where an unfamiliar material is found, a sample should be obtained. This sample should measure at least 10 cubic inches so that an ASTM E136 fire test can be conducted to determine if it is combustible.

Thickness: The thickness of all materials should be measured accurately since, under certain circumstances, the level of fire resistance is very sensitive to the material thickness.

Condition: The method of attaching the various layers and facings to one another or to the supporting structural element should be noted under the appropriate building element. The "secureness" of the attachment and the general condition of the layers and facings should be noted here.

Notes: The "Notes" column can be used for many purposes, but it might be a good idea to make specific references to other field notes or drawings.

After the building survey is completed, the data collected must be analyzed. A suggested work sheet for organizing this information is given below as Figure 2.

The required fire resistance and flame spread for each building element are normally established by the local building or rehabilitation code. The fire performance of the existing materials and assemblies should then be estimated, using

**FIGURE 2
PRELIMINARY EVALUATION WORKSHEET**

BUILDING ELEMENT		REQUIRED FIRE RESISTANCE	REQUIRED FLAME SPREAD	ESTIMATED FIRE RESISTANCE	ESTIMATED FLAME SPREAD	METHOD OF UPGRADING	ESTIMATED UPGRADED PROTECTION	NOTES
Exterior Bearing Walls								
Interior Bearing Walls								
Exterior Nonbearing Walls								
Interior Nonbearing Walls or Partitions:	A							
	B							
Structural Frame: Columns								
Beams								
Other								
Floor/Ceiling Structural System Spanning								
Roofs								
Doors (including frame and hardware): a) Enclosed vertical exitway								
b) Enclosed horizontal exitway								
c) Others								

one of the techniques described below. If the fire performance of the *existing building* element(s) is equal to or greater than that required, the materials and assemblies may remain. If the fire performance is less than required, then corrective measures must be taken.

The most common methods of upgrading the level of protection are to either remove and replace the *existing building* element(s) or to *repair* and upgrade the existing materials and assemblies. Other fire protection measures, such as automatic sprinklers or detection and alarm systems, also could be considered, though they are beyond the scope of this guideline. If the upgraded protection is still less than that required or deemed to be acceptable, additional corrective measures must be taken. This process must continue until an acceptable level of performance is obtained.

2.2
FIRE RESISTANCE OF EXISTING BUILDING ELEMENTS

The fire resistance of the *existing building* elements can be estimated from the tables and histograms contained in the Appendix. The Appendix is organized first by type of building element: walls, columns, floor/ceiling assemblies, beams, and doors. Within each building element, the tables are organized by type of construction (e.g., masonry, metal, wood frame), and then further divided by minimum dimensions or thickness of the building element.

A histogram precedes every table that has 10 or more entries. The X-axis measures fire resistance in hours; the Y-axis shows the number of entries in that table having a given level of fire resistance. The histograms also contain the location of each entry within that table for easy cross-referencing.

The histograms, because they are keyed to the tables, can speed the preliminary investigation. For example, Table 1.3.2, *Wood Frame Walls 4" to Less Than 6" Thick*, contains 96 entries. Rather than study each table entry, the histogram shows that every wall assembly listed in that table has a fire resistance of less than 2 hours. If the building code required the wall to have 2 hours fire resistance, the designer, with a minimum of effort, is made aware of a problem that requires closer study.

Suppose the code had only required a wall of 1 hour fire resistance. The histogram shows far fewer complying elements (19) than noncomplying ones (77). If the existing assembly is not one of the 19 complying entries, there is a strong possibility the existing assembly is deficient. The histograms can also be used in the converse situation. If the existing assembly is not one of the smaller number of entries with a lower than required fire resistance, there is a strong possibility the existing assembly will be acceptable.

At some point, the *existing building* component or assembly must be located within the tables. Otherwise, the fire resistance must be determined through one of the other techniques presented in the guideline. Locating the building component in the Appendix Tables not only guarantees the accuracy of the fire-resistance rating, but also provides a source of documentation for the building official.

2.3
EFFECTS OF PENETRATIONS IN FIRE-RESISTANT ASSEMBLIES

There are often many features in existing walls or floor/ceiling assemblies which were not included in the original certification or fire testing. The most common examples are pipes and utility wires passed through holes poked through an assembly. During the life of the building, many penetrations are added, and by the time a building is ready for rehabilitation it is not sufficient to just consider the fire resistance of the assembly as originally constructed. It is necessary to consider all penetrations and their relative impact upon fire performance. For instance, the fire resistance of the corridor wall may be less important than the effect of plain glass doors or transoms. In fact, doors are the most important single class of penetrations.

A fully developed fire generates substantial quantities of heat and excess gaseous fuel capable of penetrating any holes which might be present in the walls or ceiling of the fire compartment. In general, this leads to a severe degradation of the fire resistance of those building elements and to a greater potential for fire spread. This is particularly applicable to penetrations located high in a compartment where the positive pressure of the fire can force the unburned gases through the penetration.

Penetrations in a floor/ceiling assembly will generally completely negate the barrier qualities of the assembly and will lead to rapid spread of fire to the space above. It will not be a problem, however, if the penetrations are filled with noncombustible materials strongly fastened to the structure. The upper half of walls are similar to the floor/ceiling assembly in that a positive pressure can reasonably be expected in the top of the room, and this will push hot and/or burning gases through the penetration unless it is completely sealed.

Building codes require doors installed in fire resistive walls to resist the passage of fire for a specified period of time. If the door to a fully involved room is not closed, a large plume of fire will typically escape through the doorway, preventing anyone from using the space outside the door while allowing the fire to spread. This is why door closers are so important. Glass in doors and transoms can be expected to rapidly shatter unless constructed of listed or approved wire glass in a steel frame. As with other building elements, penetrations or nonrated portions of doors and transoms must be upgraded or otherwise protected.

Table 5.1 in Section V of the Appendix contains 41 entries of doors mounted in sound tight-fitting frames. Part 3.4 below outlines one procedure for evaluating and possibly upgrading existing doors.

3
FINAL EVALUATION AND DESIGN SOLUTION

The final evaluation begins after the rehabilitation project has reached the final design stage and the choice is made to keep certain archaic materials and assemblies in the rehabilitated building. The final evaluation process is essentially a more refined and detailed version of the preliminary evaluation. The specific fire resistance and flame spread requirements are determined for the project. This may involve local building and fire officials reviewing the preliminary evaluation as depicted in Figures 1 and 2 and the field drawings and notes. When necessary, provisions must be made to upgrade *existing building* elements to provide the required level of fire performance.

There are several approaches to design solutions that can make possible the continued use of archaic materials and assemblies in the rehabilitated structure. The simplest case occurs when the materials and assembly in question are found within the Appendix Tables and the fire performance properties satisfy code requirements. Other approaches must be used, though, if the assembly cannot be found within the Appendix or the fire performance needs to be upgraded. These approaches have been grouped into two classes: experimental and theoretical.

3.1
THE EXPERIMENTAL APPROACH

If a material or assembly found in a building is not listed in the Appendix Tables, there are several other ways to evaluate fire performance. One approach is to conduct the appropriate fire test(s) and thereby determine the fire-related properties directly. There are a number of laboratories in the United States which routinely conduct the various fire tests. A current list can be obtained by writing the Center for Fire Research, National Bureau of Standards, Washington, D.C. 20234.

The contract with any of these testing laboratories should require their observation of specimen preparation as well as the testing of the specimen. A complete description of where and how the specimen was obtained from the building, the transportation of the specimen, and its preparation for testing should be noted in detail so that the building official can be satisfied that the fire test is representative of the actual use.

The test report should describe the fire test procedure and the response of the material or assembly. The laboratory usually submits a cover letter with the report to describe the provisions of the fire test that were satisfied by the material or assembly under investigation. A building official will generally require this cover letter, but will also read the report to confirm that the material or assembly complies with the code requirements. Local code officials should be involved in all phases of the testing process.

The experimental approach can be costly and time consuming because specimens must be taken from the building and transported to the testing laboratory. When a load bearing assembly has continuous reinforcement, the test specimen must be removed from the building, transported, and tested in one piece. However, when the fire performance cannot be determined by other means, there may be no alternative to a full-scale test.

A "nonstandard" small-scale test can be used in special cases. Sample sizes need only be 10–25 square feet (0.93–2.3 m^2), while full-scale tests require test samples of either 100 or 180 square feet (9.3 or 17 m^2) in size. This small-scale test is best suited for testing nonload-bearing assemblies against thermal transmission only.

3.2
THE THEORETICAL APPROACH

There will be instances when materials and assemblies in a building undergoing rehabilitation cannot be found in the Appendix Tables. Even where test results are available for more or less similar construction, the proper classification may not be immediately apparent. Variations in dimensions, loading conditions, materials, or workmanship may markedly affect the performance of the individual building elements, and the extent of such a possible effect cannot be evaluated from the tables.

Theoretical methods being developed offer an alternative to the full-scale fire tests discussed above. For example, Section 4302(b) of the 1979 edition of the *Uniform Building Code* specifically allows an engineering design for fire resistance in lieu of conducting full-scale tests. These techniques draw upon computer simulation and mathematical modeling, thermodynamics, heat-flow analysis, and materials science to predict the fire performance of building materials and assemblies.

One theoretical method, known as the "Ten Rules of Fire Endurance Ratings," was published by T. Z. Harmathy in the May, 1965 edition of *Fire Technology*. (35) Harmathy's Rules provide a foundation for extending the data within the Appendix Tables to analyze or upgrade current as well as archaic building materials or assemblies.

HARMATHY'S TEN RULES

Rule 1: The "thermal"[1] fire endurance of a construction consisting of a number of parallel layers is greater than the sum of the "thermal" fire endurances characteristic of the individual layers when exposed separately to fire.

The minimum performance of an untested assembly can be estimated if the fire endurance of the individual components is known. Though the exact rating of the assembly cannot be stated, the endurance of the assembly is greater than the sum of the endurance of the components.

When a building assembly or component is found to be deficient, the fire endurance can be upgraded by providing a protective membrane. This membrane could be a new layer of brick, plaster, or drywall. The fire endurance of this membrane is called the "finish rating." Appendix Tables 1.5.1 and 1.5.2 contain the finish ratings for the most commonly employed materials. (See also the notes to Rule 2).

The test criteria for the finish rating is the same as for the thermal fire endurance of the total assembly: average temperature increases of 250°F (121°C) above ambient or 325°F (163°C) above ambient at any one place with the membrane

---

1. The "thermal" fire endurance is the time at which the average temperature on the unexposed side of a construction exceeds its initial value by 250° when the other side is exposed to the "standard" fire specified by ASTM Test Method E-19.

being exposed to the fire. The temperature is measured at the interface of the assembly and the protective membrane.

Rule 2: The fire endurance of a construction does not decrease with the addition of further layers.

Harmathy notes that this rule is a consequence of the previous rule. Its validity follows from the fact that the additional layers increase both the resistance to heat flow and the heat capacity of the construction. This, in turn, reduces the rate of temperature rise at the unexposed surface.

This rule is not just restricted to "thermal" performance but affects the other fire test criteria: direct flame passage, cotton waste ignition, and load bearing performance. This means that certain restrictions must be imposed on the materials to be added and on the loading conditions. One restriction is that a new layer, if applied to the exposed surface, must not produce additional thermal stresses in the construction, i.e., its thermal expansion characteristics must be similar to those of the adjacent layer. Each new layer must also be capable of contributing enough additional strength to the assembly to sustain the added dead load. If this requirement is not fulfilled, the allowable live load must be reduced by an amount equal to the weight of the new layer. Because of these limitations, this rule should not be applied without careful consideration.

Particular care must be taken if the material added is a good thermal insulator. Properly located, the added insulation could improve the "thermal" performance of the assembly. Improperly located, the insulation could block necessary thermal transmission through the assembly, thereby subjecting the structural elements to greater temperatures for longer periods of time, and could cause premature structural failure of the supporting members.

Rule 3: The fire endurance of constructions containing continuous air gaps or cavities is greater than the fire endurance of similar constructions of the same weight, but containing no air gaps or cavities.

By providing for voids in a construction, additional resistances are produced in the path of heat flow. Numerical heat flow analyses indicate that a 10 to 15 percent increase in fire endurance can be achieved by creating an air gap at the midplane of a brick wall. Since the gross volume is also increased by the presence of voids, the air gaps and cavities have a beneficial effect on stability as well. However, constructions containing combustible materials within an air gap may be regarded as exceptions to this rule because of the possible development of burning in the gap.

There are numerous examples of this rule in the tables. For instance:

Table 1.1.4; Item W-8-M-82: Cored concrete masonry, nominal 8 inch thick wall with one unit in wall thickness and with 62 percent minimum of solid material in each unit, load bearing (80 PSI). Fire endurance: $2^1/_2$ hours.

Table 1.1.5; Item W-10-M-11: Cored concrete masonry, nominal 10 inch thick wall with two units in wall thickness and a 2-inch (51 mm) air space, load bearing (80 PSI). The units are essentially the same as item W-8-M-82. Fire endurance: $3^1/_2$ hours.

These walls show 1 hour greater fire endurance by the addition of the 2-inch (51 mm) air space.

Rule 4: The farther an air gap or cavity is located from the exposed surface, the more beneficial is its effect on the fire endurance.

Radiation dominates the heat transfer across an air gap or cavity, and it is markedly higher where the temperature is higher.

The air gap or cavity is thus a poor insulator if it is located in a region which attains high temperatures during fire exposure.

Some of the clay tile designs take advantage of these factors. The double cell design, for instance, ensures that there is a cavity near the unexposed face. Some floor/ceiling assemblies have air gaps or cavities near the top surface and these enhance their thermal performance.

Rule 5: The fire endurance of a construction cannot be increased by increasing the thickness of a completely enclosed air layer.

Harmathy notes that there is evidence that if the thickness of the air layer is larger than about $^1/_2$ inch (12.7 mm), the heat transfer through the air layer depends only on the temperature of the bounding surfaces, and is practically independent of the distance between them. This rule is not applicable if the air layer is not completely enclosed, i.e., if there is a possibility of fresh air entering the gap at an appreciable rate.

Rule 6: Layers of materials of low thermal conductivity are better utilized on that side of the construction on which fire is more likely to happen.

As in Rule 4, the reason lies in the heat transfer process, though the conductivity of the solid is much less dependent on the ambient temperature of the materials. The low thermal conductor creates a substantial temperature differential to be established across its thickness under transient heat flow conditions. This rule may not be applicable to materials undergoing physico-chemical changes accompanied by significant heat absorption or heat evolution.

Rule 7: The fire endurance of asymmetrical constructions depends on the direction of heat flow.

This rule is a consequence of Rules 4 and 6, as well as other factors. This rule is useful in determining the relative protection of corridors and walls enclosing an exit stairway from the surrounding spaces. In addition, there are often situations where a fire is more likely, or potentially more severe, from one side or the other.

Rule 8: The presence of moisture, if it does not result in explosive spalling, increases the fire endurance.

The flow of heat into an assembly is greatly hindered by the release and evaporation of the moisture found within cementitious materials such as gypsum, Portland cement, or magnesium oxychloride. Harmathy has shown that the gain in fire endurance may be as high as 8 percent for each percent (by volume) of moisture in the construction. It is the moisture chemically bound within the construction material at the time of manufacture or processing that leads to increased fire endurance. There is no direct relationship between the relative humidity of the air in the pores of the material and the increase in fire endurance.

Under certain conditions there may be explosive spalling of low permeability cementitious materials such as dense concrete. In general, one can assume that extremely old concrete has developed enough minor cracking that this factor should not be significant.

Rule 9: Load-supporting elements, such as beams, girders and joists, yield higher fire endurances when subjected to fire endurance tests as parts of floor, roof, or ceiling assemblies than they would when tested separately.

One of the fire endurance test criteria is the ability of a load-supporting element to carry its design load. The element will be deemed to have failed when the load can no longer be supported.

Failure usually results for two reasons. Some materials, particularly steel and other metals, lose much of their structural strength at elevated temperatures. Physical deflection of the supporting element, due to decreased strength or thermal expansion, causes a redistribution of the load forces and stresses throughout the element. Structural failure often results because the supporting element is not designed to carry the redistributed load.

Roof, floor, and ceiling assemblies have primary (e.g., beams) and secondary (e.g., floor joists) structural members. Since the primary load-supporting elements span the largest distances, their deflection becomes significant at a stage when the strength of the secondary members (including the roof or floor surface) is hardly affected by the heat. As the secondary members follow the deflection of the primary load-supporting element, an increasingly larger portion of the load is transferred to the secondary members.

When load-supporting elements are tested separately, the imposed load is constant and equal to the design load throughout the test. By definition, no distribution of the load is possible because the element is being tested by itself. Without any other structural members to which the load could be transferred, the individual elements cannot yield a higher fire endurance than they do when tested as parts of a floor, roof or ceiling assembly.

Rule 10: The load-supporting elements (beams, girders, joists, etc.) of a floor, roof, or ceiling assembly can be replaced by such other load-supporting elements which, when tested separately, yielded fire endurances not less than that of the assembly.

This rule depends on Rule 9 for its validity. A beam or girder, if capable of yielding a certain performance when tested separately, will yield an equally good or better performance when it forms a part of a floor, roof, or ceiling assembly. It must be emphasized that the supporting element of one assembly must not be replaced by the supporting element of another assembly if the performance of this latter element is not known from a separate (beam) test. Because of the load-reducing effect of the secondary elements that results from a test performed on an assembly, the performance of the supporting element alone cannot be evaluated by simple arithmetic. This rule also indicates the advantage of performing separate fire tests on primary load-supporting elements.

ILLUSTRATION OF HARMATHY'S RULES

Harmathy provided one schematic figure which illustrated his Rules.[1] It should be useful as a quick reference to assist in applying his Rules.

EXAMPLE APPLICATION OF HARMATHY'S RULES

The following examples, based in whole or in part upon those presented in Harmathy's paper (35), show how the Rules can be applied to practical cases.

Example 1

Problem

A contractor would like to keep a partition which consists of a $3^3/_4$ inch (95 mm) thick layer of red clay brick, a $1^1/_4$ inch (32 mm) thick layer of plywood, and a $^3/_8$ inch (9.5 mm) thick layer of gypsum wallboard, at a location where 2-hour fire endurance is required. Is this assembly capable of providing a 2-hour protection?

Solution

(1) This partition does not appear in the Appendix Tables.

(2) Bricks of this thickness yield fire endurances of approximately 75 minutes (Table 1.1.2, Item W-4-M-2).

(3) The $1^1/_4$ inch (32 mm) thick plywood has a finish rating of 30 minutes.

(4) The $^3/_8$ inch (9.5 mm) gypsum wallboard has a finish rating of 10 minutes.

(5) Using the recommended values from the tables and applying Rule 1, the fire endurance (FI) of the assembly is larger than the sum of the individual layers, or

FI > 75 + 30 + 10 = 115 minutes

Discussion

This example illustrates how the Appendix Tables can be utilized to determine the fire resistance of assemblies not explicitly listed.

Example 2

Problem

(1) A number of buildings to be rehabilitated have the same type of roof slab which is supported with different structural elements.

(2) The designer and contractor would like to determine whether or not this roof slab is capable of yielding a 2-hour fire endurance. According to a rigorous interpretation of ASTM E119, however, only the roof assembly, including the roof slab as well as the cover and the supporting elements, can be subjected to a fire test.

---

1. Reproduced from the May 1065 *Fire Technology* (Vol. 1, No. 2). Copyright National Fire Protection Association, Boston. Reproduced by permission.

Therefore, a fire endurance classification cannot be issued for the slabs separately.

(3) The designer and contractor believe this slab will yield a 2-hour fire endurance even without the cover, and any beam of at least 2-hour fire endurance will provide satisfactory support. Is it possible to obtain a classification for the slab separately?

Solution

(1) The answer to the question is yes.
(2) According to Rule 10 it is not contrary to common sense to test and classify roofs and supporting elements separately. Furthermore, according to Rule 2, if the roof slabs actually yield a 2-hour fire endurance, the endurance of an assembly, including the slabs, cannot be less than 2 hours.
(3) The recommended procedure would be to review the tables to see if the slab appears as part of any tested roof or floor/ceiling assembly. The supporting system can be regarded as separate from the slab specimen, and the fire endurance of the assembly listed in the table is at least the fire endurance of the slab. There would have to be an adjustment for the weight of the roof cover in the allowable load if the test specimen did not contain a cover.
(4) The supporting structure or element would have to have at least a 2-hour fire endurance when tested separately.

Discussion

If the tables did not include tests on assemblies which contained the slab, one procedure would be to assemble the roof slabs on any convenient supporting system (not regarded as part of the specimen) and to subject them to a load which, besides the usually required superimposed load, includes some allowances for the weight of the cover.

Example 3

Problem

A steel-joisted floor and ceiling assembly is known to have yielded a fire endurance of 1 hour and 35 minutes. At a certain location, a 2-hour endurance is required. What is the most economical way of increasing the fire endurance by at least 25 minutes?

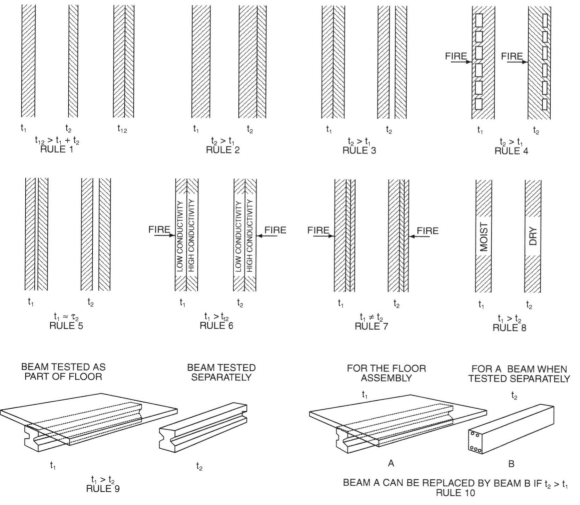

Diagrammatic illustration of 10 rules.
t = fire endurance

Solution

(1) The most effective technique would be to increase the ceiling plaster thickness. Existing coats of paint would have to be removed and the surface properly prepared before the new plaster could be applied. Other materials (e.g., gypsum wallboard) could also be considered.

(2) There may be other techniques based on other principles, but an examination of the drawings would be necessary.

Discussion

(1) The additional plaster has at least three effects:

   a) The layer of plaster is increased and thus there is a gain of fire endurance (Rule 1).

   b) There is a gain due to shifting the air gap farther from the exposed surface (Rule 4).

   c) There is more moisture in the path of heat flow to the structural elements (Rules 7 and 8).

(2) The increase in fire endurance would be at least as large as that of the finish rating for the added thickness of plaster. The combined effects in (1) above would further increase this by a factor of 2 or more, depending upon the geometry of the assembly.

Example 4

Problem

The fire endurance of item W-10-M-1 in Table 1.1.5 is 4 hours. This wall consists of two $3^3/_4$ inch (95 mm) thick layers of structural tiles separated by a 2-inch (51 mm) air gap and $^3/_4$ inch (19 mm) Portland cement plaster or stucco on both sides. If the actual wall in the building is identical to item W-10-M-1 except that it has a 4-inch (102 mm) air gap, can the fire endurance be estimated at 5 hours?

Solution

The answer to the question is no for the reasons contained in Rule 5.

Example 5

Problem

In order to increase the insulating value of its precast roof slabs, a company has decided to use two layers of different concretes. The lower layer of the slabs, where the strength of the concrete is immaterial (all the tensile load is carried by the steel reinforcement), would be made with a concrete of low strength but good insulating value. The upper layer, where the concrete is supposed to carry the compressive load, would remain the original high strength, high thermal conductivity concrete. How will the fire endurance of the slabs be affected by the change?

Solution

The effect on the thermal fire endurance is beneficial:

(1) The total resistance to heat flow of the new slabs has been increased due to the replacement of a layer of high thermal conductivity by one of low conductivity.

(2) The layer of low conductivity is on the side more likely to be exposed to fire, where it is more effectively utilized according to Rule 6. The layer of low thermal conductivity also provides better protection for the steel reinforcement, thereby extending the time before reaching the temperature at which the creep of steel becomes significant.

3.3 "THICKNESS DESIGN" STRATEGY

The "thickness design" strategy is based upon Harmathy's Rules 1 and 2. This design approach can be used when the construction materials have been identified and measured, but the specific assembly cannot be located within the tables. The tables should be surveyed again for thinner walls of like material and construction detail that have yielded the desired or greater fire endurance. If such an assembly can be found, then the thicker walls in the building have more than enough fire resistance. The thickness of the walls thus becomes the principal concern.

This approach can also be used for floor/ceiling assemblies, except that the thickness of the cover[1] and the slab become the central concern. The fire resistance of the untested assembly will be at least the fire resistance of an assembly listed in the table having a similar design but with less cover and/or thinner slabs. For other structural elements (e.g., beams and columns), the element listed in the table must also be of a similar design but with less cover thickness.

3.4 EVALUATION OF DOORS

A separate section on doors has been included because the process for evaluation presented below differs from those suggested previously for other building elements. The impact of unprotected openings or penetrations in fire resistant assemblies has been detailed in Part 2.3 above. It is sufficient to note here that openings left unprotected will likely lead to failure of the barrier under actual fire conditions.

For other types of building elements (e.g., beams, columns), the Appendix Tables can be used to establish a minimum level of fire performance. The benefit to rehabilitation is that the need for a full-scale fire test is then eliminated. For doors, however, this cannot be done. The data contained in Appendix Table 5.1, Resistance of Doors to Fire Exposure, can only provide guidance as to whether a successful fire test is even feasible.

For example, a door required to have 1 hour fire resistance is noted in the tables as providing only 5 minutes. The likelihood of achieving the required 1 hour, even if the door is upgraded, is remote. The ultimate need for replacement of the doors is reasonably clear, and the expense and time needed for testing can be saved. However, if the performance documented in the table is near or in excess of what is being required, then a fire test should be conducted. The test docu-

---

1. Cover: the protective layer or membrane of material which slows the flow of heat to the structural elements.

mentation can then be used as evidence of compliance with the required level of performance.

The table entries cannot be used as the sole proof of performance of the door in question because there are too many unknown variables which could measurably affect fire performance. The wood may have dried over the years; coats of flammable varnish could have been added. Minor deviations in the internal construction of a door can result in significant differences in performance. Methods of securing inserts in panel doors can vary. The major non-destructive method of analysis, an x-ray, often cannot provide the necessary detail. It is for these, and similar reasons, that a fire test is still felt to be necessary.

It is often possible to upgrade the fire performance of an existing door. Sometimes, "as is" and modified doors are evaluated in a single series of tests when failure of the unmodified door is expected. Because doors upgraded after an initial failure must be tested again, there is a potential savings of time and money.

The most common problems encountered are plain glass, panel inserts of insufficient thickness, and improper fit of a door in its frame. The latter problem can be significant because a fire can develop a substantial positive pressure, and the fire will work its way through otherwise innocent-looking gaps between door and frame.

One approach to solving these problems is as follows. The plain glass is replaced with approved or listed wire glass in a steel frame. The panel inserts can be upgraded by adding an additional layer of material. Gypsum wallboard is often used for this purpose. Intumescent paint applied to the edges of the door and frame will expand when exposed to fire, forming an effective seal around the edges. This seal, coupled with the generally even thermal expansion of a wood door in a wood frame, can prevent the passage of flames and other fire gases. Figure 3 below illustrates these solutions.

Because the interior construction of a door cannot be determined by a visual inspection, there is no absolute guarantee that the remaining doors are identical to the one(s) removed from the building and tested. But the same is true for doors constructed today, and reason and judgment must be applied. Doors that appear identical upon visual inspection can be weighed. If the weights are reasonably close, the doors can be assumed to be identical and therefore provide the same level of fire performance. Another approach is to fire test more than one door or to dismantle doors selected at random to see if they had been constructed in the same manner. Original building plans showing door details or other records showing that doors were purchased at one time or obtained from a single supplier can also be evidence of similar construction.

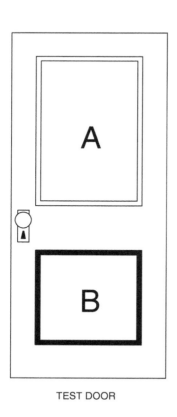
TEST DOOR

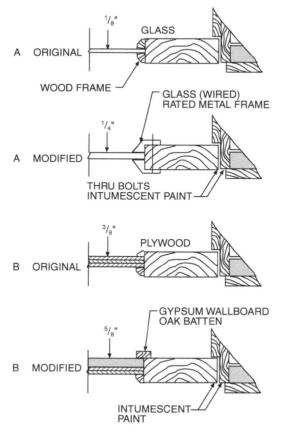

For SI: 1 inch = 25.4 mm.

**FIGURE 3
MODIFICATION DETAILS**

More often though, it is what is visible to the eye that is most significant. The investigator should carefully check the condition and fit of the door and frame, and for frames out of plumb or separating from the wall. Door closers, latches, and hinges must be examined to see that they function properly and are tightly secured. If these are in order and the door and frame have passed a full-scale test, there can be a reasonable basis for allowing the existing doors to remain.

4
SUMMARY

This section summarizes the various approaches and design solutions discussed in the preceding sections of the guideline. The term "structural system" includes: frames, beams, columns, and other structural elements. "Cover" is a protective layer(s) of materials or membrane which slows the flow of heat to the structural elements. It cannot be stressed too strongly that the fire endurance of actual building elements can be greatly reduced or totally negated by removing part of the cover to allow pipes, ducts, or conduits to pass through the element. This must be repaired in the rehabilitation process.

The following approaches shall be considered equivalent.

4.1 The fire resistance of a building element can be established from the Appendix Tables. This is subject to the following limitations:

The building element in the rehabilitated building shall be constructed of the same materials with the same nominal dimensions as stated in the tables.

All penetrations in the building element or its cover for services such as electricity, plumbing, and HVAC shall be packed with noncombustible cementitious materials and so fixed that the packing material will not fall out when it loses its water of hydration.

The effects of age and wear and tear shall be repaired so that the building element is sound and the original thickness of all components, particularly covers and floor slabs, is maintained.

This approach essentially follows the approach taken by model building codes. The assembly must appear in a table either published in or accepted by the code for a given fire-resistance rating to be recognized and accepted.

4.2 The fire resistance of a building element which does not explicitly appear in the Appendix Tables can be established if one or more elements of same design but different dimensions have been listed in the tables. For walls, the existing element must be thicker than the one listed. For floor/ceiling assemblies, the assembly listed in the table must have the same or less cover and the same or thinner slab constructed of the same material as the actual floor/ceiling assembly. For other structural elements, the element listed in the table must be of a similar design but with less cover thickness. The fire resistance in all instances shall be the fire resistance recommended in the table. This is subject to the following limitations:

The actual element in the rehabilitated building shall be constructed of the same materials as listed in the table. Only the following dimensions may vary from those specified: for walls, the overall thickness must exceed that specified in the table; for floor/ceiling assemblies, the thickness of the cover and the slab must be greater than, or equal to, that specified in the table; for other structural elements, the thickness of the cover must be greater than that specified in the table.

All penetrations in the building element or its cover for services such as electricity, plumbing, or HVAC shall be packed with noncombustible cementitious materials and so fixed that the packing material will not fall out when it loses its water of hydration.

The effects of age and wear and tear shall be repaired so that the building element is sound and the original thickness of all components, particularly covers and floor slabs, is maintained.

This approach is an application of the "thickness design" concept presented in Part 3.3 of the guideline. There should be many instances when a thicker building element was utilized than the one listed in the Appendix Tables. This guideline recognizes the inherent superiority of a thicker design. Note: "thickness design" for floor/ceiling assemblies and structural elements refers to cover and slab thickness rather than total thickness.

The "thickness design" concept is essentially a special case of Harmathy's Rules (specifically Rules 1 and 2). It should be recognized that the only source of data is the Appendix Tables. If other data are used, it must be in connection with the approach below.

4.3 The fire resistance of building elements can be established by applying Harmathy's Ten Rules of Fire Resistance Ratings as set forth in Part 3.2 of the guideline. This is subject to the following limitations:

The data from the tables can be utilized subject to the limitations in 4.2 above.

Test reports from recognized journals or published papers can be used to support data utilized in applying Harmathy's Rules.

Calculations utilizing recognized and well established computational techniques can be used in applying Harmathy's Rules. These include, but are not limited to, analysis of heat flow, mechanical properties, deflections, and load bearing capacity.

APPENDIX

INTRODUCTION

The fire-resistance tables that follow are a part of Resource A and provide a tabular form of assigning fire-resistance ratings to various archaic building elements and assemblies.

These tables for archaic materials and assemblies do for archaic materials what Tables 721.1(1) through 721.1(3) of the *International Building Code®* do for more modern building elements and assemblies. The fire-resistance tables of Resource A should be used as described in the "Purpose and Procedure" that follows the table of contents for these tables.

RESOURCE A TABLE OF CONTENTS

Purpose and Procedure   RESOURCE A-17

Section I—Walls

1.1.1	Masonry	0 in. to less than 4 in. thick	RESOURCE A-18
1.1.2	Masonry	4 in. to less than 6 in. thick	RESOURCE A-21
1.1.3	Masonry	6 in. to less than 8 in. thick	RESOURCE A-28
1.1.4	Masonry	8 in. to less than 10 in. thick	RESOURCE A-33
1.1.5	Masonry	10 in. to less than 12 in. thick	RESOURCE A-41
1.1.6	Masonry	12 in. to less than 14 in. thick	RESOURCE A-45
1.1.7	Masonry	14 in. or more thick	RESOURCE A-51
1.2.1	Metal Frame	0 in. to less than 4 in. thick	RESOURCE A-54
1.2.2	Metal Frame	4 in. to less than 6 in. thick	RESOURCE A-58
1.2.3	Metal Frame	6 in. to less than 8 in. thick	RESOURCE A-60
1.2.4	Metal Frame	8 in. to less than 10 in. thick	RESOURCE A-61
1.3.1	Wood Frame	0 in. to less than 4 in. thick	RESOURCE A-62
1.3.2	Wood Frame	4 in. to less than 6 in. thick	RESOURCE A-63
1.3.3	Wood Frame	6 in. to less than 8 in. thick	RESOURCE A-71
1.4.1	Miscellaneous Materials	0 in. to less than 4 in. thick	RESOURCE A-71
1.4.2	Miscellaneous Materials	4 in. to less than 6 in. thick	RESOURCE A-72
1.5.1	Finish Ratings—Inorganic Materials	Thickness	RESOURCE A-73
1.5.2	Finish Ratings—Organic Materials	Thickness	RESOURCE A-74

Section II—Columns

2.1.1	Reinforced Concrete	Minimum Dimension 0 in. to less than 6 in.	RESOURCE A-75
2.1.2	Reinforced Concrete	Minimum Dimension 10 in. to less than 12 in.	RESOURCE A-76
2.1.3	Reinforced Concrete	Minimum Dimension 12 in. to less than 14 in.	RESOURCE A-79
2.1.4	Reinforced Concrete	Minimum Dimension 14 in. to less than 16 in.	RESOURCE A-80
2.1.5	Reinforced Concrete	Minimum Dimension 16 in. to less than 18 in.	RESOURCE A-81
2.1.6	Reinforced Concrete	Minimum Dimension 18 in. to less than 20 in.	RESOURCE A-83
2.1.7	Reinforced Concrete	Minimum Dimension 20 in. to less than 22 in.	RESOURCE A-84

RESOURCE A—GUIDELINES ON FIRE RATINGS OF ARCHAIC MATERIALS AND ASSEMBLIES

2.1.8	Hexagonal Reinforced Concrete	Minimum Dimension 12 in. to less than 14 in.	RESOURCE A-85
2.1.9	Hexagonal Reinforced Concrete	Minimum Dimension 14 in. to less than 16 in.	RESOURCE A-86
2.1.10	Hexagonal Reinforced Concrete	Diameter—16 in. to less than 18 in.	RESOURCE A-86
2.1.11	Hexagonal Reinforced Concrete	Diameter—20 in. to less than 22 in.	RESOURCE A-86
2.2	Round Cast Iron Columns	Minimum Dimension	RESOURCE A-87
2.3	Steel—Gypsum Encasements	Minimum Area of Solid Material	RESOURCE A-88
2.4	Timber	Minimum Dimension	RESOURCE A-89
2.5.1.1	Steel/Concrete Encasements	Minimum Dimension less than 6 in.	RESOURCE A-89
2.5.1.2	Steel/Concrete Encasements	Minimum Dimension 6 in. to less than 8 in.	RESOURCE A-90
2.5.1.3	Steel/Concrete Encasements	Minimum Dimension 8 in. to less than 10 in.	RESOURCE A-91
2.5.1.4	Steel/Concrete Encasements	Minimum Dimension 10 in. to less than 12 in.	RESOURCE A-93
2.5.1.5	Steel/Concrete Encasements	Minimum Dimension 12 in. to less than 14 in.	RESOURCE A-97
2.5.1.6	Steel/Concrete Encasements	Minimum Dimension 14 in. to less than 16 in.	RESOURCE A-99
2.5.1.7	Steel/Concrete Encasements	Minimum Dimension 16 in. to less than 18 in.	RESOURCE A-101
2.5.2.1	Steel/Brick and Block Encasements	Minimum Dimension 10 in. to less than 12 in.	RESOURCE A-101
2.5.2.2	Steel/Brick and Block Encasements	Minimum Dimension 12 in. to less than 14 in.	RESOURCE A-102
2.5.2.3	Steel/Brick and Block Encasements	Minimum Dimension 14 in. to less than 16 in.	RESOURCE A-102
2.5.3.1	Steel/Plaster Encasements	Minimum Dimension 6 in. to less than 8 in.	RESOURCE A-103
2.5.3.2	Steel/Plaster Encasements	Minimum Dimension 8 in. to less than 10 in.	RESOURCE A-103
2.5.4.1	Steel/Miscellaneous Encasements	Minimum Dimension 6 in. to less than 8 in.	RESOURCE A-103
2.5.4.2	Steel/Miscellaneous Encasements	Minimum Dimension 8 in. to less than 10 in.	RESOURCE A-104
2.5.4.3	Steel/Miscellaneous Encasements	Minimum Dimension 10 in. to less than 12 in.	RESOURCE A-104
2.5.4.4	Steel/Miscellaneous Encasements	Minimum Dimension 12 in. to less than 14 in.	RESOURCE A-104
Section III—Floor/Ceiling Assemblies			
3.1	Reinforced Concrete	Assembly Thickness	RESOURCE A-105
3.2	Steel Structural Elements	Membrane Thickness	RESOURCE A-111
3.3	Wood Joist	Membrane Thickness	RESOURCE A-117
3.4	Hollow Clay Tile with Reinforced Concrete	Membrane Thickness	RESOURCE A-121
Section IV—Beams			
4.1.1	Reinforced Concrete	Depth—10 in. to less than 12 in.	RESOURCE A-124
4.1.2	Reinforced Concrete	Depth—12 in. to less than 14 in.	RESOURCE A-127
4.1.3	Reinforced Concrete	Depth—14 in. to less than 16 in.	RESOURCE A-129
4.2.1	Reinforced Concrete/Unprotected	Depth—10 in. to less than 12 in.	RESOURCE A-130
4.2.2	Steel/Concrete Protection	Depth—10 in. to less than 12 in.	RESOURCE A-130
Section V—Doors			
5.1	Resistance of Doors to Fire Exposure	Thickness	RESOURCE A-131

PURPOSE AND PROCEDURE

The tables and histograms which follow are to be used only within the analytical framework detailed in the main body of this guideline.

Histograms precede any table with 10 or more entries. The use and interpretation of these histograms is explained in Part 2 of the guideline. The tables are in a format similar to that found in the model building codes. The following example, taken from an entry in Table 1.1.2, best explains the table format.

1. Item Code: The item code consists of a four place series in the general form w-x-y-z in which each member of the series denotes the following:

    w = Type of building element (e.g., W=Walls; F=Floors, etc.)

    x = The building element thickness rounded down to the nearest 1-inch increment (e.g., $4^5/_8$ inches is rounded off to 4 inches)

    y = The general type of material from which the building element is constructed (e.g., M=Masonry; W=Wood, etc.)

    z = The item number of the particular building element in a given table

    The item code shown in the example W-4-M-50 denotes the following:

    W = Wall, as the building element

    4 = Wall thickness in the range of 4 inches (102 mm) to less than 5 inches (127 mm)

    M = Masonry construction

    50 = The 50th entry in Table 1.1.2

2. The specific name or heading of this column identifies the dimensions which, if varied, has the greatest impact on fire resistance. The critical dimension for walls, the example here, is thickness. It is different for other building elements (e.g., depth for beams; membrane thickness for some floor/ceiling assemblies). The table entry is the named dimension of the building element measured at the time of actual testing to within $\pm^1/_8$ inch (3.2 mm) tolerance. The thickness tabulated includes facings where facings are a part of the wall construction.

3. Construction Details: The construction details provide a brief description of the manner in which the building element was constructed.

4. Performance: This heading is subdivided into two columns. The column labeled "Load" will either list the load that the building element was subjected to during the fire test or it will contain a note number which will list the load and any other significant details. If the building element was not subjected to a load during the test, this column will contain "n/a," which means "not applicable."

    The second column under performance is labeled "Time" and denotes the actual fire endurance time observed in the fire test.

5. Reference Number: This heading is subdivided into three columns: Pre-BMS-92; BMS-92; and Post-BMS-92. The table entry under this column is the number in the Bibliography of the original source reference for the test data.

6. Notes: Notes are provided at the end of each table to allow a more detailed explanation of certain aspects of the test. In certain tables the notes given to this column have also been listed under the "Construction Details" and/or "Load" columns.

7. Rec Hours: This column lists the recommended fire endurance rating, in hours, of a building element. In some cases, the recommended fire endurance will be less than that listed under the "Time" column. In no case is the "Rec Hours" greater than given in the "Time" column.

ITEM CODE	THICKNESS	CONSTRUCTION DETAILS	PERFORMANCE		REFERENCE NUMBER			NOTES	REC. HOURS
			LOAD	TIME	PRE-BMS-92	BMS-92	POST-BMS-92		
W-4-M-50	$4^5/_8''$	Core: structural clay tile, See notes 12, 16, 21; Facings on unexposed side only, see note 18	N/A	25 min.		1		3, 4, 24	$^1/_3$

RESOURCE A—GUIDELINES ON FIRE RATINGS OF ARCHAIC MATERIALS AND ASSEMBLIES

SECTION I - WALLS

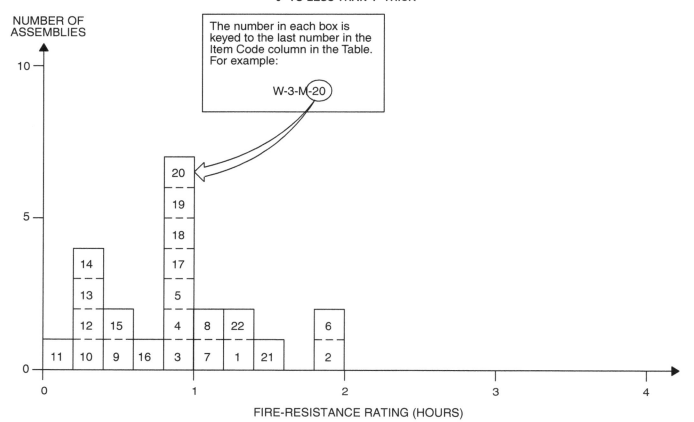

FIGURE 1.1.1
MASONRY WALLS
0″ TO LESS THAN 4″ THICK

TABLE 1.1.1
MASONRY WALLS
0″ TO LESS THAN 4″ THICK

ITEM CODE	THICKNESS	CONSTRUCTION DETAILS	PERFORMANCE		REFERENCE NUMBER			NOTES	REC. HOURS
			LOAD	TIME	PRE-BMS-92	BMS-92	POST-BMS-92		
W-2-M-1	$2\frac{1}{4}''$	Solid partition; $\frac{3}{4}''$ gypsum plank- 10′ ×1′6″; $\frac{3}{4}''$ plus gypsum plaster each side.	N/A	1 hr. 22 min.			7	1	$1\frac{1}{4}$
W-3-M-2	3″	Concrete block (18″ × 9″ × 3″) of fuel ash, Portland cement and plasticizer; cement/sand mortar.	N/A	2 hrs.			7	2, 3	2
W-2-M-3	2″	Solid gypsum block wall; No facings	N/A	1 hr.		1		4	1
W-3-M-4	3″	Solid gypsum blocks, laid in 1:3 sanded gypsum mortar.	N/A	1 hr.		1		4	1
W-3-M-5	3″	Magnesium oxysulfate wood fiber blocks; 2″ thick, laid in Portland cement-lime mortar; Facings: $\frac{1}{2}''$ of 1:3 sanded gypsum plaster on both sides.	N/A	1 hr.		1		4	1
W-3-M-6	3″	Magnesium oxysulfate bound wood fiber blocks; 3″ thick; laid in Portland cement-lime mortar; Facings: $\frac{1}{2}''$ of 1:3 sanded gypsum plaster on both sides.	N/A	2 hrs.		1		4	2

(continued)

TABLE 1.1.1—continued
MASONRY WALLS
0″ TO LESS THAN 4″ THICK

ITEM CODE	THICKNESS	CONSTRUCTION DETAILS	PERFORMANCE		REFERENCE NUMBER			NOTES	REC. HOURS
			LOAD	TIME	PRE-BMS-92	BMS-92	POST-BMS-92		
W-3-M-7	3″	Clay tile; Ohio fire clay; single cell thick; Face plaster: $^5/_8$″ (both sides) 1:3 sanded gypsum; Design "E," Construction "A."	N/A	1 hr. 6 min.	0		2	5, 6, 7, 11, 12, 39	1
W-3-M-8	3″	Clay tile; Illinois surface clay; single cell thick; Face plaster: $^5/_8$″ (both sides) 1:3 sanded gypsum; Design "A," Construction "E."	N/A	1 hr. 1 min			2	5, 8, 9, 11, 12, 39	1
W-3-M-9	3″	Clay tile; Illinois surface clay; single cell thick; No face plaster; Design "A," Construction "C."	N/A	25 min.			2	5, 10, 11, 12, 39	$^1/_3$
W-3-M-10	$3^7/_8$″	8″ × $4^7/_8$″ glass blocks; weight 4 lbs. each; Portland cement-lime mortar; horizontal mortar joints reinforced with metal lath.	N/A	15 min.		1		4	$^1/_4$
W-3-M-11	3″	Core: structural clay tile; see Notes 14, 18, 13; No facings.	N/A	10 min.		1		5, 11, 26	$^1/_6$
W-3-M-12	3″	Core: structural clay tile; see Notes 14, 19, 23; No facings.	N/A	20 min.		1		5, 11, 26	$^1/_3$
W-3-M-13	$3^5/_8$″	Core: structural clay tile; see Notes 14, 18, 23; Facings: unexposed side; see Note 20.	N/A	20 min.		1		5, 11, 26	$^1/_3$
W-3-M-14	$3^5/_8$″	Core: structural clay tile; see Notes 14, 19, 23; Facings: unexposed side only; see Note 20.	N/A	20 min.		1		5, 11, 26	$^1/_3$
W-3-M-15	$3^5/_8$″	Core: clay structural tile; see Notes 14, 18, 23; Facings: side exposed to fire; see Note 20.	N/A	30 min.		1		5, 11, 26	$^1/_2$
W-3-M-16	$3^5/_8$″	Core: clay structural tile; see Notes 14, 19, 23; Facings: side exposed to fire; see Note 20.	N/A	45 min.		1		5, 11, 26	$^3/_4$
W-2-M-17	2″	2″ thick solid gypsum blocks; see Note 27.	N/A	1 hr.		1		27	1
W-3-M-18	3″	Core: 3″ thick gypsum blocks 70% solid; see Note 2; No facings.	N/A	1 hr.		1		27	1
W-3-M-19	3″	Core: hollow concrete units; see Notes 29, 35, 36, 38; No facings.	N/A	1 hr.		1		27	1
W-3-M-20	3″	Core: hollow concrete units; see Notes 28, 35, 36, 37, 38; No facings.	N/A	1 hr.		1			1
W-3-M-21	$3^1/_2$″	Core: hollow concrete units; see Notes 28, 35, 36, 37, 38; Facings: one side; see Note 37.	N/A	$1^1/_2$ hrs.		1			$1^1/_2$
W-3-M-22	$3^1/_2$″	Core: hollow concrete units; see Notes 29, 35, 36, 38; Facings: one side, see Note 37.	N/A	$1^1/_4$ hrs.		1			$1^1/_4$

For SI: 1 inch = 25.4 mm, 1 pound per square inch = 0.00689 MPa, °C = [(°F) - 32]/1.8.

Notes:
1. Failure mode—flame thru.
2. Passed 2-hour fire test (Grade "C" fire res. - British).
3. Passed hose stream test.
4. Tested at NBS under ASA Spec. No. A2-1934. As nonload bearing partitions.
5. Tested at NBS under ASA Spec. No. 42-1934 (ASTM C19-33) except that hose stream testing where carried was run on test specimens exposed for full test duration, not for a reduced period as is contemporarily done.
6. Failure by thermal criteria—maximum temperature rise 325°F.
7. Hose stream failure.
8. Hose stream—pass.
9. Specimen removed prior to any failure occurring.
10. Failure mode—collapse.
11. For clay tile walls, unless the source or density of the clay can be positively identified or determined, it is suggested that the lowest hourly rating for the fire endurance of a clay tile partition of that thickness be followed. Identified sources of clay showing longer fire endurance can lead to longer time recommendations.

(continued)

RESOURCE A—GUIDELINES ON FIRE RATINGS OF ARCHAIC MATERIALS AND ASSEMBLIES

TABLE 1.1.1—continued
MASONRY WALLS
0" TO LESS THAN 4" THICK

12. See appendix for construction and design details for clay tile walls.
13. Load: 80 psi for gross wall area.
14. One cell in wall thickness.
15. Two cells in wall thickness.
16. Double shells plus one cell in wall thickness.
17. One cell in wall thickness, cells filled with broken tile, crushed stone, slag cinders or sand mixed with mortar.
18. Dense hard-burned clay or shale tile.
19. Medium-burned clay tile.
20. Not less than $5/8$ inch thickness of 1:3 sanded gypsum plaster.
21. Units of not less than 30 percent solid material.
22. Units of not less than 40 percent solid material.
23. Units of not less than 50 percent solid material.
24. Units of not less than 45 percent solid material.
25. Units of not less than 60 percent solid material.
26. All tiles laid in Portland cement-lime mortar.
27. Blocks laid in 1:3 sanded gypsum mortar voids in blocks not to exceed 30 percent.
28. Units of expanded slag or pumice aggregate.
29. Units of crushed limestone, blast furnace, slag, cinders and expanded clay or shale.
30. Units of calcareous sand and gravel. Coarse aggregate, 60 percent or more calcite and dolomite.
31. Units of siliceous sand and gravel. Ninety percent or more quartz, chert or flint.
32. Unit at least 49 percent solid.
33. Unit at least 62 percent solid.
34. Unit at least 65 percent solid.
35. Unit at least 73 percent solid.
36. Ratings based on one unit and one cell in wall thickness.
37. Minimum of $1/2$ inch—1:3 sanded gypsum plaster.
38. Nonload bearing.
39. See Clay Tile Partition Design Construction drawings below.

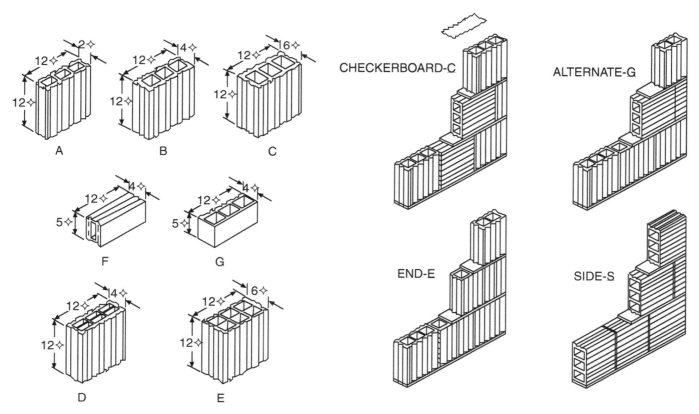

DESIGNS OF TILES USED IN FIRE-TEST PARTITIONS

THE FOUR TYPES OF CONSTRUCTION USED IN FIRE-TEST PARTITIONS

RESOURCE A—GUIDELINES ON FIRE RATINGS OF ARCHAIC MATERIALS AND ASSEMBLIES

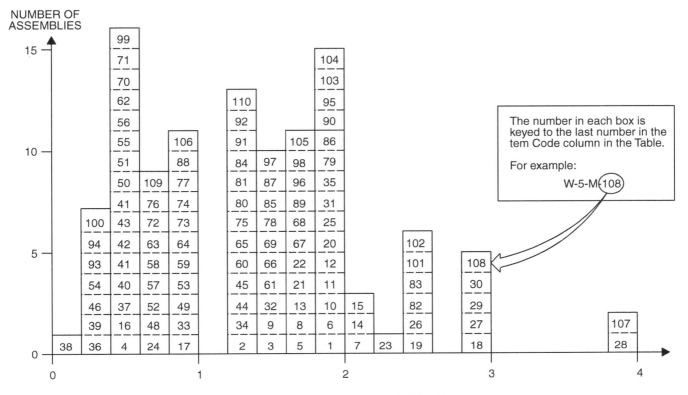

FIGURE 1.1.2
MASONRY WALLS
4″ TO LESS THAN 6″ THICK

TABLE 1.1.2
MASONRY WALLS
4″ TO LESS THAN 6″ THICK

ITEM CODE	THICKNESS	CONSTRUCTION DETAILS	PERFORMANCE		REFERENCE NUMBER			NOTES	REC. HOURS
			LOAD	TIME	PRE-BMS-92	BMS-92	POST-BMS-92		
W-4-M-1	4″	Solid 3″ thick, gypsum blocks laid in 1:3 sanded gypsum mortar; Facings: $\frac{1}{2}″$ of 1:3 sanded gypsum plaster (both sides).	N/A	2 hrs.		1		1	2
W-4-M-2	4″	Solid clay or shale brick.	N/A	1 hr. 15 min		1		1, 2	$1\frac{1}{4}$
W-4-M-3	4″	Concrete; No facings.	N/A	1 hr. 30 min.		1		1	$1\frac{1}{2}$
W-4-M-4	4″	Clay tile; Illinois surface clay; single cell thick; No face plaster; Design "B," Construction "C."	N/A	25 min.			2	3-7, 36	$\frac{1}{3}$
W-4-M-5	4″	Solid sand-lime brick.	N/A	1 hr. 45 min.		1		1	$1\frac{3}{4}$
W-4-M-6	4″	Solid wall; 3″ thick block; $\frac{1}{2}″$ plaster each side; $17\frac{3}{4}″ \times 8\frac{3}{4}″ \times 4″$ "Breeze Blocks"; Portland cement/sand mortar.	N/A	1 hr. 52 min.			7	2	$1\frac{3}{4}$
W-4-M-7	4″	Concrete (4020 psi); Reinforcement: vertical $\frac{3}{8}″$; horizontal $\frac{1}{4}″$; 6″ × 6″ grid.	N/A	2 hrs. 10 min.			7	2	2
W-4-M-8	4″	Concrete wall (4340 psi crush); reinforcement $\frac{1}{4}″$ diameter rebar on 8″ centers (vertical and horizontal).	N/A	1 hr. 40 min.			7	2	$1\frac{2}{3}$

(continued)

RESOURCE A—GUIDELINES ON FIRE RATINGS OF ARCHAIC MATERIALS AND ASSEMBLIES

TABLE 1.1.2—continued
MASONRY WALLS
4″ TO LESS THAN 6″ THICK

ITEM CODE	THICKNESS	CONSTRUCTION DETAILS	PERFORMANCE		REFERENCE NUMBER			NOTES	REC. HOURS
			LOAD	TIME	PRE-BMS-92	BMS-92	POST-BMS-92		
W-4-M-9	$4^{3}/_{16}″$	$4^{3}/_{16}″ \times 2^{5}/_{8}″$ cellular fletton brick (1873 psi) with $^{1}/_{2}″$ sand mortar; bricks are U-shaped yielding hollow cover (approx. $2″ \times 4″$) in final cross-section configuration.	N/A	1 hr. 25 min.			7	2	$1^{1}/_{3}$
W-4-M-10	$4^{1}/_{4}″$	$4^{1}/_{4}″ \times 2^{1}/_{2}″$ fletton (1831 psi) brick in $^{1}/_{2}″$ sand mortar.	N/A	1 hr. 53 min			7	2	$1^{3}/_{4}$
W-4-M-11	$4^{1}/_{4}″$	$4^{1}/_{4}″ \times 2^{1}/_{2}″$ London stock (683 psi) brick; $^{1}/_{2}″$ grout.	N/A	1 hr. 52 min.			7	2	$1^{3}/_{4}$
W-4-M-12	$4^{1}/_{2}″$	$4^{1}/_{4}″ \times 2^{1}/_{2}″$ Leicester red, wire-cut brick (4465 psi) in $^{1}/_{2}″$ sand mortar.	N/A	1 hr. 56 min.			7	6	$1^{3}/_{4}$
W-4-M-13	$4^{1}/_{4}″$	$4^{1}/_{4}″ \times 2^{1}/_{2}″$ stairfoot brick (7527 psi) $^{1}/_{2}″$ sand mortar.	N/A	1 hr. 37 min.			7	2	$1^{1}/_{2}$
W-4-M-14	$4^{1}/_{4}″$	$4^{1}/_{4}″ \times 2^{1}/_{2}″$ sand-lime brick (2603 psi) $^{1}/_{2}″$ sand mortar.	N/A	2 hrs. 6 min.			7	2	2
W-4-M-15	$4^{1}/_{4}″$	$4^{1}/_{4}″ \times 2^{1}/_{2}″$ concrete brick (2527 psi) $^{1}/_{2}″$ sand mortar.	N/A	2 hrs. 10 min.			7	2	2
W-4-M-16	$4^{1}/_{2}″$	4″ thick clay tile; Ohio fire clay; single cell thick; No plaster exposed face; $^{1}/_{2}″$ 1:2 gypsum back face; Design "F," Construction "S."	N/A	31 min.			2	3-6, 36	$^{1}/_{2}$
W-4-M-17	$4^{1}/_{2}″$	4″ thick clay tile; Ohio fire clay; single cell thick; Plaster exposed face; $^{1}/_{2}″$ 1:2 sanded gypsum; Back Face: none; Construction "S," Design "F."	80 psi	50 min.			2	3-5, 8, 36	$^{3}/_{4}$
W-4-M-18	$4^{1}/_{2}″$	Core: solid sand-lime brick; $^{1}/_{2}″$ sanded gypsum plaster facings on both sides.	80 psi	3 hrs.		1		1, 11	3
W-4-M-19	$4^{1}/_{2}″$	Core: solid sand-lime brick; $^{1}/_{2}″$ sanded gypsum plaster facings on both sides.	80 psi	2 hrs. 30 min.		1		1, 11	$2^{1}/_{2}$
W-4-M-20	$4^{1}/_{2}″$	Core: concrete brick $^{1}/_{2}″$ of 1:3 sanded gypsum plaster facings on both sides.	80 psi	2 hrs.		1		1, 11	2
W-4-M-21	$4^{1}/_{2}″$	Core: solid clay or shale brick; $^{1}/_{2}″$ thick, 1:3 sanded gypsum plaster facings on fire sides.	80 psi	1 hr. 45 min.		1		1, 2, 11	$1^{3}/_{4}$
W-4-M-22	$4^{3}/_{4}″$	4″ thick clay tile; Ohio fire clay; single cell thick; cells filled with cement and broken tile concrete; Plaster on exposed face; none on unexposed face; $^{3}/_{4}″$ 1:3 sanded gypsum; Design "G," Construction "E."	N/A	1 hr. 48 min.			2	2, 3-5, 9, 36	$1^{3}/_{4}$
W-4-M-23	$4^{3}/_{4}″$	4″ thick clay tile; Ohio fire clay; single cell thick; cells filled with cement and broken tile concrete; No plaster exposed faced; $^{3}/_{4}″$ neat gypsum plaster on unexposed face; Design "G," Construction "E."	N/A	2 hrs. 14 min.			2	2, 3-5, 9, 36	2
W-5-M-24	5″	$3″ \times 13″$ air space; 1″ thick metal reinforced concrete facings on both sides; faces connected with wood splines.	2,250 lbs./ft.	45 min.		1		1	$^{3}/_{4}$
W-5-M-25	5″	Core: 3″ thick void filled with "nondulated" mineral wool weighing 10 lbs./ft.^{3}; 1″ thick metal reinforced concrete facings on both sides.	2,250 lbs./ft.	2 hrs.		1		1	2
W-5-M-26	5″	Core: solid clay or shale brick; $^{1}/_{2}″$ thick, 1:3 sanded gypsum plaster facings on both sides.	40 psi	2 hrs. 30 min.		1		1, 2, 11	$2^{1}/_{2}$
W-5-M-27	5″	Core: solid 4″ thick gypsum blocks, laid in 1:3 sanded gypsum mortar; $^{1}/_{2}″$ of 1:3 sanded gypsum plaster facings on both sides.	N/A	3 hrs.		1		1	3

(continued)

TABLE 1.1.2—continued
MASONRY WALLS
4″ TO LESS THAN 6″ THICK

ITEM CODE	THICKNESS	CONSTRUCTION DETAILS	PERFORMANCE		REFERENCE NUMBER			NOTES	REC. HOURS
			LOAD	TIME	PRE-BMS-92	BMS-92	POST-BMS-92		
W-5-M-28	5″	Core: 4″ thick hollow gypsum blocks with 30% voids; blocks laid in 1:3 sanded gypsum mortar; No facings.	N/A	4 hrs.		1		1	4
W-5-M-29	5″	Core: concrete brick; $^1/_2$″ of 1:3 sanded gypsum plaster facings on both sides.	160 psi	3 hrs.		1		1	3
W-5-M-30	$5^1/_4$″	4″ thick clay tile; Illinois surface clay; double cell thick; Plaster: $^5/_8$″ sanded gypsum 1:3 both faces; Design "D," Construction "S."	N/A	2 hrs. 53 min.			2	2-5, 9, 36	$2^3/_4$
W-5-M-31	$5^1/_4$″	4″ thick clay tile; New Jersey fire clay; double cell thick; Plaster: $^5/_8$″ sanded gypsum 1:3 both faces; Design "D," Construction "S."	N/A	1 hr. 52 min.			2	2-5, 9, 36	$1^3/_4$
W-5-M-32	$5^1/_4$″	4″ thick clay tile; New Jersey fire clay; single cell thick; Plaster: $^5/_8$″ sanded gypsum 1:3 both faces; Design "D," Construction "S."	N/A	1 hr. 34 min.	2		2	2-5, 9, 36	$1^1/_2$
W-5-M-33	$5^1/_4$″	4″ thick clay tile; New Jersey fire clay; single cell thick; Face plaster: $^5/_8$″ both sides; 1:3 sanded gypsum; Design "B," Construction "S."	N/A	50 min.			2	3-5, 8, 36	$^3/_4$
W-5-M-34	$5^1/_4$″	4″ thick clay tile; Ohio fire clay; single cell thick; Face plaster: $^5/_8$″ both sides; 1:3 sanded gypsum; Design "B," Construction "A."	N/A	1 hr. 19 min.			2	2-5, 9, 36	$1^1/_4$
W-5-M-35	$5^1/_4$″	4″ thick clay tile; Illinois surface clay; single cell thick; Face plaster: $^5/_8$″ both sides; 1:3 sanded gypsum; Design "B," Construction "S."	N/A	1 hr. 59 min.			2	2-5, 10, 36	$1^3/_4$
W-5-M-36	4″	Core: structural clay tile; see Notes 12, 16, 21; No facings.	N/A	15 min.		1		3, 4, 24	$^1/_4$
W-4-M-37	4″	Core: structural clay tile; see Notes 12, 17, 21; No facings.	N/A	25 min.		1		3, 4, 24	$^1/_3$
W-4-M-38	4″	Core: structural clay tile; see Notes 12, 16, 20; No facings.	N/A	10 min.		1		3, 4, 24	$^1/_6$
W-4-M-39	4″	Core: structural clay tile; see Notes 12, 17, 20; No facings.	N/A	20 min.		1		3, 4, 24	$^1/_3$
W-4-M-40	4″	Core: structural clay tile; see Notes 13, 16, 23; No facings.	N/A	30 min.		1		3, 4, 24	$^1/_2$
W-4-M-41	4″	Core: structural clay tile; see Notes 13, 17, 23; No facings.	N/A	35 min.		1		3, 4, 24	$^1/_2$
W-4-M-42	4″	Core: structural clay tile; see Notes 13, 16, 21; No facings.	N/A	25 min.		1		3, 4, 24	$^1/_3$
W-4-M-43	4″	Core: structural clay tile; see Notes 13, 17, 21; No facings.	N/A	30 min.		1		3, 4, 24	$^1/_2$
W-4-M-44	4″	Core: structural clay tile; see Notes 15, 16, 20; No facings	N/A	1 hr. 15 min.		1		3, 4, 24	$1^1/_4$
W-4-M-45	4″	Core: structural clay tile; see Notes 15, 17, 20; No facings.	N/A	1 hr. 15 min.		1		3, 4, 24	$1^1/_4$
W-4-M-46	4″	Core: structural clay tile; see Notes 14, 16, 22; No facings.	N/A	20 min.		1		3, 4, 24	$^1/_3$
W-4-M-47	4″	Core: structural clay tile; see Notes 14, 17, 22; No facings.	N/A	25 min.		1		3, 4, 24	$^1/_3$
W-4-M-48	$4^1/_4$″	Core: structural clay tile; see Notes 12, 16, 21; Facings: both sides; see Note 18.	N/A	45 min.		1		3, 4, 24	$^3/_4$

(continued)

RESOURCE A—GUIDELINES ON FIRE RATINGS OF ARCHAIC MATERIALS AND ASSEMBLIES

TABLE 1.1.2—continued
MASONRY WALLS
4″ TO LESS THAN 6″ THICK

ITEM CODE	THICKNESS	CONSTRUCTION DETAILS	PERFORMANCE		REFERENCE NUMBER			NOTES	REC. HOURS
			LOAD	TIME	PRE-BMS-92	BMS-92	POST-BMS-92		
W-4-M-49	4¼″	Core: structural clay tile; see Notes 12, 17, 21; Facings: both sides; see Note 18.	N/A	1 hr.		1		3, 4, 24	1
W-4-M-50	4⅝″	Core: structural clay tile; see Notes 12, 16, 21; Facings: unexposed side only; see Note 18.	N/A	25 min.		1		3, 4, 24	⅓
W-4-M-51	4⅝″	Core: structural clay tile; see Notes 12, 17, 21; Facings: unexposed side only; see Note 18.	N/A	30 min.		1		3, 4, 24	½
W-4-M-52	4⅝″	Core: structural clay tile; see Notes 12, 16, 21; Facings: unexposed side only; see Note 18.	N/A	45 min.		1		3, 4, 24	¾
W-4-M-53	4⅝″	Core: structural clay tile; see Notes 12, 17, 21; Facings: fire side only; see Note 18.	N/A	1 hr.		1		3, 4, 24	1
W-4-M-54	4⅝″	Core: structural clay tile; see Notes 12, 16, 20; Facings: unexposed side; see Note 18.	N/A	20 min.		1		3, 4, 24	⅓
W-4-M-55	4⅝″	Core: structural clay tile; see Notes 12, 17, 20; Facings: exposed side; see Note 18.	N/A	25 min.		1		3, 4, 24	⅓
W-4-M-56	4⅝″	Core: structural clay tile; see Notes 12, 16, 20; Facings: fire side only; see Note 18.	N/A	30 min.		1		3, 4, 24	½
W-4-M-57	4⅝″	Core: structural clay tile; see Notes 12, 17, 20; Facings: fire side only; see Note 18.	N/A	45 min.		1		3, 4, 24	¾
W-4-M-58	4⅝″	Core: structural clay tile; see Notes 13, 16, 23; Facings: unexposed side only; see Note 18.	N/A	40 min.		1		3, 4, 24	⅔
W-4-M-59	4⅝″	Core: structural clay tile; see Notes 13, 17, 23; Facings: unexposed side only; see Note 18.	N/A	1 hr.		1		3, 4, 24	1
W-4-M-60	4⅝″	Core: structural clay tile; see Notes 13, 16, 23; Facings: fire side only; see Note 18.	N/A	1 hr. 15 min.		1		3, 4, 24	1¼
W-4-M-61	4⅝″	Core: structural clay tile; see Notes 13, 17, 23; Facings: fire side only; see Note 18.	N/A	1 hr. 30 min.		1		3, 4, 24	1½
W-4-M-62	4⅝″	Core: structural clay tile; see Notes 13, 16, 21; Facings: unexposed side only; see Note 18.	N/A	35 min.		1		3, 4, 24	½
W-4-M-63	4⅝″	Core: structural clay tile; see Notes 13, 17, 21; Facings: unexposed face only; see Note 18.	N/A	45 min.		1		3, 4, 24	¾
W-4-M-64	4⅝″	Core: structural clay tile; see Notes 13, 16, 23; Facings: exposed face only; see Note 18.	N/A	1 hr.		1		3, 4, 24	1
W-4-M-65	4⅝″	Core: structural clay tile; see Notes 13, 17, 21; Facings: exposed side only; see Note 18.	N/A	1 hr. 15 min.		1		3, 4, 24	1¼
W-4-M-66	4⅝″	Core: structural clay tile; see Notes 15, 17, 20; Facings: unexposed side only; see Note 18	N/A	1 hr. 30 min.		1		3, 4, 24	1½
W-4-M-67	4⅝″	Core: structural clay tile; see Notes 15, 16, 20; Facings: exposed side only; see Note 18.	N/A	1 hr. 45 min.		1		3, 4, 24	1¾
W-4-M-68	4⅝″	Core: structural clay tile; see Notes 15, 17, 20; Facings: exposed side only; see Note 18.	N/A	1 hr. 45 min.		1		3, 4, 24	1¾
W-4-M-69	4⅝″	Core: structural clay tile; see Notes 15, 16, 20; Facings: unexposed side only; see Note 18.	N/A	1 hr. 30 min.		1		3, 4, 24	1¾
W-4-M-70	4⅝″	Core: structural clay tile; see Notes 14, 16, 22; Facings: unexposed side only; see Note 18.	N/A	30 min.		1		3, 4, 24	½

(continued)

TABLE 1.1.2—continued
MASONRY WALLS
4″ TO LESS THAN 6″ THICK

ITEM CODE	THICKNESS	CONSTRUCTION DETAILS	PERFORMANCE		REFERENCE NUMBER			NOTES	REC. HOURS
			LOAD	TIME	PRE-BMS-92	BMS-92	POST-BMS-92		
W-4-M-71	$4^5/_8''$	Core: structural clay tile; see Notes 14, 17, 22; Facings: exposed side only; see Note 18.	N/A	35 min.		1		3, 4, 24	$^1/_2$
W-4-M-72	$4^5/_8''$	Core: structural clay tile; see Notes 14, 16, 22; Facings: fire side of wall only; see Note 18.	N/A	45 min.		1		3, 4, 24	$^3/_4$
W-4-M-73	$4^5/_8''$	Core: structural clay tile; see Notes 14, 17, 22; Facings: fire side of wall only; see Note 18.	N/A	1 hr.		1		3, 4, 24	1
W-4-M-74	$5^1/_4''$	Core: structural clay tile; see Notes 12, 16, 21; Facings: both sides; see Note 18.	N/A	1 hr.		1		3, 4, 24	1
W-5-M-75	$5^1/_4''$	Core: structural clay tile; see Notes 12, 17, 21; Facings: both sides; see Note 18	N/A	1 hr. 15 min.		1		3, 4, 24	$1^1/_4$
W-5-M-76	$5^1/_4''$	Core: structural clay tile; see Notes 12, 16, 20; Facings: both sides; see Note 18.	N/A	45 min.		1		3, 4, 24	$^3/_4$
W-5-M-77	$5^1/_4''$	Core: structural clay tile; see Notes 12, 17, 20; Facings: both sides; see Note 18.	N/A	1 hr.		1		3, 4, 24	1
W-5-M-78	$5^1/_4''$	Core: structural clay tile; see Notes 13, 16, 23; Facings: both sides of wall; see Note 18.	N/A	1 hr. 30 min.		1		3, 4, 24	$1^1/_2$
W-5-M-79	$5^1/_4''$	Core: structural clay tile; see Notes 13, 17, 23; Facings: both sides of wall; see Note 18.	N/A	2 hrs.		1		3, 4, 24	2
W-5-M-80	$5^1/_4''$	Core: structural clay tile; see Notes 13, 16, 21; Facings: both sides of wall; see Note 18.	N/A	1 hr. 15 min.		1		3, 4, 24	$1^1/_4$
W-5-M-81	$5^1/_4''$	Core: structural clay tile; see Notes 13, 16, 21; Facings: both sides of wall; see Note 18.	N/A	1 hr. 30 min.		1		3, 4, 24	$1^1/_2$
W-5-M-82	$5^1/_4''$	Core: structural clay tile; see Notes 15, 16, 20; Facings: both sides; see Note 18.	N/A	2 hrs. 30 min.		1		3, 4, 24	$2^1/_2$
W-5-M-83	$5^1/_4''$	Core: structural clay tile; see Notes 15, 17, 20; Facings: both sides; see Note 18.	N/A	2 hrs. 30 min.		1		3, 4, 24	$2^1/_2$
W-5-M-84	$5^1/_4''$	Core: structural clay tile; see Notes 14, 16, 22; Facings: both sides of wall; see Note 18.	N/A	1 hr. 15 min.		1		3, 4, 24	$1^1/_4$
W-5-M-85	$5^1/_4''$	Core: structural clay tile; see Notes 14, 17, 22; Facings: both sides of wall; see Note 18.	N/A	1 hr. 30 min.		1		3, 4, 24	$1^1/_2$
W-4-M-86	4″	Core: 3″ thick gypsum blocks 70% solid; see Note 26; Facings: both sides; see Note 25.	N/A	2 hrs.		1			2
W-4-M-87	4″	Core: hollow concrete units; see Notes 27, 34, 35; No facings.	N/A	1 hr. 30 min.		1			$1^1/_2$
W-4-M-88	4″	Core: hollow concrete units; see Notes 28, 33, 35; No facings.	N/A	1 hr.		1			1
W-4-M-89	4″	Core: hollow concrete units; see Notes 28, 34, 35; Facings: both sides; see Note 25.	N/A	1 hr. 45 min.		1			$1^3/_4$
W-4-M-90	4″	Core: hollow concrete units; see Notes 27, 34, 35; Facings: both sides; see Note 25.	N/A	2 hrs.		1			2
W-4-M-91	4″	Core: hollow concrete units; see Notes 27, 32, 35; No facings.	N/A	1 hr. 15 min.		1			$1^1/_4$
W-4-M-92	4″	Core: hollow concrete units; see Notes 28, 34, 35; No facings.	N/A	1 hr. 15 min.		1			$1^1/_4$
W-4-M-93	4″	Core: hollow concrete units; see Notes 29, 32, 35; No facings.	N/A	20 min.		1			$^1/_3$

(continued)

RESOURCE A—GUIDELINES ON FIRE RATINGS OF ARCHAIC MATERIALS AND ASSEMBLIES

TABLE 1.1.2—continued
MASONRY WALLS
4″ TO LESS THAN 6″ THICK

ITEM CODE	THICKNESS	CONSTRUCTION DETAILS	PERFORMANCE		REFERENCE NUMBER			NOTES	REC. HOURS
			LOAD	TIME	PRE-BMS-92	BMS-92	POST-BMS-92		
W-4-M-94	4″	Core: hollow concrete units; see Notes 30, 34, 35; No facings.	N/A	15 min.		1			1/4
W-4-M-95	4 1/2″	Core: hollow concrete units; see Notes 27, 34, 35; Facings: one side only; see Note 25.	N/A	2 hrs.		1			2
W-4-M-96	4 1/2″	Core: hollow concrete units; see Notes 27, 32, 35; Facings: one side only; see Note 25.	N/A	1 hr. 45 min.		1			1 3/4
W-4-M-97	4 1/2″	Core: hollow concrete units; see Notes 28, 33, 35; Facings: one side; see Note 25.	N/A	1 hr. 30 min.		1			1 1/2
W-4-M-98	4 1/2″	Core: hollow concrete units; see Notes 28, 34, 35; Facings: one side only; see Note 25.	N/A	1 hr. 45 min.		1			1 3/4
W-4-M-99	4 1/2″	Core: hollow concrete units; see Notes 29, 32, 35; Facings: one side; see Note 25.	N/A	30 min.		1			1/2
W-4-M-100	4 1/2″	Core: hollow concrete units; see Notes 30, 34, 35; Facings: one side; see Note 25.	N/A	20 min.		1			1/3
W-5-M-101	5″	Core: hollow concrete units; see Notes 27, 34, 35; Facings: both sides; see Note 25.	N/A	2 hrs. 30 min.		1			2 1/2
W-5-M-102	5″	Core: hollow concrete units; see Notes 27, 32, 35; Facings: both sides; see Note 25.	N/A	2 hrs. 30 min.		1			2 1/2
W-5-M-103	5″	Core: hollow concrete units; see Notes 28, 33, 35; Facings: both sides; see Note 25.	N/A	2 hrs.		1			2
W-5-M-104	5″	Core: hollow concrete units; see Notes 28, 31, 35; Facings: both sides; see Note 25.	N/A	2 hrs.		1			2
W-5-M-105	5″	Core: hollow concrete units; see Notes 29, 32, 35; Facings: both sides; see Note 25.	N/A	1 hr. 45 min.		1			1 3/4
W-5-M-106	5″	Core: hollow concrete units; see Notes 30, 34, 35; Facings: both sides; see Note 25.	N/A	1 hr.		1			1
W-5-M-107	5″	Core: 5″ thick solid gypsum blocks; see Note 26; No facings.	N/A	4 hrs.		1			4
W-5-M-108	5″	Core: 4″ thick hollow gypsum blocks; see Note 26; Facings: both sides; see Note 25.	N/A	3 hrs.		1			3
W-5-M-109	4″	Concrete with 4″ × 4″ No. 6 welded wire mesh at wall center.	100 psi	45 min.			43	2	3/4
W-4-M-110	4″	Concrete with 4″ × 4″ No. 6 welded wire mesh at wall center.	N/A	1 hr. 15 min.			43	2	1 1/4

For SI: 1 inch = 25.4 mm, 1 pound per square inch = 0.00689 MPa.

Notes:
1. Tested as NBS under ASA Spec. No. A 2-1934.
2. Failure mode—maximum temperature rise.
3. Treated at NBS under ASA Spec. No. 42-1934 (ASTM C19-53) except that hose stream testing where carried out was run on test specimens exposed for full test duration, not for or reduced period as is contemporarily done.
4. For clay tile walls, unless the source the clay can be positively identified, it is suggested that the most pessimistic hour rating for the fire endurance of a clay tile partition of that thickness to be followed. Identified sources of clay showing longer fire endurance can lead to longer time recommendations.
5. See appendix for construction and design details for clay tile walls.
6. Failure mode—flame thru or crack formation showing flames.
7. Hole formed at 25 minutes; partition collapsed at 42 minutes or removal from furnace.
8. Failure mode—collapse.
9. Hose stream pass.
10. Hose stream hole formed in specimen.
11. Load: 80 psi for gross wall cross sectional area.
12. One cell in wall thickness.
13. Two cells in wall thickness.

(continued)

RESOURCE A—GUIDELINES ON FIRE RATINGS OF ARCHAIC MATERIALS AND ASSEMBLIES

TABLE 1.1.2—continued
MASONRY WALLS
4″ TO LESS THAN 6″ THICK

14. Double cells plus one cell in wall thickness.
15. One cell in wall thickness, cells filled with broken tile, crushed stone, slag, cinders or sand mixed with mortar.
16. Dense hard-burned clay or shale tile.
17. Medium-burned clay tile.
18. Not less than $^5/_8$ inch thickness of 1:3 sanded gypsum plaster.
19. Units of not less than 30 percent solid material.
20. Units of not less than 40 percent solid material.
21. Units of not less than 50 percent solid material.
22. Units of not less than 45 percent solid material.
23. Units of not less than 60 percent solid material.
24. All tiles laid in Portland cement-lime mortar.
25. Minimum $^1/_2$ inch—1:3 sanded gypsum plaster.
26. Laid in 1:3 sanded gypsum mortar. Voids in hollow units not to exceed 30 percent.
27. Units of expanded slag or pumice aggregate.
28. Units of crushed limestone, blast furnace slag, cinders and expanded clay or shale.
29. Units of calcareous sand and gravel. Coarse aggregate, 60 percent or more calcite and dolomite.
30. Units of siliceous sand and gravel. Ninety percent or more quartz, chert or flint.
31. Unit at least 49 percent solid.
32. Unit at least 62 percent solid.
33. Unit at least 65 percent solid.
34. Unit at least 73 percent solid.
35. Ratings based on one unit and one cell in wall thickness.
36. See Clay Tile Partition Design Construction drawings below.

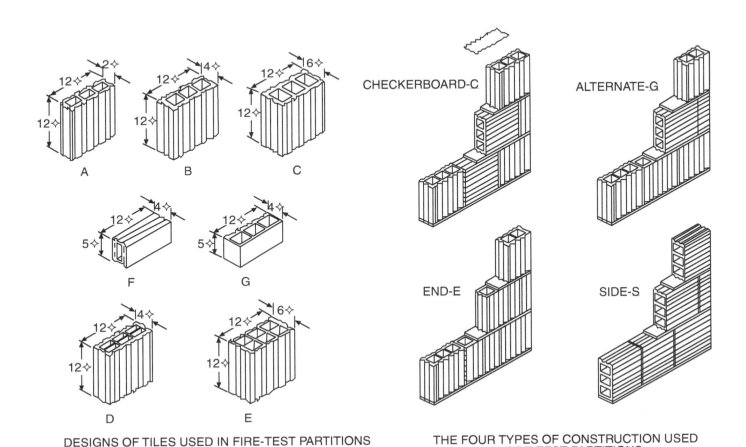

DESIGNS OF TILES USED IN FIRE-TEST PARTITIONS

THE FOUR TYPES OF CONSTRUCTION USED IN FIRE-TEST PARTITIONS

RESOURCE A—GUIDELINES ON FIRE RATINGS OF ARCHAIC MATERIALS AND ASSEMBLIES

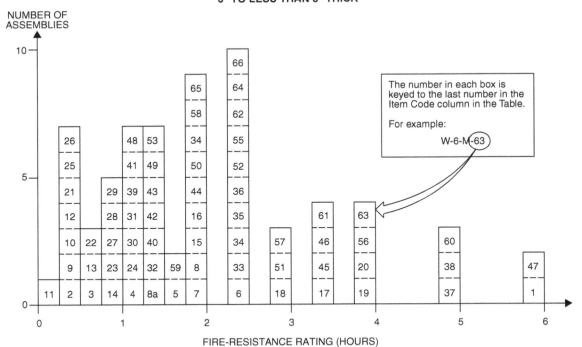

FIGURE 1.1.3
MASONRY WALLS
6″ TO LESS THAN 8″ THICK

TABLE 1.1.3
MASONRY WALLS
6″ TO LESS THAN 8″ THICK

ITEM CODE	THICKNESS	CONSTRUCTION DETAILS	PERFORMANCE LOAD	PERFORMANCE TIME	REFERENCE NUMBER PRE-BMS-92	REFERENCE NUMBER BMS-92	REFERENCE NUMBER POST-BMS-92	NOTES	REC. HOURS
W-6-M-1	6″	Core: 5″ thick, solid gypsum blocks laid in 1:3 sanded gypsum mortar; $^1/_2$″ of 1:3 sanded gypsum plaster facings on both sides.	N/A	6 hrs.		1			6
W-6-M-2	6″	6″ clay tile; Ohio fire clay; single cell thick; No plaster; Design "C," Construction "A."	N/A	17 min.			2	1, 3, 4, 6, 55	$^1/_4$
W-6-M-3	6″	6″ clay tile; Illinois surface clay; double cell thick; No plaster; Design "E," Construction "C."	N/A	45 min.			2	1-4, 7, 55	$^3/_4$
W-6-M-4	6″	6″ clay tile; New Jersey fire clay; double cell thick; No plaster; Design "E," Construction "S."	N/A	1 hr. 1 min.			2	1-4, 8, 55	1
W-7-M-5	7$^1/_4$″	6″ clay tile; Illinois surface clay; double cell thick; Plaster: $^5/_8$″—1:3 sanded gypsum both faces; Design "E," Construction "A."	N/A	1 hr. 41 min.			2	1-4, 55	1$^2/_3$
W-7-M-6	7$^1/_4$″	6″ clay tile; New Jersey fire clay; double cell thick; Plaster: $^5/_8$″—1:3 sanded gypsum both faces; Design "E," Construction "S."	N/A	2 hrs. 23 min.			2	1-4, 9, 55	2$^1/_3$
W-7-M-7	7$^1/_4$″	6″ clay tile; Ohio fire clay; single cell thick; Plaster: $^5/_8$″ sanded gypsum; 1:3 both faces; Design "C," Construction "A."	N/A	1 hr. 54 min.			2	1-4, 9, 55	2$^3/_4$
W-7-M-8	7$^1/_4$″	6″ clay tile; Illinois surface clay; single cell thick; Plaster: $^5/_8$″ sanded gypsum 1:3 both faces; Design "C," Construction "S."	N/A	2 hrs.			2	1, 3, 4, 9, 10, 55	2
W-7-M-8a	7$^1/_4$″	6″ clay tile; Illinois surface clay; single cell thick; Plaster: $^5/_8$″ sanded gypsum 1:3 both faces; Design "C," Construction "E."	N/A	1 hr. 23 min			2	1-4, 9, 10, 55	1$^3/_4$

(continued)

TABLE 1.1.3—continued
MASONRY WALLS
6″ TO LESS THAN 8″ THICK

ITEM CODE	THICKNESS	CONSTRUCTION DETAILS	PERFORMANCE		REFERENCE NUMBER			NOTES	REC. HOURS
			LOAD	TIME	PRE-BMS-92	BMS-92	POST-BMS-92		
W-6-M-9	6″	Core: structural clay tile; see Notes 12, 16, 20; No facings.	N/A	20 min.		1		3, 5, 24	$1/3$
W-6-M-10	6″	Core: structural clay tile; see Notes 12, 17, 20; No facings.	N/A	25 min.		1		3, 5, 24	$1/3$
W-6-M-11	6″	Core: structural clay tile; see Notes 12, 16, 19; No facings.	N/A	15 min.		1		3, 5, 24	$1/4$
W-6-M-12	6″	Core: structural clay tile; see Notes 12, 17, 19; No facings.	N/A	20 min.		1		3, 5, 24	$1/3$
W-6-M-13	6″	Core: structural clay tile; see Notes 13, 16, 22; No facings.	N/A	45 min.		1		3, 5, 24	$3/4$
W-6-M-14	6″	Core: structural clay tile; see Notes 13, 17, 22; No facings.	N/A	1 hr.		1		3, 5, 24	1
W-6-M-15	6″	Core: structural clay tile; see Notes 15, 17, 19; No facings.	N/A	2 hrs.		1		3, 5, 24	2
W-6-M-16	6″	Core: structural clay tile; see Notes 15, 16, 19; No facings.	N/A	2 hrs.		1		3, 5, 24	2
W-6-M-17	6″	Cored concrete masonry; see Notes 12, 34, 36, 38, 41; No facings.	80 psi	3 hrs. 30 min.		1		5, 25	$3 1/2$
W-6-M-18	6″	Cored concrete masonry; see Notes 12, 33, 36, 38, 41; No facings.	80 psi	3 hrs.		1		5, 25	3
W-6-M-19	$6 1/2$″	Cored concrete masonry; see Notes 12, 34, 36, 38, 41; Facings: side 1; see Note 35.	80 psi	4 hrs.		1		5, 25	4
W-6-M-20	$6 1/2$″	Cored concrete masonry; see Notes 12, 33, 36, 38, 41; Facings: side 1; see Note 35.	80 psi	4 hrs.		1		5, 25	4
W-6-M-21	$6 5/8$″	Core: structural clay tile; see Notes 12, 16, 20; Facings: unexposed face only; see Note 18.	N/A	30 min.		1		3, 5, 24	$1/2$
W-6-M-22	$6 5/8$″	Core: structural clay tile; see Notes 12, 17, 20; Facings: unexposed face only; see Note 18.	N/A	40 min.		1		3, 5, 24	$2/3$
W-6-M-23	$6 5/8$″	Core: structural clay tile; see Notes 12, 16, 20; Facings: exposed face only; see Note 18.	N/A	1 hr.		1		3, 5, 24	1
W-6-M-24	$6 5/8$″	Core: structural clay tile; see Notes 12, 17, 20; Facings: exposed face only; see Note 18.	N/A	1 hr. 5 min.		1		3, 5, 24	1
W-6-M-25	$6 5/8$″	Core: structural clay tile; see Notes 12, 16, 19; Facings: unexposed side only; see Note 18.	N/A	25 min.		1		3, 5, 24	$1/3$
W-6-M-26	$6 5/8$″	Core: structural clay tile; see Notes 12, 7, 19; Facings: unexposed face only; see Note 18.	N/A	30 min.		1		3, 5, 24	$1/2$
W-6-M-27	$6 5/8$″	Core: structural clay tile; see Notes 12, 16, 19; Facings: exposed side only; see Note 18.	N/A	1 hr.		1		3, 5, 24	1
W-6-M-28	$6 5/8$″	Core: structural clay tile; see Notes 12, 17, 19; Facings: fire side only; see Note 18.	N/A	1 hr.		1		3, 5, 24	1
W-6-M-29	$6 5/8$″	Core: structural clay tile; see Notes 13, 16, 22; Facings: unexposed side only; see Note 18.	N/A	1 hr.		1		3, 5, 24	1
W-6-M-30	$6 5/8$″	Core: structural clay tile; see Notes 13, 17, 22; Facings: unexposed side only; see Note 18.	N/A	1 hr. 15 min.		1		3, 5, 24	$1 1/4$
W-6-M-31	$6 5/8$″	Core: structural clay tile; see Notes 13, 16, 22; Facings: fire side only; see Note 18.	N/A	1 hr. 15 min.		1		3, 5, 24	$1 1/4$
W-6-M-32	$6 5/8$″	Core: structural clay tile; see Notes 13, 17, 22; Facings: fire side only; see Note 18.	N/A	1 hr. 30 min.		1		3, 5, 24	$1 1/2$

(continued)

RESOURCE A—GUIDELINES ON FIRE RATINGS OF ARCHAIC MATERIALS AND ASSEMBLIES

TABLE 1.1.3—continued
MASONRY WALLS
6″ TO LESS THAN 8″ THICK

ITEM CODE	THICKNESS	CONSTRUCTION DETAILS	PERFORMANCE		REFERENCE NUMBER			NOTES	REC. HOURS
			LOAD	TIME	PRE-BMS-92	BMS-92	POST-BMS-92		
W-6-M-33	6⅝″	Core: structural clay tile; see Notes 15, 16, 19; Facings: unexposed side only; see Note 18.	N/A	2 hrs. 30 min.		1		3, 5, 24	2½
W-6-M-34	6⅝″	Core: structural clay tile; see Notes 15, 17, 19; Facings: unexposed side only; see Note 18.	N/A	2 hrs. 30 min.		1		3, 5, 24	2½
W-6-M-35	6⅝″	Core: structural clay tile; see Notes 15, 16, 19; Facings: fire side only; see Note 18.	N/A	2 hrs. 30 min.		1		3, 5, 24	2½
W-6-M-36	6⅝″	Core: structural clay tile; see Notes 15, 17, 19; Facings: fire side only; see Note 18.	N/A	2 hrs. 30 min.		1		3, 5, 24	2½
W-6-M-37	7″	Cored concrete masonry; see Notes 12, 34, 36, 38, 41; see Note 35 for facings on both sides.	80 psi	5 hrs.		1		5, 25	5
W-6-M-38	7″	Cored concrete masonry; see Notes 12, 33, 36, 38, 41; see Note 35 for facings.	80 psi	5 hrs.		1		5, 25	5
W-6-M-39	7¼″	Core: structural clay tile; see Notes 12, 16, 20; Facings: both sides; see Note 18.	N/A	1 hr. 15 min.		1		3, 5, 24	1¼
W-6-M-40	7¼″	Core: structural clay tile; see Notes 12, 17, 20; Facings: both sides; see Note 18.	N/A	1 hr. 30 min.		1		3, 5, 24	1½
W-6-M-41	7¼″	Core: structural clay tile; see Notes 12, 16, 19; Facings: both sides; see Note 18.	N/A	1 hr. 15 min.		1		3, 5, 24	1¼
W-6-M-42	7¼″	Core: structural clay tile; see Notes 12, 17, 19; Facings: both sides; see Note 18.	N/A	1 hr. 30 min.		1		3, 5, 24	1½
W-7-M-43	7¼″	Core: structural clay tile; see Notes 13, 16, 22; Facings: both sides of wall; see Note 18.	N/A	1 hr. 30 min.		1		3, 5, 24	1½
W-7-M-44	7¼″	Core: structural clay tile; see Notes 13, 17, 22; Facings: both sides of wall; see Note 18.	N/A	2 hrs.		1		3, 5, 24	1½
W-7-M-45	7¼″	Core: structural clay tile; see Notes 15, 16, 19; Facings: both sides; see Note 18.	N/A	3 hrs. 30 min.		1		3, 5, 24	3½
W-7-M-46	7¼″	Core: structural clay tile; see Notes 15, 17, 19; Facings: both sides; see Note 18.	N/A	3 hrs. 30 min.		1		3, 5, 24	3½
W-6-M-47	6″	Core: 5″ thick solid gypsum blocks; see Note 45; Facings: both sides; see Note 45.	N/A	6 hrs.		1			6
W-6-M-48	6″	Core: hollow concrete units; see Notes 47, 50, 54; No facings.	N/A	1 hr. 15 min.		1			1¼
W-6-M-49	6″	Core: hollow concrete units; see Notes 46, 50, 54; No facings.	N/A	1 hr. 30 min.		1			1½
W-6-M-50	6″	Core: hollow concrete units; see Notes 46, 41, 54; No facings.	N/A	2 hrs.		1			2
W-6-M-51	6″	Core: hollow concrete units; see Notes 46, 53, 54; No facings.	N/A	3 hrs.		1			3
W-6-M-52	6″	Core: hollow concrete units; see Notes 47, 53, 54; No facings.	N/A	2 hrs. 30 min.		1			2½
W-6-M-53	6″	Core: hollow concrete units; see Notes 47, 51, 54; No facings.	N/A	1 hr. 30 min.		1			1½
W-6-M-54	6½″	Core: hollow concrete units; see Notes 46, 50, 54; Facings: one side only; see Note 35.	N/A	2 hrs.		1			2
W-6-M-55	6½″	Core: hollow concrete units; see Notes 4, 51, 54; Facings: one side; see Note 35.	N/A	2 hrs. 30 min.		1			2½
W-6-M-56	6½″	Core: hollow concrete units; see Notes 46, 53, 54; Facings: one side; see Note 35.	N/A	4 hrs.		1			4

(continued)

TABLE 1.1.3—continued
MASONRY WALLS
6″ TO LESS THAN 8″ THICK

ITEM CODE	THICKNESS	CONSTRUCTION DETAILS	PERFORMANCE		REFERENCE NUMBER			NOTES	REC. HOURS
			LOAD	TIME	PRE-BMS-92	BMS-92	POST-BMS-92		
W-6-M-57	6¹/₂″	Core: hollow concrete units; see Notes 47, 53, 54; Facings: one side; see Note 35.	N/A	3 hrs.		1			3
W-6-M-58	6¹/₂″	Core: hollow concrete units; see Notes 47, 51, 54; Facings: one side; see Note 35.	N/A	2 hrs.		1			2
W-6-M-59	6¹/₂″	Core: hollow concrete units; see Notes 47, 50, 54; Facings: one side; see Note 35.	N/A	1 hr. 45 min.		1			1³/₄
W-7-M-60	7″	Core: hollow concrete units; see Notes 46, 53, 54; Facings: both sides; see Note 35.	N/A	5 hrs.		1			5
W-7-M-61	7″	Core: hollow concrete units; see Notes 46, 51, 54; Facings: both sides; see Note 35.	N/A	3 hrs. 30 min.		1			3¹/₂
W-7-M-62	7″	Core: hollow concrete units; see Notes 46, 50, 54; Facings: both sides; see Note 35.	N/A	2 hrs. 30 min.		1			2¹/₂
W-7-M-63	7″	Core: hollow concrete units; see Notes 47, 53, 54; Facings: both sides; see Note 35.	N/A	4 hrs.		1			4
W-7-M-64	7″	Core: hollow concrete units; see Notes 47, 51, 54; Facings: both sides; see Note 35.	N/A	2 hrs. 30 min.		1			2¹/₂
W-7-M-65	7″	Core: hollow concrete units; see Notes 47, 50, 54; Facings: both sides; see Note 35.	N/A	2 hrs.		1			2
W-6-M-66	6″	Concrete wall with 4″ × 4″ No. 6 wire fabric (welded) near wall center for reinforcement.	N/A	2 hrs. 30 min.			43	2	2¹/₂

For SI: 1 inch = 25.4 mm, 1 pound per square inch = 0.00689 MPa.

Notes:
1. Tested at NBS under ASA Spec. No. 43-1934 (ASTM C19-53) except that hose stream testing where carried out was run on test specimens exposed for full test duration, not for a reduced period as is contemporarily done.
2. Failure by thermal criteria—maximum temperature rise.
3. For clay tile walls, unless the source or density of the clay can be positively identified or determined, it is suggested that the lowest hourly rating for the fire endurance of a clay tile partition of that thickness be followed. Identified sources of clay showing longer fire endurance can lead to longer time recommendations.
4. See Note 55 for construction and design details for clay tile walls.
5. Tested at NBS under ASA Spec. No. A2-1934.
6. Failure mode—collapse.
7. Collapsed on removal from furnace at 1 hour 9 minutes.
8. Hose stream—failed.
9. Hose stream—passed.
10. No end point met in test.
11. Wall collapsed at 1 hour 28 minutes.
12. One cell in wall thickness.
13. Two cells in wall thickness.
14. Double shells plus one cell in wall thickness.
15. One cell in wall thickness, cells filled with broken tile, crushed stone, slag, cinders or sand mixed with mortar.
16. Dense hard-burned clay or shale tile.
17. Medium-burned clay tile.
18. Not less than ⁵/₈ inch thickness of 1:3 sanded gypsum plaster.
19. Units of not less than 30 percent solid material.
20. Units of not less than 40 percent solid material.
21. Units of not less than 50 percent solid material.
22. Units of not less than 45 percent solid material.
23. Units of not less than 60 percent solid material.
24. All tiles laid in Portland cement-lime mortar.
25. Load: 80 psi for gross cross sectional area of wall.
26. Three cells in wall thickness.
27. Minimum percent of solid material in concrete units = 52.
28. Minimum percent of solid material in concrete units = 54.
29. Minimum percent of solid material in concrete units = 55.
30. Minimum percent of solid material in concrete units = 57.

(continued)

RESOURCE A—GUIDELINES ON FIRE RATINGS OF ARCHAIC MATERIALS AND ASSEMBLIES

TABLE 1.1.3—continued
MASONRY WALLS
6″ TO LESS THAN 8″ THICK

31. Minimum percent of solid material in concrete units = 62.
32. Minimum percent of solid material in concrete units = 65.
33. Minimum percent of solid material in concrete units = 70.
34. Minimum percent of solid material in concrete units = 76.
35. Not less than $^1/_2$ inch of 1:3 sanded gypsum plaster.
36. Noncombustible or no members framed into wall.
37. Combustible members framed into wall.
38. One unit in wall thickness.
39. Two units in wall thickness.
40. Three units in wall thickness.
41. Concrete units made with expanded slag or pumice aggregates.
42. Concrete units made with expanded burned clay or shale, crushed limestone, air cooled slag or cinders.
43. Concrete units made with calcareous sand and gravel. Coarse aggregate, 60 percent or more calcite and dolomite.
44. Concrete units made with siliceous sand and gravel. Ninety percent or more quartz, chert or flint.
45. Laid in 1:3 sanded gypsum mortar.
46. Units of expanded slag or pumice aggregate.
47. Units of crushed limestone, blast furnace, slag, cinder and expanded clay or shale.
48. Units of calcareous sand and gravel. Coarse aggregate, 60 percent or more calcite and dolomite.
49. Units of siliceous sand and gravel. Ninety percent or more quartz, chert or flint.
50. Unit minimum 49 percent solid.
51. Unit minimum 62 percent solid.
52. Unit minimum 65 percent solid.
53. Unit minimum 73 percent solid.
54. Ratings based on one unit and one cell in wall section.
55. See Clay Tile Partition Design Construction drawings below.

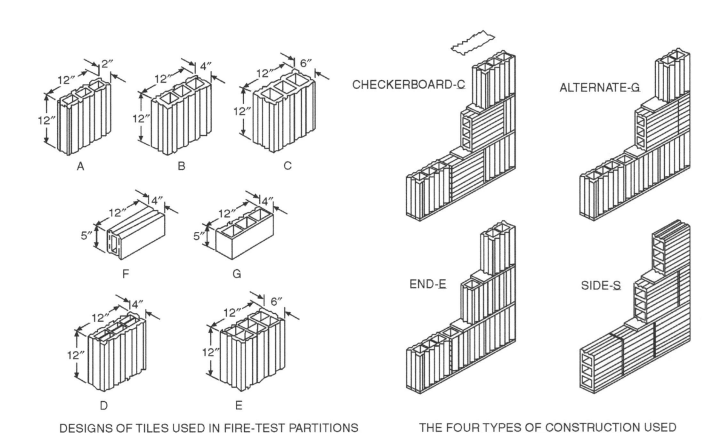

DESIGNS OF TILES USED IN FIRE-TEST PARTITIONS

THE FOUR TYPES OF CONSTRUCTION USED IN FIRE-TEST PARTITIONS

RESOURCE A—GUIDELINES ON FIRE RATINGS OF ARCHAIC MATERIALS AND ASSEMBLIES

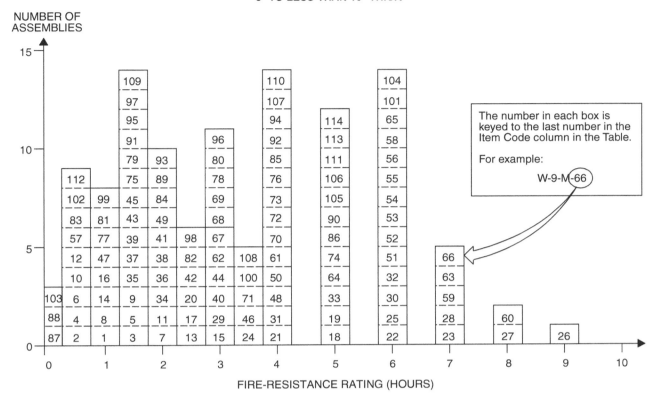

FIGURE 1.1.4
MASONRY WALLS
8″ TO LESS THAN 10″ THICK

TABLE 1.1.4
MASONRY WALLS
8″ TO LESS THAN 10″ THICK

ITEM CODE	THICKNESS	CONSTRUCTION DETAILS	PERFORMANCE		REFERENCE NUMBER			NOTES	REC. HOURS
			LOAD	TIME	PRE-BMS-92	BMS-92	POST-BMS-92		
W-8-M-1	8″	Core: clay or shale structural tile; Units in wall thickness: 1; Cells in wall thickness: 2; Minimum % solids in units: 40.	80 psi	1 hr. 15 min.		1		1, 20	$1\,^{1}/_{4}$
W-8-M-2	8″	Core: clay or shale structural tile; Units in wall thickness: 1; Cells in wall thickness: 2; Minimum % solids in units: 40; No facings; Result for wall with combustible members framed into interior.	80 psi	45 min.		1		1, 20	$^{3}/_{4}$
W-8-M-3	8″	Core: clay or shale structural tile; Units in wall thickness: 1; Cells in wall thickness: 2; Minimum % solids in units: 43.	80 psi	1 hr. 30 min.		1		1, 20	$1\,^{1}/_{2}$
W-8-M-4	8″	Core: clay or shale structural tile; Units in wall thickness: 1; Cells in wall thickness: 2; Minimum % solids in units: 43; No facings; Combustible members framed into wall.	80 psi	45 min.		1		1, 20	$^{3}/_{4}$
W-8-M-5	8″	Core: clay or shale structural tile; No facings.	See Notes	1 hr. 30 min.		1		1, 2, 5, 10, 18, 20, 21	$1\,^{1}/_{2}$
W-8-M-6	8″	Core: clay or shale structural tile; No facings.	See Notes	45 min.		1		1, 2, 5, 10, 19, 20, 21	$^{3}/_{4}$

(continued)

TABLE 1.1.4—continued
MASONRY WALLS
8″ TO LESS THAN 10″ THICK

ITEM CODE	THICKNESS	CONSTRUCTION DETAILS	PERFORMANCE		REFERENCE NUMBER			NOTES	REC. HOURS
			LOAD	TIME	PRE-BMS-92	BMS-92	POST-BMS-92		
W-8-M-7	8″	Core: clay or shale structural tile; No facings	See Notes	2 hrs.		1		1, 2, 5, 13, 18, 20, 21	2
W-8-M-8	8″	Core: clay or shale structural tile; No facings.	See Notes	1 hr. 45 min.		1		1, 2, 5, 13, 19, 20, 21	$1^1/_4$
W-8-M-9	8″	Core: clay or shale structural tile; No facings.	See Notes	1 hr. 15 min.		1		1, 2, 6, 9, 18, 20, 21	$1^3/_4$
W-8-M-10	8″	Core: clay or shale structural tile; No facings.	See Notes	45 min.		1		1, 2, 6, 9, 19, 20, 21	$^3/_4$
W-8-M-11	8″	Core: clay or shale structural tile; No facings.	See Notes	2 hrs.		1		1, 2, 6, 10, 18, 20, 21	2
W-8-M-12	8″	Core: clay or shale structural tile; No facings.	See Notes	45 min.		1		1, 2, 6, 10, 19, 20, 21	$^3/_4$
W-8-M-13	8″	Core: clay or shale structural tile; No facings.	See Notes	2 hrs. 30 min.		1		1, 3, 6, 12, 18, 20, 21	$2^1/_2$
W-8-M-14	8″	Core: clay or shale structural tile; No facings.	See Notes	1 hr.		1		1, 2, 6, 12, 19, 20, 21	1
W-8-M-15	8″	Core: clay or shale structural tile; No facings.	See Notes	3 hrs.		1		1, 2, 6, 16, 18, 20, 21	3
W-8-M-16	8″	Core: clay or shale structural tile; No facings.	See Notes	1 hr. 15 min.		1		1, 2, 6, 16, 19, 20, 21	$1^1/_4$
W-8-M-17	8″	Cored clay or shale brick; Units in wall thickness: 1; Cells in wall thickness: 1; Minimum % solids: 70; No facings.	See Notes	2 hrs. 30 min.		1		1, 44	$2^1/_2$
W-8-M-18	8″	Cored clay or shale brick; Units in wall thickness: 2; Cells in wall thickness: 2; Minimum % solids: 87; No facings.	See Notes	5 hrs.		1		1, 45	5
W-8-M-19	8″	Core: solid clay or shale brick; No facings.	See Notes	5 hrs.		1		1, 22, 45	5
W-8-M-20	8″	Core: hollow rolok of clay or shale.	See Notes	2 hrs. 30 min.		1		1, 22, 45	$2^1/_2$
W-8-M-21	8″	Core: hollow rolok bak of clay or shale; No facings.	See Notes	4 hrs.		1		1, 45	4
W-8-M-22	8″	Core: concrete brick; No facings.	See Notes	6 hrs.		1		1, 45	6
W-8-M-23	8″	Core: sand-lime brick; No facings.	See Notes	7 hrs.		1		1, 45	7
W-8-M-24	8″	Core: 4″, 40% solid clay or shale structural tile; 1 side 4″ brick facing.	See Notes	3 hrs. 30 min.		1		1, 20	$3^1/_2$
W-8-M-25	8″	Concrete wall (3220 psi); Reinforcing vertical rods 1″ from each face and 1″ diameter; horizontal rods $^5/_8$″ diameter.	22,200 lbs./ft.	6 hrs.			7		6
W-8-M-26	8″	Core: sand-line brick; $^1/_2$″ of 1:3 sanded gypsum plaster facings on one side.	See Notes	9 hrs.		1		1, 45	9
W-8-M-27	$8^1/_2$″	Core: sand-line brick; $^1/_2$″ of 1:3 sanded gypsum plaster facings on one side.	See Notes	8 hrs.		1		1, 45	8
W-8-M-28	$8^1/_2$″	Core: concrete; $^1/_2$″ of 1:3 sanded gypsum plaster facings on one side.	See Notes	7 hrs.		1		1, 45	7

(continued)

TABLE 1.1.4—continued
MASONRY WALLS
8″ TO LESS THAN 10″ THICK

ITEM CODE	THICKNESS	CONSTRUCTION DETAILS	PERFORMANCE		REFERENCE NUMBER			NOTES	REC. HOURS
			LOAD	TIME	PRE-BMS-92	BMS-92	POST-BMS-92		
W-8-M-29	$8\frac{1}{2}''$	Core: hollow rolok of clay or shale; $\frac{1}{2}''$ of 1:3 sanded gypsum plaster facings on one side.	See Notes	3 hrs.		1		1, 45	3
W-8-M-30	$8\frac{1}{2}''$	Core: solid clay or shale brick $\frac{1}{2}''$ thick, 1:3 sanded gypsum plaster facings on one side.	See Notes	6 hrs.		1		1, 22, 45,	6
W-8-M-31	$8\frac{1}{2}''$	Core: cored clay or shale brick; Units in wall thickness: 1; Cells in wall thickness: 1; Minimum % solids: 70; $\frac{1}{2}''$ of 1:3 sanded gypsum plaster facings on both sides.	See Notes	4 hrs.		1		1, 44	4
W-8-M-32	$8\frac{1}{2}''$	Core: cored clay or shale brick; Units in wall thickness: 2; Cells in wall thickness: 2; Minimum % solids: 87; $\frac{1}{2}''$ of 1:3 sanded gypsum plaster facings on one side.	See Notes	6 hrs.		1		1, 45	6
W-8-M-33	$8\frac{1}{2}''$	Core: hollow rolok bak of clay or shale; $\frac{1}{2}''$ of 1:3 sanded gypsum plaster facings on one side.	See Notes	5 hrs.		1		1, 45	5
W-8-M-34	$8\frac{5}{8}''$	Core: clay or shale structural tile; Units in wall thickness: 1; Cells in wall thickness: 2; Minimum % solids in units: 40; $\frac{5}{8}''$ of 1:3 sanded gypsum plaster facings on one side.	See Notes	2 hrs.		1		1, 20 21	2
W-8-M-35	$8\frac{5}{8}''$	Core: clay or shale structural tile; Units in wall thickness: 1; Cells in wall thickness: 2; Minimum % solids in units: 40; Exposed face: $\frac{5}{8}''$ of 1:3 sanded gypsum plaster.	See Notes	1 hr. 30 min.		1		1, 20, 21	$1\frac{1}{2}$
W-8-M-36	$8\frac{5}{8}''$	Core: clay or shale structural tile; Units in wall thickness: 1; Cells in wall thickness: 2; Minimum % solids in units: 43; $\frac{5}{8}''$ of 1:3 sanded gypsum plaster facings on one side.	See Notes	2 hrs.			1	1, 20,21	2
W-8-M-37	$8\frac{5}{8}''$	Core: clay or shale structural tile; Units in wall thickness: 1; Cells in wall thickness: 2; Minimum % solids in units: 43; $\frac{5}{8}''$ of 1:3 sanded gypsum plaster of the exposed face only.	See Notes	1 hr. 30 min.			1	1, 20, 21	$1\frac{1}{2}$
W-8-M-38	$8\frac{5}{8}''$	Core: clay or shale structural tile; Facings: side 1; see Note 17.	See Notes	2 hrs.		1		1, 2, 5, 10, 18, 20, 21	2
W-8-M-39	$8\frac{5}{8}''$	Core: clay or shale structural tile; Facings: exposed side only; see Note 17.	See Notes	1 hr. 30 min.		1		1, 2, 5, 10, 19, 20, 21	$1\frac{1}{2}$
W-8-M-40	$8\frac{5}{8}''$	Core: clay or shale structural tile; Facings: exposed side only; see Note 17.	See Notes	3 hrs.		1		1, 2, 5, 13, 18, 20, 21	3
W-8-M-41	$8\frac{5}{8}''$	Core: clay or shale structural tile; Facings: exposed side only; see Note 17.	See Notes	2 hrs.		1		1, 2, 5, 13, 19, 20, 21	2
W-8-M-42	$8\frac{5}{8}''$	Core: clay or shale structural tile; Facings: side 1; see Note 17.	See Notes	2 hrs. 30 min.		1		1, 2, 9, 18, 20, 21	$2\frac{1}{2}$
W-8-M-43	$8\frac{5}{8}''$	Core: clay or shale structural tile; Facings: exposed side only; see Note 17.	See Notes	1 hr. 30 min.		1		1, 2, 6, 9, 19, 20, 21	$1\frac{1}{2}$

(continued)

TABLE 1.1.4—continued
MASONRY WALLS
8″ TO LESS THAN 10″ THICK

ITEM CODE	THICKNESS	CONSTRUCTION DETAILS	PERFORMANCE LOAD	PERFORMANCE TIME	REFERENCE NUMBER PRE-BMS-92	REFERENCE NUMBER BMS-92	REFERENCE NUMBER POST-BMS-92	NOTES	REC. HOURS
W-8-M-44	$8^5/_8″$	Core: clay or shale structural tile; Facings: side 1, see Note 17; side 2, none.	See Notes	3 hrs.		1		1, 2, 10, 18, 20, 21	3
W-8-M-45	$8^5/_8″$	Core: clay or shale structural tile; Facings: fire side only; see Note 17.	See Notes	1 hr. 30 min.		1		1, 2, 6, 10, 19, 20, 21	$1^1/_2$
W-8-M-46	$8^5/_8″$	Core: clay or shale structural tile; Facings: side 1, see Note 17; side 2, none.	See Notes	3 hrs. 30 min.		1		1, 2, 6, 12, 18, 20, 21	$3^1/_2$
W-8-M-47	$8^5/_8″$	Core: clay or shale structural tile; Facings: exposed side only; see Note 17.	See Notes	1 hr. 45 min.		1		1, 2, 6, 12, 19, 20, 21	$1^3/_4$
W-8-M-48	$8^5/_8″$	Core: clay or shale structural tile; Facings: side 1, see Note 17; side 2, none.	See Notes	4 hrs.		1		1, 2, 6, 16, 18, 20, 21	4
W-8-M-49	$8^5/_8″$	Core: clay or shale structural tile; Facings: fire side only; see Note 17.	See Notes	2 hrs.		1		1, 2, 6, 16, 19, 20, 21	2
W-8-M-50	$8^5/_8″$	Core: 4″, 40% solid clay or shale clay structural tile; 4″ brick plus $^5/_8″$ of 1:3 sanded gypsum plaster facings on one side.	See Notes	4 hrs.		1		1, 20	4
W-8-M-51	$8^3/_4″$	$8^3/_4″ \times 2^1/_2″$ and $4″ \times 2^1/_2″$ cellular fletton (1873 psi) single and triple cell hollow brick set in $^1/_2″$ sand mortar in alternate courses.	3.6 tons/ft.	6 hrs.			7	23, 29	6
W-8-M-52	$8^3/_4″$	$8^3/_4″$ thick cement brick (2527 psi) with P.C. and sand mortar.	3.6 tons/ft.	6 hrs.			7	23, 24	6
W-8-M-53	$8^3/_4″$	$8^3/_4″ \times 2^1/_2″$ fletton brick (1831 psi) in $^1/_2″$ sand mortar.	3.6 tons/ft.	6 hrs.			7	23, 24	6
W-8-M-54	$8^3/_4″$	$8^3/_4″ \times 2^1/_2″$ London stock brick (683 psi) in $^1/_2″$ P.C. - sand mortar.	7.2 tons/ft.	6 hrs.			7	23, 24	6
W-9-M-55	9″	$9″ \times 2^1/_2″$ Leicester red wire-cut brick (4465 psi) in $^1/_2″$ P.C. - sand mortar.	6.0 tons/ft.	6 hrs.			7	23, 24	6
W-9-M-56	9″	$9″ \times 3″$ sand-lime brick (2603 psi) in $^1/_2″$ P.C. - sand mortar.	3.6 tons/ft.	6 hrs.			7	23, 24	6
W-9-M-57	9″	2 layers $2^7/_8″$ fletton brick (1910 psi) with $3^1/_4″$ air space; Cement and sand mortar.	1.5 tons/ft.	32 min.			7	23, 25	$^1/_3$
W-9-M-58	9″	$9″ \times 3″$ stairfoot brick (7527 psi) in $^1/_2″$ sand-cement mortar.	7.2 tons/ft.	6 hrs.			7	23, 24	6
W-9-M-59	9″	Core: solid clay or shale brick; $^1/_2″$ thick; 1:3 sanded gypsum plaster facings on both sides.	See Notes	7 hrs.		1		1, 22, 45	7
W-9-M-60	9″	Core: concrete brick; $^1/_2″$ of 1:3 sanded gypsum plaster facings on both sides.	See Notes	8 hrs.		1		1, 45	8
W-9-M-61	9″	Core: hollow rolok of clay or shale; $^1/_2″$ of 1:3 sanded gypsum plaster facings on both sides.	See Notes	4 hrs.		1		1, 45	4
W-9-M-62	9″	Cored clay or shale brick; Units in wall thickness: 1; Cells in wall thickness: 1; Minimum % solids: 70; $^1/_2″$ of 1:3 sanded gypsum plaster facings on one side.	See Notes	3 hrs.		1		1, 44	3

(continued)

TABLE 1.1.4—continued
MASONRY WALLS
8″ TO LESS THAN 10″ THICK

ITEM CODE	THICKNESS	CONSTRUCTION DETAILS	PERFORMANCE		REFERENCE NUMBER			NOTES	REC. HOURS
			LOAD	TIME	PRE-BMS-92	BMS-92	POST-BMS-92		
W-9-M-63	9″	Cored clay or shale brick; Units in wall thickness: 2; Cells in wall thickness: 2; Minimum % solids: 87; $^1/_2$″ of 1:3 sanded gypsum plaster facings on both sides.	See Notes	7 hrs.		1		1, 45	7
W-9-M-64	9-10″	Core: cavity wall of clay or shale brick; No facings.	See Notes	5 hrs.		1		1, 45	5
W-9-M-65	9-10″	Core: cavity construction of clay or shale brick; $^1/_2$″ of 1:3 sanded gypsum plaster facings on one side.	See Notes	6 hrs.		1		1, 45	6
W-9-M-66	9-10″	Core: cavity construction of clay or shale brick; $^1/_2$″ of 1:3 sanded gypsum plaster facings on both sides.	See Notes	7 hrs.		1		1, 45	7
W-9-M-67	$9^1/_4$″	Core: clay or shale structural tile; Units in wall thickness: 1; Cells in wall thickness: 2; Minimum % solids in units: 40; $^5/_8$″ of 1:3 sanded gypsum plaster facings on both sides.	See Notes	3 hrs.		1		1, 20, 21	3
W-9-M-68	$9^1/_4$″	Core: clay or shale structural tile; Units in wall thickness: 1; Cells in wall thickness: 2; Minimum % solids in units: 43; $^5/_8$″ of 1:3 sanded gypsum plaster facings on both sides.	See Notes	3 hrs.		1		1, 20, 21	3
W-9-M-69	$9^1/_4$″	Core: clay or shale structural tile; Facings: sides 1 and 2; see Note 17.	See Notes	3 hrs.		1		1, 2, 5, 10, 18, 20, 21	3
W-9-M-70	$9^1/_4$″	Core: clay or shale structural tile; Facings: sides 1 and 2; see Note 17.	See Notes	4 hrs.		1		1, 2, 5, 13, 18, 20, 21	4
W-9-M-71	$9^1/_4$″	Core: clay or shale structural tile; Facings: sides 1 and 2; see Note 17.	See Notes	3 hrs. 30 min.		1		1, 2, 6, 9, 18, 20, 21	$3^1/_2$
W-9-M-72	$9^1/_4$″	Core: clay or shale structural tile; Facings: sides 1 and 2; see Note 17.	See Notes	4 hrs.		1		1, 2, 6, 10, 18, 20, 21	4
W-9-M-73	$9^1/_4$″	Core: clay or shale structural tile; Facings: sides 1 and 2; see Note 17.	See Notes	4 hrs.		1		1, 2, 6, 12, 18, 20, 21	4
W-9-M-74	$9^1/_4$″	Core: clay or shale structural tile; Facings: sides 1 and 2; see Note 17.	See Notes	5 hrs.		1		1, 2, 6, 16, 18, 20, 21	5
W-9-M-75	8″	Cored concrete masonry; see Notes 2, 19, 26, 34, 40; No facings.	80 psi	1 hr. 30 min.		1		1, 20	$1^1/_2$
W-8-M-76	8″	Cored concrete masonry; see Notes 2, 18, 26, 34, 40; No facings	80 psi	4 hrs.		1		1, 20	4
W-8-M-77	8″	Cored concrete masonry; see Notes 2, 19, 26, 31, 40; No facings.	80 psi	1 hr. 15 min.		1		1, 20	$1^1/_4$
W-8-M-78	8″	Cored concrete masonry; see Notes 2, 18, 26, 31, 40; No facings.	80 psi	3 hrs.		1		1, 20	3
W-8-M-79	8″	Cored concrete masonry; see Notes 2, 19, 26, 36, 42; No facings.	80 psi	1 hr. 30 min.		1		1, 20	$1^1/_2$

(continued)

RESOURCE A—GUIDELINES ON FIRE RATINGS OF ARCHAIC MATERIALS AND ASSEMBLIES

TABLE 1.1.4—continued
MASONRY WALLS
8″ TO LESS THAN 10″ THICK

ITEM CODE	THICKNESS	CONSTRUCTION DETAILS	PERFORMANCE		REFERENCE NUMBER			NOTES	REC. HOURS
			LOAD	TIME	PRE-BMS-92	BMS-92	POST-BMS-92		
W-8-M-80	8″	Cored concrete masonry; see Notes 2, 18, 26, 36, 41; No facings.	80 psi	3 hrs.		1		1, 20	3
W-8-M-81	8″	Cored concrete masonry; see Notes 2, 19, 26, 34, 41; No facings.	80 psi	1 hr.		1		1, 20	1
W-8-M-82	8″	Cored concrete masonry; see Notes 2, 18, 26, 34, 41; No facings.	80 psi	2 hrs. 30 min.		1		1, 20	$2^{1}/_{2}$
W-8-M-83	8″	Cored concrete masonry; see Notes 2, 19, 26, 29, 41; No facings.	80 psi	45 min.		1		1, 20	$^{3}/_{4}$
W-8-M-84	8″	Cored concrete masonry; see Notes 2, 18, 26, 29, 41; No facings.	80 psi	2 hrs.		1		1, 20	2
W-8-M-85	$8^{1}/_{2}$″	Cored concrete masonry; see Notes 3, 18, 26, 34, 41; Facings: $2^{1}/_{4}$″ brick.	80 psi	4 hrs.		1		1, 20	4
W-8-M-86	8″	Cored concrete masonry; see Notes 3, 18, 26, 34, 41; Facings: $3^{3}/_{4}$″ brick face.	80 psi	5 hrs.		1		1, 20	5
W-8-M-87	8″	Cored concrete masonry; see Notes 2, 19, 26, 30, 43; No facings.	80 psi	12 min.		1		1, 20	$^{1}/_{5}$
W-8-M-88	8″	Cored concrete masonry; see Notes 2, 18, 26, 30, 43; No facings.	80 psi	12 min.		1		1, 20	$^{1}/_{5}$
W-8-M-89	$8^{1}/_{2}$″	Cored concrete masonry; see Notes 2, 19, 26, 34, 40; Facings: fire side only; see Note 38.	80 psi	2 hrs.		1		1, 20	2
W-8-M-90	$8^{1}/_{2}$″	Cored concrete masonry; see Notes 2, 18, 26, 34, 40; Facings: side 1; see Note 38.	80 psi	5 hrs.		1		1, 20	5
W-8-M-91	$8^{1}/_{2}$″	Cored concrete masonry; see Notes 2, 19, 26, 31, 40; Facings: fire side only; see Note 38.	80 psi	1 hr. 45 min.		1		1, 20	$1^{3}/_{4}$
W-8-M-92	$8^{1}/_{2}$″	Cored concrete masonry; see Notes 2, 18, 26, 31, 40; Facings: one side; see Note 38.	80 psi	4 hrs.		1		1, 20	4
W-8-M-93	$8^{1}/_{2}$″	Cored concrete masonry; see Notes 2, 19, 26, 36, 41; Facings: fire side only; see Note 38.	80 psi	2 hrs.		1		1, 20	2
W-8-M-94	$8^{1}/_{2}$″	Cored concrete masonry; see Notes 2, 18, 26, 36, 41; Facings: fire side only; see Note 38.	80 psi	4 hrs.		1		1, 20	4
W-8-M-95	$8^{1}/_{2}$″	Cored concrete masonry; see Notes 2, 19, 26, 34, 41; Facings: fire side only; see Note 38.	80 psi	1 hr. 30 min.		1		1, 20	$1^{1}/_{2}$
W-8-M-96	$8^{1}/_{2}$″	Cored concrete masonry; see Notes 2, 18, 26, 34, 41; Facings: one side; see Note 38.	80 psi	3 hrs.				1, 20	3
W-8-M-97	$8^{1}/_{2}$″	Cored concrete masonry; see Notes 2, 19, 26, 29, 41; Facings: fire side only; see Note 38.	80 psi	1 hr. 30 min.		1		1, 20	$1^{1}/_{2}$
W-8-M-98	$8^{1}/_{2}$″	Cored concrete masonry; see Notes 2, 18, 26, 29, 41; Facings: one side; see Note 38.	80 psi	2 hrs. 30 min.		1		1, 20	$2^{1}/_{2}$
W-8-M-99	$8^{1}/_{2}$″	Cored concrete masonry; see Notes 3, 19, 23, 27, 41; No facings.	80 psi	1 hr. 15 min.		1		1, 20	$1^{1}/_{4}$

(continued)

TABLE 1.1.4—continued
MASONRY WALLS
8″ TO LESS THAN 10″ THICK

ITEM CODE	THICKNESS	CONSTRUCTION DETAILS	PERFORMANCE		REFERENCE NUMBER			NOTES	REC. HOURS
			LOAD	TIME	PRE-BMS-92	BMS-92	POST-BMS-92		
W-8-M-100	$8^1/_2″$	Cored concrete masonry; see Notes 3, 18, 23, 27, 41; No facings.	80 psi	3 hrs. 30 min.		1		1, 20	$3^1/_2$
W-8-M-101	$8^1/_2″$	Cored concrete masonry; see Notes 3, 18, 26, 34, 41; Facings: $3^3/_4″$ brick face; one side only; see Note 38.	80 psi	6 hrs.		1		1, 20	6
W-8-M-102	$8^1/_2″$	Cored concrete masonry; see Notes 2, 19, 26, 30, 43; Facings: fire side only; see Note 38.	80 psi	30 min.		1		1, 20	$^1/_2$
W-8-M-103	$8^1/_2″$	Cored concrete masonry; see Notes 2, 18, 26, 30, 43; Facings: one side only; see Note 38.	80 psi	12 min.		1		1, 20	$^1/_5$
W-8-M-104	9″	Cored concrete masonry; see Notes 2, 18, 26, 34, 40; Facings: both sides; see Note 38.	80 psi	6 hrs.		1		1, 20	6
W-8-M-105	9″	Cored concrete masonry; see Notes 2, 18, 26, 31, 40; Facings: both sides; see Note 38.	80 psi	5 hrs.		1		1, 20	5
W-8-M-106	9″	Cored concrete masonry; see Notes 2, 18, 26, 36, 41; Facings: both sides of wall; see Note 38.	80 psi	5 hrs.		1		1, 20	5
W-8-M-107	9″	Cored concrete masonry; see Notes 2, 18, 26, 34, 41; Facings: both sides; see Note 38.	80 psi	4 hrs.		1		1, 20	4
W-8-M-108	9″	Cored concrete masonry; see Notes 2, 18, 26, 29, 41; Facings: both sides; see Note 38.	80 psi	3 hrs. 30 min.		1		1, 20	$3^1/_2$
W-8-M-109	9″	Cored concrete masonry; see Notes 3, 19, 23, 27, 40; Facings: fire side only; see Note 38.	80 psi	1 hr. 45 min.		1		1, 20	$1^3/_4$
W-8-M-110	9″	Cored concrete masonry; see Notes 3, 18, 23, 27, 41; Facings: one side only; see Note 38.	80 psi	4 hrs.		1		1, 20	4
W-8-M-111	9″	Cored concrete masonry; see Notes 3, 18, 26, 34, 41; $2^1/_4″$ brick face on one side only; see Note 38.	80 psi	5 hrs.		1		1, 20	5
W-8-M-112	9″	Cored concrete masonry; see Notes 2, 18, 26, 30, 43; Facings: both sides; see Note 38.	80 psi	30 min.		1		1, 20	$^1/_2$
W-9-M-113	$9^1/_2″$	Cored concrete masonry; see Notes 3, 18, 23, 27, 41; Facings: both sides; see Note 38.	80 psi	5 hrs.		1		1, 20	5
W-8-M-114	8″		200 psi	5 hrs.			43	22	5

For SI: 1 inch = 25.4 mm, 1 pound per square inch = 0.00689 MPa.

Notes:
1. Tested at NBS under ASA Spec. No. 43-1934 (ASTM C19-53).
2. One unit in wall thickness.
3. Two units in wall thickness.
4. Two or three units in wall thickness.
5. Two cells in wall thickness.
6. Three or four cells in wall thickness.
7. Four or five cells in wall thickness.
8. Five or six cells in wall thickness.
9. Minimum percent of solid materials in units = 40%.
10. Minimum percent of solid materials in units = 43%.
11. Minimum percent of solid materials in units = 46%.
12. Minimum percent of solid materials in units = 48%.
13. Minimum percent of solid materials in units = 49%.
14. Minimum percent of solid materials in units = 45%.
15. Minimum percent of solid materials in units = 51%.
16. Minimum percent of solid materials in units = 53%.
17. Not less than $^5/_8$ inch thickness of 1:3 sanded gypsum plaster.
18. Noncombustible or no members framed into wall.

(continued)

TABLE 1.1.4—continued
MASONRY WALLS
8″ TO LESS THAN 10″ THICK

19. Combustible members framed into wall.
20. Load: 80 psi for gross cross-sectional area of wall.
21. Portland cement-lime mortar.
22. Failure mode thermal.
23. British test.
24. Passed all criteria.
25. Failed by sudden collapse with no preceding signs of impending failure.
26. One cell in wall thickness.
27. Two cells in wall thickness.
28. Three cells in wall thickness.
29. Minimum percent of solid material in concrete units = 52.
30. Minimum percent of solid material in concrete units = 54.
31. Minimum percent of solid material in concrete units = 55.
32. Minimum percent of solid material in concrete units = 57.
33. Minimum percent of solid material in concrete units = 60.
34. Minimum percent of solid material in concrete units = 62.
35. Minimum percent of solid material in concrete units = 65.
36. Minimum percent of solid material in concrete units = 70.
37. Minimum percent of solid material in concrete units = 76.
38. Not less than $1/2$ inch of 1:3 sanded gypsum plaster.
39. Three units in wall thickness.
40. Concrete units made with expanded slag or pumice aggregates.
41. Concrete units made with expanded burned clay or shale, crushed limestone, air cooled slag or cinders.
42. Concrete units made with calcareous sand and gravel. Coarse aggregate, 60 percent or more calcite and dolomite.
43. Concrete units made with siliceous sand and gravel. Ninety percent or more quartz, chert and dolomite.
44. Load: 120 psi for gross cross-sectional area of wall.
45. Load: 160 psi for gross cross-sectional area of wall.

RESOURCE A—GUIDELINES ON FIRE RATINGS OF ARCHAIC MATERIALS AND ASSEMBLIES

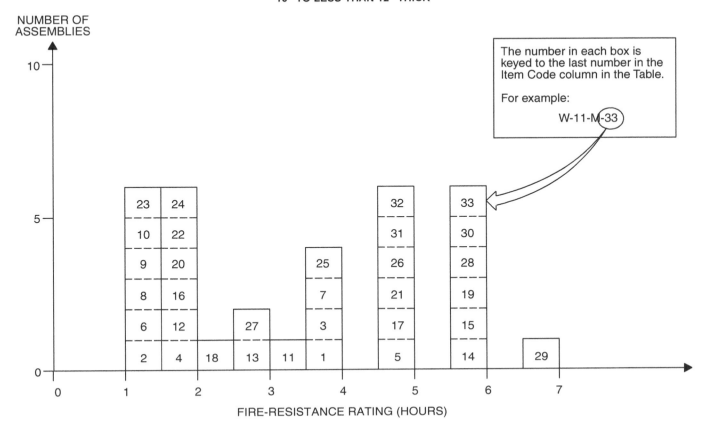

FIGURE 1.1.5
MASONRY WALLS
10″ TO LESS THAN 12″ THICK

TABLE 1.1.5
MASONRY WALLS
10″ TO LESS THAN 12″ THICK

ITEM CODE	THICKNESS	CONSTRUCTION DETAILS	PERFORMANCE		REFERENCE NUMBER			NOTES	REC. HOURS
			LOAD	TIME	PRE-BMS-92	BMS-92	POST-BMS-92		
W-10-M-1	10″	Core: two $3^3/_4″$, 40% solid clay or shale structural tiles with 2″ air space between; Facings: $^3/_4″$ Portland cement plaster on stucco on both sides.	80 psi	4 hrs.		1		1, 20	4
W-10-M-2	10″	Core: cored concrete masonry, 2″ air cavity; see Notes 3, 19, 27, 34, 40; No facings.	80 psi	1 hr. 30 min.		1		1, 20	$1^1/_2$
W-10-M-3	10″	Cored concrete masonry; see Notes 3, 18, 27, 34, 40; No facings.	80 psi	4 hrs.		1		1, 20	4
W-10-M-4	10″	Cored concrete masonry; see Notes 2, 19, 26, 34, 40; No facings.	80 psi	2 hrs.		1		1, 20	2
W-10-M-5	10″	Cored concrete masonry; see Notes 2, 18, 26, 33, 40; No facings.	80 psi	5 hrs.		1		1, 20	5
W-10-M-6	10″	Cored concrete masonry; see Notes 2, 19, 26, 33, 41; No facings.	80 psi	1 hr. 30 min.		1		1, 20	$1^1/_2$
W-10-M-7	10″	Cored concrete masonry; see Notes 2, 18, 26, 33, 41; No facings.	80 psi	4 hrs.		1		1, 20	4
W-10-M-8	10″	Cored concrete masonry (cavity type 2″ air space); see Notes 3, 19, 27, 34, 42; No facings.	80 psi	1 hr. 15 min.		1		1, 20	$1^1/_4$

(continued)

RESOURCE A—GUIDELINES ON FIRE RATINGS OF ARCHAIC MATERIALS AND ASSEMBLIES

TABLE 1.1.5—continued
MASONRY WALLS
10″ TO LESS THAN 12″ THICK

ITEM CODE	THICKNESS	CONSTRUCTION DETAILS	PERFORMANCE		REFERENCE NUMBER			NOTES	REC. HOURS
			LOAD	TIME	PRE-BMS-92	BMS-92	POST-BMS-92		
W-10-M-9	10″	Cored concrete masonry (cavity type 2″ air space); see Notes 3, 18, 27, 34, 42; No facings.	80 psi	1 hr. 15 min.		1		1, 20	$1\frac{1}{4}$
W-10-M-10	10″	Cored concrete masonry (cavity type 2″ air space); see Notes 3, 19, 27, 34, 41; No facings.	80 psi	1 hr. 15 min.		1		1, 20	$1\frac{1}{4}$
W-10-M-11	10″	Cored concrete masonry (cavity type 2″ air space); see Notes 3, 18, 27, 34, 41; No facings.	80 psi	3 hrs. 30 min.		1		1, 20	$3\frac{1}{2}$
W-10-M-12	10″	9″ thick concrete block ($11\frac{3}{4}″ \times 9″ \times 4\frac{1}{4}″$) with two 2″ thick voids included; $\frac{3}{8}″$ P.C. plaster $\frac{1}{8}″$ neat gypsum.	N/A	1 hr. 53 min.			7	23, 44	$1\frac{3}{4}$
W-10-M-13	10″	Holly clay tile block wall - $8\frac{1}{2}″$ block with two 3″ voids in each $8\frac{1}{2}″$ section; $\frac{3}{4}″$ gypsum plaster - each face.	N/A	2 hrs. 42 min.			7	23, 25	$2\frac{1}{2}$
W-10-M-14	10″	Two layers $4\frac{1}{4}″$ brick with $1\frac{1}{2}″$ air space; No ties sand cement mortar. (Fletton brick - 1910 psi).	N/A	6 hrs.			7	23, 24	6
W-10-M-15	10″	Two layers $4\frac{1}{4}″$ thick Fletton brick (1910 psi); $1\frac{1}{2}″$ air space; Ties: 18″ o.c. vertical; 3′ o.c. horizontal.	N/A	6 hrs.			7	23, 24	6
W-10-M-16	$10\frac{1}{2}″$	Cored concrete masonry; 2″ air cavity; see Notes 3, 19, 27, 34, 40; Facings: fire side only; see Note 38.	80 psi	2 hrs.		1		1, 20	2
W-10-M-17	$10\frac{1}{2}″$	Cored concrete masonry; see Notes 3, 18, 27, 34, 40; Facings: side 1 only; see Note 38.	80 psi	5 hrs.		1		1, 20	5
W-10-M-18	$10\frac{1}{2}″$	Cored concrete masonry; see Notes 2, 19, 26, 33, 40; Facings: fire side only; see Note 38.	80 psi	2 hrs. 30 min.		1		1, 20	$2\frac{1}{2}$
W-10-M-19	$10\frac{1}{2}″$	Cored concrete masonry; see Notes 2, 18, 26, 33, 40; Facings: one side; see Note 38.	80 psi	6 hrs.		1		1, 20	6
W-10-M-20	$10\frac{1}{2}″$	Cored concrete masonry; see Notes 2, 19, 26, 33, 41; Facings: fire side of wall only; see Note 38.	80 psi	2 hrs.		1		1, 20	2
W-10-M-21	$10\frac{1}{2}″$	Cored concrete masonry; see Notes 2, 18, 26, 33, 41; Facings: one side only; see Note 38.	80 psi	5 hrs.		1		1, 20	5
W-10-M-22	$10\frac{1}{2}″$	Cored concrete masonry (cavity type 2″ air space); see Notes 3,19, 27, 34, 42; Facings: fire side only; see Note 38.	80 psi	1 hr. 45 min.		1		1, 20	$1\frac{3}{4}$
W-10-M-23	$10\frac{1}{2}″$	Cored concrete masonry (cavity type 2″ air space); see Notes 3, 18, 27, 34, 42; Facings: one side only; see Note 38.	80 psi	1 hr. 15 min.		1		1, 20	$1\frac{1}{4}$
W-10-M-24	$10\frac{1}{2}″$	Cored concrete masonry (cavity type 2″ air space); see Notes 3, 19, 27, 34, 41; Facings: fire side only; see Note 38.	80 psi	2 hrs.		1		1, 20	2
W-10-M-25	$10\frac{1}{2}″$	Cored concrete masonry (cavity type 2″ air space); see Notes 3, 18, 27, 34, 41; Facings: one side only; see Note 38.	80 psi	4 hrs.		1		1, 20	4
W-10-M-26	$10\frac{5}{8}″$	Core: 8″, 40% solid tile plus 2″ furring tile; $\frac{5}{8}″$ sanded gypsum plaster between tile types; Facings: both sides $\frac{3}{4}″$ Portland cement plaster or stucco.	80 psi	5 hrs.		1		1, 20	5

(continued)

TABLE 1.1.5—continued
MASONRY WALLS
10″ TO LESS THAN 12″ THICK

ITEM CODE	THICKNESS	CONSTRUCTION DETAILS	PERFORMANCE		REFERENCE NUMBER			NOTES	REC. HOURS
			LOAD	TIME	PRE-BMS-92	BMS-92	POST-BMS-92		
W-10-M-27	$10^5/_8''$	Core: 8″, 40% solid tile plus 2″ furring tile; $^5/_8''$ sanded gypsum plaster between tile types; Facings: one side $^3/_4''$ Portland cement plaster or stucco.	80 psi	3 hrs. 30 min.		1		1, 20	$3^1/_2$
W-11-M-28	11″	Cored concrete masonry; see Notes 3, 18, 27, 34, 40; Facings: both sides; see Note 38.	80 psi	6 hrs.		1		1, 20	6
W-11-M-29	11″	Cored concrete masonry; see Notes 2, 18, 26, 33, 40; Facings: both sides; see Note 38.	80 psi	7 hrs.		1		1, 20	7
W-11-M-30	11″	Cored concrete masonry; see Notes 2, 18, 26, 33, 41; Facings: both sides of wall; see Note 38.	80 psi	6 hrs.		1		1, 20	6
W-11-M-31	11″	Cored concrete masonry (cavity type 2″ air space); see Notes 3, 18, 27, 34, 42; Facings: both sides; see Note 38.	80 psi	5 hrs.		1		1, 20	5
W-11-M-32	11″	Cored concrete masonry (cavity type 2″ air space); see Notes 3, 18, 27, 34, 41; Facings: both sides; see Note 38.	80 psi	5 hrs.		1		1, 20	5
W-11-M-33	11″	Two layers brick ($4^1/_2''$ Fletton, 2,428 psi) 2″ air space; galvanized ties; 18″ o.c. - horizontal; 3′ o.c. - vertical.	3 tons/ft.	6 hrs.			7	23, 24	6

For SI: 1 inch = 25.4 mm, 1 pound per square inch = 0.00689 MPa.

Notes:
1. Tested at NBS - ASA Spec. No. A2-1934.
2. One unit in wall thickness.
3. Two units in wall thickness.
4. Two or three units in wall thickness.
5. Two cells in wall thickness.
6. Three or four cells in wall thickness.
7. Four or five cells in wall thickness.
8. Five or six cells in wall thickness.
9. Minimum percent of solid materials in units = 40%.
10. Minimum percent of solid materials in units = 43%.
11. Minimum percent of solid materials in units = 46%.
12. Minimum percent of solid materials in units = 48%.
13. Minimum percent of solid materials in units = 49%.
14. Minimum percent of solid materials in units = 45%.
15. Minimum percent of solid materials in units = 51%.
16. Minimum percent of solid materials in units = 53%.
17. Not less than $^5/_8$ inch thickness of 1:3 sanded gypsum plaster.
18. Noncombustible or no members framed into wall.
19. Combustible members framed into wall.
20. Load: 80 psi for gross cross sectional area of wall.
21. Portland cement-lime mortar.
22. Failure mode—thermal.
23. British test.
24. Passed all criteria.
25. Failed by sudden collapse with no preceding signs of impending failure.
26. One cell in wall thickness.
27. Two cells in wall thickness.
28. Three cells in wall thickness.
29. Minimum percent of solid material in concrete units = 52%.
30. Minimum percent of solid material in concrete units = 54%.
31. Minimum percent of solid material in concrete units = 55%.
32. Minimum percent of solid material in concrete units = 57%.
33. Minimum percent of solid material in concrete units = 60%.
34. Minimum percent of solid material in concrete units = 62%.
35. Minimum percent of solid material in concrete units = 65%.

(continued)

**TABLE 1.1.5—continued
MASONRY WALLS
10″ TO LESS THAN 12″ THICK**

36. Minimum percent of solid material in concrete units = 70%.
37. Minimum percent of solid material in concrete units = 76%.
38. Not less than $1/2$ inch of 1:3 sanded gypsum plaster.
39. Three units in wall thickness.
40. Concrete units made with expanded slag or pumice aggregates.
41. Concrete units made with expanded burned clay or shale, crushed limestone, air cooled slag or cinders.
42. Concrete units made with calcareous sand and gravel. Coarse aggregate, 60 percent or more calcite and dolomite.

RESOURCE A—GUIDELINES ON FIRE RATINGS OF ARCHAIC MATERIALS AND ASSEMBLIES

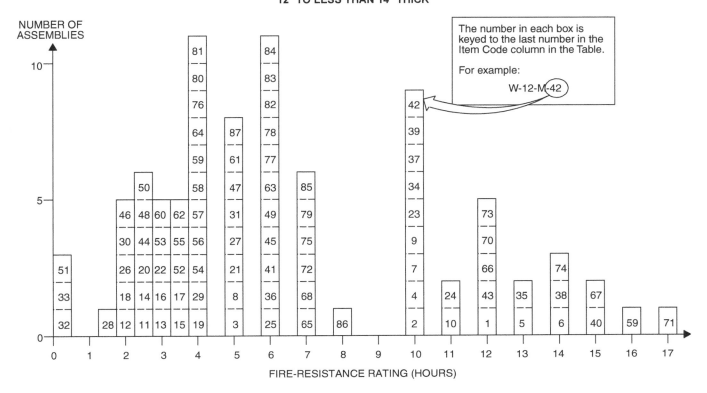

FIGURE 1.1.6
MASONRY WALLS
12″ TO LESS THAN 14″ THICK

TABLE 1.1.6
MASONRY WALLS
12″ TO LESS THAN 14″ THICK

ITEM CODE	THICKNESS	CONSTRUCTION DETAILS	PERFORMANCE		REFERENCE NUMBER			NOTES	REC. HOURS
			LOAD	TIME	PRE-BMS-92	BMS-92	POST-BMS-92		
W-12-M-1	12″	Core: solid clay or shale brick; No facings.	N/A	12 hrs.		1		1	12
W-12-M-2	12″	Core: solid clay or shale brick; No facings.	160 psi	10 hrs.		1		1, 44	10
W-12-M-3	12″	Core: hollow rolok of clay or shale; No facings.	160 psi	5 hrs.		1		1, 44	5
W-12-M-4	12″	Core: hollow rolok bak of clay or shale; No facings.	160 psi	10 hrs.		1		1, 44	10
W-12-M-5	12″	Core: concrete brick; No facings.	160 psi	13 hrs.		1		1, 44	13
W-12-M-6	12″	Core: sand-lime brick; No facings.	N/A	14 hrs.		1		1	14
W-12-M-7	12″	Core: sand-lime brick; No facings.	160 psi	10 hrs.		1		1, 44	10
W-12-M-8	12″	Cored clay or shale brick; Units in wall thickness: 1; Cells in wall thickness: 2; Minimum % solids: 70; No facings.	120 psi	5 hrs.		1		1, 45	5
W-12-M-9	12″	Cored clay or shale brick; Units in wall thickness: 3; Cells in wall thickness: 3; Minimum % solids: 87; No facings.	160 psi	10 hrs.		1		1, 44	10
W-12-M-10	12″	Cored clay or shale brick; Units in wall thickness: 3; Cells in wall thickness: 3; Minimum % solids: 87; No facings.	N/A	11 hrs.		1		1	11

(continued)

RESOURCE A—GUIDELINES ON FIRE RATINGS OF ARCHAIC MATERIALS AND ASSEMBLIES

TABLE 1.1.6—continued
MASONRY WALLS
12″ TO LESS THAN 14″ THICK

ITEM CODE	THICKNESS	CONSTRUCTION DETAILS	PERFORMANCE		REFERENCE NUMBER			NOTES	REC. HOURS
			LOAD	TIME	PRE-BMS-92	BMS-92	POST-BMS-92		
W-12-M-11	12″	Core: clay or shale structural tile; see Notes 2, 6, 9, 18; No facings.	80 psi	2 hrs.		1		1, 20	2$\frac{1}{2}$
W-12-M-12	12″	Core: clay or shale structural tile; see Notes 2, 4, 9, 19; No facings.	80 psi	2 hrs.		1		1, 20	2
W-12-M-13	12″	Core: clay or shale structural tile; see Notes 2, 6, 14, 19; No facings.	80 psi	3 hrs.		1		1, 20	3
W-12-M-14	12″	Core: clay or shale structural tile; see Notes 2, 6, 14, 18; No facings.	80 psi	2 hrs. 30 min.		1		1, 20	2$\frac{1}{2}$
W-12-M-15	12″	Core: clay or shale structural tile; see Notes 2, 4, 13, 18; No facings.	80 psi	3 hrs. 30 min.		1		1, 20	3$\frac{1}{2}$
W-12-M-16	12″	Core: clay or shale structural tile; see Notes 2, 4, 13, 19; No facings.	80 psi	3 hrs.		1		1, 20	3
W-12-M-17	12″	Core: clay or shale structural tile; see Notes 3, 6, 9, 18; No facings.	80 psi	3 hrs. 30 min.		1		1, 20	3$\frac{1}{2}$
W-12-M-18	12″	Core: clay or shale structural tile; see Notes 3, 6, 9, 19; No facings.	80 psi	2 hrs.		1		1, 20	2
W-12-M-19	12″	Core: clay or shale structural tile; see Notes 3, 6, 14, 18; No facings.	80 psi	4 hrs.		1		1, 20	4
W-12-M-20	12″	Core: clay or shale structural tile; see Notes 3, 6, 14, 19; No facings.	80 psi	2 hrs. 30 min.		1		1, 20	2$\frac{1}{2}$
W-12-M-21	12″	Core: clay or shale structural tile; see Notes 3, 6, 16, 18; No facings.	80 psi	5 hrs.		1		1, 20	5
W-12-M-22	12″	Core: clay or shale structural tile; see Notes 3, 6, 16, 19; No facings.	80 psi	3 hrs.		1		1, 20	3
W-12-M-23	12″	Core: 8″, 70% solid clay or shale structural tile; 4″ brick facings on one side.	80 psi	10 hrs.		1		1, 20	10
W-12-M-24	12″	Core: 8″, 70% solid clay or shale structural tile; 4″ brick facings on one side.	N/A	11 hrs.		1		1	11
W-12-M-25	12″	Core: 8″, 40% solid clay or shale structural tile; 4″ brick facings on one side.	80 psi	6 hrs.		1		1, 20	6
W-12-M-26	12″	Cored concrete masonry; see Notes 1, 9, 15, 16, 20; No facings.	80 psi	2 hrs.		1		1, 20	2
W-12-M-27	12″	Cored concrete masonry; see Notes 2, 18, 26, 34, 41; No facings.	80 psi	5 hrs.		1		1, 20	5
W-12-M-28	12″	Cored concrete masonry; see Notes 2, 19, 26, 31, 41; No facings.	80 psi	1 hr. 30 min.		1		1, 20	1$\frac{1}{2}$
W-12-M-29	12″	Cored concrete masonry; see Notes 2, 18, 26, 31, 41; No facings.	80 psi	4 hrs.		1		1, 20	4
W-12-M-30	12″	Cored concrete masonry; see Notes 3, 19, 27, 31, 43; No facings.	80 psi	2 hrs.		1		1, 20	2
W-12-M-31	12″	Cored concrete masonry; see Notes 3, 18, 27, 31, 43; No facings.	80 psi	5 hrs.		1		1, 20	5
W-12-M-32	12″	Cored concrete masonry; see Notes 2, 19, 26, 32, 43; No facings.	80 psi	25 min.		1		1, 20	$\frac{1}{3}$
W-12-M-33	12″	Cored concrete masonry; see Notes 2, 18, 26, 32, 43; No facings.	80 psi	25 min.		1		1, 20	$\frac{1}{3}$

(continued)

TABLE 1.1.6—continued
MASONRY WALLS
12″ TO LESS THAN 14″ THICK

ITEM CODE	THICKNESS	CONSTRUCTION DETAILS	PERFORMANCE		REFERENCE NUMBER			NOTES	REC. HOURS
			LOAD	TIME	PRE-BMS-92	BMS-92	POST-BMS-92		
W-12-M-34	12½″	Core: solid clay or shale brick; ½″ of 1:3 sanded gypsum plaster facings on one side.	160 psi	10 hrs.		1		1, 44	10
W-12-M-35	12½″	Core: solid clay or shale brick; ½″ of 1:3 sanded gypsum plaster facings on one side.	N/A	13 hrs.		1		1	13
W-12-M-36	12½″	Core: hollow rolok of clay or shale; ½″ of 1:3 sanded gypsum plaster facings on one side.	160 psi	6 hrs.		1		1, 44	6
W-12-M-37	12½″	Core: hollow rolok bak of clay or shale; ½″ of 1:3 sanded gypsum plaster facings on one side.	160 psi	10 hrs.		1		1, 44	10
W-12-M-38	12½″	Core: concrete; ½″ of 1:3 sanded gypsum plaster facings on one side.	160 psi	14 hrs.		1		1, 44	14
W-12-M-39	12½″	Core: sand-lime brick; ½″ of 1:3 sanded gypsum plaster facings on one side.	160 psi	10 hrs.		1		1, 44	10
W-12-M-40	12½″	Core: sand-lime brick; ½″ of 1:3 sanded gypsum plaster facings on one side.	N/A	15 hrs.		1		1	15
W-12-M-41	12½″	Cored clay or shale brick; Units in wall thickness: 1; Cells in wall thickness: 2; Minimum % solids: 70; ½″ of 1:3 sanded gypsum plaster facings on one side.	120 psi	6 hrs.		1		1, 45	6
W-12-M-42	12½″	Cored clay or shale brick; Units in wall thickness: 3; Cells in wall thickness: 3; Minimum % solids: 87; ½″ of 1:3 sanded gypsum plaster facings on one side.	160 psi	10 hrs.		1		1, 44	10
W-12-M-43	12½″	Cored clay or shale brick; Units in wall thickness: 3; Cells in wall thickness: 3; Minimum % solids: 87; ½″ of 1:3 sanded gypsum plaster facings on one side.	N/A	12 hrs.		1		1	12
W-12-M-44	12½″	Cored concrete masonry; see Notes 2, 19, 26, 34, 41; Facings: fire side only; see Note 38.	80 psi	2 hrs. 30 min.		1		1, 20	2½
W-12-M-45	12½″	Cored concrete masonry; see Notes 2, 18, 26, 34, 39, 41; Facings: one side only; see Note 38.	80 psi	6 hrs.		1		1, 20	6
W-12-M-46	12½″	Cored concrete masonry; see Notes 2, 19, 26, 31, 41; Facings: fire side only; see Note 38.	80 psi	2 hrs.		1		1, 20	2
W-12-M-47	12½″	Cored concrete masonry; see Notes 2, 18, 26, 31, 41; Facings: one side of wall only; see Note 38.	80 psi	5 hrs.		1		1, 20	5
W-12-M-48	12½″	Cored concrete masonry; see Notes 3, 19, 27, 31, 43; Facings: fire side only; see Note 38.	80 psi	2 hrs. 30 min.		1		1, 20	2½
W-12-M-49	12½″	Cored concrete masonry; see Notes 3, 18, 27, 31, 43; Facings: one side only; see Note 38.	80 psi	6 hrs.		1		1, 20	6
W-12-M-50	12½″	Cored concrete masonry; see Notes 2, 19, 26, 32, 43; Facings: fire side only; see Note 38.	80 psi	2 hrs. 30 min.		1		1, 20	2½

(continued)

RESOURCE A—GUIDELINES ON FIRE RATINGS OF ARCHAIC MATERIALS AND ASSEMBLIES

TABLE 1.1.6—continued
MASONRY WALLS
12″ TO LESS THAN 14″ THICK

ITEM CODE	THICKNESS	CONSTRUCTION DETAILS	PERFORMANCE		REFERENCE NUMBER			NOTES	REC. HOURS
			LOAD	TIME	PRE-BMS-92	BMS-92	POST-BMS-92		
W-12-M-51	12 1/2″	Cored concrete masonry; see Notes 2, 18, 26, 32, 43; Facings: one side only; see Note 38.	80 psi	25 min.		1		1, 20	1/3
W-12-M-52	12 5/8″	Clay or shale structural tile; see Notes 2, 6, 9, 18; Facings: side 1, see Note 17; side 2, none.	80 psi	3 hrs. 30 min.		1		1, 20	3 1/2
W-12-M-53	12 5/8″	Clay or shale structural tile; see Notes 2, 6, 9, 19; Facings: fire side only; see Note 17.	80 psi	3 hrs.		1		1, 20	3
W-12-M-54	12 5/8″	Clay or shale structural tile; see Notes 2, 6, 14, 19; Facings: side 1, see Note 17; side 2, none.	80 psi	4 hrs.		1		1, 20	4
W-12-M-55	12 5/8″	Clay or shale structural tile; see Notes 2, 6, 14, 18; Facings: exposed side only; see Note 17.	80 psi	3 hrs. 30 min.		1		1, 20	3 1/2
W-12-M-56	12 5/8″	Clay or shale structural tile; see Notes 2, 4, 13, 18; Facings: side 1, see Note 17; side 2, none.	80 psi	4 hrs.		1		1, 20	4
W-12-M-57	12 5/8″	Clay or shale structural tile; see Notes 1, 4, 13, 19; Facings: fire side only; see Note 17.	80 psi	4 hrs.		1		1, 20	4
W-12-M-58	12 5/8″	Clay or shale structural tile; see Notes 3, 6, 9, 18; Facings: side 1, see Note 17; side 2, none.	80 psi	4 hrs.		1		1, 20	4
W-12-M-59	12 5/8″	Clay or shale structural tile; see Notes 3, 6, 9, 19; Facings: fire side only; see Note 17.	80 psi	3 hrs.		1		1, 20	3
W-12-M-60	12 5/8″	Clay or shale structural tile; see Notes 3, 6, 14, 18; Facings: side 1, see Note 17; side 2, none.	80 psi	5 hrs.		1		1, 20	5
W-12-M-61	12 5/8″	Clay or shale structural tile; see Notes 3, 6, 14, 19; Facings: fire side only; see Note 17.	80 psi	3 hrs. 30 min.		1		1, 20	3 1/2
W-12-M-62	12 5/8″	Clay or shale structural tile; see Notes 3, 6, 16, 18; Facings: side 1, see Note 17; side 2, none.	80 psi	6 hrs.		1		1, 20	6
W-12-M-63	12 5/8″	Clay or shale structural tile; see Notes 3, 6, 16, 19; Facings: fire side only; see Note 17.	80 psi	4 hrs.		1		1, 20	4
W-12-M-64	12 5/8″	Core: 8″, 40% solid clay or shale structural tile; Facings: 4″ brick plus 5/8″ of 1:3 sanded gypsum plaster on one side.	80 psi	7 hrs.		1		1, 20	7
W-13-M-65	13″	Core: solid clay or shale brick; 1/2″ of 1:3 sanded gypsum plaster facings on both sides.	160 psi	12 hrs.		1		1, 44	12
W-13-M-66	13″	Core: solid clay or shale brick; 1/2″ of 1:3 sanded gypsum plaster facings on both sides.	N/A	15 hrs.		1		1, 20	15
W-13-M-67	13″	Core: solid clay or shale brick; 1/2″ of 1:3 sanded gypsum plaster facings on both sides.	N/A	15 hrs.		1		1	15
W-13-M-68	13″	Core: hollow rolok of clay or shale; 1/2″ of 1:3 sanded gypsum plaster facings on both sides.	80 psi	7 hrs.		1		1, 20	7
W-13-M-69	13″	Core: concrete brick; 1/2″ of 1:3 sanded gypsum plaster facings on both sides.	160 psi	16 hrs.		1		1, 44	16

(continued)

TABLE 1.1.6—continued
MASONRY WALLS
12″ TO LESS THAN 14″ THICK

ITEM CODE	THICKNESS	CONSTRUCTION DETAILS	PERFORMANCE		REFERENCE NUMBER			NOTES	REC. HOURS
			LOAD	TIME	PRE-BMS-92	BMS-92	POST-BMS-92		
W-13-M-70	13″	Core: sand-lime brick; $\frac{1}{2}$″ of 1:3 sanded gypsum plaster facings on both sides.	160 psi	12 hrs.		1		1, 44	12
W-13-M-71	13″	Core: sand-lime brick; $\frac{1}{2}$″ of 1:3 sanded gypsum plaster facings on both sides.	N/A	17 hrs.		1		1	17
W-13-M-72	13″	Cored clay or shale brick; Units in wall thickness: 1; Cells in wall thickness: 2; Minimum % solids: 70; $\frac{1}{2}$″ of 1:3 sanded gypsum plaster facings on both sides.	120 psi	7 hrs.		1		1, 45	7
W-13-M-73	13″	Cored clay or shale brick; Units in wall thickness: 3; Cells in wall thickness: 3; Minimum % solids: 87; $\frac{1}{2}$″ of 1:3 sanded gypsum plaster facings on both sides.	160 psi	12 hrs.		1		1, 44	12
W-13-M-74	13″	Cored clay or shale brick; Units in wall thickness: 3; Cells in wall thickness: 2; Minimum % solids: 87; $\frac{1}{2}$″ of 1:3 sanded gypsum plaster facings on both sides.	N/A	14 hrs.		1		1	14
W-13-M-75	13″	Cored concrete masonry; see Notes 18, 23, 28, 39, 41; No facings.	80 psi	7 hrs.		1		1, 20	7
W-13-M-76	13″	Cored concrete masonry; see Notes 19, 23, 28, 39, 41; No facings.	80 psi	4 hrs.		1		1, 20	4
W-13-M-77	13″	Cored concrete masonry; see Notes 3, 18, 27, 31, 43; Facings: both sides; see Note 38.	80 psi	6 hrs.		1		1, 20	6
W-13-M-78	13″	Cored concrete masonry; see Notes 2, 18, 26, 31, 41; Facings: both sides; see Note 38.	80 psi	6 hrs.		1		1, 20	6
W-13-M-79	13″	Cored concrete masonry; see Notes 2, 18, 26, 34, 41; Facings: both sides of wall; see Note 38.	80 psi	7 hrs.		1		1, 20	7
W-13-M-80	$13\frac{1}{4}$″	Core: clay or shale structural tile; see Notes 2, 6, 9, 18; Facings: both sides; see Note 17.	80 psi	4 hrs.		1		1, 20	4
W-13-M-81	$13\frac{1}{4}$″	Core: clay or shale structural tile; see Notes 2, 6, 14, 19; Facings: both sides; see Note 17.	80 psi	4 hrs.		1		1, 20	4
W-13-M-82	$13\frac{1}{4}$″	Core: clay or shale structural tile; see Notes 2, 4, 13, 18; Facings: both sides; see Note 17.	80 psi	6 hrs.		1		1, 20	6
W-13-M-83	$13\frac{1}{4}$″	Core: clay or shale structural tile; see Notes 3, 6, 9, 18; Facings: both sides; see Note 17.	80 psi	6 hrs.		1		1, 20	6
W-13-M-84	$13\frac{1}{4}$″	Core: clay or shale structural tile; see Notes 3, 6, 14, 18; Facings: both sides; see Note 17.	80 psi	6 hrs.		1		1, 20	6
W-13-M-85	$13\frac{1}{4}$″	Core: clay or shale structural tile; see Notes 3, 6, 16, 18; Facings: both sides; see Note 17.	80 psi	7 hrs.		1		1, 20	7

(continued)

RESOURCE A—GUIDELINES ON FIRE RATINGS OF ARCHAIC MATERIALS AND ASSEMBLIES

TABLE 1.1.6—continued
MASONRY WALLS
12″ TO LESS THAN 14″ THICK

ITEM CODE	THICKNESS	CONSTRUCTION DETAILS	PERFORMANCE		REFERENCE NUMBER			NOTES	REC. HOURS
			LOAD	TIME	PRE-BMS-92	BMS-92	POST-BMS-92		
W-13-M-86	13 1/2″	Cored concrete masonry; see Notes 18, 23, 28, 39, 41; Facings: one side only; see Note 38.	80 psi	8 hrs.		1		1, 20	8
W-13-M-87	13 1/2″	Cored concrete masonry; see Notes 19, 23, 28, 39, 41; Facings: fire side only; see Note 38.	80 psi	5 hrs.		1		1, 20	5

For SI: 1 inch = 25.4 mm, 1 pound per square inch = 0.00689 MPa.

Notes:
1. Tested at NBS - ASA Spec. No. A2-1934.
2. One unit in wall thickness.
3. Two units in wall thickness.
4. Two or three units in wall thickness.
5. Two cells in wall thickness.
6. Three or four cells in wall thickness.
7. Four or five cells in wall thickness.
8. Five or six cells in wall thickness.
9. Minimum percent of solid materials in units = 40%.
10. Minimum percent of solid materials in units = 43%.
11. Minimum percent of solid materials in units = 46%.
12. Minimum percent of solid materials in units = 48%.
13. Minimum percent of solid materials in units = 49%.
14. Minimum percent of solid materials in units = 45%.
15. Minimum percent of solid materials in units = 51%.
16. Minimum percent of solid materials in units = 53%.
17. Not less than $5/8$ inch thickness of 1:3 sanded gypsum plaster.
18. Noncombustible or no members framed into wall.
19. Combustible members framed into wall.
20. Load: 80 psi for gross area.
21. Portland cement-lime mortar.
22. Failure mode-thermal.
23. British test.
24. Passed all criteria.
25. Failed by sudden collapse with no preceding signs of impending failure.
26. One cell in wall thickness.
27. Two cells in wall thickness.
28. Three cells in wall thickness.
29. Minimum percent of solid material in concrete units = 52%.
30. Minimum percent of solid material in concrete units = 54%.
31. Minimum percent of solid material in concrete units = 55%.
32. Minimum percent of solid material in concrete units = 57%.
33. Minimum percent of solid material in concrete units = 60%.
34. Minimum percent of solid material in concrete units = 62%.
35. Minimum percent of solid material in concrete units = 65%.
36. Minimum percent of solid material in concrete units = 70%.
37. Minimum percent of solid material in concrete units = 76%.
38. Not less than $1/2$ inch of 1:3 sanded gypsum plaster.
39. Three units in wall thickness.
40. Concrete units made with expanded slag or pumice aggregates.
41. Concrete units made with expanded burned clay or shale, crushed limestone, air cooled slag or cinders.
42. Concrete units made with calcareous sand and gravel. Coarse aggregate, 60 percent or more calcite and dolomite.
43. Concrete units made with siliceous sand and gravel. Ninety percent or more quartz, chert or flint.
44. Load: 160 psi of gross wall cross sectional area.
45. Load: 120 psi of gross wall cross sectional area.

RESOURCE A—GUIDELINES ON FIRE RATINGS OF ARCHAIC MATERIALS AND ASSEMBLIES

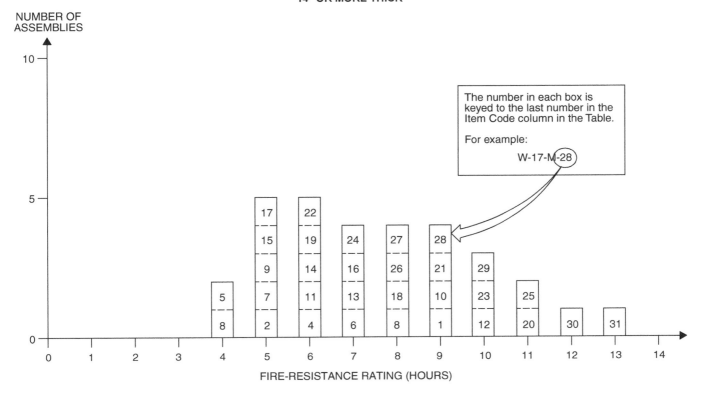

FIGURE 1.1.7
MASONRY WALLS
14″ OR MORE THICK

TABLE 1.1.7
MASONRY WALLS
14″ OR MORE THICK

ITEM CODE	THICKNESS	CONSTRUCTION DETAILS	PERFORMANCE		REFERENCE NUMBER			NOTES	REC. HOURS
			LOAD	TIME	PRE-BMS-92	BMS-92	POST-BMS-92		
W-14-M-1	14″	Core: cored masonry; see Notes 18, 28, 33, 39, 41; Facings: both sides; see Note 38.	80 psi	9 hrs.		1		1, 20	9
W-16-M-2	16″	Core: clay or shale structural tile; see Notes 4, 7, 9, 19; No facings.	80 psi	5 hrs.		1		1, 20	5
W-16-M-3	16″	Core: clay or shale structural tile; see Notes 4, 7, 9, 19; No facings.	80 psi	4 hrs.		1		1, 20	4
W-16-M-4	16″	Core: clay or shale structural tile; see Notes 4, 7, 10, 18; No facings.	80 psi	6 hrs.		1		1, 20	6
W-16-M-5	16″	Core: clay or shale structural tile; see Notes 4, 7, 10, 19; No facings.	80 psi	4 hrs.		1		1, 20	4
W-16-M-6	16″	Core: clay or shale structural tile; see Notes 4, 7, 11, 18; No facings.	80 psi	7 hrs.		1		1, 20	7
W-16-M-7	16″	Core: clay or shale structural tile; see Notes 4, 7, 11, 19; No facings.	80 psi	5 hrs.		1		1, 20	5
W-16-M-8	16″	Core: clay or shale structural tile; see Notes 4, 8, 13, 18; No facings.	80 psi	8 hrs.		1		1, 20	8
W-16-M-9	16″	Core: clay or shale structural tile; see Notes 4, 8, 13, 19; No facings.	80 psi	5 hrs.		1		1, 20	5

(continued)

RESOURCE A—GUIDELINES ON FIRE RATINGS OF ARCHAIC MATERIALS AND ASSEMBLIES

TABLE 1.1.7—continued
MASONRY WALLS
14″ OR MORE THICK

ITEM CODE	THICKNESS	CONSTRUCTION DETAILS	PERFORMANCE		REFERENCE NUMBER			NOTES	REC. HOURS
			LOAD	TIME	PRE-BMS-92	BMS-92	POST-BMS-92		
W-16-M-10	16″	Core: clay or shale structural tile; see Notes 4, 8, 15, 18; No facings.	80 psi	9 hrs.		1		1, 20	9
W-16-M-11	16″	Core: clay or shale structural tile; see Notes 3, 7, 14, 18; No facings.	80 psi	6 hrs.		1		1, 20	6
W-16-M-12	16″	Core: clay or shale structural tile; see Notes 4, 8, 16, 18; No facings.	80 psi	10 hrs.		1		1, 20	10
W-16-M-13	16″	Core: clay or shale structural tile; see Notes 4, 6, 16, 19; No facings.	80 psi	7 hrs.		1		1, 20	7
W-16-M-14	16 5/8″	Core: clay or shale structural tile; see Notes 4, 7, 9, 18; Facings: side 1, see Note 17; side 2, none.	80 psi	6 hrs.		1		1, 20	6
W-16-M-15	16 5/8″	Core: clay or shale structural tile; see Notes 4, 7, 9, 19; Facings: fire side only; see Note 17.	80 psi	5 hrs.		1		1, 20	5
W-16-M-16	16 5/8″	Core: clay or shale structural tile; see Notes 4, 7, 10, 18; Facings: side 1, see Note 17; side 2, none.	80 psi	7 hrs.		1		1, 20	7
W-16-M-17	16 5/8″	Core: clay or shale structural tile; see Notes 4, 7, 10, 19; Facings: fire side only; see Note 17.	80 psi	5 hrs.		1		1, 20	5
W-16-M-18	16 5/8″	Core: clay or shale structural tile; see Notes 4, 7, 11, 18; Facings: side 1, see Note 17; side 2, none.	80 psi	5 hrs.		1		1, 20	5
W-16-M-19	16 5/8″	Core: clay or shale structural tile; see Notes 4, 7, 11, 19; Facings: fire side only; see Note 17.	80 psi	6 hrs.		1		1, 20	6
W-16-M-20	16 5/8″	Core: clay or shale structural tile; see Notes 4, 8, 13, 18; Facings: sides 1 and 2; see Note 17.	80 psi	11 hrs.		1		1, 20	11
W-16-M-21	16 5/8″	Core: clay or shale structural tile; see Notes 4, 8, 13 18; Facings: side 1, see Note 17; side 2, none.	80 psi	9 hrs.		1		1, 20	9
W-16-M-22	16 5/8″	Core: clay or shale structural tile; see Notes 4, 8, 13, 19; Facings: fire side only; see Note 17.	80 psi	6 hrs.		1		1, 20	6
W-16-M-23	16 5/8″	Core: clay or shale structural tile; see Notes 4, 8, 15, 18; Facings: side 1, see Note 17; side 2, none.	80 psi	10 hrs.		1		1, 20	10
W-16-M-24	16 5/8″	Core: clay or shale structural tile; see Notes 4, 8, 15, 19; Facings: fire side only; see Note 17.	80 psi	7 hrs.		1		1, 20	7
W-16-M-25	16 5/8″	Core: clay or shale structural tile; see Notes 4, 6, 16, 18; Facings: side 1, see Note 17; side 2, none.	80 psi	11 hrs.		1		1, 20	11
W-16-M-26	16 5/8″	Core: clay or shale structural tile; see Notes 4, 6, 16, 19; Facings: fire side only; see Note 17.	80 psi	8 hrs.		1		1, 20	8

(continued)

RESOURCE A—GUIDELINES ON FIRE RATINGS OF ARCHAIC MATERIALS AND ASSEMBLIES

TABLE 1.1.7—continued
MASONRY WALLS
14″ OR MORE THICK

ITEM CODE	THICKNESS	CONSTRUCTION DETAILS	PERFORMANCE		REFERENCE NUMBER			NOTES	REC. HOURS
			LOAD	TIME	PRE-BMS-92	BMS-92	POST-BMS-92		
W-17-M-27	17¼″	Core: clay or shale structural tile; see Notes 4, 7, 9, 18; Facings: sides 1 and 2; see Note 17.	80 psi	8 hrs.		1		1, 20	8
W-17-M-28	17¼″	Core: clay or shale structural tile; see Notes 4, 7, 10, 18; Facings: sides 1 and 2; see Note 17.	80 psi	9 hrs.		1		1, 20	9
W-17-M-29	17¼″	Core: clay or shale structural tile; see Notes 4, 7, 11, 18; Facings: sides 1 and 2; see Note 17.	80 psi	10 hrs.		1		1, 20	10
W-17-M-30	17¼″	Core: clay or shale structural tile; see Notes 4, 8, 15, 18; Facings: sides 1 and 2; see Note 17.	80 psi	12 hrs.		1		1, 20	12
W-17-M-31	17¼″	Core: clay or shale structural tile; see Notes 4, 6, 16, 18; Facings: sides 1 and 2; see Note 17.	80 psi	13 hrs.		1		1, 20	13

For SI: 1 inch = 25.4 mm, 1 pound per square inch = 0.00689 MPa.

Notes:
1. Tested at NBS - ASA Spec. No. A2-1934.
2. One unit in wall thickness.
3. Two units in wall thickness.
4. Two or three units in wall thickness.
5. Two cells in wall thickness.
6. Three or four cells in wall thickness.
7. Four or five cells in wall thickness.
8. Five or six cells in wall thickness.
9. Minimum percent of solid materials in units = 40%.
10. Minimum percent of solid materials in units = 43%.
11. Minimum percent of solid materials in units = 46%.
12. Minimum percent of solid materials in units = 48%.
13. Minimum percent of solid materials in units = 49%.
14. Minimum percent of solid materials in units = 45%.
15. Minimum percent of solid materials in units = 51%.
16. Minimum percent of solid materials in units = 53%.
17. Not less than ⅝ inch thickness of 1:3 sanded gypsum plaster.
18. Noncombustible or no members framed into wall.
19. Combustible members framed into wall.
20. Load: 80 psi for gross area.
21. Portland cement-lime mortar.
22. Failure mode—thermal.
23. British test.
24. Passed all criteria.
25. Failed by sudden collapse with no preceding signs of impending failure.
26. One cell in wall thickness.
27. Two cells in wall thickness.
28. Three cells in wall thickness.
29. Minimum percent of solid material in concrete units = 52%.
30. Minimum percent of solid material in concrete units = 54%.
31. Minimum percent of solid material in concrete units = 55%.
32. Minimum percent of solid material in concrete units = 57%.
33. Minimum percent of solid material in concrete units = 60%.
34. Minimum percent of solid material in concrete units = 62%.
35. Minimum percent of solid material in concrete units = 65%.
36. Minimum percent of solid material in concrete units = 70%.
37. Minimum percent of solid material in concrete units = 76%.
38. Not less than ½ inch of 1:3 sanded gypsum plaster.
39. Three units in wall thickness.
40. Concrete units made with expanded slag or pumice aggregates.
41. Concrete units made with expanded burned clay or shale, crushed limestone, air cooled slag or cinders.
42. Concrete units made with calcareous sand and gravel. Coarse aggregate, 60 percent or more calcite and dolomite.
43. Concrete units made with siliceous sand and gravel. Ninety percent or more quartz, chert or flint.

RESOURCE A—GUIDELINES ON FIRE RATINGS OF ARCHAIC MATERIALS AND ASSEMBLIES

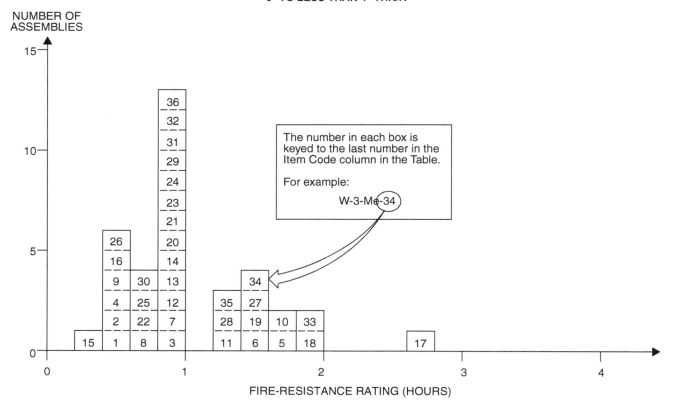

FIGURE 1.2.1
METAL FRAME WALLS
0″ TO LESS THAN 4″ THICK

TABLE 1.2.1
METAL FRAME WALLS
0″ TO LESS THAN 4″ THICK

ITEM CODE	THICKNESS	CONSTRUCTION DETAILS	PERFORMANCE		REFERENCE NUMBER			NOTES	REC. HOURS
			LOAD	TIME	PRE-BMS-92	BMS-92	POST-BMS-92		
W-3-Me-1	3″	Core: steel channels having three rows of 4″ × $^1/_8$″ staggered slots in web; core filled with heat expanded vermiculite weighing 1.5 lbs./ft.^{2} of wall area; Facings: sides 1 and 2, 18 gage steel, spot welded to core.	N/A	25 min.		1			$^1/_3$
W-3-Me-2	3″	Core: steel channels having three rows of 4″ × $^1/_8$″ staggered slots in web; core filled with heat expanded vermiculite weighing 2 lbs./ft.^{2} of wall area; Facings: sides 1 and 2, 18 gage steel, spot welded to core.	N/A	30 min.		1			$^1/_2$
W-3-Me-3	$2^1/_2$″	Solid partition: $^3/_8$″ tension rods (vertical) 3′ o.c. with metal lath; Scratch coat: cement/sand/lime plaster; Float coats: cement/sand/lime plaster; Finish coats: neat gypsum plaster.	N/A	1 hr.			7	1	1
W-2-Me-4	2″	Solid wall: steel channel per Note 1; 2″ thickness of 1:2; 1:3 Portland cement on metal lath.	N/A	30 min.		1			$^1/_2$

(continued)

**TABLE 1.2.1—continued
METAL FRAME WALLS
0″ TO LESS THAN 4″ THICK**

ITEM CODE	THICKNESS	CONSTRUCTION DETAILS	PERFORMANCE		REFERENCE NUMBER			NOTES	REC. HOURS
			LOAD	TIME	PRE-BMS-92	BMS-92	POST-BMS-92		
W-2-Me-5	2″	Solid wall: steel channel per Note 1; 2″ thickness of neat gypsum plaster on metal lath.	N/A	1 hr. 45 min.		1			$1^{3}/_{4}$
W-2-Me-6	2″	Solid wall: steel channel per Note 1; 2″ thickness of 1:1$^{1}/_{2}$; 1:1$^{1}/_{2}$ gypsum plaster on metal lath.	N/A	1 hr. 30 min.		1			$1^{1}/_{2}$
W-2-Me-7	2″	Solid wall: steel channel per Note 2; 2″ thickness of 1:1; 1:1 gypsum plaster on metal lath.	N/A	1 hr.		1			1
W-2-Me-8	2″	Solid wall: steel channel per Note 1; 2″ thickness of 1:2; 1:2 gypsum plaster on metal lath.	N/A	45 min.		1			$^{3}/_{4}$
W-2-Me-9	$2^{1}/_{4}″$	Solid wall: steel channel per Note 2; $2^{1}/_{4}″$ thickness of 1:2; 1:3 Portland cement on metal lath.	N/A	30 min.		1			$^{1}/_{2}$
W-2-Me-10	$2^{1}/_{4}″$	Solid wall: steel channel per Note 2; $2^{1}/_{4}″$ thickness of neat gypsum plaster on metal lath.	N/A	2 hrs.		1			2
W-2-Me-11	$2^{1}/_{4}″$	Solid wall: steel channel per Note 2; $2^{1}/_{4}″$ thickness of 1:$^{1}/_{2}$; 1:$^{1}/_{2}$ gypsum plaster on metal lath.	N/A	1 hr. 45 min.		1			$1^{3}/_{4}$
W-2-Me-12	$2^{1}/_{4}″$	Solid wall: steel channel per Note 2; $2^{1}/_{4}″$ thickness of 1:1; 1:1 gypsum plaster on metal lath.	N/A	1 hr. 15 min.		1			$1^{1}/_{4}$
W-2-Me-13	$2^{1}/_{4}″$	Solid wall: steel channel per Note 2; $2^{1}/_{4}″$ thickness of 1:2; 1:2 gypsum plaster on metal lath.	N/A	1 hr.		1			1
W-2-Me-14	$2^{1}/_{2}″$	Solid wall: steel channel per Note 1; $2^{1}/_{2}″$ thickness of 4.5:1:7; 4.5:1:7 Portland cement, sawdust and sand sprayed on wire mesh; see Note 3.	N/A	1 hr.		1			1
W-2-Me-15	$2^{1}/_{2}″$	Solid wall: steel channel per Note 2; $2^{1}/_{2}″$ thickness of 1:4; 1:4 Portland cement sprayed on wire mesh; see Note 3.	N/A	20 min.		1			$^{1}/_{3}$
W-2-Me-16	$2^{1}/_{2}″$	Solid wall: steel channel per Note 2; $2^{1}/_{2}″$ thickness of 1:2; 1:3 Portland cement on metal lath.	N/A	30 min.		1			$^{1}/_{2}$
W-2-Me-17	$2^{1}/_{2}″$	Solid wall: steel channel per Note 2; $2^{1}/_{2}″$ thickness of neat gypsum plaster on metal lath.	N/A	2 hrs. 30 min.		1			$2^{1}/_{2}$
W-2-Me-18	$2^{1}/_{2}″$	Solid wall: steel channel per Note 2; $2^{1}/_{2}″$ thickness of 1:$^{1}/_{2}$; 1:$^{1}/_{2}$ gypsum plaster on metal lath.	N/A	2 hrs.		1			2
W-2-Me-19	$2^{1}/_{2}″$	Solid wall: steel channel per Note 2; $2^{1}/_{2}″$ thickness of 1:1; 1:1 gypsum plaster on metal lath.	N/A	1 hr. 30 min.		1			$1^{1}/_{2}$

(continued)

RESOURCE A—GUIDELINES ON FIRE RATINGS OF ARCHAIC MATERIALS AND ASSEMBLIES

TABLE 1.2.1—continued
METAL FRAME WALLS
0″ TO LESS THAN 4″ THICK

ITEM CODE	THICKNESS	CONSTRUCTION DETAILS	PERFORMANCE		REFERENCE NUMBER			NOTES	REC. HOURS
			LOAD	TIME	PRE-BMS-92	BMS-92	POST-BMS-92		
W-2-Me-20	$2\frac{1}{2}$″	Solid wall: steel channel per Note 2; $2\frac{1}{2}$″ thickness of 1:2; 1:2 gypsum plaster on metal lath.	N/A	1 hr.		1			1
W-2-Me-21	$2\frac{1}{2}$″	Solid wall: steel channel per Note 2; $2\frac{1}{2}$″ thickness of 1:2; 1:3 gypsum plaster on metal lath.	N/A	1 hr.		1			1
W-3-Me-22	3″	Core: steel channel per Note 2; 1:2; 1:2 gypsum plaster on $\frac{3}{4}$″ soft asbestos lath; plaster thickness 2″.	N/A	45 min.		1			$\frac{3}{4}$
W-3-Me-23	$3\frac{1}{2}$″	Solid wall: steel channel per Note 2; $2\frac{1}{2}$″ thickness of 1:2; 1:2 gypsum plaster on $\frac{3}{4}$″ asbestos lath.	N/A	1 hr.		1			1
W-3-Me-24	$3\frac{1}{2}$″	Solid wall: steel channel per Note 2; lath over and $1:2\frac{1}{2}$; $1:2\frac{1}{2}$ gypsum plaster on 1″ magnesium oxysulfate wood fiberboard; plaster thickness $2\frac{1}{2}$″.	N/A	1 hr.		1			1
W-3-Me-25	$3\frac{1}{2}$″	Core: steel studs; see Note 4; Facings: $\frac{3}{4}$″ thickness of $1:\frac{1}{30}:2$; $1:\frac{1}{30}:3$ Portland cement and asbestos fiber plaster.	N/A	45 min.		1			$\frac{3}{4}$
W-3-Me-26	$3\frac{1}{2}$″	Core: steel studs; see Note 4; Facings: both sides $\frac{3}{4}$″ thickness of 1:2; 1:3 Portland cement.	N/A	30 min.		1			$\frac{1}{2}$
W-3-Me-27	$3\frac{1}{2}$″	Core: steel studs; see Note 4; Facings: both sides $\frac{3}{4}$″ thickness of neat gypsum plaster.	N/A	1 hr. 30 min.		1			$1\frac{1}{2}$
W-3-Me-28	$3\frac{1}{2}$″	Core: steel studs; see Note 4; Facings: both sides $\frac{3}{4}$″ thickness of $1:\frac{1}{2}$; $1:\frac{1}{2}$ gypsum plaster.	N/A	1 hr. 15 min.		1			$1\frac{1}{4}$
W-3-Me-29	$3\frac{1}{2}$″	Core: steel studs; see Note 4; Facings: both sides $\frac{3}{4}$″ thickness of 1:2; 1:2 gypsum plaster.	N/A	1 hr.		1			1
W-3-Me-30	$3\frac{1}{2}$″	Core: steel studs; see Note 4; Facings: both sides $\frac{3}{4}$″ thickness of 1:2; 1:3 gypsum plaster.	N/A	45 min.		1			$\frac{3}{4}$
W-3-Me-31	$3\frac{3}{4}$″	Core: steel studs; see Note 4; Facings: both sides $\frac{7}{8}$″ thickness of $1:\frac{1}{30}:2$; $1:\frac{1}{30}:3$ Portland cement and asbestos fiber plaster.	N/A	1 hr.		1			1
W-3-Me-32	$3\frac{3}{4}$″	Core: steel studs; see Note 4; Facings: both sides $\frac{7}{8}$″ thickness of 1:2; 1:3 Portland cement.	N/A	45 min.		1			$\frac{3}{4}$
W-3-Me-33	$3\frac{3}{4}$″	Core: steel studs; see Note 4; Facings: both sides $\frac{7}{8}$″ thickness of neat gypsum plaster.	N/A	2 hrs.		1			2
W-3-Me-34	$3\frac{3}{4}$″	Core: steel studs; see Note 4; Facings: both sides $\frac{7}{8}$″ thickness of $1:\frac{1}{2}$; $1:\frac{1}{2}$ gypsum plaster.	N/A	1 hr. 30 min.		1			$1\frac{1}{2}$

(continued)

TABLE 1.2.1—continued
METAL FRAME WALLS
0″ TO LESS THAN 4″ THICK

ITEM CODE	THICKNESS	CONSTRUCTION DETAILS	PERFORMANCE		REFERENCE NUMBER			NOTES	REC. HOURS
			LOAD	TIME	PRE-BMS-92	BMS-92	POST-BMS-92		
W-3-Me-35	$3^3/_4″$	Core: steel studs; see Note 4; Facings: both sides $^7/_8″$ thickness of 1:2; 1:2 gypsum plaster.	N/A	1 hr. 15 min.	1				$1^1/_4$
W-3-Me-36	$3^3/_4″$	Core: steel; see Note 4; Facings: $^7/_8″$ thickness of 1:2; 1:3 gypsum plaster on both sides.	N/A	1 hr.	1				1

For SI: 1 inch = 25.4 mm.

Notes:
1. Failure mode—local temperature rise—back face.
2. Three-fourths inch or 1 inch channel framing—hot-rolled or strip-steel channels.
3. Reinforcement is 4-inch square mesh of No. 6 wire welded at intersections (no channels).
4. Ratings are for any usual type of nonload-bearing metal framing providing 2 inches (or more) air space.

General Note:
The construction details of the wall assemblies are as complete as the source documentation will permit. Data on the method of attachment of facings and the gauge of steel studs was provided when known. The cross-sectional area of the steel stud can be computed, thereby permitting a reasoned estimate of actual loading conditions. For load-bearing assemblies, the maximum allowable stress for the steel studs has been provided in the table "Notes." More often, it is the thermal properties of the facing materials, rather than the specific gauge of the steel, that will determine the degree of fire resistance. This is particularly true for nonbearing wall assemblies.

RESOURCE A—GUIDELINES ON FIRE RATINGS OF ARCHAIC MATERIALS AND ASSEMBLIES

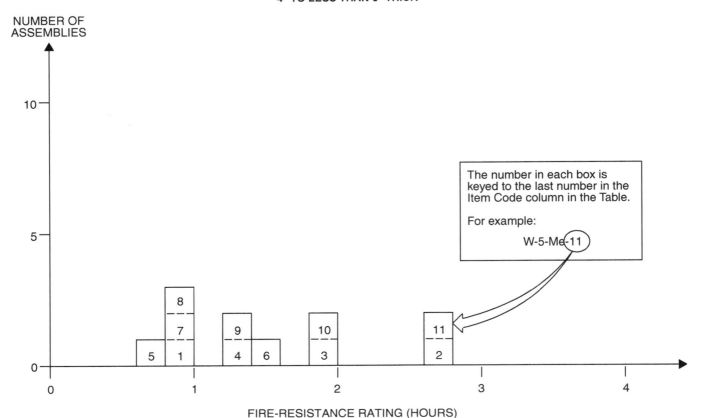

FIGURE 1.2.2
METAL FRAME WALLS
4″ TO LESS THAN 6″ THICK

TABLE 1.2.2
METAL FRAME WALLS
4″ TO LESS THAN 6″ THICK

ITEM CODE	THICKNESS	CONSTRUCTION DETAILS	PERFORMANCE		REFERENCE NUMBER			NOTES	REC. HOURS
			LOAD	TIME	PRE-BMS-92	BMS-92	POST-BMS-92		
W-5-Me-1	$5\frac{1}{2}''$	3″ cavity with 16 ga. channel studs ($3\frac{1}{2}''$ o.c.) of $\frac{1}{2}'' \times \frac{1}{2}''$ channel and 3″ spacer; Metal lath on ribs with plaster (three coats) $\frac{3}{4}''$ over face of lath; Plaster (each side): scratch coat, cement/lime/sand with hair; float coat, cement/lime/sand; finish coat, neat gypsum.	N/A	1 hr. 11 min.			7	1	1
W-4-Me-2	4″	Core: steel studs; see Note 2; Facings: both sides 1″ thickness of neat gypsum plaster.	N/A	2 hrs. 30 min.		1			$2\frac{1}{2}$
W-4-Me-3	4″	Core: steel studs; see Note 2; Facings: both sides 1″ thickness of $1:1\frac{1}{2}$; $1:1\frac{1}{2}$ gypsum plaster.	N/A	2 hrs.		1			2
W-4-Me-4	4″	Core: steel; see Note 2; Facings: both sides 1″ thickness of 1:2; 1:3 gypsum plaster.	N/A	1 hr. 15 min.		1			$1\frac{1}{4}$
W-4-Me-5	$4\frac{1}{2}''$	Core: lightweight steel studs 3″ in depth; Facings: both sides $\frac{3}{4}''$ thick sanded gypsum plaster, 1:2 scratch coat, 1:3 brown coat applied on metal lath.	See Note 4	45 min.		1		5	$\frac{3}{4}$

(continued)

TABLE 1.2.2—continued
METAL FRAME WALLS
4″ TO LESS THAN 6″ THICK

ITEM CODE	THICKNESS	CONSTRUCTION DETAILS	PERFORMANCE		REFERENCE NUMBER			NOTES	REC. HOURS
			LOAD	TIME	PRE-BMS-92	BMS-92	POST-BMS-92		
W-4-Me-6	4$\frac{1}{2}$″	Core: lightweight steel studs 3″ in depth; Facings: both sides $\frac{3}{4}$″ thick neat gypsum plaster on metal lath.	See Note 4	1 hr. 30 min.		1		5	1$\frac{1}{2}$
W-4-Me-7	4$\frac{1}{2}$″	Core: lightweight steel studs 3″ in depth; Facings: both sides $\frac{3}{4}$″ thick sanded gypsum plaster, 1:2 scratch and brown coats applied on metal lath.	See Note 4	1 hr.		1		5	1
W-4-Me-8	4$\frac{3}{4}$″	Core: lightweight steel studs 3″ in depth; Facings: both sides $\frac{7}{8}$″ thick sanded gypsum plaster, 1:2 scratch coat, 1:3 brown coat, applied on metal lath.	See Note 4	1 hr.		1		5	1
W-4-Me-9	4$\frac{3}{4}$″	Core: lightweight steel studs 3″ in depth; Facings: both sides $\frac{7}{8}$″ thick sanded gypsum plaster, 1:2 scratch and 1:3 brown coats applied on metal lath.	See Note 4	1 hr. 15 min.		1		5	1$\frac{1}{4}$
W-5-Me-10	5″	Core: lightweight steel studs 3″ in depth; Facings: both sides 1″ thick neat gypsum plaster on metal lath.	See Note 4	2 hrs.		1		5	2
W-5-Me-11	5″	Core: lightweight steel studs 3″ in depth; Facings: both sides 1″ thick neat gypsum plaster on metal lath.	See Note 4	2 hrs. 30 min.		1		5, 6	2$\frac{1}{2}$

For SI: 1 inch = 25.4 mm, 1 pound per square inch = 0.00689 MPa.

Notes:
1. Failure mode—local back face temperature rise.
2. Ratings are for any usual type of nonbearing metal framing providing a minimum 2 inches air space.
3. Facing materials secured to lightweight steel studs not less than 3 inches deep.
4. Rating based on loading to develop a maximum stress of 7270 psi for net area of each stud.
5. Spacing of steel studs must be sufficient to develop adequate rigidity in the metal-lath or gypsum-plaster base.
6. As per Note 4 but load/stud not to exceed 5120 psi.

General Note:
The construction details of the wall assemblies are as complete as the source documentation will permit. Data on the method of attachment of facings and the gauge of steel studs was provided when known. The cross sectional area of the steel stud can be computed, thereby permitting a reasoned estimate of actual loading conditions. For load-bearing assemblies, the maximum allowable stress for the steel studs has been provided in the table "Notes." More often, it is the thermal properties of the facing materials, rather than the specific gauge of the steel, that will determine the degree of fire resistance. This is particularly true for nonbearing wall assemblies.

RESOURCE A—GUIDELINES ON FIRE RATINGS OF ARCHAIC MATERIALS AND ASSEMBLIES

**TABLE 1.2.3
METAL FRAME WALLS
6″ TO LESS THAN 8″ THICK**

ITEM CODE	THICKNESS	CONSTRUCTION DETAILS	PERFORMANCE		REFERENCE NUMBER			NOTES	REC HOURS
			LOAD	TIME	PRE-BMS-92	BMS-92	POST-BMS-92		
W-6-Me-1	$6^5/_8''$	On one side of 1″ magnesium oxysulfate wood fiberboard sheathing attached to steel studs (see Notes 1 and 2), 1″ air space, $3^3/_4''$ brick secured with metal ties to steel frame every fifth course; Inside facing of $^7/_8''$ 1:2 sanded gypsum plaster on metal lath secured directly to studs; Plaster side exposed to fire.	See Note 2	1 hr. 45 min.		1		1	$1^3/_4$
W-6-Me-2	$6^5/_8''$	On one side of 1″ magnesium oxysulfate wood fiberboard sheathing attached to steel studs (see Notes 1 and 2), 1″ air space, $3^3/_4''$ brick secured with metal ties to steel frame every fifth course; Inside facing of $^7/_8''$ 1:2 sanded gypsum plaster on metal lath secured directly to studs; Brick face exposed to fire.	See Note 2	4 hrs.		1		1	4
W-6-Me-3	$6^5/_8''$	On one side of 1″ magnesium oxysulfate wood fiberboard sheathing attached to steel studs (see Notes 1 and 2), 1″ air space, $3^3/_4''$ brick secured with metal ties to steel frame every fifth course; Inside facing of $^7/_8''$ vermiculite plaster on metal lath secured directly to studs; Plaster side exposed to fire.	See Note 2	2 hrs.		1		1	2

For SI: 1 inch = 25.4 mm, 1 pound per square inch = 0.00689 MPa.

Notes:
1. Lightweight steel studs (minimum 3 inches deep) used. Stud spacing dependent on loading, but in each case, spacing is to be such that adequate rigidity is provided to the metal lath plaster base.
2. Load is such that stress developed in studs is not greater than 5120 psi calculated from net stud area.

General Note:
The construction details of the wall assemblies are as complete as the source documentation will permit. Data on the method of attachment of facings and the gauge of steel studs was provided when known. The cross sectional area of the steel stud can be computed, thereby permitting a reasoned estimate of actual loading conditions. For load-bearing assemblies, the maximum allowable stress for the steel studs has been provided in the table "Notes." More often, it is the thermal properties of the facing materials, rather than the specific gauge of the steel, that will determine the degree of fire resistance. This is particularly true for nonbearing wall assemblies.

**TABLE 1.2.4
METAL FRAME WALLS
8″ TO LESS THAN 10″ THICK**

ITEM CODE	THICKNESS	CONSTRUCTION DETAILS	PERFORMANCE		REFERENCE NUMBER			NOTES	REC. HOURS
			LOAD	TIME	PRE-BMS-92	BMS-92	POST-BMS-92		
W-9-Me-1	$9^1/_{16}″$	On one side of $^1/_2″$ wood fiberboard sheathing next to studs, $^3/_4″$ air space formed with $^3/_4″ \times 1^5/_8″$ wood strips placed over the fiberboard and secured to the studs, paper backed wire lath nailed to strips $3^3/_4″$ brick veneer held in place by filling a $^3/_4″$ space between the brick and paper backed lath with mortar; Inside facing of $^3/_4″$ neat gypsum plaster on metal lath attached to $^5/_{16}″$ plywood strips secured to edges of steel studs; Rated as combustible because of the sheathing; See Notes 1 and 2; Plaster exposed.	See Note 2	1 hr. 45 min.	1			1	$1^3/_4$
W-9-Me-2	$9^1/_{16}″$	Same as above with brick exposed.	See Note 2	4 hrs.	1			1	4
W-8-Me-3	$8^1/_2″$	On one side of paper backed wire lath attached to studs and $3^3/_4″$ brick veneer held in place by filling a 1″ space between the brick and lath with mortar; Inside facing of 1″ paper-enclosed mineral wool blanket weighing 0.6 lb./ft.^{2} attached to studs, metal lath or paper backed wire lath laid over the blanket and attached to the studs, $^3/_4″$ sanded gypsum plaster 1:2 for the scratch coat and 1:3 for the brown coat; See Notes 1 and 2; Plaster face exposed.	See Note 2	4 hrs.	1			1	4
W-8-Me-4	$8^1/_2″$	Same as above with brick exposed.	See Note 2	5 hrs.	1			1	5

For SI: 1 inch = 25.4 mm, 1 pound per square inch = 0.00689 MPa.

Notes:
1. Lightweight steel studs ≥ 3 inches in depth. Stud spacing dependent on loading, but in any case, the spacing is to be such that adequate rigidity is provided to the metal-lath plaster base.
2. Load is such that stress developed in studs is ≤ 5120 psi calculated from the net area of the stud.

General Note:
The construction details of the wall assemblies are as complete as the source documentation will permit. Data on the method of attachment of facings and the gauge of steel studs was provided when known. The cross sectional area of the steel stud can be computed, thereby permitting a reasoned estimate of actual loading conditions. For load-bearing assemblies, the maximum allowable stress for the steel studs has been provided in the table "Notes." More often, it is the thermal properties of the facing materials, rather than the specific gauge of the steel, that will determine the degree of fire resistance. This is particularly true for nonbearing wall assemblies.

RESOURCE A—GUIDELINES ON FIRE RATINGS OF ARCHAIC MATERIALS AND ASSEMBLIES

**TABLE 1.3.1
WOOD FRAME WALLS
0″ TO LESS THAN 4″ THICK**

ITEM CODE	THICKNESS	CONSTRUCTION DETAILS	PERFORMANCE		REFERENCE NUMBER			NOTES	REC. HOURS
			LOAD	TIME	PRE-BMS-92	BMS-92	POST-BMS-92		
W-3-W-1	$3\frac{3}{4}″$	Solid wall: $2\frac{1}{4}″$ wood-wool slab core; $\frac{3}{4}″$ gypsum plaster each side.	N/A	2 hrs.			7	1, 6	2
W-3-W-2	$3\frac{7}{8}″$	2 × 4 stud wall; $\frac{3}{16}″$ thick cement asbestos board on both sides of wall.	360 psi net area	10 min.	1			2-5	$\frac{1}{6}$
W-3-W-3	$3\frac{7}{8}″$	Same as W-3-W-2 but stud cavities filled with 1 lb./ft.^{2} mineral wool batts.	360 psi net area	40 min.	1			2-5	$\frac{2}{3}$

For SI: 1 inch = 25.4 mm, 1 pound per square inch = 0.00689 MPa.

Notes:
1. Achieved "Grade C" fire resistance (British).
2. Nominal 2 × 4 wood studs of No. 1 common or better lumber set edgewise, 2 × 4 plates at top and bottom and blocking at mid height of wall.
3. All horizontal joints in facing material backed by 2 × 4 blocking in wall.
4. Load: 360 psi of net stud cross sectional area.
5. Facings secured with 6d casing nails. Nail holes predrilled and 0.02 inch to 0.03 inch smaller than nail diameter.
6. The wood-wool core is a pressed excelsior slab which possesses insulating properties similar to cellulosic insulation.

RESOURCE A—GUIDELINES ON FIRE RATINGS OF ARCHAIC MATERIALS AND ASSEMBLIES

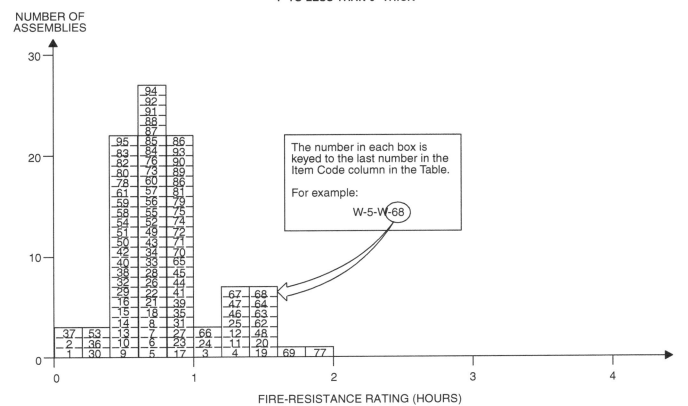

FIGURE 1.3.2
WOOD FRAME WALLS
4″ TO LESS THAN 6″ THICK

TABLE 1.3.2
WOOD FRAME WALLS
4″ TO LESS THAN 6″ THICK

ITEM CODE	THICKNESS	CONSTRUCTION DETAILS	PERFORMANCE LOAD	PERFORMANCE TIME	REFERENCE NUMBER PRE-BMS-92	REFERENCE NUMBER BMS-92	REFERENCE NUMBER POST-BMS-92	NOTES	REC. HOURS
W-4-W-1	4″	2″ × 4″ stud wall; $3/16$″ CAB; no insulation; Design A.	35 min.	10 min.			4	1-10	$1/6$
W-4-W-2	$4 1/8$″	2″ × 4″ stud wall; $3/16$″ CAB; no insulation; Design A.	38 min.	9 min.			4	1-10	$1/6$
W-4-W-3	$4 3/4$″	2″ × 4″ stud wall; $3/16$″ CAB and $3/8$″ gypsum board face (both sides); Design B.	62 min.	64 min.			4	1-10	1
W-5-W-4	5″	2″ × 4″ stud wall; $3/16$″ CAB and $1/2$″ gypsum board (both sides); Design B.	79 min.	Greater than 90 min.			4	1-10	1
W-4-W-5	$4 3/4$″	2″ × 4″ stud wall; $3/16$″ CAB and $3/8$″ gypsum board (both sides); Design B.	45 min.	45 min.			4	1-12	—
W-5-W-6	5″	2″ × 4″ stud wall; $3/16$″ CAB and $1/2$″ gypsum board face (both sides); Design B.	45 min.	45 min.			4	1-10, 12, 13	—
W-4-W-7	4″	2″ × 4″ stud wall; $3/16$″ CAB face; $3 1/2$″ mineral wool insulation; Design C.	40 min.	42 min.			4	1-10	$2/3$
W-4-W-8	4″	2″ × 4″ stud wall; $3/16$″ CAB face; $3 1/2$″ mineral wool insulation; Design C.	46 min.	46 min.			4	1-10, 43	$2/3$
W-4-W-9	4″	2″ × 4″ stud wall; $3/16$″ CAB face; $3 1/2$″ mineral wool insulation; Design C.	30 min.	30 min.			4	1-10, 12, 14	—

(continued)

RESOURCE A—GUIDELINES ON FIRE RATINGS OF ARCHAIC MATERIALS AND ASSEMBLIES

TABLE 1.3.2—continued
WOOD FRAME WALLS
4″ TO LESS THAN 6″ THICK

ITEM CODE	THICKNESS	CONSTRUCTION DETAILS	PERFORMANCE		REFERENCE NUMBER			NOTES	REC. HOURS
			LOAD	TIME	PRE-BMS-92	BMS-92	POST-BMS-92		
W-4-W-10	4$\frac{1}{8}$″	2″ × 4″ stud wall; $\frac{3}{16}$″ CAB face; 3$\frac{1}{2}$″ mineral wool insulation; Design C.	—	30 min.			4	1-8, 12, 14	—
W-4-W-11	4$\frac{3}{4}$″	2″ × 4″ stud wall; $\frac{3}{16}$″ CAB face; $\frac{3}{8}$″ gypsum strips over studs; 5$\frac{1}{2}$″ mineral wool insulation; Design D.	79 min.	79 min.			4	1-10	1
W-4-W-12	4$\frac{3}{4}$″	2″ × 4″ stud wall; $\frac{3}{16}$″ CAB face; $\frac{3}{8}$″ gypsum strips at stud edges; 7$\frac{1}{2}$″ mineral wool insulation; Design D.	82 min.	82 min.			4	1-10	1
W-4-W-13	4$\frac{3}{4}$″	2″ × 4″ stud wall; $\frac{3}{16}$″ CAB face; $\frac{3}{8}$″ gypsum board strips over studs; 5$\frac{1}{2}$″ mineral wool insulation; Design D.	30 min.	30 min.			4	1-12	—
W-4-W-14	4$\frac{3}{4}$″	2″ × 4″ stud wall; $\frac{3}{16}$″ CAB face; $\frac{3}{8}$″ gypsum board strips over studs; 7″ mineral wool insulation; Design D.	30 min.	30 min.			4	1-12	—
W-5-W-15	5$\frac{1}{2}$″	2″ × 4″ stud wall; Exposed face: CAB shingles over 1″ × 6″; Unexposed face: $\frac{1}{8}$″ CAB sheet; $\frac{7}{16}$″ fiberboard (wood); Design E.	34 min.	—			4	1-10	$\frac{1}{2}$
W-5-W-16	5$\frac{1}{2}$″	2″ × 4″ stud wall; Exposed face: $\frac{1}{8}$″ CAB sheet; $\frac{7}{16}$″ fiberboard; Unexposed face: CAB shingles over 1″ × 6″; Design E.	32 min.	33 min.			4	1-10	$\frac{1}{2}$
W-5-W-17	5$\frac{1}{2}$″	2″ × 4″ stud wall; Exposed face: CAB shingles over 1″ × 6″; Unexposed face: $\frac{1}{8}$″ CAB sheet; gypsum at stud edges; 3$\frac{1}{2}$″ mineral wood insulation; Design F.	51 min.	—			4	1-10	$\frac{3}{4}$
W-5-W-18	5$\frac{1}{2}$″	2″ × 4″ stud wall; Exposed face: $\frac{1}{8}$″ CAB sheet; gypsum board at stud edges; Unexposed face: CAB shingles over 1″ × 6″; 3$\frac{1}{2}$″ mineral wool insulation; Design F.	42 min.	—			4	1-10	$\frac{2}{3}$
W-5-W-19	5$\frac{5}{8}$″	2″ × 4″ stud wall; Exposed face: CAB shingles over 1″ × 6″; Unexposed face: $\frac{1}{8}$″ CAB sheet; gypsum board at stud edges; 5$\frac{1}{2}$″ mineral wool insulation; Design G.	74 min.	85 min.			4	1-10	1
W-5-W-20	5$\frac{5}{8}$″	2″ × 4″ stud wall; Exposed face: $\frac{1}{8}$″ CAB sheet; gypsum board at $\frac{3}{16}$″ stud edges; $\frac{7}{16}$″ fiberboard; Unexposed face: CAB shingles over 1″ × 6″; 5$\frac{1}{2}$″ mineral wool insulation; Design G.	79 min.	85 min.			4	1-10	1$\frac{1}{4}$
W-5-W-21	5$\frac{5}{8}$″	2″ × 4″ stud wall; Exposed face: CAB shingles 1″ × 6″ sheathing; Unexposed face: CAB sheet; gypsum board at stud edges; 5$\frac{1}{2}$″ mineral wool insulation; Design G.	38 min.	38 min.			4	1-10, 12, 14	—
W-5-W-22	5$\frac{5}{8}$″	2″ × 4″ stud wall; Exposed face: CAB sheet; gypsum board at stud edges; Unexposed face: CAB shingles 1″ × 6″ sheathing; 5$\frac{1}{2}$″ mineral wool insulation; Design G.	38 min.	38 min.			4	1-12	—
W-6-W-23	6″	2″ × 4″ stud wall; 16″ o.c.; $\frac{1}{2}$″ gypsum board each side; $\frac{1}{2}$″ gypsum plaster each side.	N/A	60 min.			7	15	1

(continued)

TABLE 1.3.2—continued
WOOD FRAME WALLS
4″ TO LESS THAN 6″ THICK.

ITEM CODE	THICKNESS	CONSTRUCTION DETAILS	PERFORMANCE		REFERENCE NUMBER			NOTES	REC. HOURS
			LOAD	TIME	PRE-BMS-92	BMS-92	POST-BMS-92		
W-6-W-24	6″	2″ × 4″ stud wall; 16″ o.c.; $1/2$″ gypsum board each side; $1/2$″ gypsum plaster each side.	N/A	68 min.			7	16	1
W-6-W-25	$6^{7}/_{8}$″	2″ × 4″ stud wall; 18″ o.c.; $3/4$″ gypsum plank each side; $3/16$″ gypsum plaster each side.	N/A	80 min.			7	15	$1^{1}/_{3}$
W-5-W-26	$5^{1}/_{8}$″	2″ × 4″ stud wall; 16″ o.c.; $3/8$″ gypsum board each side; $3/16$″ gypsum plaster each side.	N/A	37 min.			7	15	$1/2$
W-5-W-27	$5^{3}/_{4}$″	2″ × 4″ stud wall; 16″ o.c.; $3/8$″ gypsum lath each side; $1/2$″ gypsum plaster each side.	N/A	52 min.			7	15	$3/4$
W-5-W-28	5″	2″ × 4″ stud wall; 16″ o.c.; $1/2$″ gypsum board each side.	N/A	37 min.			7	16	$1/2$
W-5-W-29	5″	2″ × 4″ stud wall; $1/2$″ fiberboard both sides 14% M.C. with F.R. paint at 35 gm./ft.^{2}.	N/A	28 min.			7	15	$1/3$
W-4-W-30	$4^{3}/_{4}$″	2″ × 4″ stud wall; Fire side: $1/2$″ (wood) fiberboard; Back side: $1/4$″ CAB; 16″ o.c.	N/A	17 min.			7	15, 16	$1/4$
W-5-W-31	$5^{1}/_{8}$″	2″ × 4″ stud wall; 16″ o.c.; $1/2$″ fiberboard insulation with $1/32$″ asbestos (both sides of each board).	N/A	50 min.			7	16	$3/4$
W-4-W-32	$4^{1}/_{4}$″	2″ × 4″ stud wall; $3/8$″ thick gypsum wallboard on both faces; insulated cavities.	See Note 23	25 min.		1		17, 18, 23	$1/3$
W-4-W-33	$4^{1}/_{2}$″	2″ × 4″ stud wall; $1/2$″ thick gypsum wallboard on both faces.	See Note 17	40 min.		1		17, 23	$1/3$
W-4-W-34	$4^{1}/_{2}$″	2″ × 4″ stud wall; $1/2$″ thick gypsum wallboard on both faces; insulated cavities.	See Note 17	45 min.		1		17, 18, 23	$3/4$
W-4-W-35	$4^{1}/_{2}$″	2″ × 4″ stud wall; $1/2$″ thick gypsum wallboard on both faces; insulated cavities.	N/A	1 hr.		1		17, 18, 24	1
W-4-W-36	$4^{1}/_{2}$″	2″ × 4″ stud wall; $1/2$″ thick, 1.1 lbs./ft.^{2} wood fiberboard sheathing on both faces.	See Note 23	15 min.		1		17, 23	$1/4$
W-4-W-37	$4^{1}/_{2}$″	2″ × 4″ stud wall; $1/2$″ thick, 0.7 lb./ft.^{2} wood fiberboard sheathing on both faces.	See Note 23	10 min.		1		17, 23	$1/6$
W-4-W-38	$4^{1}/_{2}$″	2″ × 4″ stud wall; $1/2$″ thick, flameproofed 1.6 lbs./ft.^{2} wood fiberboard sheathing on both faces.	See Note 23	30 min.		1		17, 23	$1/2$
W-4-W-39	$4^{1}/_{2}$″	2″ × 4″ stud wall; $1/2$″ thick gypsum wallboard on both faces; insulated cavities.	See Note 23	1 hr.		1		17, 18, 23	1
W-4-W-40	$4^{1}/_{2}$″	2″ × 4″ stud wall; $1/2$″ thick, 1:2; 1:3 gypsum plaster on wood lath on both faces.	See Note 23	30 min.		1		17, 21, 23	$1/2$
W-4-W-41	$4^{1}/_{2}$″	2″ × 4″ stud wall; $1/2$″, 1:2; 1:3 gypsum plaster on wood lath on both faces; insulated cavities.	See Note 23	1 hr.		1		17, 18, 21, 24	1

(continued)

TABLE 1.3.2—continued
WOOD FRAME WALLS
4″ TO LESS THAN 6″ THICK

ITEM CODE	THICKNESS	CONSTRUCTION DETAILS	PERFORMANCE		REFERENCE NUMBER			NOTES	REC. HOURS
			LOAD	TIME	PRE-BMS-92	BMS-92	POST-BMS-92		
W-4-W-42	4½″	2″ × 4″ stud wall; ½″, 1:5; 1:7.5 lime plaster on wood lath on both wall faces.	See Note 23	30 min.		1		17, 21, 23	½
W-4-W-43	4½″	2″ × 4″ stud wall; ½″ thick 1:5; 1:7.5 lime plaster on wood lath on both faces; insulated cavities.	See Note 23	45 min.		1		17, 18, 21, 23	¾
W-4-W-44	4⅝″	2″ × 4″ stud wall; 3/16″ thick cement-asbestos over ⅜″ thick gypsum board on both faces.	See Note 23	1 hr.		1		23, 25, 26, 27	1
W-4-W-45	4⅝″	2″ × 4″ stud wall; studs faced with 4″ wide strips of ⅜″ thick gypsum board; 3/16″ thick gypsum cement-asbestos board on both faces; insulated cavities.	See Note 23	1 hr.		1		23, 25, 27, 28	1
W-4-W-46	4⅝″	Same as W-4-W-45 but nonload bearing.	N/A	1 hr. 15 min.		1		24, 28	1¼
W-4-W-47	4⅞″	2″ × 4″ stud wall; 3/16″ thick cement-asbestos board over ½″ thick gypsum sheathing on both faces.	See Note 23	1 hr. 15 min.		1		23, 25, 26, 27	1¼
W-4-W-48	4⅞″	Same as W-4-W-47 but nonload bearing.	N/A	1 hr. 30 min.		1		24, 27	1½
W-5-W-49	5″	2″ × 4″ stud wall; Exterior face: ¾″ wood sheathing; asbestos felt 14 lbs./100 ft.² and 5/32″ cement-asbestos shingles; Interior face: 4″ wide strips of ⅜″ gypsum board over studs; wall faced with 3/16″ thick cement-asbestos board.	See Note 23	40 min.		1		18, 23, 25, 26, 29	⅔
W-5-W-50	5″	2″ × 4″ stud wall; Exterior face: as per W-5-W-49; Interior face: 9/16″ composite board consisting of 7/16″ thick wood fiberboard faced with ⅛″ thick cement-asbestos board; Exterior side exposed to fire.	See Note 23	30 min.		1		23, 25, 26, 30	½
W-5-W-51	5″	Same as W-5-W-50 but interior side exposed to fire.	See Note 23	30 min.		1		23, 25, 26	½
W-5-W-52	5″	Same as W-5-W-49 but exterior side exposed to fire.	See Note 23	45 min.		1		18, 23, 25, 26	¾
W-5-W-53	5″	2″ × 4″ stud wall; ¾″ thick T&G wood boards on both sides.	See Note 23	20 min.		1		17, 23	⅓
W-5-W-54	5″	Same as W-5-W-53 but with insulated cavities.	See Note 23	35 min.		1		17, 18, 23	½
W-5-W-55	5″	2″ × 4″ stud wall; ¾″ thick T&G wood boards on both sides with 30 lbs./100 ft.² asbestos; paper, between studs and boards.	See Note 23	45 min.		1		17, 23	¾
W-5-W-56	5″	2″ × 4″ stud wall; ½″ thick, 1:2; 1:3 gypsum plaster on metal lath on both sides of wall.	See Note 23	45 min.		1		17, 21, 34	¾

(continued)

TABLE 1.3.2—continued
WOOD FRAME WALLS
4″ TO LESS THAN 6″ THICK

ITEM CODE	THICKNESS	CONSTRUCTION DETAILS	PERFORMANCE		REFERENCE NUMBER			NOTES	REC. HOURS
			LOAD	TIME	PRE-BMS-92	BMS-92	POST-BMS-92		
W-5-W-57	5″	2″ × 4″ stud wall; $3/4$″ thick 2:1:8; 2:1:12 lime and Keene's cement plaster over metal lath on both sides of wall.	See Note 23	45 min.		1		17, 21, 23	$1/2$
W-5-W-58	5″	2″ × 4″ stud wall; $3/4$″ thick 2:1:8; 2:1:10 lime Portland cement plaster over metal lath on both sides of wall.	See Note 23	30 min.		1		17, 21, 23	$1/2$
W-5-W-59	5″	2″ × 4″ stud wall; $3/4$″ thick 1:5; 1:7.5 lime plaster on metal lath on both sides of wall.	See Note 23	30 min.		1		17, 21, 23	$1/2$
W-5-W-60	5″	2″ × 4″ stud wall; $3/4$″ thick 1:$1/30$:2; 1:$1/30$:3 Portland cement, asbestos fiber plaster on metal lath on both sides of wall.	See Note 23	45 min.		1		17, 21, 23	$3/4$
W-5-W-61	5″	2″ × 4″ stud wall; $3/4$″ thick 1:2; 1:3 Portland cement plaster on metal lath on both sides of wall.	See Note 23	30 min.		1		17, 21, 23	$1/2$
W-5-W-62	5″	2″ × 4″ stud wall; $3/4$″ thick neat gypsum plaster on metal lath on both sides of wall.	N/A	1 hr. 30 min.		1		17, 22, 24	$1 1/2$
W-5-W-63	5″	2″ × 4″ stud wall; $3/4$″ thick neat gypsum plaster on metal lath on both sides of wall.	See Note 23	1 hr. 30 min.		1		17, 21, 23	$1 1/2$
W-5-W-64	5″	2″ × 4″ stud wall; $3/4$″ thick 1:2; 1:2 gypsum plaster on metal lath on both sides of wall; insulated cavities.	See Note 23	1 hr. 30 min.		1		17, 18, 21, 23	$1 1/2$
W-5-W-65	5″	2″ × 4″ stud wall; same as W-5-W-64 but cavities not insulated.	See Note 23	1 hr.		1		17, 21, 23	1
W-5-W-66	5″	2″ × 4″ stud wall; $3/4$″ thick 1:2; 1:3 gypsum plaster on metal lath on both sides of wall; insulated cavities.	See Note 23	1 hr. 15 min.		1		17, 18, 21, 23	$1 1/4$
W-5-W-67	$5 1/16$″	Same as W-5-W-49 except cavity insulation of 1.75 lbs./ft.^{2} mineral wool bats; rating applies when either wall side exposed to fire.	See Note 23	1 hr. 15 min.		1		23, 26, 25	$1 1/4$
W-5-W-68	$5 1/4$″	2″ × 4″ stud wall, $7/8$″ thick 1:2; 1:3 gypsum plaster on metal lath on both sides of wall; insulated cavities.	See Note 23	1 hr. 30 min.		1		17, 18, 21, 23	$1 1/2$
W-5-W-69	$5 1/4$″	2″ × 4″ stud wall; $7/8$″ thick neat gypsum plaster applied on metal lath on both sides of wall.	N/A	1 hr. 45 min.		1		17, 22, 24	$1 3/4$
W-5-W-70	$5 1/4$″	2″ × 4″ stud wall; $1/2$″ thick neat gypsum plaster on $3/8$″ plain gypsum lath on both sides of wall.	See Note 23	1 hr.		1		17, 22, 23	1
W-5-W-71	$5 1/4$″	2″ × 4″ stud wall; $1/2$″ thick of 1:2; 1:2 gypsum plaster on $3/8$″ thick plain gypsum lath with $1 3/4$″ × $1 3/4$″ metal lath pads nailed 8″ o.c. vertically and 16″ o.c. horizontally on both sides of wall.	See Note 23	1 hr.		1		17, 21, 23	1
W-5-W-72	$5 1/4$″	2″ × 4″ stud wall; $1/2$″ thick of 1:2; 1:2 gypsum plaster on $3/8$″ perforated gypsum lath, one $3/4$″ diameter hole or larger per 16″ square of lath surface, on both sides of wall.	See Note 23	1 hr.		1		17, 21, 23	1

(continued)

RESOURCE A—GUIDELINES ON FIRE RATINGS OF ARCHAIC MATERIALS AND ASSEMBLIES

TABLE 1.3.2—continued
WOOD FRAME WALLS
4″ TO LESS THAN 6″ THICK

ITEM CODE	THICKNESS	CONSTRUCTION DETAILS	PERFORMANCE LOAD	PERFORMANCE TIME	PRE-BMS-92	BMS-92	POST-BMS-92	NOTES	REC. HOURS
W-5-W-73	$5\frac{1}{4}″$	2″ × 4″ stud wall; $\frac{1}{2}″$ thick of 1:2; 1:2 gypsum plaster on $\frac{3}{8}″$ gypsum lath (plain, indented or perforated) on both sides of wall.	See Note 23	45 min.		1		17, 21, 23	$\frac{3}{4}$
W-5-W-74	$5\frac{1}{4}″$	2″ × 4″ stud wall; $\frac{7}{8}″$ thick of 1:2; 1:3 gypsum plaster over metal lath on both sides of wall.	See Note 23	1 hr.		1		17, 21, 23	1
W-5-W-75	$5\frac{1}{4}″$	2″ × 4″ stud wall; $\frac{7}{8}″$ thick of $1:\frac{1}{30}:2$; $1:\frac{1}{30}:3$ Portland cement, asbestos plaster applied over metal lath on both sides of wall.	See Note 23	1 hr.		1		17, 21, 23	1
W-5-W-76	$5\frac{1}{4}″$	2″ × 4″ stud wall; $\frac{7}{8}″$ thick of 1:2; 1:3 Portland cement plaster over metal lath on both sides of wall.	See Note 23	45 min.		1		17, 21, 23	$\frac{3}{4}$
W-5-W-77	$5\frac{1}{2}″$	2″ × 4″ stud wall; 1″ thick neat gypsum plaster over metal lath on both sides of wall; nonload bearing.	N/A	2 hrs.		1		17, 22, 24	2
W-5-W-78	$5\frac{1}{2}″$	2″ × 4″ stud wall; $\frac{1}{2}″$ thick of 1:2; 1:2 gypsum plaster on $\frac{1}{2}″$ thick, 0.7 lb./ft.^{2} wood fiberboard on both sides of wall.	See Note 23	35 min.		1		17, 21, 23	$\frac{1}{2}$
W-4-W-79	$4\frac{3}{4}″$	2″ × 4″ wood stud wall; $\frac{1}{2}″$ thick of 1:2; 1:2 gypsum plaster over wood lath on both sides of wall; mineral wool insulation.	N/A	1 hr.			43	21, 31, 35, 38	1
W-4-W-80	$4\frac{3}{4}″$	Same as W-4-W-79 but uninsulated.	N/A	35 min.			43	21, 31, 35	$\frac{1}{2}$
W-4-W-81	$4\frac{3}{4}″$	2″ × 4″ wood stud wall; $\frac{1}{2}″$ thick of 3:1:8; 3:1:12 lime, Keene's cement, sand plaster over wood lath on both sides of wall; mineral wool insulation.	N/A	1 hr.			43	21, 31, 35, 40	1
W-4-W-82	$4\frac{3}{4}″$	2″ × 4″ wood stud wall; $\frac{1}{2}″$ thick of $1:6\frac{1}{4}$; $1:6\frac{1}{4}$ lime Keene's cement plaster over wood lath on both sides of wall; mineral wool insulation.	N/A	30 min.			43	21, 31, 35, 40	$\frac{1}{2}$
W-4-W-83	$4\frac{3}{4}″$	2″ × 4″ wood stud wall; $\frac{1}{2}″$ thick of 1:5; 1:7.5 lime plaster over wood lath on both sides of wall.	N/A	30 min.			43	21, 31, 35	$\frac{1}{2}$
W-5-W-84	$5\frac{1}{8}″$	2″ × 4″ wood stud wall; $\frac{11}{16}″$ thick of 1:5; 1:7.5 lime plaster over wood lath on both sides of wall; mineral wool insulation.	N/A	45 min.			43	21, 31, 35, 39	$\frac{3}{4}$
W-5-W-85	$5\frac{1}{4}″$	2″ × 4″ wood stud wall; $\frac{3}{4}″$ thick of 1:5; 1:7 lime plaster over wood lath on both sides of wall; mineral wool insulation.	N/A	40 min.			43	21, 31, 35, 40	$\frac{2}{3}$
W-5-W-86	$5\frac{1}{4}″$	2″ × 4″ wood stud wall; $\frac{1}{2}″$ thick of 2:1:12 lime, Keene's cement and sand scratch coat; $\frac{1}{2}″$ thick 2:1:18 lime, Keene's cement and sand brown coat over wood lath on both sides of wall; mineral wool insulation.	N/A	1 hr.			43	21, 31, 35, 40	1
W-5-W-87	$5\frac{1}{4}″$	2″ × 4″ wood stud wall; $\frac{1}{2}″$ thick of 1:2; 1:2 gypsum plaster over $\frac{3}{8}″$ plaster board on both sides of wall.	N/A	45 min.			43	21, 31	$\frac{3}{4}$

(continued)

TABLE 1.3.2—continued
WOOD FRAME WALLS
4″ TO LESS THAN 6″ THICK

ITEM CODE	THICKNESS	CONSTRUCTION DETAILS	PERFORMANCE		REFERENCE NUMBER			NOTES	REC. HOURS
			LOAD	TIME	PRE-BMS-92	BMS-92	POST-BMS-92		
W-5-W-88	5 1/4″	2″ × 4″ wood stud wall; 1/2″ thick of 1:2; 1:2 gypsum plaster over 3/8″ gypsum lath on both sides of wall.	N/A	45 min.			43	21, 31	3/4
W-5-W-89	5 1/4″	2″ × 4″ wood stud wall; 1/2″ thick of 1:2; 1:2 gypsum plaster over 3/8″ gypsum lath on both sides of wall.	N/A	1 hr.			43	21, 31, 33	1
W-5-W-90	5 1/4″	2″ × 4″ wood stud wall; 1/2″ thick neat plaster over 3/8″ thick gypsum lath on both sides of wall.	N/A	1 hr.			43	21, 22, 31	1
W-5-W-91	5 1/4″	2″ × 4″ wood stud wall; 1/2″ thick of 1:2; 1:2 gypsum plaster over 3/8″ thick indented gypsum lath on both sides of wall.	N/A	45 min.			43	21, 31	3/4
W-5-W-92	5 1/4″	2″ × 4″ wood stud wall; 1/2″ thick of 1:2; 1:2 gypsum plaster over 3/8″ thick perforated gypsum lath on both sides of wall.	N/A	45 min.			43	21, 31, 34	3/4
W-5-W-93	5 1/4″	2″ × 4″ wood stud wall; 1/2″ thick of 1:2; 1:2 gypsum plaster over 3/8″ perforated gypsum lath on both sides of wall.	N/A	1 hr.			43	21, 31	1
W-5-W-94	5 1/4″	2″ × 4″ wood stud wall; 1/2″ thick of 1:2; 1:2 gypsum plaster over 3/8″ thick perforated gypsum lath on both sides of wall.	N/A	45 min.			43	21, 31, 34	3/4
W-5-W-95	5 1/2″	2″ × 4″ wood stud wall; 1/2″ thick of 1:2; 1:2 gypsum plaster over 1/2″ thick wood fiberboard plaster base on both sides of wall.	N/A	35 min.			43	21, 31, 36	1/2
W-5-W-96	5 3/4″	2″ × 4″ wood stud wall; 1/2″ thick of 1:2; 1:2 gypsum plaster over 7/8″ thick flameproofed wood fiberboard on both sides of wall.	N/A	1 hr.			43	21, 31, 37	1

For SI: 1 inch = 25.4 mm, 1 foot = 305 mm, 1 pound = 0.004448 kN, 1 pound per square inch = 0.00689 MPa, 1 pound per square foot = 47.9 N/m^2.

Notes:
1. All specimens 8 feet or 8 feet 8 inches by 10 feet 4 inches, i.e. one-half of furnace size. See Note 42 for design cross section.
2. Specimens tested in tandem (two per exposure).
3. Test per ASA No. A2-1934 except where unloaded. Also, panels were of "half" size of furnace opening. Time value signifies a thermal failure time.
4. Two-inch by 4-inch studs: 16 inches on center.; where 10 feet 4 inches, blocking at 2-foot 4-inch height.
5. Facing 4 feet by 8 feet, cement-asbestos board sheets, 3/16 inch thick.
6. Sheathing (diagonal): 25/22 inch by 5 1/2 inch, 1 inch by 6 inches pine.
7. Facing shingles: 24 inches by 12 inches by 5/32 inch where used.
8. Asbestos felt: asphalt sat between sheathing and shingles.
9. Load: 30,500 pounds or 360 psi/stud where load was tested.
10. Walls were tested beyond achievement of first test end point. A load-bearing time in excess of performance time indicates that although thermal criteria were exceeded, load-bearing ability continued.
11. Wall was rated for one hour combustible use in original source.
12. Hose steam test specimen. See table entry of similar design above for recommended rating.
13. Rated one and one-fourth hour load bearing. Rated one and one-half hour nonload bearing.
14. Failed hose stream.
15. Test terminated due to flame penetration.
16. Test terminated—local back face temperature rise.
17. Nominal 2-inch by 4-inch wood studs of No. 1 common or better lumber set edgewise. Two-inch by four-inch plates at top and bottom and blocking at mid height of wall.
18. Cavity insulation consists of rock wool bats 1.0 lb./ft.^{2} of filled cavity area.
19. Cavity insulation consists of glass wool bats 0.6 lb./ft.^{2} of filled cavity area.
20. Cavity insulation consists of blown-in rock wool 2.0 lbs./ft.^{2} of filled cavity area
21. Mix proportions for plastered walls as follows: first ratio indicates scratch coat mix, weight of dry plaster: dry sand; second ratio indicates brown coat mix.
22. "Neat" plaster is taken to mean unsanded wood-fiber gypsum plaster.
23. Load: 360 psi of net stud cross sectional area.
24. Rated as nonload bearing.

(continued)

RESOURCE A—GUIDELINES ON FIRE RATINGS OF ARCHAIC MATERIALS AND ASSEMBLIES

**TABLE 1.3.2—continued
WOOD FRAME WALLS
4″ TO LESS THAN 6″ THICK**

25. Nominal 2-inch by 4-inch studs per Note 17, spaced at 16 inches on center.
26. Horizontal joints in facing material supported by 2-inch by 4-inch blocking within wall.
27. Facings secured with 6d casing nails. Nail holes predrilled and were 0.02 to 0.03 inch smaller than nail diameter.
28. Cavity insulation consists of mineral wool bats weighing 2 lbs./ft.^{2} of filled cavity area.
29. Interior wall face exposed to fire.
30. Exterior wall faced exposed to fire.
31. Nominal 2-inch by 4-inch studs of yellow pine or Douglas-fir spaced 16 inches on center in a single row.
32. Studs as in Note 31 except double row, with studs in rows staggered.
33. Six roofing nails with metal-lath pads around heats to each 16-inch by 48-inch lath.
34. Areas of holes less than $2^3/_4$ percent of area of lath.
35. Wood laths were nailed with either 3d or 4d nails, one nail to each bearing, and the end joining broken every seventh course.
36. One-half-inch thick fiberboard plaster base nailed with 3d or 4d common wire nails spaced 4 to 6 inches on center.
37. Seven-eighths-inch thick fiberboard plaster base nailed with 5d common wire nails spaced 4 to 6 inches on center.
38. Mineral wood bats 1.05 to 1.25 lbs./ft.^{2} with waterproofed-paper backing.
39. Blown-in mineral wool insulation, 2.2 lbs./ft.^{2}.
40. Mineral wool bats, 1.4 lbs./ft.^{2} with waterproofed-paper backing.
41. Mineral wood bats, 0.9 lb./ft.^{2}.
42. See wall design diagram below.

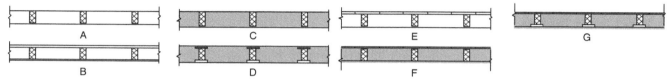

43. Duplicate specimen of W-4-W-7, tested simultaneously with W-4-W-7 in 18-foot test furnace.

RESOURCE A—GUIDELINES ON FIRE RATINGS OF ARCHAIC MATERIALS AND ASSEMBLIES

TABLE 1.3.3
WOOD FRAME WALLS
6″ TO LESS THAN 8″ THICK

ITEM CODE	THICKNESS	CONSTRUCTION DETAILS	PERFORMANCE		REFERENCE NUMBER			NOTES	REC. HOURS
			LOAD	TIME	PRE-BMS-92	BMS-92	POST-BMS-92		
W-6-W-1	6 1/4″	2 × 4 stud wall; 1/2″ thick, 1:2; 1:2 gypsum plaster on 7/8″ flameproofed wood fiberboard weighing 2.8 lbs./ft.² on both sides of wall.	See Note 3	1 hr.		1		1-3	1
W-6-W-2	6 1/2″	2 × 4 stud wall; 1/2″ thick, 1:3; 1:3 gypsum plaster on 1″ thick magnesium oxysulfate wood fiberboard on both sides of wall.	See Note 3	45 min.		1		1-3	3/4
W-7-W-3	7 1/4″	Double row of 2 × 4 studs, 1/2″ thick of 1:2; 1:2 gypsum plaster applied over 3/8″ thick perforated gypsum lath on both sides of wall; mineral wool insulation.	N/A	1 hr.			43	2, 4, 5	1
W-7-W-4	7 1/2″	Double row of 2 × 4 studs, 5/8″ thick of 1:2; 1:2 gypsum plaster applied over 3/8″ thick perforated gypsum lath over laid with 2″ × 2″, 16 gage wire fabric, on both sides of wall.	N/A	1 hr. 15 min.			43	2, 4	1 1/4

For SI: 1 inch = 25.4 mm, 1 pound = 0.004448 kN, 1 pound per square inch = 0.00689 MPa, 1 pound per square foot = 47.9 N/m².

Notes:
1. Nominal 2-inch by 4-inch wood studs of No. 1 common or better lumber set edgewise. Two-inch by 4-inch plates at top and bottom and blocking at mid height of wall.
2. Mix proportions for plastered walls as follows: first ratio indicates scratch coat mix, weight of dry plaster: dry sand; second ratio indicates brown coat mix.
3. Load: 360 psi of net stud cross sectional area.
4. Nominal 2-inch by 4-inch studs of yellow pine of Douglas-fir spaced 16 inches in a double row, with studs in rows staggered.
5. Mineral wool bats, 0.19 lb./ft.²

TABLE 1.4.1
MISCELLANEOUS MATERIALS WALLS
0″ TO LESS THAN 4″ THICK

ITEM CODE	THICKNESS	CONSTRUCTION DETAILS	PERFORMANCE		REFERENCE NUMBER			NOTES	REC. HOURS
			LOAD	TIME	PRE-BMS-92	BMS-92	POST-BMS-92		
W-3-Mi-1	3 7/8″	Glass brick wall: (bricks 5 3/4″ × 5 3/4″ × 3 7/8″) 1/4″ mortar bed, cement/lime/sand; mounted in brick (9″) wall with mastic and 1/2″ asbestos rope.	N/A	1 hr.			7	1, 2	1
W-3-Mi-2	3″	Core: 2″ magnesium oxysulfate wood-fiber blocks; laid in Portland cement-lime mortar; Facings: on both sides; see Note 3.	N/A	1 hr.		1		3	1
W-3-Mi-3	3 7/8″	Core: 8″ × 4 7/8″ glass blocks 3 7/8″ thick weighing 4 lbs. each; laid in Portland cement-lime mortar; horizontal mortar joints reinforced with metal lath.	N/A	15 min.		1			1/4

For SI: 1 inch = 25.4 mm, 1 pound = 0.004448 kN.

Notes:
1. No failure reached at 1 hour.
2. These glass blocks are assumed to be solid based on other test data available for similar but hollow units which show significantly reduced fire endurance.
3. Minimum of 1/2 inch of 1:3 sanded gypsum plaster required to develop this rating.

TABLE 1.4.2
MISCELLANEOUS MATERIALS WALLS
4″ TO LESS THAN 6″ THICK

ITEM CODE	THICKNESS	CONSTRUCTION DETAILS	PERFORMANCE		REFERENCE NUMBER			NOTES	REC. HOURS
			LOAD	TIME	PRE-BMS-92	BMS-92	POST-BMS-92		
W-4-Mi-1	4″	Core: 3″ magnesium oxysulfate wood-fiber blocks; laid in Portland cement mortar; Facings: both sides; see Note 1.	N/A	2 hrs.		1			2

For SI: 1 inch = 25.4 mm.

Notes:
1. One-half inch sanded gypsum plaster. Voids in hollow blocks to be not more than 30 percent.

RESOURCE A—GUIDELINES ON FIRE RATINGS OF ARCHAIC MATERIALS AND ASSEMBLIES

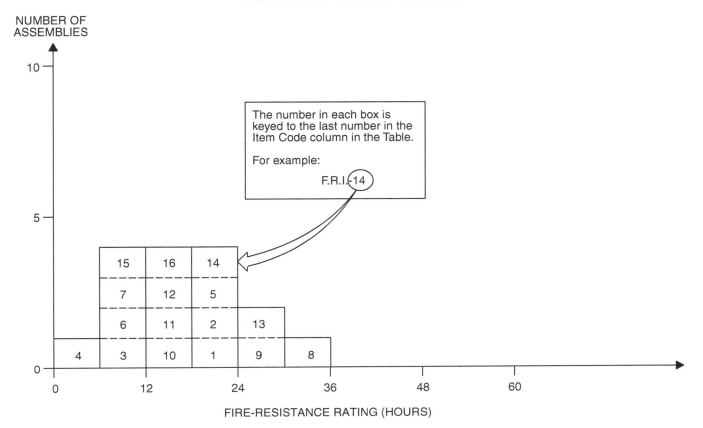

FIGURE 1.5.1
FINISH RATINGS—INORGANIC MATERIALS

TABLE 1.5.1
FINISH RATINGS—INORGANIC MATERIALS

ITEM CODE	THICKNESS	CONSTRUCTION DETAILS	PERFORMANCE FINISH RATING	REFERENCE NUMBER PRE-BMS-92	BMS-92	POST-BMS-92	NOTES	REC. F.R. (MIN.)
F.R.-I-1	$9/16''$	$3/8''$ gypsum wallboard faced with $3/16''$ cement-asbestos board.	20 minutes	1			1, 2	15
F.R.-I-2	$11/16''$	$1/2''$ gypsum sheathing faced with $3/16''$ cement-asbestos board.	20 minutes	1			1, 2	20
F.R.-I-3	$3/16''$	$3/16''$ cement-asbestos board over uninsulated cavity.	10 minutes	1			1, 2	5
F.R.-I-4	$3/16''$	$3/16''$ cement-asbestos board over insulated cavities.	5 minutes	1			1, 2	5
F.R.-I-5	$3/4''$	$3/4''$ thick 1:2; 1:3 gypsum plaster over paper backed metal lath.	20 minutes	1			1, 2, 3	20
F.R.-I-6	$3/4''$	$3/4''$ thick Portland cement plaster on metal lath.	10 minutes	1			1, 2	10
F.R.-I-7	$3/4''$	$3/4''$ thick 1:5; 1:7.5 lime plaster on metal lath.	10 minutes	1			1, 2	10
F.R.-I-8	$1''$	$1''$ thick neat gypsum plaster on metal lath.	35 minutes	1			1, 2, 4	35
F.R.-I-9	$3/4''$	$3/4''$ thick neat gypsum plaster on metal lath.	30 minutes	1			1, 2, 4	30

(continued)

RESOURCE A—GUIDELINES ON FIRE RATINGS OF ARCHAIC MATERIALS AND ASSEMBLIES

TABLE 1.5.1—continued
FINISH RATINGS—INORGANIC MATERIALS

ITEM CODE	THICKNESS	CONSTRUCTION DETAILS	PERFORMANCE FINISH RATING	REFERENCE NUMBER PRE-BMS-92	BMS-92	POST-BMS-92	NOTES	REC. F.R. (MIN.)
F.R.-I-10	$3/4''$	$3/4''$ thick 1:2; 1:2 gypsum plaster on metal lath.	15 minutes	1			1, 2, 3	15
F.R.-I-11	$1/2''$	Same as F.R.-1-7, except $1/2''$ thick on wood lath.	15 minutes	1			1, 2, 3	15
F.R.-I-12	$1/2''$	$1/2''$ thick 1:2; 1:3 gypsum plaster on wood lath.	15 minutes	1			1, 2, 3	15
F.R.-I-13	$7/8''$	$1/2''$ thick 1:2; 1:2 gypsum plaster on $3/8''$ perforated gypsum lath.	30 minutes	1			1, 2, 3	30
F.R.-I-14	$7/8''$	$1/2''$ thick 1:2; 1:2 gypsum plaster on $3/8''$ thick plain or indented gypsum plaster.	20 minutes	1			1, 2, 3	20
F.R.-I-15	$3/8''$	$3/8''$ gypsum wallboard.	10 minutes	1			1, 2	10
F.R.-I-16	$1/2''$	$1/2''$ gypsum wallboard.	15 minutes	1			1, 2	15

For SI: 1 inch = 25.4 mm, °C = [(°F) - 32]/1.8.

Notes:
1. The finish rating is the time required to obtain an average temperature rise of 250°F, or a single point rise of 325°F, at the interface between the material being rated and the substrate being protected.
2. Tested in accordance with the Standard Specifications for Fire Tests of Building Construction and Materials, ASA No. A2-1932.
3. Mix proportions for plasters as follows: first ratio, dry weight of plaster: dry weight of sand for scratch coat; second ratio, plaster: sand for brown coat.
4. Neat plaster means unsanded wood-fiber gypsum plaster.

General Note:
The finish rating of modern building materials can be found in the current literature.

TABLE 1.5.2
FINISH RATINGS—ORGANIC MATERIALS

ITEM CODE	THICKNESS	CONSTRUCTION DETAILS	PERFORMANCE FINISH RATING	REFERENCE NUMBER PRE-BMS-92	BMS-92	POST-BMS-92	NOTES	REC. F.R. (MIN.)
F.R.-O-1	$9/16''$	$7/16''$ wood fiberboard faced with $1/8''$ cement-asbestos board.	15 minutes	1			1, 2	15
F.R.-O-2	$29/32''$	$3/4''$ wood sheathing, asbestos felt weighing 14 lbs./100 ft.^{2} and $5/32''$ cement-asbestos shingles.	20 minutes	1			1, 2	20
F.R.-O-3	$1 1/2''$	1'' thick magnesium oxysulfate wood fiberboard faced with 1:3; 1:3 gypsum plaster, $1/2''$ thick.	20 minutes	1			1, 2, 3	20
F.R.-O-4	$1/2''$	$1/2''$ thick wood fiberboard.	5 minutes	1			1, 2	5
F.R.-O-5	$1/2''$	$1/2''$ thick flameproofed wood fiberboard.	10 minutes	1			1, 2	10
F.R.-O-6	1''	$1/2''$ thick wood fiberboard faced with $1/2''$ thick 1:2; 1:2 gypsum plaster.	15 minutes	1			1, 2, 3	30
F.R.-O-7	$1 3/8''$	$7/8''$ thick flameproofed wood fiberboard faced with $1/2''$ thick 1:2; 1:2 gypsum plaster.	30 minutes	1			1, 2, 3	30
F.R.-O-8	$1 1/4''$	$1 1/4''$ thick plywood.	30 minutes			35		30

For SI: 1 inch = 25.4 mm, 1 pound = 0.004448 kN, 1 pound per square foot = 47.9 N/m^2, °C = [(°F) - 32]/1.8.

Notes:
1. The finish rating is the time required to obtain an average temperature rise of 250°F, or a single point rise of 325°F, at the interface between the material being rated and he substrate being protected.
2. Tested in accordance with the Standard Specifications for Fire Tests of Building Construction and Materials, ASA No. A2-1932.
3. Plaster ratios as follows: first ratio is for scratch coat, weight of dry plaster: weight of dry sand; second ratio is for the brown coat.

General Note:
The finish rating of thinner materials, particularly thinner woods, have not been listed because the possible effects of shrinkage, warpage and aging cannot be predicted.

SECTION II
COLUMNS

TABLE 2.1.1
REINFORCED CONCRETE COLUMNS
MINIMUM DIMENSION 0″ TO LESS THAN 6″

ITEM CODE	MINIMUM DIMENSION	CONSTRUCTION DETAILS	PERFORMANCE		REFERENCE NUMBER			NOTES	REC. HOURS
			LOAD	TIME	PRE-BMS-92	BMS-92	POST-BMS-92		
C-6-RC-1	6″	6″ × 6″ square columns; gravel aggregate concrete (4030 psi); Reinforcement: vertical, four $^7/_8$″ rebars; horizontal, $^5/_{16}$″ ties at 6″ pitch; Cover: 1″.	34.7 tons	62 min.			7	1, 2	1
C-6-RC-2	6″	6″ × 6″ square columns; gravel aggregate concrete (4200 psi); Reinforcement: vertical, four $^1/_2$″ rebars; horizontal, $^5/_{16}$″ ties at 6″ pitch; Cover: 1″.	21 tons	69 min.			7	1, 2	1

Notes:
1. Collapse.
2. British test.

RESOURCE A—GUIDELINES ON FIRE RATINGS OF ARCHAIC MATERIALS AND ASSEMBLIES

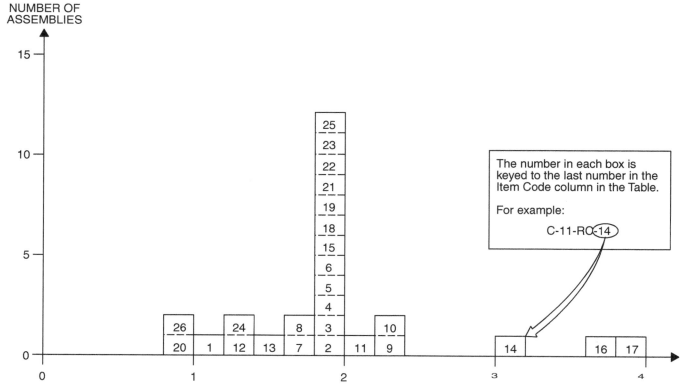

FIGURE 2.1.2
REINFORCED CONCRETE COLUMNS
MINIMUM DIMENSION 10″ TO LESS THAN 12″

TABLE 2.1.2
REINFORCED CONCRETE COLUMNS
MINIMUM DIMENSION 10″ TO LESS THAN 12″

ITEM CODE	MINIMUM DIMENSION	CONSTRUCTION DETAILS	PERFORMANCE		REFERENCE NUMBER			NOTES	REC. HOURS
			LOAD	TIME	PRE-BMS-92	BMS-92	POST-BMS-92		
C-10-RC-1	10″	10″ square columns; aggregate concrete (4260 psi); Reinforcement: vertical, four $1\frac{1}{4}″$ rebars; horizontal, $\frac{3}{8}″$ ties at 6″ pitch; Cover: $1\frac{1}{4}″$.	92.2 tons	1 hr. 2 min.			7	1	1
C-10-RC-2	10″	10″ square columns; aggregate concrete (2325 psi); Reinforcement: vertical, four $\frac{1}{2}″$ rebars; horizontal, $\frac{5}{16}″$ ties at 6″ pitch; Cover: 1″.	46.7 tons	1 hr. 52 min.			7	1	$1\frac{3}{4}$
C-10-RC-3	10″	10″ square columns; aggregate concrete (5370 psi); Reinforcement: vertical, four $\frac{1}{2}″$ rebars; horizontal, $\frac{5}{16}″$ ties at 6″ pitch; Cover: 1″.	46.5 tons	2 hrs.			7	2, 3, 11	2
C-10-RC-4	10″	10″ square columns; aggregate concrete (5206 psi); Reinforcement: vertical, four $\frac{1}{2}″$ rebars; horizontal, $\frac{5}{16}″$ ties at 6″ pitch; Cover: 1″.	46.5 tons	2 hrs.			7	2, 7	2
C-10-RC-5	10″	10″ square columns; aggregate concrete (5674 psi); Reinforcement: vertical, four $\frac{1}{2}″$ rebars; horizontal, $\frac{5}{16}″$ ties at 6″ pitch; Cover: 1″.	46.7 tons	2 hrs.			7	1	2

(continued)

TABLE 2.1.2—continued
REINFORCED CONCRETE COLUMNS
MINIMUM DIMENSION 10″ TO LESS THAN 12″

ITEM CODE	MINIMUM DIMENSION	CONSTRUCTION DETAILS	PERFORMANCE		REFERENCE NUMBER			NOTES	REC. HOURS
			LOAD	TIME	PRE-BMS-92	BMS-92	POST-BMS-92		
C-10-RC-6	10″	10″ square columns; aggregate concrete (5150 psi); Reinforcement: vertical, four $1^1/_2$″ rebars; horizontal, $^5/_{16}$″ ties at 6″ pitch; Cover: 1″.	66 tons	1 hr. 43 min.			7	1	$1^3/_4$
C-10-RC-7	10″	10″ square columns; aggregate concrete (5580 psi); Reinforcement: vertical, four $^1/_2$″ rebars; horizontal, $^5/_{16}$″ ties at 6″ pitch; Cover: $1^1/_8$″.	62.5 tons	1 hr. 38 min.			7	1	$1^1/_2$
C-10-RC-8	10″	10″ square columns; aggregate concrete (4080 psi); Reinforcement: vertical, four $1^1/_8$″ rebars; horizontal, $^5/_{16}$″ ties at 6″ pitch; Cover: $1^1/_8$″.	72.8 tons	1 hr. 48 min.			7	1	$1^3/_4$
C-10-RC-9	10″	10″ square columns; aggregate concrete (2510 psi); Reinforcement: vertical, four $^1/_2$″ rebars; horizontal, $^5/_{16}$″ ties at 6″ pitch; Cover: 1″.	51 tons	2 hrs. 16 min.			7	1	$2^1/_4$
C-10-RC-10	10″	10″ square columns; aggregate concrete (2170 psi); Reinforcement: vertical, four $^1/_2$″ rebars; horizontal, $^5/_{16}$″ ties at 6″ pitch; Cover: 1″.	45 tons	2 hrs. 14 min.			7	12	$2^1/_4$
C-10-RC-11	10″	10″ square columns; gravel aggregate concrete (4015 psi); Reinforcement: vertical, four $^1/_2$″ rebars; horizontal, $^5/_{16}$″ ties at 6″ pitch; Cover: $1^1/_8$″.	46.5 tons	2 hrs. 6 min.			7	1	2
C-11-RC-12	11″	11″ square columns; gravel aggregate concrete (4150 psi); Reinforcement: vertical, four $1^1/_4$″ rebars; horizontal, $^3/_8$″ ties at $7^1/_2$″ pitch; Cover: $1^1/_2$″.	61 tons	1 hr. 23 min.			7	1	$1^1/_4$
C-11-RC-13	11″	11″ square columns; gravel aggregate concrete (4380 psi); Reinforcement: vertical, four $1^1/_4$″ rebars; horizontal, $^3/_8$″ ties at $7^1/_2$″ pitch; Cover: $1^1/_2$″.	61 tons	1 hr. 26 min.			7	1	$1^1/_4$
C-11-RC-14	11″	11″ square columns; gravel aggregate concrete (4140 psi); Reinforcement: vertical, four $1^1/_4$″ rebars; horizontal, $^3/_8$″ ties at $7^1/_2$″ pitch; steel mesh around reinforcement; Cover: $1^1/_2$″.	61 tons	3 hrs. 9 min.			7	1	3
C-11-RC-15	11″	11″ square columns; slag aggregate concrete (3690 psi); Reinforcement: vertical, four $1^1/_4$″ rebars; horizontal, $^3/_8$″ ties at $7^1/_2$″ pitch; Cover: $1^1/_2$″.	91 tons	2 hrs.			7	2, 3, 4, 5	2
C-11-RC-16	11″	11″ square columns; limestone aggregate concrete (5230 psi); Reinforcement: vertical, four $1^1/_4$″ rebars; horizontal, $^3/_8$″ ties at $7^1/_2$″ pitch; Cover: $1^1/_2$″.	91.5 tons	3 hrs. 41 min.			7	1	$3^1/_2$
C-11-RC-17	11″	11″ square columns; limestone aggregate concrete (5530 psi); Reinforcement: vertical, four $1^1/_4$″ rebars; horizontal, $^3/_8$″ ties at $7^1/_2$″ pitch; Cover: $1^1/_2$″.	91.5 tons	3 hrs. 47 min.			7	1	$3^1/_2$

(continued)

TABLE 2.1.2—continued
REINFORCED CONCRETE COLUMNS
MINIMUM DIMENSION 10″ TO LESS THAN 12″

ITEM CODE	MINIMUM DIMENSION	CONSTRUCTION DETAILS	PERFORMANCE		REFERENCE NUMBER			NOTES	REC. HOURS
			LOAD	TIME	PRE-BMS-92	BMS-92	POST- BMS-92		
C-11-RC-18	11″	11″ square columns; limestone aggregate concrete (5280 psi); Reinforcement: vertical, four $1^{1}/_{4}$″ rebars; horizontal, $^{3}/_{8}$″ ties at $7^{1}/_{2}$″ pitch; Cover: $1^{1}/_{2}$″.	91.5 tons	2 hrs.			7	2, 3, 4, 6	2
C-11-RC-19	11″	11″ square columns; limestone aggregate concrete (4180 psi); Reinforcement: vertical, four $^{5}/_{8}$″ rebars; horizontal, $^{3}/_{8}$″ ties at 7″ pitch; Cover: $1^{1}/_{2}$″.	71.4 tons	2 hrs.			7	2, 7	2
C-11-RC-20	11″	11″ square columns; gravel concrete (4530 psi); Reinforcement: vertical, four $^{5}/_{8}$″ rebars; horizontal, $^{3}/_{8}$″ ties at 7″ pitch; Cover: $1^{1}/_{2}$″ with $^{1}/_{2}$″ plaster.	58.8 tons	2 hrs.			7	2, 3, 9	$1^{1}/_{4}$
C-11-RC-21	11″	11″ square columns; gravel concrete (3520 psi); Reinforcement: vertical, four $^{5}/_{8}$″ rebars; horizontal, $^{3}/_{8}$″ ties at 7″ pitch; Cover: $1^{1}/_{2}$″.	Variable	1 hr. 24 min.			7	1, 8	2
C-11-RC-22	11″	11″ square columns; aggregate concrete (3710 psi); Reinforcement: vertical, four $^{5}/_{8}$″ rebars; horizontal, $^{3}/_{8}$″ ties at 7″ pitch; Cover: $1^{1}/_{2}$″.	58.8 tons	2 hrs.			7	2, 3, 10	2
C-11-RC-23	11″	11″ square columns; aggregate concrete (3190 psi); Reinforcement: vertical, four $^{5}/_{8}$″ rebars; horizontal, $^{3}/_{8}$″ ties at 7″ pitch; Cover: $1^{1}/_{2}$″.	58.8 tons	2 hrs.			7	2, 3, 10	2
C-11-RC-24	11″	11″ square columns; aggregate concrete (4860 psi); Reinforcement: vertical, four $^{5}/_{8}$″ rebars; horizontal, $^{3}/_{8}$″ ties at 7″ pitch; Cover: $1^{1}/_{2}$″.	86.1 tons	1 hr. 20 min.			7	1	$1^{1}/_{3}$
C-11-RC-25	11″	11″ square columns; aggregate concrete (4850 psi); Reinforcement: vertical, four $^{5}/_{8}$″ rebars; horizontal, $^{3}/_{8}$″ ties at 7″ pitch; Cover: $1^{1}/_{2}$″.	58.8 tons	1 hr. 59 min.			7	1	$1^{3}/_{4}$
C-11-RC-26	11″	11″ square columns; aggregate concrete (3834 psi); Reinforcement: vertical, four $^{5}/_{8}$″ rebars; horizontal, $^{5}/_{16}$″ ties at $4^{1}/_{2}$″ pitch; Cover: $1^{1}/_{2}$″.	71.4 tons	53 min.			7	1	$^{3}/_{4}$

For SI: 1 inch = 25.4 mm, 1 pound per square inch = 0.00689 MPa, 1 ton = 8.896 kN.

Notes:
1. Failure mode—collapse.
2. Passed 2 hour fire exposure.
3. Passed hose stream test.
4. Reloaded effectively after 48 hours but collapsed at load in excess of original test load.
5. Failing load was 150 tons.
6. Failing load was 112 tons.
7. Failed during hose stream test.
8. Range of load 58.8 tons (initial) to 92 tons (92 minutes) to 60 tons (80 minutes).
9. Collapsed at 44 tons in reload after 96 hours.
10. Withstood reload after 72 hours.
11. Collapsed on reload after 48 hours.

TABLE 2.1.3
REINFORCED CONCRETE COLUMNS
MINIMUM DIMENSION 12″ TO LESS THAN 14″

ITEM CODE	MINIMUM DIMENSION	CONSTRUCTION DETAILS	PERFORMANCE LOAD	PERFORMANCE TIME	REFERENCE NUMBER PRE-BMS-92	REFERENCE NUMBER BMS-92	REFERENCE NUMBER POST-BMS-92	NOTES	REC. HOURS
C-12-RC-1	12″	12″ square columns; gravel aggregate concrete (2647 psi); Reinforcement: vertical, four $5/8$″ rebars; horizontal, $5/16$″ ties at $4^1/_2$″ pitch; Cover: 2″.	78.2 tons	38 min.		1	7	1	$^1/_2$
C-12-RC-2	12″	Reinforced columns with $1^1/_2$″ concrete outside of reinforced steel; Gross diameter or side of column: 12″; Group I, Column A.	—	6 hrs.		1		2, 3	6
C-12-RC-3	12″	Description as per C-12-RC-2; Group I, Column B.	—	4 hrs.		1		2, 3	4
C-12-RC-4	12″	Description as per C-12-RC-2; Group II, Column A.	—	4 hrs.		1		2, 3	4
C-12-RC-5	12″	Description as per C-12-RC-2; Group II, Column B.	—	2 hrs. 30 min.		1		2, 3	$2^1/_2$
C-12-RC-6	12″	Description as per C-12-RC-2; Group III, Column A.	—	3 hrs.		1		2, 3	3
C-12-RC-7	12″	Description as per C-12-RC-2; Group III, Column B.	—	2 hrs.		1		2, 3	2
C-12-RC-8	12″	Description as per C-12-RC-2; Group IV, Column A.	—	2 hrs.		1		2, 3	2
C-12-RC-9	12″	Description as per C-12-RC-2; Group IV, Column B.	—	1 hr. 30 min.		1		2, 3	$1^1/_2$

For SI: 1 inch = 25.4 mm, 1 pound per square inch = 0.00689 MPa, 1 pound per square yard = 5.3 N/m^2.

Notes:
1. Failure mode—unspecified structural.
2. Group I: includes concrete having calcareous aggregate containing a combined total of not more than 10 percent of quartz, chert and flint for the coarse aggregate.
   Group II: includes concrete having trap-rock aggregate applied without metal ties and also concrete having cinder, sandstone or granite aggregate, if held in place with wire mesh or expanded metal having not larger than 4-inch mesh, weighing not less than 1.7 lbs./yd.^{2}, placed not more than 1 inch from the surface of the concrete.
   Group III: includes concrete having cinder, sandstone or granite aggregate tied with No. 5 gage steel wire, wound spirally over the column section on a pitch of 8 inches, or equivalent ties, and concrete having siliceous aggregates containing a combined total of 60 percent or more of quartz, chert and flint, if held in place with wire mesh or expanded metal having not larger than 4-inch mesh, weighing not less than 1.7 lbs./yd.^{2}, placed not more than 1 inch from the surface of the concrete.
   Group IV: includes concrete having siliceous aggregates containing a combined total of 60 percent or more of quartz, chert and flint, and tied with No. 5 gage steel wire wound spirally over the column section on a pitch of 8 inches, or equivalent ties.
3. Groupings of aggregates and ties are the same as for structural steel columns protected solidly with concrete, the ties to be placed over the vertical reinforcing bars and the mesh where required, to be placed within 1 inch from the surface of the column.
   Column A: working loads are assumed as carried by the area of the column inside of the lines circumscribing the reinforcing steel.
   Column B: working loads are assumed as carried by the gross area of the column.

RESOURCE A—GUIDELINES ON FIRE RATINGS OF ARCHAIC MATERIALS AND ASSEMBLIES

TABLE 2.1.4
REINFORCED CONCRETE COLUMNS
MINIMUM DIMENSION 14″ TO LESS THAN 16″

ITEM CODE	MINIMUM DIMENSION	CONSTRUCTION DETAILS	PERFORMANCE LOAD	PERFORMANCE TIME	REFERENCE NUMBER PRE-BMS-92	REFERENCE NUMBER BMS-92	REFERENCE NUMBER POST-BMS-92	NOTES	REC. HOURS
C-14-RC-1	14″	14″ square columns; gravel aggregate concrete (4295 psi); Reinforcement: vertical four $3/4$″ rebars; horizontal: $1/4$″ ties at 9″ pitch; Cover: $1 1/2$″	86 tons	1 hr. 22 min.			7	1	$1 1/4$
C-14-RC-2	14″	Reinforced concrete columns with $1 1/2$″ concrete outside reinforcing steel; Gross diameter or side of column: 12″; Group I, Column A.	—	7 hrs.	1			2, 3	7
C-14-RC-3	14″	Description as per C-14-RC-2; Group II, Column B.	—	5 hrs.	1			2, 3	5
C-14-RC-4	14″	Description as per C-14-RC-2; Group III, Column A.	—	5 hrs.	1			2, 3	5
C-14-RC-5	14″	Description as per C-14-RC-2; Group IV, Column B.	—	3 hrs. 30 min.	1			2, 3	$3 1/2$
C-14-RC-6	14″	Description as per C-14-RC-2; Group III, Column A.	—	4 hrs.	1			2, 3	4
C-14-RC-7	14″	Description as per C-14-RC-2; Group III, Column B.	—	2 hrs. 30 min.	1			2, 3	$2 1/2$
C-14-RC-8	14″	Description as per C-14-RC-2; Group IV, Column A.	—	2 hrs. 30 min.	1			2, 3	$2 1/2$
C-14-RC-9	14″	Description as per C-14-RC-2; Group IV, Column B.	—	1 hr. 30 min.	1			2, 3	$1 1/2$

For SI: 1 inch = 25.4 mm, 1 pound per square inch = 0.00689 MPa, 1 pound per square yard = 5.3 N/m^2.

Notes:
1. Failure mode—main rebars buckled between links at various points.
2. Group I: includes concrete having calcareous aggregate containing a combined total of not more than 10 percent of quartz, chert and flint for the coarse aggregate.
   Group II: includes concrete having trap-rock aggregate applied without metal ties and also concrete having cinder, sandstone or granite aggregate, if held in place with wire mesh or expanded metal having not larger than 4-inch mesh, weighing not less than 1.7 lbs./yd.^{2}, placed not more than 1 inch from the surface of the concrete.
   Group III: includes concrete having cinder, sandstone or granite aggregate tied with No. 5 gage steel wire, wound spirally over the column section on a pitch of 8 inches, or equivalent ties, and concrete having siliceous aggregates containing a combined total of 60 percent or more of quartz, chert and flint, if held in place with wire mesh or expanded metal having not larger than 4-inch mesh, weighing not less than 1.7 lbs./yd.^{2}, placed not more than 1 inch from the surface of the concrete.
   Group IV: includes concrete having siliceous aggregates containing a combined total of 60 percent or more of quartz, chert and flint, and tied with No. 5 gage steel wire wound spirally over the column section on a pitch of 8 inches, or equivalent ties.
3. Groupings of aggregates and ties are the same as for structural steel columns protected solidly with concrete, the ties to be placed over the vertical reinforcing bars and the mesh where required, to be placed within 1 inch from the surface of the column.
   Column A: working loads are assumed as carried by the area of the column inside of the lines circumscribing the reinforcing steel.
   Column B: working loads are assumed as carried by the gross area of the column.

RESOURCE A—GUIDELINES ON FIRE RATINGS OF ARCHAIC MATERIALS AND ASSEMBLIES

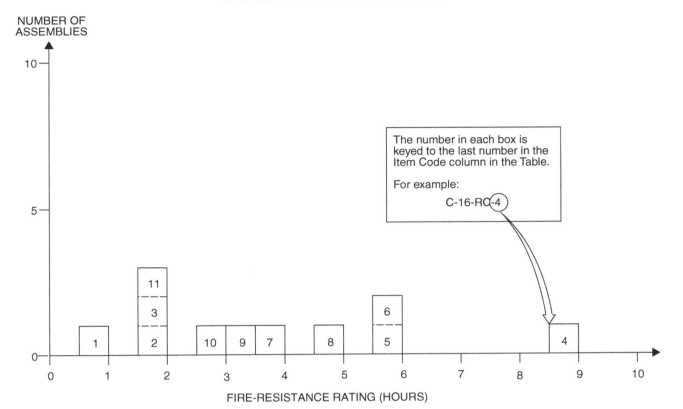

FIGURE 2.1.5
REINFORCED CONCRETE COLUMNS
MINIMUM DIMENSION 16″ TO LESS THAN 18″

TABLE 2.1.5
REINFORCED CONCRETE COLUMNS
MINIMUM DIMENSION 16″ TO LESS THAN 18″

ITEM CODE	MINIMUM DIMENSION	CONSTRUCTION DETAILS	PERFORMANCE LOAD	PERFORMANCE TIME	REFERENCE NUMBER PRE-BMS-92	REFERENCE NUMBER BMS-92	REFERENCE NUMBER POST-BMS-92	NOTES	REC. HOURS
C-16-RC-1	16″	16″ square columns; gravel aggregate concrete (4550 psi); Reinforcement: vertical, eight $1^3/_8$″ rebars; horizontal, $^5/_{16}$″ ties at 6″ pitch $1^3/_8$″ below column surface and $^5/_{16}$″ ties at 6″ pitch linking center rebars of each face forming a smaller square in column cross section.	237 tons	1 hr			7	1, 2, 3	1
C-16-RC-2	16″	16″ square columns; gravel aggregate concrete (3360 psi); Reinforcement: vertical, eight $1^3/_8$″ rebars; horizontal, $^5/_{16}$″ ties at 6″ pitch; Cover: $1^3/_8$″.	210 tons	2 hrs.			7	2, 4, 5, 6	2
C-16-RC-3	16″	16″ square columns; gravel aggregate concrete (3980 psi); Reinforcement: vertical, four $^7/_8$″ rebars; horizontal, $^3/_8$″ ties at 6″ pitch; Cover: 1″.	123.5 tons	2 hrs.			7	2, 4, 7	2
C-16-RC-4	16″	Reinforced concrete columns with $1^1/_2$″ concrete outside reinforcing steel; Gross diameter or side of column: 16″ ; Group I, Column A.	—	9 hrs.	1			8, 9	9
C-16-RC-5	16″	Description as per C-16-RC-4; Group I, Column B.	—	6 hrs.	1			8, 9	6
C-16-RC-6	16″	Description as per C-16-RC-4; Group II, Column A.	—	6 hrs.	1			8, 9	6

(continued)

RESOURCE A—GUIDELINES ON FIRE RATINGS OF ARCHAIC MATERIALS AND ASSEMBLIES

TABLE 2.1.5—continued
REINFORCED CONCRETE COLUMNS
MINIMUM DIMENSION 16″ TO LESS THAN 18″

ITEM CODE	MINIMUM DIMENSION	CONSTRUCTION DETAILS	PERFORMANCE LOAD	PERFORMANCE TIME	REFERENCE NUMBER PRE-BMS-92	REFERENCE NUMBER BMS-92	REFERENCE NUMBER POST-BMS-92	NOTES	REC. HOURS
C-16-RC-7	16″	Description as per C-16-RC-4; Group II, Column B.	—	4 hrs.		1		8, 9	4
C-16-RC-8	16″	Description as per C-16-RC-4; Group III, Column A.	—	5 hrs.		1		8, 9	5
C-16-RC-9	16″	Description as per C-16-RC-4; Group III, Column B.	—	3 hrs. 30 min.		1		8, 9	$3^1/_2$
C-16-RC-10	16″	Description as per C-16-RC-4; Group IV, Column A.	—	3 hrs.		1		8, 9	3
C-16-RC-11	16″	Description as per C-16-RC-4; Group IV, Column B.	—	2 hrs.		1		8, 9	2

For SI: 1 inch = 25.4 mm, 1 pound per square inch = 0.00689 MPa, 1 pound per square yard = 5.3 N/m^2.

Notes:
1. Column passed 1-hour fire test.
2. Column passed hose stream test.
3. No reload specified.
4. Column passed 2-hour fire test.
5. Column reloaded successfully after 24 hours.
6. Reinforcing details same as C-16-RC-1.
7. Column passed reload after 72 hours.
8. Group I: includes concrete having calcareous aggregate containing a combined total of not more than 10 percent of quartz, chert and flint for the coarse aggregate.
    Group II: includes concrete having trap-rock aggregate applied without metal ties and also concrete having cinder, sandstone or granite aggregate, if held in place with wire mesh or expanded metal having not larger than 4-inch mesh, weighing not less than 1.7 lbs./yd.^{2}, placed not more than 1 inch from the surface of the concrete.
    Group III: includes concrete having cinder, sandstone or granite aggregate tied with No. 5 gage steel wire, wound spirally over the column section on a pitch of 8 inches, or equivalent ties, and concrete having siliceous aggregates containing a combined total of 60 percent or more of quartz, chert and flint, if held in place with wire mesh or expanded metal having not larger than 4-inch mesh, weighing not less than 1.7 lbs./yd.^{2}, placed not more than 1 inch from the surface of the concrete.
    Group IV: includes concrete having siliceous aggregates containing a combined total of 60 percent or more of quartz, chert and flint, and tied with No. 5 gage steel wire wound spirally over the column section on a pitch of 8 inches, or equivalent ties.
9. Groupings of aggregates and ties are the same as for structural steel columns protected solidly with concrete, the ties to be placed over the vertical reinforcing bars and the mesh where required, to be placed within 1 inch from the surface of the column.
    Column A: working loads are assumed as carried by the area of the column inside of the lines circumscribing the reinforcing steel.
    Column B: working loads are assumed as carried by the gross area of the column.

RESOURCE A—GUIDELINES ON FIRE RATINGS OF ARCHAIC MATERIALS AND ASSEMBLIES

TABLE 2.1.6
REINFORCED CONCRETE COLUMNS
MINIMUM DIMENSION 18″ TO LESS THAN 20″

ITEM CODE	MINIMUM DIMENSION	CONSTRUCTION DETAILS	PERFORMANCE		REFERENCE NUMBER			NOTES	REC. HOURS
			LOAD	TIME	PRE-BMS-92	BMS-92	POST-BMS-92		
C-18-RC-1	18″	Reinforced concrete columns with $1^1/_2$″ concrete outside reinforced steel; Gross diameter or side of column: 18″ ; Group I, Column A.	—	11 hrs.		1		1, 2	11
C-18-RC-2	18″	Description as per C-18-RC-1; Group I, Column B.	—	8 hrs.		1		1, 2	8
C-18-RC-3	18″	Description as per C-18-RC-1; Group II, Column A.	—	7 hrs.		1		1, 2	7
C-18-RC-4	18″	Description as per C-18-RC-1; Group II, Column B.	—	5 hrs.		1		1, 2	5
C-18-RC-5	18″	Description as per C-18-RC-1; Group III, Column A.	—	6 hrs.		1		1, 2	6
C-18-RC-6	18″	Description as per C-18-RC-1; Group III, Column B.	—	4 hrs.		1		1, 2	4
C-18-RC-7	18″	Description as per C-18-RC-1; Group IV, Column A.	—	3 hrs. 30 min.		1		1, 2	$3^1/_2$
C-18-RC-8	18″	Description as per C-18-RC-1; Group IV, Column B.	—	2 hrs. 30 min.		1		1, 2	$2^1/_2$

For SI: 1 inch = 25.4 mm, 1 pound per square yard = 5.3 N/m².

Notes:
1. Group I: includes concrete having calcareous aggregate containing a combined total of not more than 10 percent of quartz, chert and flint for the coarse aggregate.
   Group II: includes concrete having trap-rock aggregate applied without metal ties and also concrete having cinder, sandstone or granite aggregate, if held in place with wire mesh or expanded metal having not larger than 4-inch mesh, weighing not less than 1.7 lbs./yd.², placed not more than 1 inch from the surface of the concrete.
   Group III: includes concrete having cinder, sandstone or granite aggregate tied with No. 5 gage steel wire, wound spirally over the column section on a pitch of 8 inches, or equivalent ties, and concrete having siliceous aggregates containing a combined total of 60 percent or more of quartz, chert and flint, if held in place with wire mesh or expanded metal having not larger than 4-inch mesh, weighing not less than 1.7 lbs./yd.², placed not more than 1 inch from the surface of the concrete.
   Group IV: includes concrete having siliceous aggregates containing a combined total of 60 percent or more of quartz, chert and flint and, tied with No. 5 gage steel wire wound spirally over the column section on a pitch of 8 inches, or equivalent ties.
2. Groupings of aggregates and ties are the same as for structural steel columns protected solidly with concrete, the ties to be placed over the vertical reinforcing bars and the mesh where required, to be placed within 1 inch from the surface of the column.
   Column A: working loads are assumed as carried by the area of the column inside of the lines circumscribing the reinforcing steel.
   Column B: working loads are assumed as carried by the gross area of the column.

RESOURCE A—GUIDELINES ON FIRE RATINGS OF ARCHAIC MATERIALS AND ASSEMBLIES

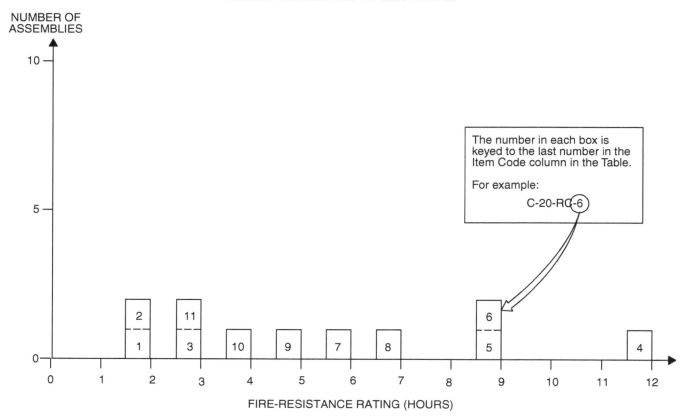

FIGURE 2.1.7
REINFORCED CONCRETE COLUMNS
MINIMUM DIMENSION 20″ TO LESS THAN 22″

TABLE 2.1.7
REINFORCED CONCRETE COLUMNS
MINIMUM DIMENSION 20″ TO LESS THAN 22″

ITEM CODE	MINIMUM DIMENSION	CONSTRUCTION DETAILS	PERFORMANCE		REFERENCE NUMBER			NOTES	REC. HOURS
			LOAD	TIME	PRE-BMS-92	BMS-92	POST-BMS-92		
C-20-RC-1	20″	20″ square columns; gravel aggregate concrete (6690 psi); Reinforcement: vertical, four $1^3/_4$″ rebars; horizontal, $^3/_8$″ wire at 6″ pitch; Cover $1^3/_4$″.	367 tons	2 hrs.			7	1, 2, 3	2
C-20-RC-2	20″	20″ square columns; gravel aggregate concrete (4330 psi); Reinforcement: vertical, four $1^3/_4$″ rebars; horizontal, $^3/_8$″ ties at 6″ pitch; Cover $1^3/_4$″.	327 tons	2 hrs.			7	1, 2, 4	2
C-20-RC-3	$20^1/_4$″	20″ square columns; gravel aggregate concrete (4230 psi); Reinforcement: vertical, four $1^1/_8$″ rebars; horizontal, $^3/_8$″ wire at 5″ pitch; Cover $1^1/_8$″.	199 tons	2 hrs. 56 min.			7	5	$2^3/_4$
C-20-RC-4	20″	Reinforced concrete columns with $1^1/_2$″ concrete outside of reinforcing steel; Gross diameter or side of column: 20″; Group I, Column A.	—	12 hrs.	1			6, 7	12
C-20-RC-5	20″	Description as per C-20-RC-4; Group I, Column B.	—	9 hrs.	1			6, 7	9

(continued)

TABLE 2.1.7—continued
REINFORCED CONCRETE COLUMNS
MINIMUM DIMENSION 20″ TO LESS THAN 22″

ITEM CODE	MINIMUM DIMENSION	CONSTRUCTION DETAILS	PERFORMANCE LOAD	PERFORMANCE TIME	REFERENCE NUMBER PRE-BMS-92	REFERENCE NUMBER BMS-92	REFERENCE NUMBER POST-BMS-92	NOTES	REC. HOURS
C-20-RC-6	20″	Description as per C-20-RC-4; Group II, Column A.	—	9 hrs.	1			6, 7	9
C-20-RC-7	20″	Description as per C-20-RC-4; Group II, Column B.	—	6 hrs.	1			6, 7	6
C-20-RC-8	20″	Description as per C-20-RC-4; Group III, Column A.	—	7 hrs.	1			6, 7	7
C-20-RC-9	20″	Description as per C-20-RC-4; Group III, Column B.	—	5 hrs.	1			6, 7	5
C-20-RC-10	20″	Description as per C-20-RC-4; Group IV, Column A.	—	4 hrs.	1			6, 7	4
C-20-RC-11	20″	Description as per C-20-RC-4; Group IV, Column B.	—	3 hrs.	1			6, 7	3

For SI: 1 inch = 25.4 mm, 1 pound per square yard = 5.3 N/m^2, 1 ton = 8.896 kN.

Notes:
1. Passed 2-hour fire test.
2. Passed hose stream test.
3. Failed during reload at 300 tons.
4. Passed reload after 72 hours.
5. Failure mode—collapse.
6. Group I: includes concrete having calcareous aggregate containing a combined total of not more than 10 percent of quartz, chert and flint for the coarse aggregate.
   Group II: includes concrete having trap-rock aggregate applied without metal ties and also concrete having cinder, sandstone or granite aggregate, if held in place with wire mesh or expanded metal having not larger than 4-inch mesh, weighing not less than 1.7 lbs./yd.^{2}, placed not more than 1 inch from the surface of the concrete.
   Group III: includes concrete having cinder, sandstone or granite aggregate tied with No. 5 gage steel wire, wound spirally over the column section on a pitch of 8 inches, or equivalent ties, and concrete having siliceous aggregates containing a combined total of 60 percent or more of quartz, chert and flint, if held in place with wire mesh or expanded metal having not larger than 4-inch mesh, weighing not less than 1.7 lbs./yd.^{2}, placed not more than 1 inch from the surface of the concrete.
   Group IV: includes concrete having siliceous aggregates containing a combined total of 60 percent or more of quartz, chert and flint, and tied with No. 5 gage steel wire wound spirally over the column section on a pitch of 8 inches, or equivalent ties.
7. Groupings of aggregates and ties are the same as for structural steel columns protected solidly with concrete, the ties to be placed over the vertical reinforcing bars and the mesh where required, to be placed within 1 inch from the surface of the column.
   Column A: working loads are assumed as carried by the area of the column inside of the lines circumscribing the reinforcing steel.
   Column B: working loads are assumed as carried by the gross area of the column.

TABLE 2.1.8
HEXAGONAL REINFORCED CONCRETE COLUMNS
MINIMUM DIMENSION 12″ TO LESS THAN 14″

ITEM CODE	MINIMUM DIMENSION	CONSTRUCTION DETAILS	PERFORMANCE LOAD	PERFORMANCE TIME	REFERENCE NUMBER PRE-BMS-92	REFERENCE NUMBER BMS-92	REFERENCE NUMBER POST-BMS-92	NOTES	REC. HOURS
C-12-HRC-1	12″	12″ hexagonal columns; gravel aggregate concrete (4420 psi); Reinforcement: vertical, eight $1/2$″ rebars; horizontal, $5/16$″ helical winding at $1 1/2$″ pitch; Cover: $1/2$″.	88 tons	58 min.			7	1	$3/4$
C-12-HRC-2	12″	12″ hexagonal columns; gravel aggregate concrete (3460 psi); Reinforcement: vertical, eight $1/2$″ rebars; horizontal, $5/16$″ helical winding at $1 1/2$″ pitch; Cover: $1/2$″.	78.7 tons	1 hr.			7	2	1

For SI: 1 inch = 25.4 mm, 1 pound per square inch = 0.00689 MPa, 1 ton = 8.896 kN.

Notes:
1. Failure mode—collapse.
2. Test stopped at 1 hour.

RESOURCE A—GUIDELINES ON FIRE RATINGS OF ARCHAIC MATERIALS AND ASSEMBLIES

TABLE 2.1.9
HEXAGONAL REINFORCED CONCRETE COLUMNS
MINIMUM DIMENSION 14″ TO LESS THAN 16″

ITEM CODE	MINIMUM DIMENSION	CONSTRUCTION DETAILS	PERFORMANCE LOAD	PERFORMANCE TIME	REFERENCE NUMBER PRE-BMS-92	REFERENCE NUMBER BMS-92	REFERENCE NUMBER POST-BMS-92	NOTES	REC. HOURS
C-14-HRC-1	14″	14″ hexagonal columns; gravel aggregate concrete (4970 psi); Reinforcement: vertical, eight $\frac{1}{2}$″ rebars; horizontal, $\frac{5}{16}$″ helical winding on 2″ pitch; Cover: $\frac{1}{2}$″.	90 tons	2 hrs.			7	1, 2, 3	2

For SI: 1 inch = 25.4 mm, 1 pound per square inch = 0.00689 MPa, 1 ton = 8.896 kN.

Notes:
1. Withstood 2-hour fire test.
2. Withstood hose stream test.
3. Withstood reload after 48 hours.

TABLE 2.1.10
HEXAGONAL REINFORCED CONCRETE COLUMNS
DIAMETER—16″ TO LESS THAN 18″

ITEM CODE	MINIMUM DIMENSION	CONSTRUCTION DETAILS	PERFORMANCE LOAD	PERFORMANCE TIME	REFERENCE NUMBER PRE-BMS-92	REFERENCE NUMBER BMS-92	REFERENCE NUMBER POST-BMS-92	NOTES	REC. HOURS
C-16-HRC-1	16″	16″ hexagonal columns; gravel concrete (6320 psi); Reinforcement: vertical, eight $\frac{5}{8}$″ rebars; horizontal, $\frac{5}{16}$″ helical winding on $\frac{3}{4}$″ pitch; Cover: $\frac{1}{2}$″.	140 tons	1 hr. 55 min.			7	1	$1\frac{3}{4}$
C-16-HRC-2	16″	16″ hexagonal columns; gravel aggregate concrete (5580 psi); Reinforcement: vertical, eight $\frac{5}{8}$″ rebars; horizontal, $\frac{5}{16}$″ helical winding on $1\frac{3}{4}$″ pitch; Cover: $\frac{1}{2}$″.	124 tons	2 hrs.			7	2	2

For SI: 1 inch = 25.4 mm, 1 pound per square inch = 0.00689 MPa, 1 ton = 8.896 kN.

Notes:
1. Failure mode—collapse.
2. Failed on furnace removal.

TABLE 2.1.11
HEXAGONAL REINFORCED CONCRETE COLUMNS
DIAMETER—20″ TO LESS THAN 22″

ITEM CODE	MINIMUM DIMENSION	CONSTRUCTION DETAILS	PERFORMANCE LOAD	PERFORMANCE TIME	REFERENCE NUMBER PRE-BMS-92	REFERENCE NUMBER BMS-92	REFERENCE NUMBER POST-BMS-92	NOTES	REC. HOURS
C-20-HRC-1	20″	20″ hexagonal columns; gravel concrete (6080 psi); Reinforcement: vertical, $\frac{3}{4}$″ rebars; horizontal, $\frac{5}{6}$″ helical winding on $1\frac{3}{4}$″ pitch; Cover: $\frac{1}{2}$″.	211 tons	2 hrs.			7	1	2
C-20-HRC-2	20″	20″ hexagonal columns; gravel concrete (5080 psi); Reinforcement: vertical, $\frac{3}{4}$″ rebars; horizontal, $\frac{5}{16}$″ wire on $1\frac{3}{4}$″ pitch; Cover: $\frac{1}{2}$″.	184 tons	2 hrs. 15 min.			7	2, 3, 4	$2\frac{1}{4}$

For SI: 1 inch = 25.4 mm, 1 pound per square inch = 0.00689 MPa, 1 ton = 8.896 kN.

Notes:
1. Column collapsed on furnace removal.
2. Passed $2\frac{1}{4}$-hour fire test.
3. Passed hose stream test.
4. Withstood reload after 48 hours.

TABLE 2.2
ROUND CAST IRON COLUMNS

ITEM CODE	MINIMUM DIMENSION	CONSTRUCTION DETAILS	PERFORMANCE		REFERENCE NUMBER			NOTES	REC. HOURS
			LOAD	TIME	PRE-BMS-92	BMS-92	POST-BMS-92		
C-7-CI-1	7″ O.D.	Column: 0.6″ minimum metal thickness; unprotected.	—	30 min.		1			$^1/_2$
C-7-CI-2	7″ O.D.	Column: 0.6″ minimum metal thickness concrete filled, outside unprotected.	—	45 min.		1			$^3/_4$
C-11-CI-3	11″ O.D.	Column: 0.6″ minimum metal thickness; Protection: $1^1/_2$″ Portland cement plaster on high ribbed metal lath, $^1/_2$″ broken air space.	—	3 hrs.		1			3
C-11-CI-4	11″ O.D.	Column: 0.6″ minimum metal thickness; Protection: 2″ concrete other than siliceous aggregate.	—	2 hrs. 30 min.		1			$2^1/_2$
C-12-CI-5	12.5″ O.D.	Column: 7″ O.D. 0.6″ minimum metal thickness; Protection: 2″ porous hollow tile, $^3/_4$″ mortar between tile and column, outside wire ties.	—	3 hrs.		1			3
C-7-CI-6	7.6″ O.D.	Column: 7″ I.D., $^3/_{10}$″ minimum metal thickness, concrete filled unprotected.	—	30 min.		1			$^1/_2$
C-8-CI-7	8.6″ O.D.	Column: 8″ I.D., $^3/_{10}$″ minimum metal thickness; concrete filled reinforced with four $3^1/_2$″ × $^3/_8$″ angles, in fill; unprotected outside.	—	1 hr.		1			1

For SI: 1 inch = 25.4 mm.

RESOURCE A—GUIDELINES ON FIRE RATINGS OF ARCHAIC MATERIALS AND ASSEMBLIES

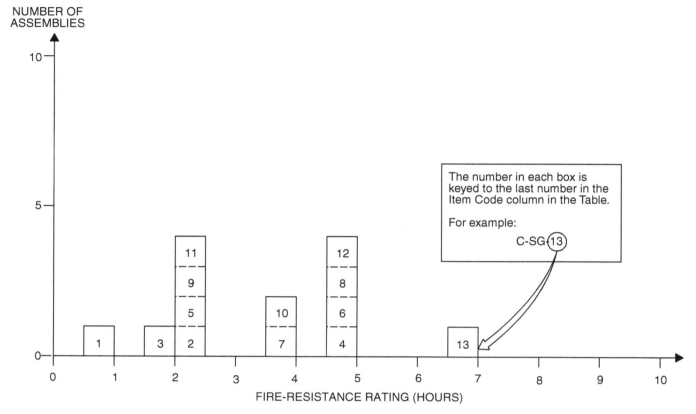

FIGURE 2.3
STEEL COLUMNS—GYPSUM ENCASEMENTS

TABLE 2.3
STEEL COLUMNS—GYPSUM ENCASEMENTS

ITEM CODE	MINIMUM AREA OF SOLID MATERIAL	CONSTRUCTION DETAILS	PERFORMANCE		REFERENCE NUMBER			NOTES	REC. HOURS
			LOAD	TIME	PRE-BMS-92	BMS-92	POST-BMS-92		
C-SG-1	—	Steel protected with $^3/_4$" 1:3 sanded gypsum or 1" 1:2$^1/_2$ Portland cement plaster on wire or lath; one layer.	—	1 hr.		1			1
C-SG-2	—	Same as C-SG-1; two layers.	—	2 hrs. 30 min.		1			2$^1/_2$
C-SG-3	130 in.^{2}	2" solid blocks with wire mesh in horizontal joints; 1" mortar on flange; reentrant space filled with block and mortar.	—	2 hrs.		1			2
C-SG-4	150 in.^{2}	Same as C-130-SG-3 with $^1/_2$" sanded gypsum plaster.	—	5 hrs.		1			5
C-SG-5	130 in.^{2}	2" solid blocks with wire mesh in horizontal joints; 1" mortar on flange; reentrant space filled with gypsum concrete.	—	2 hrs. 30 min.		1			2$^1/_2$
C-SG-6	150 in.^{2}	Same as C-130-SG-5 with $^1/_2$" sanded gypsum plaster.	—	5 hrs.		1			5
C-SG-7	300 in.^{2}	4" solid blocks with wire mesh in horizontal joints; 1" mortar on flange; reentrant space filled with block and mortar.	—	4 hrs.		1			4

(continued)

TABLE 2.3—continued
STEEL COLUMNS—GYPSUM ENCASEMENTS

ITEM CODE	MINIMUM AREA OF SOLID MATERIAL	CONSTRUCTION DETAILS	PERFORMANCE		REFERENCE NUMBER			NOTES	REC. HOURS
			LOAD	TIME	PRE-BMS-92	BMS-92	POST-BMS-92		
C-SG-8	300 in.²	Same as C-300-SG-7 with reentrant space filled with gypsum concrete.	—	5 hrs.		1			5
C-SG-9	85 in.²	2″ solid blocks with cramps at horizontal joints; mortar on flange only at horizontal joints; reentrant space not filled.	—	2 hrs. 30 min.		1			2 $^1/_2$
C-SG-10	105 in.²	Same as C-85-SG-9 with $^1/_2$″ sanded gypsum plaster.	—	4 hrs.		1			4
C-SG-11	95 in.²	3″ hollow blocks with cramps at horizontal joints; mortar on flange only at horizontal joints; reentrant space not filled.	—	2 hrs. 30 min.		1			2 $^1/_2$
C-SG-12	120 in.²	Same as C-95-SG-11 with $^1/_2$″ sanded gypsum plaster.	—	5 hrs.		1			5
C-SG-13	130 in.²	2″ neat fibered gypsum reentrant space filled poured solid and reinforced with 4″ × 4″ wire mesh $^1/_2$″ sanded gypsum plaster.	—	7 hrs.		1			7

For SI: 1 inch = 25.4 mm, 1 square inch = 645 mm².

TABLE 2.4
TIMBER COLUMNS MINIMUM DIMENSION

ITEM CODE	MINIMUM DIMENSION	CONSTRUCTION DETAILS	PERFORMANCE		REFERENCE NUMBER			NOTES	REC. HOURS
			LOAD	TIME	PRE-BMS-92	BMS-92	POST-BMS-92		
C-11-TC-1	11″	With unprotected steel plate cap.	—	30 min.		1		1, 2	$^1/_2$
C-11-TC-2	11″	With unprotected cast iron cap and pintle.	—	45 min.		1		1, 2	$^3/_4$
C-11-TC-3	11″	With concrete or protected steel or cast iron cap.	—	1 hr. 15 min.		1		1, 2	1 $^1/_4$
C-11-TC-4	11″	With $^3/_8$″ gypsum wallboard over column and over cast iron or steel cap.	—	1 hr. 15 min.		1		1, 2	1 $^1/_4$
C-11-TC-5	11″	With 1″ Portland cement plaster on wire lath over column and over cast iron or steel cap; $^3/_4$″ air space.	—	2 hrs.		1		1, 2	2

For SI: 1 inch = 25.4 mm, 1 square inch = 645 mm².

Notes:
1. Minimum area: 120 square inches.
2. Type of wood: long leaf pine or Douglas fir.

TABLE 2.5.1.1
STEEL COLUMNS—CONCRETE ENCASEMENTS
MINIMUM DIMENSION LESS THAN 6″

ITEM CODE	MINIMUM DIMENSION	CONSTRUCTION DETAILS	PERFORMANCE		REFERENCE NUMBER			NOTES	REC. HOURS
			LOAD	TIME	PRE-BMS-92	BMS-92	POST-BMS-92		
C-5-SC-1	5″	5″ × 6″ outer dimensions; 4″ × 3″ × 10 lbs. "H" beam; Protection: gravel concrete (4900 psi) 6″ × 4″ - 13 SWG mesh.	12 tons	1 hr. 29 min.			7	1	1 $^1/_4$

For SI: 1 inch = 25.4 mm, 1 pound per square inch = 0.00689 MPa, 1 ton = 8.896 kN.

Notes:
1. Failure mode—collapse.

**TABLE 2.5.1.2
STEEL COLUMNS—CONCRETE ENCASEMENTS
6″ TO LESS THAN 8″ THICK**

ITEM CODE	MINIMUM DIMENSION	CONSTRUCTION DETAILS	PERFORMANCE		REFERENCE NUMBER			NOTES	REC. HOURS
			LOAD	TIME	PRE-BMS-92	BMS-92	POST-BMS-92		
C-7-SC-1	7″	7″ × 8″ column; 4″ × 3″ × 10 lbs. "H" beam; Protection: brick filled concrete (6220 psi); 6″ × 4″ mesh - 13 SWG; 1″ below column surface.	12 tons	2 hrs. 46 min.			7	1	$2^3/_4$
C-7-SC-2	7″	7″ × 8″ column; 4″ × 3″ × 10 lbs. "H" beam; Protection: gravel concrete (5140 psi); 6″ × 4″ 13 SWG mesh 1″ below surface.	12 tons	3 hrs. 1 min.			7	1	3
C-7-SC-3	7″	7″ × 8″ column; 4″ × 3″ × 10 lbs. "H" beam; Protection: concrete (4540 psi); 6″ × 4″ - 13 SWG mesh; 1″ below column surface.	12 tons	3 hrs. 9 min.			7	1	3
C-7-SC-4	7″	7″ × 8″ column; 4″ × 3″ × 10 lbs. "H" beam; Protection: gravel concrete (5520 psi); 4″ × 4″ mesh; 16 SWG.	12 tons	2 hrs. 50 min.			7	1	$2^3/_4$

For SI: 1 inch = 25.4 mm, 1 pound per square inch = 0.00689 MPa, 1 ton = 8.896 kN.

Notes:
1. Failure mode—collapse.

RESOURCE A—GUIDELINES ON FIRE RATINGS OF ARCHAIC MATERIALS AND ASSEMBLIES

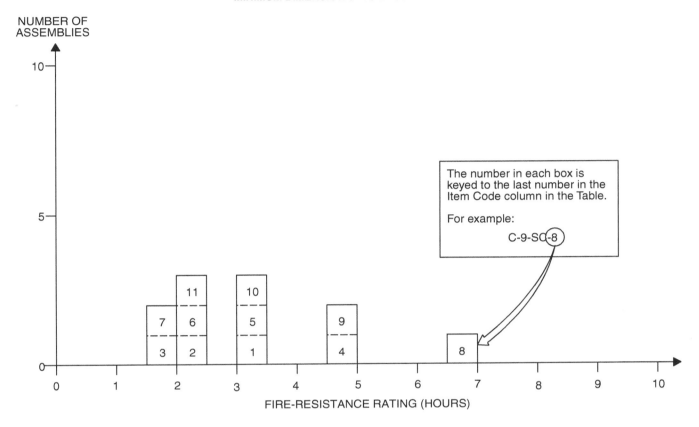

FIGURE 2.5.1.3
STEEL COLUMNS—CONCRETE ENCASEMENTS
MINIMUM DIMENSION 8″ TO LESS THAN 10″

TABLE 2.5.1.3
STEEL COLUMNS—CONCRETE ENCASEMENTS
MINIMUM DIMENSION 8″ TO LESS THAN 10″

ITEM CODE	MINIMUM DIMENSION	CONSTRUCTION DETAILS	PERFORMANCE LOAD	PERFORMANCE TIME	REFERENCE NUMBER PRE-BMS-92	REFERENCE NUMBER BMS-92	REFERENCE NUMBER POST-BMS-92	NOTES	REC. HOURS
C-8-SC-1	$8^1/_2″$	$8^1/_2″ \times 10″$ column; $6″ \times 4^1/_2″ \times 20$ lbs. "H" beam; Protection: gravel concrete (5140 psi); $6″ \times 4″$ - 13 SWG mesh.	39 tons	3 hrs. 8 min.			7	1	3
C-8-SC-2	8″	$8″ \times 10″$ column; $8″ \times 6″ \times 35$ lbs. "I" beam; Protection: gravel concrete (4240 psi); $6″ \times 4″$ - 13 SWG mesh; $^1/_2″$ cover.	90 tons	2 hrs. 1 min.			7	1	2
C-8-SC-3	8″	$8″ \times 10″$ concrete encased column; $8″ \times 6″ \times 35$ lbs. "H" beam; protection: aggregate concrete (3750 psi); 4″ mesh - 16 SWG reinforcing $^1/_2″$ below column surface.	90 tons	1 hr. 58 min.			7	1	$1^3/_4$
C-8-SC-4	8″	$6″ \times 6″$ steel column; 2″ outside protection; Group I.	—	5 hrs.	1			2	5
C-8-SC-5	8″	$6″ \times 6″$ steel column; 2″ outside protection; Group II.	—	3 hrs. 30 min.	1			2	$3^1/_2$
C-8-SC-6	8″	$6″ \times 6″$ steel column; 2″ outside protection; Group III.	—	2 hrs. 30 min.	1			2	$2^1/_2$

(continued)

TABLE 2.5.1.3—continued
STEEL COLUMNS—CONCRETE ENCASEMENTS
MINIMUM DIMENSION 8″ TO LESS THAN 10″

ITEM CODE	MINIMUM DIMENSION	CONSTRUCTION DETAILS	PERFORMANCE LOAD	PERFORMANCE TIME	REFERENCE NUMBER PRE-BMS-92	REFERENCE NUMBER BMS-92	REFERENCE NUMBER POST-BMS-92	NOTES	REC. HOURS
C-8-SC-7	8″	6″ × 6″ steel column; 2″ outside protection; Group IV.	—	1 hr. 45 min.		1		2	$1^3/_4$
C-9-SC-8	9″	6″ × 6″ steel column; 3″ outside protection; Group I.	—	7 hrs.		1		2	7
C-9-SC-9	9″	6″ × 6″ steel column; 3″ outside protection; Group II.	—	5 hrs.		1		2	5
C-9-SC-10	9″	6″ × 6″ steel column; 3″ outside protection; Group III.	—	3 hrs. 30 min.		1		2	$3^1/_2$
C-9-SC-11	9″	6″ × 6″ steel column; 3″ outside protection; Group IV.	—	2 hrs. 30 min.		1		2	$2^1/_2$

For SI: 1 inch = 25.4 mm, 1 pound = 0.004448 kN, 1 pound per square inch = 0.00689 MPa, 1 pound per square yard = 5.3 N/m^2, 1 ton = 8.896 kN.

Notes:
1. Failure mode—collapse.
2. Group I: includes concrete having calcareous aggregate containing a combined total of not more than 10 percent of quartz, chert and flint for the coarse aggregate.

   Group II: includes concrete having trap-rock aggregate applied without metal ties and also concrete having cinder, sandstone or granite aggregate, if held in place with wire mesh or expanded metal having not larger than 4-inch mesh, weighing not less than 1.7 lbs./yd.^{2}, placed not more than 1 inch from the surface of the concrete.

   Group III: includes concrete having cinder, sandstone or granite aggregate tied with No. 5 gage steel wire, wound spirally over the column section on a pitch of 8 inches, or equivalent ties, and concrete having siliceous aggregates containing a combined total of 60 percent or more of quartz, chert and flint, if held in place with wire mesh or expanded metal having not larger than 4-inch mesh, weighing not less than 1.7 lbs./yd.^{2}, placed not more than 1 inch from the surface of the concrete.

   Group IV: includes concrete having siliceous aggregates containing a combined total of 60 percent or more of quartz, chert and flint, and tied with No. 5 gage steel wire wound spirally over the column section on a pitch of 8 inches, or equivalent ties.

RESOURCE A—GUIDELINES ON FIRE RATINGS OF ARCHAIC MATERIALS AND ASSEMBLIES

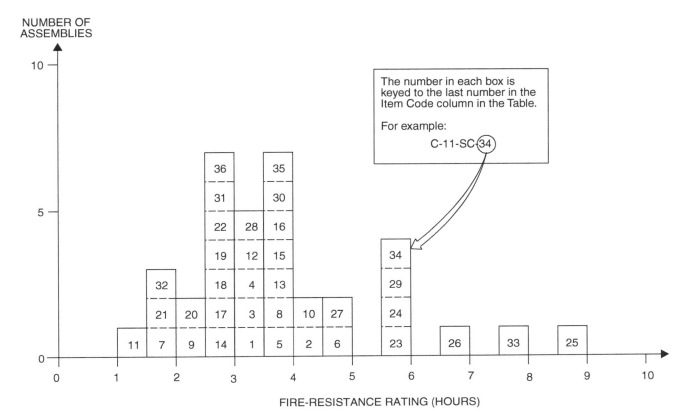

FIGURE 2.5.1.4
STEEL COLUMNS—CONCRETE ENCASEMENTS
MINIMUM DIMENSION 10″ TO LESS THAN 12″

TABLE 2.5.1.4
STEEL COLUMNS—CONCRETE ENCASEMENTS
MINIMUM DIMENSION 10″ TO LESS THAN 12″

ITEM CODE	MINIMUM DIMENSION	CONSTRUCTION DETAILS	PERFORMANCE LOAD	PERFORMANCE TIME	REFERENCE NUMBER PRE-BMS-92	REFERENCE NUMBER BMS-92	REFERENCE NUMBER POST-BMS-92	NOTES	REC. HOURS
C-10-SC-1	10″	10″ × 12″ concrete encased steel column; 8″ × 6″ × 35 lbs. "H" beam; Protection: gravel aggregate concrete (3640 psi); Mesh 6″ × 4″ 13 SWG, 1″ below column surface.	90 tons	3 hrs. 7 min.			7	1,2	3
C-10-SC-2	10″	10″ × 16″ column; 8″ × 6″ × 35 lbs. "H" beam; Protection: clay brick concrete (3630 psi); 6″ × 4″ mesh; 13 SWG, 1″ below column surface.	90 tons	4 hrs. 6 min.			7	2	4
C-10-SC-3	10″	10″ × 12″ column; 8″ × 6″ × 35 lbs. "H" beam; Protection: crushed stone and sand concrete (3930 psi); 6″ × 4″ - 13 SWG mesh; 1″ below column surface.	90 tons	3 hrs. 17 min.			7	2	$3^{1}/_{4}$
C-10-SC-4	10″	10″ × 12″ column; 8″ × 6″ × 35 lbs. "H" beam; Protection: crushed basalt and sand concrete (4350 psi); 6″ × 4″ - 13 SWG mesh; 1″ below column surface.	90 tons	3 hrs. 22 min.			7	2	$3^{1}/_{3}$
C-10-SC-5	10″	10″ × 12″ column; 8″ × 6″ × 35 lbs. "H" beam; Protection: gravel aggregate concrete (5570 psi); 6″ × 4″ mesh; 13 SWG.	90 tons	3 hrs. 39 min.			7	2	$3^{1}/_{2}$
C-10-SC-6	10″	10″ × 16″ column; 8″ × 6″ × 35 lbs. "I" beam; Protection: gravel concrete (4950 psi); mesh; 6″ × 4″ 13 SWG 1″ below column surface.	90 tons	4 hrs. 32 min.			7	2	$4^{1}/_{2}$

(continued)

RESOURCE A—GUIDELINES ON FIRE RATINGS OF ARCHAIC MATERIALS AND ASSEMBLIES

TABLE 2.5.1.4—continued
STEEL COLUMNS—CONCRETE ENCASEMENTS
MINIMUM DIMENSION 10″ TO LESS THAN 12″

ITEM CODE	MINIMUM DIMENSION	CONSTRUCTION DETAILS	PERFORMANCE		REFERENCE NUMBER			NOTES	REC. HOURS
			LOAD	TIME	PRE-BMS-92	BMS-92	POST-BMS-92		
C-10-SC-7	10″	10″ × 12″ concrete encased steel column; 8″ × 6″ × 35 lbs. "H" beam; Protection: aggregate concrete (1370 psi); 6″ × 4″ mesh; 13 SWG reinforcing 1″ below column surface.	90 tons	2 hrs.			7	3, 4	2
C-10-SC-8	10″	10″ × 12″ concrete encased steel column; 8″ × 6″ × 35 lbs. "H" column; Protection: aggregate concrete (4000 psi); 13 SWG iron wire loosely around column at 6″ pitch about 2″ beneath column surface.	86 tons	3 hrs. 36 min.			7	2	$3\frac{1}{2}$
C-10-SC-9	10″	10″ × 12″ concrete encased steel column; 8″ × 6″ × 35 lbs. "H" beam; Protection: aggregate concrete (3290 psi); 2″ cover minimum.	86 tons	2 hrs. 8 min.			7	2	2
C-10-SC-10	10″	10″ × 14″ concrete encased steel column; 8″ × 6″ × 35 lbs. "H" column; Protection: crushed brick filled concrete (5310 psi); 6″ × 4″ mesh; 13 SWG reinforcement 1″ below column surface.	90 tons	4 hrs. 28 min.			7	2	$4\frac{1}{3}$
C-10-SC-11	10″	10″ × 14″ concrete encased column; 8″ × 6″ × 35 lbs. "H" beam; Protection: aggregate concrete (342 psi); 6″ × 4″ mesh; 13 SWG reinforcement 1″ below surface.	90 tons	1 hr. 2 min.			7	2	1
C-10-SC-12	10″	10″ × 12″ concrete encased steel column; 8″ × 6″ × 35 lbs. "H" beam; Protection: aggregate concrete (4480 psi); four $\frac{3}{8}$″ vertical bars at "H" beam edges with $\frac{3}{16}$″ spacers at beam surface at 3′ pitch and $\frac{3}{16}$″ binders at 10″ pitch; 2″ concrete cover.	90 tons	3 hrs. 2 min.			7	2	3
C-10-SC-13	10″	10″ × 12″ concrete encased steel column; 8″ × 6″ × 35 lbs. "H" beam; Protection: aggregate concrete (5070 psi); 6″ × 4″ mesh; 13 SWG reinforcing at 6″ beam sides wrapped and held by wire ties across (open) 8″ beam face; reinforcements wrapped in 6″ × 4″ mesh; 13 SWG throughout; $\frac{1}{2}$″ cover to column surface.	90 tons	3 hrs. 59 min.			7	2	$3\frac{3}{4}$
C-10-SC-14	10″	10″ × 12″ concrete encased steel column; 8″ × 6″ × 35 lbs. "H" beam; Protection: aggregate concrete (4410 psi); 6″ × 4″ mesh; 13 SWG reinforcement $1\frac{1}{4}$″ below column surface; $\frac{1}{2}$″ limestone cement plaster with $\frac{3}{8}$″ gypsum plaster finish.	90 tons	2 hrs. 50 min.			7	2	$2\frac{3}{4}$
C-10-SC-15	10″	10″ × 12″ concrete encased steel column; 8″ × 6″ × 35 lbs. "H" beam; Protection: crushed clay brick filled concrete (4260 psi); 6″ × 4″ mesh; 13 SWG reinforcing 1″ below column surface.	90 tons	3 hrs. 54 min.			7	2	$3\frac{3}{4}$
C-10-SC-16	10″	10″ × 12″ concrete encased steel column; 8″ × 6″ × 35 lbs. "H" beam; Protection: limestone aggregate concrete (4350 psi); 6″ × 4″ mesh; 13 SWG reinforcing 1″ below column surface.	90 tons	3 hrs. 54 min.			7	2	$3\frac{3}{4}$

(continued)

TABLE 2.5.1.4—continued
STEEL COLUMNS—CONCRETE ENCASEMENTS
MINIMUM DIMENSION 10″ TO LESS THAN 12″

ITEM CODE	MINIMUM DIMENSION	CONSTRUCTION DETAILS	PERFORMANCE LOAD	PERFORMANCE TIME	REFERENCE NUMBER PRE-BMS-92	REFERENCE NUMBER BMS-92	REFERENCE NUMBER POST-BMS-92	NOTES	REC. HOURS
C-10-SC-17	10″	10″ × 12″ concrete encased steel column; 8″ × 6″ × 35 lbs. "H" beam; Protection: limestone aggregate concrete (5300 psi); 6″ × 4″; 13 SWG wire mesh 1″ below column surface.	90 tons	3 hrs.			7	4, 5	3
C-10-SC-18	10″	10″ × 12″ concrete encased steel column; 8″ × 6″ × 35 lbs. "H" beam; Protection: limestone aggregate concrete (4800 psi) with 6″ × 4″; 13 SWG mesh reinforcement 1″ below surface.	90 tons	3 hrs.			7	4, 5	3
C-10-SC-19	10″	10″ × 14″ concrete encased steel column; 12″ × 8″ × 65 lbs. "H" beam; Protection: aggregate concrete (3900 psi); 4″ mesh; 16 SWG reinforcing $^1/_2$″ below column surface.	118 tons	2 hrs. 42 min.			7	2	2
C-10-SC-20	10″	10″ × 14″ concrete encased steel column; 12″ × 8″ × 65 lbs. "H" beam; Protection: aggregate concrete (4930 psi); 4″ mesh; 16 SWG reinforcing $^1/_2$″ below column surface.	177 tons	2 hrs. 8 min.			7	2	2
C-10-SC-21	$10^3/_8$″	$10^3/_8$″ × $12^3/_8$″ concrete encased steel column; 8″ × 6″ × 35 lbs. "H" beam; Protection: aggregate concrete (835 psi) with 6″ × 4″ mesh; 13 SWG reinforcing $1^3/_{16}$″ below column surface; $^3/_{16}$″ gypsum plaster finish.	90 tons	2 hrs.			7	3, 4	2
C-11-SC-22	11″	11″ × 13″ concrete encased steel column; 8″ × 6″ × 35 lbs. "H" beam; Protection: "open texture" brick filled concrete (890 psi) with 6″ × 4″ mesh; 13 SWG reinforcing $1^1/_2$″ below column surface; $^3/_8$″ lime cement plaster; $^1/_8$″ gypsum plaster finish.	90 tons	3 hrs.			7	6, 7	3
C-11-SC-23	11″	11″ × 12″ column; 4″ × 3″ × 10 lbs. "H" beam; gravel concrete (4550 psi); 6″ × 4″ - 13 SWG mesh reinforcing; 1″ below column surface.	12 tons	6 hrs.			7	7, 8	6
C-11-SC-24	11″	11″ × 12″ column; 4″ × 3″ × 10 lbs. "H" beam; Protection: gravel aggregate concrete (3830 psi); with 4″ × 4″ mesh; 16 SWG, 1″ below column surface.	16 tons	5 hrs. 32 min.			7	2	$5^1/_2$
C-10-SC-25	10″	6″ × 6″ steel column with 4″ outside protection; Group I.	—	9 hrs.	1			9	9
C-10-SC-26	10″	Description as per C-SC-25; Group II.	—	7 hrs.	1			9	7
C-10-SC-27	10″	Description as per C-10-SC-25; Group III.	—	5 hrs.	1			9	5
C-10-SC-28	10″	Description as per C-10-SC-25; Group IV.	—	3 hrs. 30 min.	1			9	$3^1/_2$
C-10-SC-29	10″	8″ × 8″ steel column with 2″ outside protection; Group I.	—	6 hrs.	1			9	6
C-10-SC-30	10″	Description as per C-10-SC-29; Group II.	—	4 hrs.	1			9	4
C-10-SC-31	10″	Description as per C-10-SC-29; Group III.	—	3 hrs.	1			9	3
C-10-SC-32	10″	Description as per C-10-SC-29; Group IV.	—	2 hrs.	1			9	2

(continued)

RESOURCE A—GUIDELINES ON FIRE RATINGS OF ARCHAIC MATERIALS AND ASSEMBLIES

TABLE 2.5.1.4—continued
STEEL COLUMNS—CONCRETE ENCASEMENTS
MINIMUM DIMENSION 10″ TO LESS THAN 12″

ITEM CODE	MINIMUM DIMENSION	CONSTRUCTION DETAILS	PERFORMANCE LOAD	PERFORMANCE TIME	REFERENCE NUMBER PRE-BMS-92	REFERENCE NUMBER BMS-92	REFERENCE NUMBER POST-BMS-92	NOTES	REC. HOURS
C-11-SC-33	11″	8″ × 8″ steel column with 3″ outside protection; Group I.	—	8 hrs.		1		9	8
C-11-SC-34	11″	Description as per C-10-SC-33; Group II.	—	6 hrs.		1		9	6
C-11-SC-35	11″	Description as per C-10-SC-33; Group III.	—	4 hrs.		1		9	4
C-11-SC-36	11″	Description as per C-10-SC-33; Group IV.	—	3 hrs.		1		9	3

For SI: 1 inch = 25.4 mm, 1 pound = 0.004448 kN, 1 pound per square inch = 0.00689 MPa, 1 pound per square yard = 5.3 N/m^2, 1 ton = 8.896 kN.

Notes:
1. Tested under total restraint load to prevent expansion—minimum load 90 tons.
2. Failure mode—collapse.
3. Passed 2-hour fire test (Grade "C," British).
4. Passed hose stream test.
5. Column tested and passed 3-hour grade fire resistance (British).
6. Column passed 3-hour fire test.
7. Column collapsed during hose stream testing.
8. Column passed 6-hour fire test.
9. Group I: includes concrete having calcareous aggregate containing a combined total of not more than 10 percent of quartz, chert and flint for the coarse aggregate.
   Group II: includes concrete having trap-rock aggregate applied without metal ties and also concrete having cinder, sandstone or granite aggregate, if held in place with wire mesh or expanded metal having not larger than 4-inch mesh, weighing not less than 1.7 lbs./yd.^{2}, placed not more than 1 inch from the surface of the concrete.
   Group III: includes concrete having cinder, sandstone or granite aggregate tied with No. 5 gage steel wire, wound spirally over the column section on a pitch of 8 inches, or equivalent ties, and concrete having siliceous aggregates containing a combined total of 60 percent or more of quartz, chert and flint, if held in place with wire mesh or expanded metal having not larger than 4-inch mesh, weighing not less than 1.7 lbs./yd.^{2}, placed not more than 1 inch from the surface of the concrete.
   Group IV: includes concrete having siliceous aggregates containing a combined total of 60 percent or more of quartz, chert and flint, and tied with No. 5 gage steel wire wound spirally over the column section on a pitch of 8 inches, or equivalent ties.

RESOURCE A—GUIDELINES ON FIRE RATINGS OF ARCHAIC MATERIALS AND ASSEMBLIES

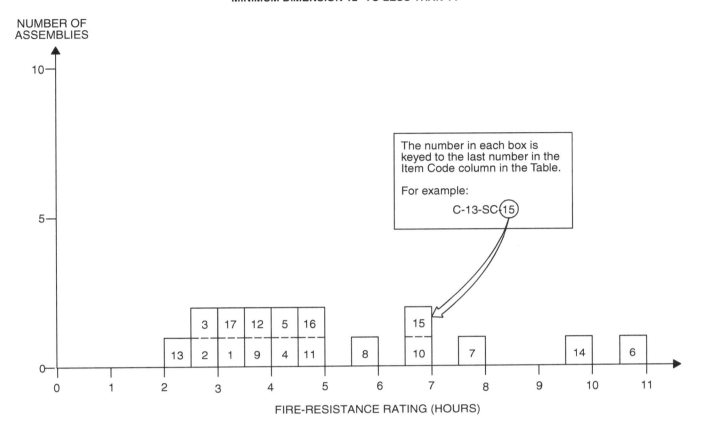

FIGURE 2.5.1.5
STEEL COLUMNS—CONCRETE ENCASEMENTS
MINIMUM DIMENSION 12″ TO LESS THAN 14″

TABLE 2.5.1.5
STEEL COLUMNS—CONCRETE ENCASEMENTS
MINIMUM DIMENSION 12″ TO LESS THAN 14″

ITEM CODE	MINIMUM DIMENSION	CONSTRUCTION DETAILS	PERFORMANCE		REFERENCE NUMBER			NOTES	REC. HOURS
			LOAD	TIME	PRE-BMS-92	BMS-92	POST-BMS-92		
C-12-SC-1	12″	12″ × 14″ concrete encased steel column; 8″ × 6″ × 35 lbs. "H" beam; Protection: aggregate concrete (4150 psi) with 4″ mesh; 16 SWG reinforcing 1″ below column surface.	120 tons	3 hrs. 24 min.			7	1	$3^1/_3$
C-12-SC-2	12″	12″ × 16″ concrete encased column; 8″ × 6″ × 35 lbs. "H" beam; Protection: aggregate concrete (4300 psi) with 4″ mesh; 16 SWG reinforcing 1″ below column surface.	90 tons	2 hrs. 52 min.			7	1	$2^3/_4$
C-12-SC-3	12″	12″ × 16″ concrete encased steel column; 12″ × 8″ × 65 lbs. "H" column; Protection: gravel aggregate concrete (3550 psi) with 4″ mesh; 16 SWG reinforcement 1″ below column surface.	177 tons	2 hrs. 31 min.			7	1	$2^1/_2$
C-12-SC-4	12″	12″ × 16″ concrete encased column; 12″ × 8″ × 65 lbs. "H" beam; Protection: aggregate concrete (3450 psi) with 4″ mesh; 16 SWG reinforcement 1″ below column surface.	118 tons	4 hrs. 4 min.			7	1	4

(continued)

TABLE 2.5.1.5—continued
STEEL COLUMNS—CONCRETE ENCASEMENTS
MINIMUM DIMENSION 12″ TO LESS THAN 14″

ITEM CODE	MINIMUM DIMENSION	CONSTRUCTION DETAILS	PERFORMANCE LOAD	PERFORMANCE TIME	REFERENCE NUMBER PRE-BMS-92	REFERENCE NUMBER BMS-92	REFERENCE NUMBER POST-BMS-92	NOTES	REC. HOURS
C-12-SC-5	$12^1/_2$″	$12^1/_2$″ × 14″ column; 6″ × $4^1/_2$″ × 20 lbs. "H" beam; Protection: gravel aggregate concrete (3750 psi) with 4″ × 4″ mesh; 16 SWG reinforcing 1″ below column surface.	52 tons	4 hrs. 29 min.			7	1	$4^1/_3$
C-12-SC-6	12″	8″ × 8″ steel column; 2″ outside protection; Group I.	—	11 hrs.			1	2	11
C-12-SC-7	12″	Description as per C-12-SC-6; Group II.	—	8 hrs.		1		2	8
C-12-SC-8	12″	Description as per C-12-SC-6; Group III.	—	6 hrs.		1		2	6
C-12-SC-9	12″	Description as per C-12-SC-6; Group IV.	—	4 hrs.		1		2	4
C-12-SC-10	12″	10″ × 10″ steel column; 2″ outside protection; Group I.	—	7 hrs.		1		2	7
C-12-SC-11	12″	Description as per C-12-SC-10; Group II.	—	5 hrs.		1		2	5
C-12-SC-12	12″	Description as per C-12-SC-10; Group III.	—	4 hrs.		1		2	4
C-12-SC-13	12″	Description as per C-12-SC-10; Group IV.	—	2 hrs. 30 min.		1		2	$2^1/_2$
C-13-SC-14	13″	10″ × 10″ steel column; 3″ outside protection; Group I.	—	10 hrs.		1		2	10
C-13-SC-15	13″	Description as per C-12-SC-14; Group II.	—	7 hrs.		1		2	7
C-13-SC-16	13″	Description as per C-12-SC-14; Group III.	—	5 hrs.		1		2	5
C-13-SC-17	13″	Description as per C-12-SC-14; Group IV.	—	3 hrs. 30 min.		1		2	$3^1/_2$

For SI: 1 inch = 25.4 mm, 1 pound = 0.004448 kN, 1 pound per square inch = 0.00689 MPa, 1 pound per square yard = 5.3 N/m^2, 1 ton = 8.896 kN.

Notes:
1. Failure mode—collapse.
2. Group I: includes concrete having calcareous aggregate containing a combined total of not more than 10 percent of quartz, chert and flint for the coarse aggregate.

 Group II: includes concrete having trap-rock aggregate applied without metal ties and also concrete having cinder, sandstone or granite aggregate, if held in place with wire mesh or expanded metal having not larger than 4-inch mesh, weighing not less than 1.7 lbs./yd.^{2}, placed not more than 1 inch from the surface of the concrete.

 Group III: includes concrete having cinder, sandstone or granite aggregate tied with No. 5 gage steel wire, wound spirally over the column section on a pitch of 8 inches, or equivalent ties, and concrete having siliceous aggregates containing a combined total of 60 percent or more of quartz, chert and flint, if held in place with wire mesh or expanded metal having not larger than 4-inch mesh, weighing not less than 1.7 lbs./yd.^{2}, placed not more than 1 inch from the surface of the concrete.

 Group IV: includes concrete having siliceous aggregates containing a combined total of 60 percent or more of quartz, chert and flint, and tied with No. 5 gage steel wire wound spirally over the column section on a pitch of 8 inches, or equivalent ties.

RESOURCE A—GUIDELINES ON FIRE RATINGS OF ARCHAIC MATERIALS AND ASSEMBLIES

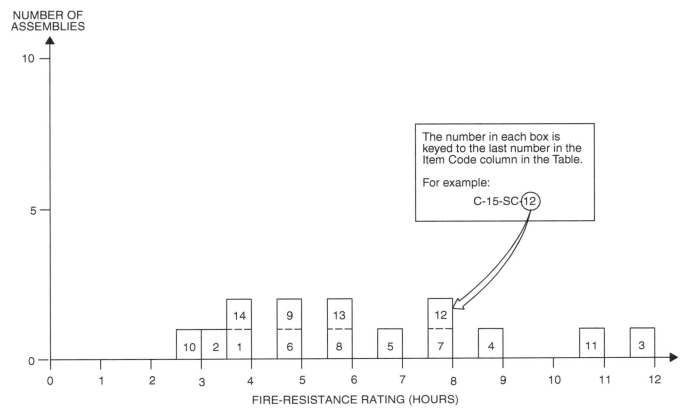

FIGURE 2.5.1.6
STEEL COLUMNS—CONCRETE ENCASEMENTS
MINIMUM DIMENSION 14″ TO LESS THAN 16″

TABLE 2.5.1.6
STEEL COLUMNS—CONCRETE ENCASEMENTS
MINIMUM DIMENSION 14″ TO LESS THAN 16″

ITEM CODE	MINIMUM DIMENSION	CONSTRUCTION DETAILS	PERFORMANCE		REFERENCE NUMBER			NOTES	REC. HOURS
			LOAD	TIME	PRE-BMS-92	BMS-92	POST-BMS-92		
C-14-SC-1	14″	24″ × 16″ concrete encased steel column; 8″ × 6″ × 35 lbs. "H" column; Protection: aggregate concrete (4240 psi); 4″ mesh - 16 SWG reinforcing 1″ below column surface.	90 tons	3 hrs. 40 min.			7	1	3
C-14-SC-2	14″	14″ × 18″ concrete encased steel column; 12″ × 8″ × 65 lbs. "H" beam; Protection: gravel aggregate concrete (4000 psi) with 4″ - 16 SWG wire mesh reinforcement 1″ below column surface.	177 tons	3 hrs. 20 min.			7	1	3
C-14-SC-3	14″	10″ × 10″ steel column; 4″ outside protection; Group I.	—	12 hrs.	1			2	12
C-14-SC-4	14″	Description as per C-14-SC-3; Group II.	—	9 hrs.	1			2	9
C-14-SC-1	14″	24″ × 16″ concrete encased steel column; 8″ × 6″ × 35 lbs. "H" column; Protection: aggregate concrete (4240 psi); 4″ mesh - 16 SWG reinforcing 1″ below column surface.	90 tons	3 hrs. 40 min.			7	1	3

(continued)

RESOURCE A—GUIDELINES ON FIRE RATINGS OF ARCHAIC MATERIALS AND ASSEMBLIES

TABLE 2.5.1.6—continued
STEEL COLUMNS—CONCRETE ENCASEMENTS
MINIMUM DIMENSION 14″ TO LESS THAN 16″

ITEM CODE	MINIMUM DIMENSION	CONSTRUCTION DETAILS	PERFORMANCE LOAD	PERFORMANCE TIME	REFERENCE NUMBER PRE-BMS-92	REFERENCE NUMBER BMS-92	REFERENCE NUMBER POST-BMS-92	NOTES	REC. HOURS
C-14-SC-2	14″	14″ × 18″ concrete encased steel column; 12″ × 8″ × 65 lbs. "H" beam; Protection: gravel aggregate concrete (4000 psi) with 4″-16 SWG wire mesh reinforcement 1″ below column surface.	177 tons	3 hrs. 20 min.			7	1	3
C-14-SC-3	14″	10″ × 10″ steel column; 4″ outside protection; Group I.	—	12 hrs.		1		2	12
C-14-SC-4	14″	Description as per C-14-SC-3; Group II.	—	9 hrs.		1		2	9
C-14-SC-5	14″	Description as per C-14-SC-3; Group III.	—	7 hrs.		1		2	7
C-14-SC-6	14″	Description as per C-14-SC-3; Group IV.	—	5 hrs.		1		2	5
C-14-SC-7	14″	12″ × 12″ steel column; 2″ outside protection; Group I.	—	8 hrs.		1		2	8
C-14-SC-8	14″	Description as per C-14-SC-7; Group II.	—	6 hrs.		1		2	6
C-14-SC-9	14″	Description as per C-14-SC-7; Group III.	—	5 hrs.		1		2	5
C-14-SC-10	14″	Description as per C-14-SC-7; Group IV	—	3 hrs.		1		2	3
C-15-SC-11	15″	12″ × 12″ steel column; 3″ outside protection; Group I.	—	11 hrs.		1		2	11
C-15-SC-12	15″	Description as per C-15-SC-11; Group II.	—	8 hrs.		1		2	8
C-15-SC-13	15″	Description as per C-15-SC-11; Group III.	—	6 hrs.		1		2	6
C-15-SC-14	15″	Description as per C-15-SC-11; Group IV.	—	4 hrs.		1		2	4

For SI: 1 inch = 25.4 mm, 1 pound = 0.004448 kN, 1 pound per square inch = 0.00689 MPa, 1 pound per square yard = 5.3 N/m^2, 1 ton = 8.896 kN.

Notes:
1. Collapse.
2. Group I: includes concrete having calcareous aggregate containing a combined total of not more than 10 percent of quartz, chert and flint for the coarse aggregate.
   Group II: includes concrete having trap-rock aggregate applied without metal ties and also concrete having cinder, sandstone or granite aggregate, if held in place with wire mesh or expanded metal having not larger than 4-inch mesh, weighing not less than 1.7 lbs./yd.^{2}, placed not more than 1 inch from the surface of the concrete.
   Group III: includes concrete having cinder, sandstone or granite aggregate tied with No. 5 gage steel wire, wound spirally over the column section on a pitch of 8 inches, or equivalent ties, and concrete having siliceous aggregates containing a combined total of 60 percent or more of quartz, chert and flint, if held in place with wire mesh or expanded metal having not larger than 4-inch mesh, weighing not less than 1.7 lbs./yd.^{2}, placed not more than 1 inch from the surface of the concrete.
   Group IV: includes concrete having siliceous aggregates containing a combined total of 60 percent or more of quartz, chert and flint, and tied with No. 5 gage steel wire wound spirally over the column section on a pitch of 8 inches, or equivalent ties.

RESOURCE A—GUIDELINES ON FIRE RATINGS OF ARCHAIC MATERIALS AND ASSEMBLIES

TABLE 2.5.1.7
STEEL COLUMNS—CONCRETE ENCASEMENTS
MINIMUM DIMENSION 16″ TO LESS THAN 18″

ITEM CODE	MINIMUM DIMENSION	CONSTRUCTION DETAILS	PERFORMANCE LOAD	PERFORMANCE TIME	REFERENCE NUMBER PRE-BMS-92	REFERENCE NUMBER BMS-92	REFERENCE NUMBER POST-BMS-92	NOTES	REC. HOURS
C-16-SC-13	16″	12″ × 12″ steel column; 4″ outside protection; Group I.	—	14 hrs.	1			1	14
C-16-SC-2	16″	Description as per C-16-SC-1; Group II.	—	10 hrs.	1			1	10
C-16-SC-3	16″	Description as per C-16-SC-1; Group III.	—	8 hrs.	1			1	8
C-16-SC-4	16″	Description as per C-16-SC-1; Group IV.	—	5 hrs.	1			1	5

For SI: 1 inch = 25.4 mm.

Notes:
1. Group I: includes concrete having calcareous aggregate containing a combined total of not more than 10 percent of quartz, chert and flint for the coarse aggregate.
Group II: includes concrete having trap-rock aggregate applied without metal ties and also concrete having cinder, sandstone or granite aggregate, if held in place with wire mesh or expanded metal having not larger than 4-inch mesh, weighing not less than 1.7 lbs./yd.^{2}, placed not more than 1 inch from the surface of the concrete.
Group III: includes concrete having cinder, sandstone or granite aggregate tied with No. 5 gage steel wire, wound spirally over the column section on a pitch of 8 inches, or equivalent ties, and concrete having siliceous aggregates containing a combined total of 60 percent or more of quartz, chert and flint, if held in place with wire mesh or expanded metal having not larger than 4-inch mesh, weighing not less than 1.7 lbs./yd.^{2}, placed not more than 1 inch from the surface of the concrete.
Group IV: includes concrete having siliceous aggregates containing a combined total of 60 percent or more of quartz, chert and flint, and tied with No. 5 gage steel wire wound spirally over the column section on a pitch of 8 inches, or equivalent ties.

TABLE 2.5.2.1
STEEL COLUMNS—BRICK AND BLOCK ENCASEMENTS
MINIMUM DIMENSION 10″ TO LESS THAN 12″

ITEM CODE	MINIMUM DIMENSION	CONSTRUCTION DETAILS	PERFORMANCE LOAD	PERFORMANCE TIME	REFERENCE NUMBER PRE-BMS-92	REFERENCE NUMBER BMS-92	REFERENCE NUMBER POST-BMS-92	NOTES	REC. HOURS
C-10-SB-1	10$\frac{1}{2}$″	10$\frac{1}{2}$″ × 13″ brick encased steel columns; 8″ × 6″ × 35 lbs. "H" beam; Protection. Fill of broken brick and mortar; 2″ brick on edge; joints broken in alternate courses; cement-sand grout; 13 SWG wire reinforcement in every third horizontal joint.	90 tons	3 hrs. 6 min.			7	1	3
C-10-SB-2	10$\frac{1}{2}$″	10$\frac{1}{2}$″ × 13″ brick encased steel columns; 8″ × 6″ × 35 lbs. "H" beam; Protection: 2″ brick; joints broken in alternate courses; cement-sand grout; 13 SWG iron wire reinforcement in alternate horizontal joints.	90 tons	2 hrs.			7	2, 3, 4	2
C-10-SB-3	10″	10″ × 12″ block encased columns; 8″ × 6″ × 35 lbs. "H" beam; Protection: 2″ foamed slag concrete blocks; 13 SWG wire at each horizontal joint; mortar at each joint.	90 tons	2 hrs.			7	5	2
C-10-SB-4	10$\frac{1}{2}$″	10$\frac{1}{2}$″ × 12″ block encased steel columns; 8″ × 6″ × 35 lbs. "H" beam; Protection: gravel aggregate concrete fill (unconsolidated) 2″ thick hollow clay tiles with mortar at edges.	86 tons	56 min.			7	1	$\frac{3}{4}$
C-10-SB-5	10$\frac{1}{2}$″	10$\frac{1}{2}$″ × 12″ block encased steel columns; 8″ × 6″ × 35 lbs. "H" beam; Protection: 2″ hollow clay tiles with mortar at edges.	86 tons	22 min.			7	1	$\frac{1}{4}$

For SI: 1 inch = 25.4 mm, 1 pound = 0.004448 kN, 1 ton = 8.896 kN.

Notes:
1. Failure mode—collapse.
2. Passed 2-hour fire test (Grade "C" - British).
3. Passed hose stream test.
4. Passed reload test.
5. Passed 2-hour fire exposure but collapsed immediately following hose stream test.

RESOURCE A—GUIDELINES ON FIRE RATINGS OF ARCHAIC MATERIALS AND ASSEMBLIES

TABLE 2.5.2.2
STEEL COLUMNS—BRICK AND BLOCK ENCASEMENTS
MINIMUM DIMENSION 12″ TO LESS THAN 14″

ITEM CODE	MINIMUM DIMENSION	CONSTRUCTION DETAILS	PERFORMANCE		REFERENCE NUMBER			NOTES	REC. HOURS
			LOAD	TIME	PRE-BMS-92	BMS-92	POST-BMS-92		
C-12-SB-1	12″	12″ × 15″ brick encased steel columns; 8″ × 6″ × 35 lbs. "H" beam; Protection: $2^{5}/_{8}$″ thick brick; joints broken in alternate courses; cement-sand grout; fill of broken brick and mortar.	90 tons	1 hr. 49 min.			7	1	$1^{3}/_{4}$

For SI: 1 inch = 25.4 mm, 1 pound = 0.004448 kN, 1 ton = 8.896 kN.

Notes:
1. Failure mode—collapse.

TABLE 2.5.2.3
STEEL COLUMNS—BRICK AND BLOCK ENCASEMENTS
MINIMUM DIMENSION 14″ TO LESS THAN 16″

ITEM CODE	MINIMUM DIMENSION	CONSTRUCTION DETAILS	PERFORMANCE		REFERENCE NUMBER			NOTES	REC. HOURS
			LOAD	TIME	PRE-BMS-92	BMS-92	POST-BMS-92		
C-15-SB-1	15″	15″ × 17″ brick encased steel columns; 8″ × 6″ × 35 lbs. "H" beam; Protection: $4^{1}/_{2}$″ thick brick; joints broken in alternate courses; cement-sand grout; fill of broken brick and mortar.	45 tons	6 hrs.			7	1	6
C-15-SB-2	15″	15″ × 17″ brick encased steel columns; 8″ × 6″ × 35 lbs. "H" beam; Protection. Fill of broken brick and mortar; $4^{1}/_{2}$″ brick; joints broken in alternate courses; cement-sand grout.	86 tons	6 hrs.			7	2, 3, 4	6
C-15-SB-3	15″	15″ × 18″ brick encased steel columns; 8″ × 6″ × 35 lbs. "H" beam; Protection: $4^{1}/_{2}$″ brick work; joints alternating; cement-sand grout.	90 tons	4 hrs.			7	5, 6	4
C-15-SB-4	14″	14″ × 16″ block encased steel columns; 8″ × 6″ × 35 lbs. "H" beam; Protection: 4″ thick foam slag concrete blocks; 13 SWG wire reinforcement in each horizontal joint; mortar in joints.	90 tons	5 hrs. 52 min.			7	7	$4^{3}/_{4}$

For SI: 1 inch = 25.4 mm, 1 pound = 0.004448 kN, 1 ton = 8.896 kN.

Notes:
1. Only a nominal load was applied to specimen.
2. Passed 6-hour fire test (Grade "A" - British).
3. Passed (6 minute) hose stream test.
4. Reload not specified.
5. Passed 4-hour fire exposure.
6. Failed by collapse between first and second minute of hose stream exposure.
7. Mode of failure-collapse.

TABLE 2.5.3.1
STEEL COLUMNS—PLASTER ENCASEMENTS
MINIMUM DIMENSION 6″ TO LESS THAN 8″

ITEM CODE	MINIMUM DIMENSION	CONSTRUCTION DETAILS	PERFORMANCE		REFERENCE NUMBER			NOTES	REC. HOURS
			LOAD	TIME	PRE-BMS-92	BMS-92	POST-BMS-92		
C-7-SP-1	$7\frac{1}{2}″$	$7\frac{1}{2}″ \times 9\frac{1}{2}″$ plaster protected steel columns; $8″ \times 6″ \times 35$ lbs. "H" beam; Protection: 24 SWG wire metal lath; $1\frac{1}{4}″$ lime plaster.	90 tons	57 min.			7	1	$\frac{3}{4}$
C-7-SP-2	$7\frac{7}{8}″$	$7\frac{7}{8}″ \times 10″$ plaster protected steel columns; $8″ \times 6″ \times 35$ lbs. "H" beam; Protection: $\frac{3}{8}″$ gypsum board; wire wound with 16 SWG wire helically wound at 4″ pitch; $\frac{1}{2}″$ gypsum plaster.	90 tons	1 hr. 13 min.			7	1	1
C-7-SP-3	$7\frac{1}{4}″$	$7\frac{1}{4}″ \times 9\frac{3}{8}″$ plaster protected steel columns; $8″ \times 6″ \times 35$ lbs. "H" beam; Protection: $\frac{3}{8}″$ gypsum board; wire helically wound 16 SWG at 4″ pitch; $\frac{1}{4}″$ gypsum plaster finish.	90 tons	1 hr. 14 min.			7	1	1

Notes:
1. Failure mode—collapse.

TABLE 2.5.3.2
STEEL COLUMNS—PLASTER ENCASEMENTS
MINIMUM DIMENSION 8″ TO LESS THAN 10″

ITEM CODE	MINIMUM DIMENSION	CONSTRUCTION DETAILS	PERFORMANCE		REFERENCE NUMBER			NOTES	REC. HOURS
			LOAD	TIME	PRE-BMS-92	BMS-92	POST-BMS-92		
C-8-SP-1	$8″$	$8″ \times 10″$ plaster protected steel columns; $8″ \times 6″ \times 35$ lbs. "H" beam; Protection: 24 SWG wire lath; 1″ gypsum plaster.	86 tons	1 hr. 23 min.			7	1	$1\frac{1}{4}$
C-8-SP-2	$8\frac{1}{2}″$	$8\frac{1}{2}″ \times 10\frac{1}{2}″$ plaster protected steel columns; $8″ \times 6″ \times 35$ lbs. "H" beam; Protection: 24 SWG metal lath wrap; $1\frac{1}{4}″$ gypsum plaster.	90 tons	1 hr. 36 min.			7	1	$1\frac{1}{2}$
C-9-SP-3	$9″$	$9″ \times 11″$ plaster protected steel columns; $8″ \times 6″ \times 35$ lbs. "H" beam; Protection: 24 SWG metal lath wrap; $\frac{1}{8}″$ M.S. ties at 12″ pitch wire netting $1\frac{1}{2}″ \times 22$ SWG between first and second plaster coats; $1\frac{1}{2}″$ gypsum plaster.	90 tons	1 hr. 33 min.			7	1	$1\frac{1}{2}$
C-8-SP-4	$8\frac{3}{4}″$	$8\frac{3}{4}″ \times 10\frac{3}{4}″$ plaster protected steel columns; $8″ \times 6″ \times 35$ lbs. "H" beam; Protection: $\frac{3}{4}″$ gypsum board; wire wound spirally (#16 SWG) at $1\frac{1}{2}″$ pitch; $\frac{1}{2}″$ gypsum plaster.	90 tons	2 hrs.			7	2, 3, 4	2

For SI: 1 inch = 25.4 mm, 1 pound = 0.004448 kN, 1 ton = 8.896 kN.

Notes:
1. Failure mode—collapse.
2. Passed 2 hour fire exposure test (Grade "C" - British).
3. Passed hose stream test.

TABLE 2.5.4.1
STEEL COLUMNS—MISCELLANEOUS ENCASEMENTS
MINIMUM DIMENSION 6″ TO LESS THAN 8″

ITEM CODE	MINIMUM DIMENSION	CONSTRUCTION DETAILS	PERFORMANCE		REFERENCE NUMBER			NOTES	REC. HOURS
			LOAD	TIME	PRE-BMS-92	BMS-92	POST-BMS-92		
C-7-SM-1	$7\frac{5}{8}″$	$7\frac{5}{8}″ \times 9\frac{1}{2}″$ (asbestos plaster) protected steel columns; $8″ \times 6″ \times 35$ lbs. "H" beam; Protection: 20 gage $\frac{1}{2}″$ metal lath; $\frac{9}{16}″$ asbestos plaster (minimum).	90 tons	1 hr. 52 min.			7	1	$1\frac{3}{4}$

For SI: 1 inch = 25.4 mm, 1 pound = 0.004448 kN, 1 ton = 8.896 kN.

Notes:
1. Failure mode—collapse.

RESOURCE A—GUIDELINES ON FIRE RATINGS OF ARCHAIC MATERIALS AND ASSEMBLIES

TABLE 2.5.4.2
STEEL COLUMNS—MISCELLANEOUS ENCASEMENTS
MINIMUM DIMENSION 8″ TO LESS THAN 10″

ITEM CODE	MINIMUM DIMENSION	CONSTRUCTION DETAILS	PERFORMANCE		REFERENCE NUMBER			NOTES	REC. HOURS
			LOAD	TIME	PRE-BMS-92	BMS-92	POST-BMS-92		
C-9-SM-1	$9^5/_8″$	$9^5/_8″ \times 11^3/_8″$ asbestos slab and cement plaster protected columns; $8″ \times 6″ \times 35$ lbs. "H" beam; Protection: 1″ asbestos slab; wire wound; $^5/_8″$ plaster.	90 tons	2 hrs.			7	1, 2	2

For SI: 1 inch = 25.4 mm, 1 pound = 0.004448 kN, 1 ton = 8.896 kN.

Notes:
1. Passed 2 hour fire exposure test.
2. Collapsed during hose stream test.

TABLE 2.5.4.3
STEEL COLUMNS—MISCELLANEOUS ENCASEMENTS
MINIMUM DIMENSION 10″ TO LESS THAN 12″

ITEM CODE	MINIMUM DIMENSION	CONSTRUCTION DETAILS	PERFORMANCE		REFERENCE NUMBER			NOTES	REC. HOURS
			LOAD	TIME	PRE-BMS-92	BMS-92	POST-BMS-92		
C-11-SM-1	$11^1/_2″$	$11^1/_2″ \times 13^1/_2″$ wood wool and plaster protected steel columns; $8″ \times 6″ \times 35$ lbs. "H" beam; Protection: wood-wool-cement paste as fill and to 2″ cover over beam; $^3/_4″$ gypsum plaster finish.	90 tons	2 hrs.			7	1, 2, 3	2
C-10-SM-1	10″	$10″ \times 12″$ asbestos protected steel columns; $8″ \times 6″ \times 35$ lbs. "H" beam; Protection: sprayed on asbestos paste to 2″ cover over column.	90 tons	4 hrs.			7	2, 3, 4	4

For SI: 1 inch = 25.4 mm, 1 pound = 0.004448 kN, 1 ton = 8.896 kN.

Notes:
1. Passed 2 hour fire exposure (Grade "C" - British).
2. Passed hose stream test.
3. Passed reload test.
4. Passed 4 hour fire exposure test.

TABLE 2.5.4.4
STEEL COLUMNS—MISCELLANEOUS ENCASEMENTS
MINIMUM DIMENSION 12″ TO LESS THAN 14″

ITEM CODE	MINIMUM DIMENSION	CONSTRUCTION DETAILS	PERFORMANCE		REFERENCE NUMBER			NOTES	REC. HOURS
			LOAD	TIME	PRE-BMS-92	BMS-92	POST-BMS-92		
C-12-SM-1	12″	$12″ \times 14^1/_4″$ cement and asbestos protected columns; $8″ \times 6″ \times 35$ lbs. "H" beam; Protection: fill of asbestos packing pieces 1″ thick 1′3″ o.c.; cover of 2″ molded asbestos inner layer; 1″ molded asbestos outer layer; held in position by 16 SWG nichrome wire ties; wash of refractory cement on outer surface.	86 tons	4 hrs. 43 min.			7	1, 2, 3	$4^2/_3$

For SI: 1 inch = 25.4 mm, 1 pound = 0.004448 kN, 1 ton = 8.896 kN.

Notes:
1. Passed 4 hour fire exposure (Grade "B" - British).
2. Passed hose stream test.
3. Passed reload test.

RESOURCE A—GUIDELINES ON FIRE RATINGS OF ARCHAIC MATERIALS AND ASSEMBLIES

SECTION III
FLOOR/CEILING ASSEMBLIES

FIGURE 3.1
FLOOR/CEILING ASSEMBLIES—REINFORCED CONCRETE

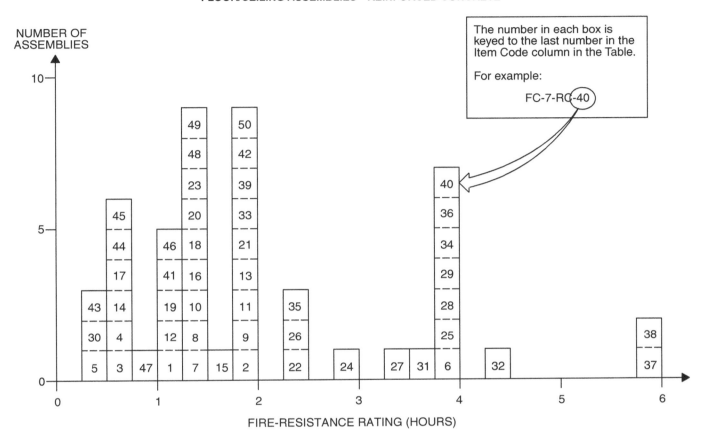

TABLE 3.1
FLOOR/CEILING ASSEMBLIES—REINFORCED CONCRETE

ITEM CODE	ASSEMBLY THICKNESS	CONSTRUCTION DETAILS	PERFORMANCE		REFERENCE NUMBER			NOTES	REC. HOURS
			LOAD	TIME	PRE-BMS-92	BMS-92	POST-BMS-92		
F/C-3-RC-1	$3^3/_4''$	$3^3/_4''$ thick floor; $3^1/_4''$ (5475 psi) concrete deck; $1/_2''$ plaster under deck; $3/_8''$ main reinforcement bars at $5^1/_2''$ pitch with $7/_8''$ concrete cover; $3/_8''$ main reinforcement bars at $4^1/_2''$ pitch perpendicular with $1/_2''$ concrete cover; 13'1'' span restrained.	195 psf	24 min.			7	1, 2	$1/_3$
F/C-3-RC-2	$3^1/_4''$	$3^1/_4''$ deep (3540 psi) concrete deck; $3/_8''$ main reinforcement bars at $5^1/_2''$ pitch with $7/_8''$ cover; $3/_8''$ main reinforcement bars at $4^1/_2''$ pitch perpendicular with $1/_2''$ cover; 13'1'' span restrained.	195 psf	2 hrs.			7	1, 3, 4	2

(continued)

RESOURCE A—GUIDELINES ON FIRE RATINGS OF ARCHAIC MATERIALS AND ASSEMBLIES

TABLE 3.1—continued
FLOOR/CEILING ASSEMBLIES—REINFORCED CONCRETE

ITEM CODE	ASSEMBLY THICKNESS	CONSTRUCTION DETAILS	PERFORMANCE		REFERENCE NUMBER			NOTES	REC. HOURS
			LOAD	TIME	PRE-BMS-92	BMS-92	POST-BMS-92		
F/C-3-RC-3	$3\frac{1}{4}''$	$3\frac{1}{4}''$ deep (4175 psi) concrete deck; $\frac{3}{8}''$ main reinforcement bars at $5\frac{1}{2}''$ pitch with $\frac{7}{8}''$ cover; $\frac{3}{8}''$ main reinforcement bars at $4\frac{1}{2}''$ pitch perpendicular with $\frac{1}{2}''$ cover; 13'1" span restrained.	195 psf	31 min.			7	1, 5	$\frac{1}{2}$
F/C-3-RC-4	$3\frac{1}{4}''$	$3\frac{1}{4}''$ deep (4355 psi) concrete deck; $\frac{3}{8}''$ main reinforcement bars at $5\frac{1}{2}''$ pitch with $\frac{7}{8}''$ cover; $\frac{3}{8}''$ main reinforcement bars at $4\frac{1}{2}''$ pitch perpendicular with $\frac{1}{2}''$ cover; 13'1" span restrained.	195 psf	41 min.			7	1, 5, 6	$\frac{1}{2}$
F/C-3-RC-5	$3\frac{1}{4}''$	$3\frac{1}{4}''$ thick (3800 psi) concrete deck; $\frac{3}{8}''$ main reinforcement bars at $5\frac{1}{2}''$ pitch with $\frac{7}{8}''$ cover; $\frac{3}{8}''$ main reinforcement bars at $4\frac{1}{2}''$ pitch perpendicular with $\frac{1}{2}''$ cover; 13'1" span restrained.	195 psf	1 hr. 5 min.			7	1, 5	1
F/C-4-RC-6	$4\frac{1}{4}''$	$4\frac{1}{4}''$ thick; $3\frac{1}{4}''$ (4000 psi) concrete deck; 1" sprayed asbestos lower surface; $\frac{3}{8}''$ main reinforcement bars at $5\frac{7}{8}''$ pitch with $\frac{7}{8}''$ concrete cover; $\frac{3}{8}''$ main reinforcement bars at $4\frac{1}{2}''$ pitch perpendicular with $\frac{1}{2}''$ concrete cover; 13'1" span restrained.	195 psf	4 hrs.			7	1, 7	4
F/C-4-RC-7	4"	4" (5025 psi) concrete deck; $\frac{1}{4}''$ reinforcement bars at $7\frac{1}{2}''$ pitch with $\frac{3}{4}''$ cover; $\frac{3}{8}''$ main reinforcement bars at $3\frac{3}{4}''$ pitch perpendicular with $\frac{1}{2}''$ cover; 13'1" span restrained.	140 psf	1 hr. 16 min.			7	1, 2	$1\frac{1}{4}$
F/C-4-RC-8	4"	4" thick (4905 psi) deck; $\frac{1}{4}''$ reinforcement bars at $7\frac{1}{2}''$ pitch with $\frac{7}{8}''$ cover; $\frac{3}{8}''$ main reinforcement bars at $3\frac{3}{4}''$ pitch perpendicular with $\frac{1}{2}''$ cover; 13'1" span restrained.	100 psf	1 hr. 23 min.			7	1, 2	$1\frac{1}{3}$
F/C-4-RC-9	4"	4" deep (4370 psi); $\frac{1}{4}''$ reinforcement bars at 6" pitch with $\frac{3}{4}''$ cover; $\frac{1}{4}''$ main reinforcement bars at 4" pitch perpendicular with $\frac{1}{2}''$ cover; 13'1" span restrained.	150 psf	2 hrs.			7	1, 3	2
F/C-4-RC-10	4"	4" thick (5140 psi) deck; $\frac{1}{4}''$ reinforcement bars at $7\frac{1}{2}''$ pitch with $\frac{7}{8}''$ cover; $\frac{3}{8}''$ main reinforcement bars at $3\frac{3}{4}''$ pitch perpendicular with $\frac{1}{2}''$ cover; 13'1" span restrained.	140 psf	1 hr. 16 min.			7	1, 5	$1\frac{1}{4}$
F/C-4-RC-11	4"	4" thick (4000 psi) concrete deck; 3" × $1\frac{1}{2}''$ × 4 lbs. R.S.J.; 2'6" C.R.S.; flush with top surface; 4" × 6" x 13 SWG mesh reinforcement 1" from bottom of slab; 6'6" span restrained.	150 psf	2 hrs.			7	1, 3	2

(continued)

TABLE 3.1—continued
FLOOR/CEILING ASSEMBLIES—REINFORCED CONCRETE

ITEM CODE	ASSEMBLY THICKNESS	CONSTRUCTION DETAILS	PERFORMANCE		REFERENCE NUMBER			NOTES	REC. HOURS
			LOAD	TIME	PRE-BMS-92	BMS-92	POST-BMS-92		
F/C-4-RC-12	4″	4″ deep (2380 psi) concrete deck; 3″ × $1^1/_2$″ × 4 lbs. R.S.J.; 2′6″ C.R.S.; flush with top surface; 4″ × 6″ x 13 SWG mesh reinforcement 1″ from bottom surface; 6′6″ span restrained.	150 psf	1 hr. 3 min.			7	1, 2	1
F/C-4-RC-13	$4^1/_2$″	$4^1/_2$″ thick (5200 psi) deck; $^1/_4$″ reinforcement bars at $7^1/_4$″ pitch with $^7/_8$″ cover; $^3/_8$″ main reinforcement bars at $3^3/_4$″ pitch perpendicular with $^1/_2$″ cover; 13′1″ span restrained.	140 psf	2 hrs.			7	1, 3	2
F/C-4-RC-14	$4^1/_2$″	$4^1/_2$″ deep (2525 psi) concrete deck; $^1/_4$″ reinforcement bars at $7^1/_2$″ pitch with $^7/_8$″ cover; $^3/_8$″ main reinforcement bars at $3^3/_8$″ pitch perpendicular with $^1/_2$″ cover; 13′1″ span restrained.	150 psf	42 min.			7	1, 5	$^2/_3$
F/C-4-RC-15	$4^1/_2$″	$4^1/_2$″ deep (4830 psi) concrete deck; $1^1/_2$″ × No. 15 gauge wire mesh; $^3/_8$″ reinforcement bars at 15″ pitch with 1″ cover; $^1/_2$″ main reinforcement bars at 6″ pitch perpendicular with $^1/_2$″ cover; 12′ span simply supported.	75 psf	1 hr. 32 min.			7	1, 8	$1^1/_2$
F/C-4-RC-16	$4^1/_2$″	$4^1/_2$″ deep (4595 psi) concrete deck; $^1/_4$″ reinforcement bars at $7^1/_2$″ pitch with $^7/_8$″ cover; $^3/_8$″ main reinforcement bars at $3^1/_2$″ pitch perpendicular with $^1/_2$″ cover; 12′ span simply supported.	75 psf	1 hr. 20 min.			7	1, 8	$1^1/_3$
F/C-4-RC-17	$4^1/_2$″	$4^1/_2$″ deep (3625 psi) concrete deck; $^1/_4$″ reinforcement bars at $7^1/_2$″ pitch with $^7/_8$″ cover; $^3/_8$″ main reinforcement bars at $3^1/_2$″ pitch perpendicular with $^1/_2$″ cover; 12′ span simply supported.	75 psf	35 min.			7	1, 8	$^1/_2$
F/C-4-RC-18	$4^1/_2$″	$4^1/_2$″ deep (4410 psi) concrete deck; $^1/_4$″ reinforcement bars at $7^1/_2$″ pitch with $^7/_8$″ cover; $^3/_8$″ main reinforcement bars at $3^1/_2$″ pitch perpendicular with $^1/_2$″ cover; 12′ span simply supported.	85 psf	1 hr. 27 min.			7	1, 8	$1^1/_3$
F/C-4-RC-19	$4^1/_2$″	$4^1/_2$″ deep (4850 psi) deck; $^3/_8$″ reinforcement bars at 15″ pitch with 1″ cover; $^1/_2$″ main reinforcement bars at 6″ pitch perpendicular with $^1/_2$″ cover; 12′ span simply supported.	75 psf	2 hrs. 15 min.			7	1, 9	$1^1/_4$
F/C-4-RC-20	$4^1/_2$″	$4^1/_2$″ deep (3610 psi) deck; $^1/_4$″ reinforcement bars at $7^1/_2$″ pitch with $^7/_8$″ cover; $^3/_8$″ main reinforcement bars at $3^1/_2$″ pitch perpendicular with $^1/_2$″ cover; 12′ span simply supported.	75 psf	1 hr. 22 min.			7	1, 8	$1^1/_3$

(continued)

RESOURCE A—GUIDELINES ON FIRE RATINGS OF ARCHAIC MATERIALS AND ASSEMBLIES

TABLE 3.1—continued
FLOOR/CEILING ASSEMBLIES—REINFORCED CONCRETE

ITEM CODE	ASSEMBLY THICKNESS	CONSTRUCTION DETAILS	PERFORMANCE		REFERENCE NUMBER			NOTES	REC. HOURS
			LOAD	TIME	PRE-BMS-92	BMS-92	POST-BMS-92		
F/C-5-RC-21	5″	5″ deep; 4½″ (5830 psi) concrete deck; ½″ plaster finish bottom of slab; ¼″ reinforcement bars at 7½″ pitch with ⅞″ cover; ⅜″ main reinforcement bars at 3½″ pitch perpendicular with ½″ cover; 12′ span simply supported.	69 psf	2 hrs.			7	1, 3	2
F/C-5-RC-22	5″	4½″ (5290 psi) concrete deck; ½″ plaster finish bottom of slab; ¼″ reinforcement bars at 7½″ pitch with ⅞″ cover; ⅜″ main reinforcement bars at 3½″ pitch perpendicular with ½″ cover; 12′ span simply supported.	No load	2 hrs. 28 min.			7	1, 10, 11	2¼
F/C-5-RC-23	5″	5″ (3020 psi) concrete deck; 3″ × 1½″ × 4 lbs. R.S.J.; 2′ C.R.S. with 1″ cover on bottom and top flanges; 8′ span restrained.	172 psf	1 hr. 24 min.			7	1, 2, 12	1½
F/C-5-RC-24	5½″	5″ (5180 psi) concrete deck; ½″ retarded plaster underneath slab; ¼″ reinforcement bars at 7½″ pitch with 1⅜″ cover; ⅜″ main reinforcement bars at 3½″ pitch perpendicular with 1″ cover; 12′ span simply supported.	60 psf	2 hrs. 48 min.			7	1, 10	2¾
F/C-6-RC-25	6″	6″ deep (4800 psi) concrete deck; ¼″ reinforcement bars at 7½″ pitch with ⅞″ cover; ⅜″ main reinforcement bars at 3½″ pitch perpendicular with ⅞″ cover; 13′1″ span restrained.	195 psf	4 hrs.			7	1, 7	4
F/C-6-RC-26	6″	6″ (4650 psi) concrete deck; ¼″ reinforcement bars at 7½″ pitch with ⅞″ cover; ⅜″ main reinforcement bars at 3½″ pitch perpendicular with ½″ cover; 13′1″ span restrained.	195 psf	2 hrs. 23 min.			7	1, 2	2¼
F/C-6-RC-27	6″	6″ deep (6050 psi) concrete deck; ¼″ reinforcement bars at 7½″ pitch ⅞″ cover; ⅜″ reinforcement bars at 3½″ pitch perpendicular with ½″ cover; 13′1″ span restrained.	195 psf	3 hrs. 30 min.			7	1, 10	3½
F/C-6-RC-28	6″	6″ deep (5180 psi) concrete deck; ¼″ reinforcement bars at 8″ pitch ¾″ cover; ¼″ reinforcement bars at 5½″ pitch perpendicular with ½″ cover; 13′1″ span restrained.	150 psf	4 hrs.			7	1, 7	4
F/C-6-RC-29	6″	6″ thick (4180 psi) concrete deck; 4″ × 3″ × 10 lbs. R.S.J.; 2′ 6″ C.R.S. with 1″ cover on both top and bottom flanges; 13′1″ span restrained.	160 psf	3 hrs. 48 min.			7	1, 10	3¾
F/C-6-RC-30	6″	6″ thick (3720 psi) concrete deck; 4″ × 3″ × 10 lbs. R.S.J.; 2′ 6″ C.R.S. with 1″ cover on both top and bottom flanges; 12′ span simply supported.	115 psf	29 min.			7	1, 5, 13	¼

(continued)

TABLE 3.1—continued
FLOOR/CEILING ASSEMBLIES—REINFORCED CONCRETE

ITEM CODE	ASSEMBLY THICKNESS	CONSTRUCTION DETAILS	PERFORMANCE		REFERENCE NUMBER			NOTES	REC. HOURS
			LOAD	TIME	PRE-BMS-92	BMS-92	POST-BMS-92		
F/C-6-RC-31	6″	6″ deep (3450 psi) concrete deck; 4″ × $1^3/_4$″ × 5 lbs. R.S.J.; 2′ 6″ C.R.S. with 1″ cover on both top and bottom flanges; 12′ span simply supported.	25 psf	3 hrs. 35 min.			7	1, 2	$3^1/_2$
F/C-6-RC-32	6″	6″ deep (4460 psi) concrete deck; 4″ × $1^3/_4$″ × 5 lbs. R.S.J.; 2′ C.R.S.; with 1″ cover on both top and bottom flanges; 12′ span simply supported.	60 psf	4 hrs. 30 min.			7	1, 10	$4^1/_2$
F/C-6-RC-33	6″	6″ deep (4360 psi) concrete deck; 4″ × $1^3/_4$″ × 5 lbs. R.S.J.; 2′ C.R.S.; with 1″ cover on both top and bottom flanges; 13′1″ span restrained.	60 psf	2 hrs.			7	1, 3	2
F/C-6-RC-34	$6^1/_4$″	$6^1/_4$″ thick; $4^3/_4$″ (5120 psi) concrete core; 1″ T&G board flooring; $1/_2$″ plaster undercoat; 4″ × 3″ × 10 lbs. R.S.J.; 3′ C.R.S. flush with top surface concrete; 12′ span simply supported; 2″ × 1′3″ clinker concrete insert.	100 psf	4 hrs.			7	1, 7	4
F/C-6-RC-35	$6^1/_4$″	$4^3/_4$″ (3600 psi) concrete core; 1″ T&G board flooring; $1/_2$″ plaster undercoat; 4″ × 3″ × 10 lbs. R.S.J.; 3′ C.R.S.; flush with top surface concrete; 12′ span simply supported; 2″ × 1′3″ clinker concrete insert.	100 psf	2 hrs. 30 min.			7	1, 5	$2^1/_2$
F/C-6-RC-36	$6^1/_4$″	$4^3/_4$″ (2800 psi) concrete core; 1″ T&G board flooring; $1/_2$″ plaster undercoat; 4″ × 3″ × 10 lbs. R.S.J.; 3′ C.R.S.; flush with top surface concrete; 12″ span simply supported; 2″ × 1′3″ clinker concrete insert.	80 psf	4 hrs.			7	1, 7	4
F/C-7-RC-37	7″	(3640 psi) concrete deck; $1/_4$″ reinforcement bars at 6″ pitch with $1^1/_2$″ cover; $1/_4$″ reinforcement bars at 5″ pitch perpendicular with $1^1/_2$″ cover; 13′1″ span restrained.	169 psf	6 hrs.			7	1, 14	6
F/C-7-RC-38	7″	(4060 psi) concrete deck; 4″ × 3″ × 10 lbs. R.S.J.; 2′ 6″ C.R.S. with $1^1/_2$″ cover on both top and bottom flanges; 4″ × 6″ × 13 SWG mesh reinforcement $1^1/_2$″ from bottom of slab; 13′1″ span restrained.	175 psf	6 hrs.			7	1, 14	6
F/C-7-RC-39	$7^1/_4$″	$5^3/_4$″ (4010 psi) concrete core; 1″ T&G board flooring; $1/_2$″ plaster undercoat; 4″ × 3″ × 10 lbs. R.S.J.; 2′ 6″ C.R.S.; 1″ down from top surface of concrete; 12′ simply supported span; 2″ × 1′ 3″ clinker concrete insert.	95 psf	2 hrs.			7	1, 3	2
F/C-7-RC-40	$7^1/_4$″	$5^3/_4$″ (3220 psi) concrete core; 1″ T&G flooring; $1/_2$″ plaster undercoat; 4″ × 3″ × 10 lbs. R.S.J.; 2′6″ C.R.S.; 1″ down from top surface of concrete; 12′ simply supported span; 2″ × 1′3″ clinker concrete insert.	95 psf	4 hrs.			7	1, 7	4

(continued)

RESOURCE A—GUIDELINES ON FIRE RATINGS OF ARCHAIC MATERIALS AND ASSEMBLIES

TABLE 3.1—continued
FLOOR/CEILING ASSEMBLIES—REINFORCED CONCRETE

ITEM CODE	ASSEMBLY THICKNESS	CONSTRUCTION DETAILS	PERFORMANCE LOAD	PERFORMANCE TIME	REFERENCE NUMBER PRE-BMS-92	REFERENCE NUMBER BMS-92	REFERENCE NUMBER POST-BMS-92	NOTES	REC. HOURS
F/C-7-RC-41	10" (2¼" Slab)	Ribbed floor, see Note 15 for details; slab 2½" deep (3020 psi); ¼" reinforcement bars at 6" pitch with ¾" cover; beams 7½" deep × 5" wide; 24" C.R.S.; ⅝" reinforcement bars two rows ½" vertically apart with 1" cover; 13'1" span restricted.	195 psf	1 hr. 4 min.			7	1, 2, 15	1
F/C-5-RC-42	5½"	Composite ribbed concrete slab assembly; see Note 17 for details.	See Note 16	2 hrs.			43	16, 17	2
F/C-3-RC-43	3"	2500 psi concrete; ⅝" cover; fully restrained at test.	See Note 16	30 min.			43	16	½
F/C-3-RC-44	3"	2000 psi concrete; ⅝" cover; free or partial restraint at test.	See Note 16	45 min.			43	16	¾
F/C-4-RC-45	4"	2500 psi concrete; ⅝" cover; fully restrained at test.	See Note 16	40 min.			43	16	⅔
F/C-4-RC-46	4"	2000 psi concrete; ¾" cover; free or partial restraint at test.	See Note 16	1 hr. 15 min.			43	16	1¼
F/C-5-RC-47	5"	2500 psi concrete; ¾" cover; fully restrained at test.	See Note 16	1 hr.			43	16	1
F/C-5-RC-48	5"	2000 psi concrete; ¾" cover; free or partial restraint at test.	See Note 16	1 hr. 30 min.			43	16	1½
F/C-6-RC-49	6"	2500 psi concrete; 1" cover; fully restrained at test.	See Note 16	1 hr. 30 min.			43	16	1½
F/C-6-RC-50	6"	2000 psi concrete; 1" cover; free or partial restraint at test.	See Note 16	2 hrs.			43	16	2

For SI: 1 inch = 25.4 mm, 1 foot = 305 mm, 1 pound per square inch = 0.00689 MPa, 1 pound per square foot = 47.9 N/m².

Notes:
1. British test.
2. Failure mode—local back face temperature rise.
3. Tested for Grade "C" (2 hour) fire resistance
4. Collapse imminent following hose stream.
5. Failure mode—flame thru.
6. Void formed with explosive force and report.
7. Achieved Grade "B" (4 hour) fire resistance (British).
8. Failure mode—collapse.
9. Test was run to 2 hours, but specimen was partially supported by the furnace at 1¼ hours.
10. Failure mode—average back face temperature.
11. Recommended endurance for nonload bearing performance only.
12. Floor maintained load bearing ability to 2 hours at which point test was terminated.
13. Test was run to 3 hours at which time failure mode 2 (above) was reached in spite of crack formation at 29 minutes.
14. Tested for Grade "A" (6 hour) fire resistance.
15.

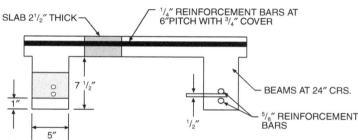

16. Load unspecified.
17. Total assembly thickness 5½ inches. Three-inch thick blocks of molded excelsior bonded with Portland cement used as inserts with 2½-inch cover (concrete) above blocks and ¾-inch gypsum plaster below. Nine-inch wide ribs containing reinforcing steel of unspecified size interrupted 20-inch wide segments of slab composite (i.e., plaster, excelsior blocks, concrete cover).

RESOURCE A—GUIDELINES ON FIRE RATINGS OF ARCHAIC MATERIALS AND ASSEMBLIES

FIGURE 3.2
FLOOR/CEILING ASSEMBLIES—STEEL STRUCTURAL ELEMENTS

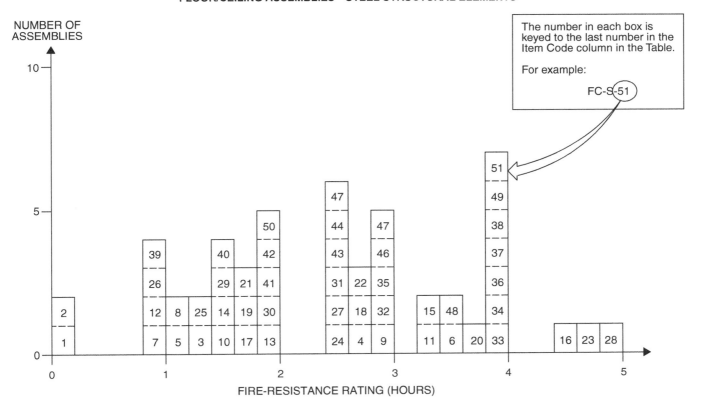

TABLE 3.2
FLOOR/CEILING ASSEMBLIES—STEEL STRUCTURAL ELEMENTS

ITEM CODE	MEMBRANE THICKNESS	CONSTRUCTION DETAILS	PERFORMANCE		REFERENCE NUMBER			NOTES	REC. HOURS
			LOAD	TIME	PRE-BMS-92	BMS-92	POST-BMS-92		
F/C-S-1	0″	10′ × 13′6″; S.J. 103 - 24″ o.c.; Deck: 2″ concrete; Membrane: none.	145 psf	7 min.			3	1, 2, 3, 8	0
F/C-S-2	0″	10′ × 13′6″; S.J. 103 - 24″ o.c.; Deck: 2″ concrete; Membrane: none	145 psf	7 min.			3	1, 2, 3, 8	0
F/C-S-3	$\frac{1}{2}″$	10′ × 13′ 6″; S.J. 103 - 24″ o.c.; Deck: 2″ concrete 1:2:4; Membrane: furring 12″ o.c.; Clips A, B, G; No extra reinforcement; $\frac{1}{2}″$ plaster - 1.5:2.5.	145 psf	1 hr. 15 min.			3	2, 3, 8	$1\frac{1}{4}$
F/C-S-4	$\frac{1}{2}″$	10′ × 13′ 6″; S.J. 103 - 24″ o.c.; Deck: 2″ concrete 1:2:4; Membrane: furring 16″ o.c.; Clips D, E, F, G; Diagonal wire reinforcement; $\frac{1}{2}″$ plaster - 1.5:2.5.	145 psf	2 hrs. 46 min.			3	3, 8	$2\frac{3}{4}$
F/C-S-5	$\frac{1}{2}″$	10′ × 13′6″; S.J. 103 - 24″ o.c.; Deck: 2″ concrete 1:2:4; Membrane: furring 16″ o.c.; Clips A, B, G; No extra reinforcement; $\frac{1}{2}″$ plaster - 1.5:2.5.	145 psf	1 hr. 4 min.			3	2, 3, 8	1
F/C-S-6	$\frac{1}{2}″$	10′ × 13′6″; S.J. 103 - 24″ o.c.; Deck: 2″ concrete 1:2:4; Membrane: furring 16″ o.c.; Clips D, E, F, G; Hexagonal mesh reinforcement; $\frac{1}{2}″$ plaster.	145 psf	3 hrs. 28 min.			3	2, 3, 8	$2\frac{1}{3}$

(continued)

TABLE 3.2—continued
FLOOR/CEILING ASSEMBLIES—STEEL STRUCTURAL ELEMENTS

ITEM CODE	MEMBRANE THICKNESS	CONSTRUCTION DETAILS	PERFORMANCE		REFERENCE NUMBER			NOTES	REC. HOURS
			LOAD	TIME	PRE-BMS-92	BMS-92	POST-BMS-92		
F/C-S-7	$1/2''$	10′ × 13′6″; S.J. 103 - 24″ o.c.; Deck: 4 lbs. rib lath; 6″ × 6″ - 10 × 10 ga. reinforcement; 2″ deck gravel concrete; Membrane: furring 16″ o.c.; Clips C, E; Reinforcement: none; $1/2''$ plaster - 1.5:2.5 mill mix.	N/A	55 min.			3	5, 8	$3/4$
F/C-S-8	$1/2''$	Spec. 9′ × 4′4″; S.J. 103 bar joists - 18″ o.c.; Deck: 4 lbs. rib lath base; 6″ × 6″ - 10 × 10 ga. reinforcement; 2″ deck 1:2:4 gravel concrete; Membrane: furring, $3/4''$ C.R.S., 16″ o.c.; Clips C, E; Reinforcement: none; $1/2''$ plaster - 1.5:2.5 mill mix.	300 psf	1 hr. 10 min.			3	2, 3, 8	1
F/C-S-9	$5/8''$	10′ × 13′6″; S.J. 103 - 24″ o.c.; Deck: 2″ concrete 1:2:4; Membrane: furring 12″ o.c.; Clips A, B, G; Extra "A" clips reinforcement; $5/8''$ plaster - 1.5:2; 1.5:3.	145 psf	3 hrs.			3	6, 8	3
F/C-S-10	$5/8''$	18′ × 13′6″; Joists, S.J. 103 - 24″ o.c.; Deck: 4 lbs. rib lath; 6″ × 6″ - 10 × 10 ga. reinforcement; 2″ deck 1:2:3.5 gravel concrete; Membrane: furring, spacing 16″ o.c.; Clips C, E; Reinforcement: none; $5/8''$ plaster - 1.5:2.5 mill mix.	145 psf	1 hr. 25 min.			3	2, 3, 8	$1 1/3$
F/C-S-11	$5/8''$	10′ × 13′6″; S.J. 103 - 24″ o.c.; Deck: 2″ concrete 1:2:4; Membrane: furring 12″ o.c.; Clips D, E, F, G; Diagonal wire reinforcement; $5/8''$ plaster - 1.5:2; 0.5:3.	145 psf	3 hrs. 15 min.			3	2, 4, 8	$3 1/4$
F/C-S-12	$5/8''$	10′ × 13′6″; Joists, S.J. 103 - 24″ o.c.; Deck: 3.4 lbs. rib lath; 6″ × 6″ - 10 × 10 ga. reinforcement; 2″ deck 1:2:4 gravel concrete; Membrane: furring 16″ o.c.; Clips D, E, F, G; Reinforcement: none; $5/8''$ plaster - 1.5:2.5.	145 psf	1 hr.			3	7, 8	1
F/C-S-13	$3/4''$	Spec. 9′ × 4′4″; S.J. 103 - 18″ o.c.; Deck: 4 lbs. rib lath; 6″ × 6″ - 10 × 10 ga. reinforcement; 2″ deck 1:2:4 gravel concrete; Membrane: furring, $3/4''$ C.R.S., 16″ o.c.; Clips C, E; Reinforcement: none; $3/4''$ plaster - 1.5:2.5 mill mix.	300 psf	1 hr. 56 min.			3	3, 8	$1 3/4$
F/C-S-14	$7/8''$	Floor finish: 1″ concrete; plate cont. weld; 4″ - 7.7 lbs. "I" beams; Ceiling: $1/4''$ rods 12″ o.c.; $7/8''$ gypsum sand plaster.	105 psf	1 hr. 35 min.			6	2, 4, 9, 10	$1 1/2$
F/C-S-15	1″	Floor finish: $1 1/2''$ L.W. concrete; $1/2''$ limestone cement; plate cont. weld; 5″ - 10 lbs. "I" beams; Ceiling: $1/4''$ rods 12″ o.c. tack welded to beams metal lath; 1″ P. C. plaster.	165 psf	3 hrs. 20 min.			6	4, 9, 11	$3 1/3$
F/C-S-16	1″	10′ × 13′6″; S.J. 103 - 24″ o.c.; Deck: 2″ concrete 1:2:4; Membrane: furring 12″ o.c.; Clips D, E, F, G; Hexagonal mesh reinforcement; 1″ thick plaster - 1.5:2; 1.5:3.	145 psf	4 hrs. 26 min.			3	2, 4, 8	$4 1/3$
F/C-S-17	1″	10′ × 13′6″; Joists - S.J. 103 - 24″ o.c.; Deck: 3.4 lbs. rib lath; 6″ × 6″ - 10 × 10 ga. reinforcement; 2″ deck 1:2:4 gravel concrete; Membrane: furring 16″ o.c.; Clips D, E, F, G; 1″ plaster.	145 psf	1 hr. 42 min.			3	2, 4, 8	$1 2/3$

(continued)

TABLE 3.2—continued
FLOOR/CEILING ASSEMBLIES—STEEL STRUCTURAL ELEMENTS

ITEM CODE	MEMBRANE THICKNESS	CONSTRUCTION DETAILS	PERFORMANCE		REFERENCE NUMBER			NOTES	REC. HOURS
			LOAD	TIME	PRE-BMS-92	BMS-92	POST-BMS-92		
F/C-S-18	$1\frac{1}{8}''$	10′ × 13′6″ ; S. J. 103 - 24″ o.c.; Deck: 2″ concrete 1:2:4; Membrane: furring 12″ o.c.; Clips C, E, F, G; Diagonal wire reinforcement; $1\frac{1}{8}''$ plaster.	145 psf	2 hrs. 44 min.			3	2, 4, 8	$2\frac{2}{3}$
F/C-S-19	$1\frac{1}{8}''$	10′ × 13′6″ ; Joists - S.J. 103 - 24″ o.c.; Deck: $1\frac{1}{2}''$ gypsum concrete over; $\frac{1}{2}''$ gypsum board; Membrane: furring 12″ o.c.; Clips D, E, F, G; $1\frac{1}{8}''$ plaster - 1.5:2; 1.5:3.	145 psf	1 hr. 40 min.			3	2, 3, 8	$1\frac{2}{3}$
F/C-S-20	$1\frac{1}{8}''$	$2\frac{1}{2}''$ cinder concrete; $\frac{1}{2}''$ topping; plate 6″ welds 12″ o.c.; 5″ - 18.9 lbs. "H" center; 5″ - 10 lbs. "I" ends; 1″ channels 18″ o.c.; $1\frac{1}{8}''$ gypsum sand plaster.	150 psf	3 hrs. 43 min.			6	2, 4, 9, 11	$3\frac{2}{3}$
F/C-S-21	$1\frac{1}{4}''$	10′ × 13′6″ ; Joists - S.J. 103 - 24″ o.c.; Deck: $1\frac{1}{2}''$ gypsum concrete over; $\frac{1}{2}''$ gypsum board base; Membrane: furring 12″ o.c.; Clips D, E, F, G; $1\frac{1}{4}''$ plaster - 1.5:2; 1.5:3.	145 psf	1 hr. 48 min.			3	2, 3, 8	$1\frac{2}{3}$
F/C-S-22	$1\frac{1}{4}''$	Floor finish: $1\frac{1}{2}''$ limestone concrete; $\frac{1}{2}''$ sand cement topping; plate to beams $3\frac{1}{2}''$; 12″ o.c. welded; 5″ - 10 lbs. "I" beams; 1″ channels 18″ o.c.; $1\frac{1}{4}''$ wood fiber gypsum sand plaster on metal lath.	292 psf	2 hrs. 45 min.			6	2, 4, 9, 10	$2\frac{3}{4}$
F/C-S-23	$1\frac{1}{2}''$	$2\frac{1}{2}''$ L.W. (gas exp.) concrete; Deck: $\frac{1}{2}''$ topping; plate $6\frac{1}{4}''$ welds 12″ o.c.; Beams: 5″ - 18.9 lbs. "H" center; 5″ - 10 lbs. "I" ends; Membrane: 1″ channels 18″ o.c.; $1\frac{1}{2}''$ gypsum sand plaster.	150 psf	4 hrs. 42 min.			6	2, 4, 9	$4\frac{2}{3}$
F/C-S-24	$1\frac{1}{2}''$	Floor finish: $1\frac{1}{2}''$ limestone concrete; $\frac{1}{2}''$ cement topping; plate $3\frac{1}{2}''$ - 12″ o.c. welded; 5″ - 10 lbs. "I" beams; Ceiling: 1″ channels 18″ o.c.; $1\frac{1}{2}''$ gypsum plaster.	292 psf	2 hrs. 34 min.			6	2, 4, 9, 10	$2\frac{1}{2}$
F/C-S-25	$1\frac{1}{2}''$	Floor finish: $1\frac{1}{2}''$ gravel concrete on exp. metal; plate cont. weld; 4″ - 7.7 lbs. "I" beams; Ceiling: $\frac{1}{4}''$ rods 12″ o.c. welded to beams; $1\frac{1}{2}''$ fiber gypsum sand plaster.	70 psf	1 hr. 24 min.			6	2, 4, 9, 10	$1\frac{1}{3}$
F/C-S-26	$2\frac{1}{2}''$	Floor finish: bare plate; $6\frac{1}{4}''$ welding - 12″ o.c.; 5″ - 18.9 lbs. "H" girders (inner); 5″ - 10 lbs "I" girders (two outer); 1″ channels 18″ o.c.; 2″ reinforced gypsum tile; $\frac{1}{2}''$ gypsum sand plaster.	122 psf	1 hr.			6	7, 9, 11	1
F/C-S-27	$2\frac{1}{2}''$	Floor finish: 2″ gravel concrete; plate to beams $3\frac{1}{2}''$ - 12″ o.c. welded; 4″ - 7.7 lbs. "I" beams; 2″ gypsum ceiling tiles; $\frac{1}{2}''$ 1:3 gypsum sand plaster.	105 psf	2 hrs. 31 min.			6	2, 4, 9, 10	$2\frac{1}{2}$
F/C-S-28	$2\frac{1}{2}''$	Floor finish: $1\frac{1}{2}''$ gravel concrete; $\frac{1}{2}''$ gypsum asphalt; plate continuous weld; 4″ - 7.7 lbs. "I" beams; 12″ - 31.8 lbs. "I" beams - girder at 5′ from one end; 1″ channels 18″ o.c.; 2″ reinforcement gypsum tile; $\frac{1}{2}''$ 1:3 gypsum sand plaster.	200 psf	4 hrs. 55 min.			6	2, 4, 9, 11	$4\frac{2}{3}$

(continued)

TABLE 3.2—continued
FLOOR/CEILING ASSEMBLIES—STEEL STRUCTURAL ELEMENTS

ITEM CODE	MEMBRANE THICKNESS	CONSTRUCTION DETAILS	PERFORMANCE LOAD	PERFORMANCE TIME	REFERENCE NUMBER PRE-BMS-92	REFERENCE NUMBER BMS-92	REFERENCE NUMBER POST-BMS-92	NOTES	REC. HOURS
F/C-S-29	$3/4''$	Floor: 2″ reinforced concrete or 2″ precast reinforced gypsum tile; Ceiling: $3/4''$ Portland cement-sand plaster 1:2 for scratch coat and 1:3 for brown coat with 15 lbs. hydrated lime and 3 lbs. of short asbestos fiber bag per cement or $3/4''$ sanded gypsum plaster 1:2 for scratch coat and 1:3 for brown coat.	See Note 12	1 hr. 30 min.		1		12, 13, 14	$1 1/2$
F/C-S-30	$3/4''$	Floor: $2 1/4''$ reinforced concrete or 2″ reinforced gypsum tile; the latter with $1/4''$ mortar finish; Ceiling: $3/4''$ sanded gypsum plaster; 1:2 for scratch coat and 1:3 for brown coat.	See Note 12	2 hrs.		1		12, 13, 14	2
F/C-S-31	$3/4''$	Floor: $2 1/2''$ reinforced concrete or 2″ reinforced gypsum tile; the latter with $1/4''$ mortar finish; Ceiling: 1″ neat gypsum plaster or $3/4''$ gypsum-vermiculite plaster, ratio of gypsum to fine vermiculite 2:1 to 3:1.	See Note 12	2 hrs. 30 min.		1		12, 13, 14	$2 1/2$
F/C-S-32	$3/4''$	Floor: $2 1/2''$ reinforced concrete or 2″ reinforced gypsum tile; the latter with $1/2''$ mortar finish; Ceiling: 1″ neat gypsum plaster or $3/4''$ gypsum-vermiculite plaster, ratio of gypsum to fine vermiculite 2:1 to 3:1.	See Note 12	3 hrs.		1		12, 13, 14	3
F/C-S-33	1″	Floor: $2 1/2''$ reinforced concrete or 2″ reinforced gypsum slabs; the latter with $1/2''$ mortar finish; Ceiling: 1″ gypsum-vermiculite plaster applied on metal lath and ratio 2:1 to 3:1 gypsum to vermiculite by weight.	See Note 12	4 hrs.		1		12, 13, 14	4
F/C-S-34	$2 1/2''$	Floor: 2″ reinforced concrete or 2″ precast reinforced Portland cement concrete or gypsum slabs; precast slabs to be finished with $1/4''$ mortar top coat; Ceiling: 2″ precast reinforced gypsum tile, anchored into beams with metal ties or clips and covered with $1/2''$ 1:3 sanded gypsum plaster.	See Note 12	4 hrs.		1		12, 13, 14	4
F/C-S-35	1″	Floor: 1:3:6 Portland cement, sand and gravel concrete applied directly to the top of steel units and $1 1/2''$ thick at top of cells, plus $1/2''$ $1:2 1/2''$ cement-sand finish, total thickness at top of cells, 2″; Ceiling: 1″ neat gypsum plaster, back of lath 2″ or more from underside of cellular steel.	See Note 15	3 hrs.		1		15, 16, 17, 18	3
F/C-S-36	1″	Floor: same as F/C-S-35; Ceiling: 1″ gypsum-vermiculite plaster (ratio of gypsum to vermiculite 2:1 to 3:1), the back of lath 2″ or more from under-side of cellular steel.	See Note 15	4 hrs.		1		15, 16, 17, 18	4
F/C-S-37	1″	Floor: same as F/C-S-35; Ceiling: 1″ neat gypsum plaster; back of lath 9″ or more from underside of cellular steel.	See Note 15	4 hrs.		1		15, 16, 17, 18	4
F/C-S-38	1″	Floor: same as F/C-S-35; Ceiling: 1″ gypsum-vermiculite plaster (ratio of gypsum to vermiculite 2:1 to 3:1), the back of lath being 9″ or more from underside of cellular steel.	See Note 15	5 hrs.		1		15, 16, 17,18	5

(continued)

RESOURCE A—GUIDELINES ON FIRE RATINGS OF ARCHAIC MATERIALS AND ASSEMBLIES

TABLE 3.2—continued
FLOOR/CEILING ASSEMBLIES—STEEL STRUCTURAL ELEMENTS

ITEM CODE	MEMBRANE THICKNESS	CONSTRUCTION DETAILS	PERFORMANCE LOAD	PERFORMANCE TIME	REFERENCE NUMBER PRE-BMS-92	REFERENCE NUMBER BMS-92	REFERENCE NUMBER POST-BMS-92	NOTES	REC. HOURS
F/C-S-39	$3/4''$	Floor: asbestos paper 14 lbs./100 ft.^{2} cemented to steel deck with waterproof linoleum cement, wood screeds and $7/8''$ wood floor; Ceiling: $3/4''$ sanded gypsum plaster 1:2 for scratch coat and 1:3 for brown coat.	See Note 19	1 hr.		1		19, 20, 21, 22	1
F/C-S-40	$3/4''$	Floor: $1^1/_2''$, 1:2:4 Portland cement concrete; Ceiling: $3/4''$ sanded gypsum plaster 1:2 for scratch coat and 1:3 for brown coat.	See Note 19	1 hr. 30 min.		1		19, 20, 21, 22	$1^1/_2$
F/C-S-41	$3/4''$	Floor: 2'', 1:2:4 Portland cement concrete; Ceiling: $3/4''$ sanded gypsum plaster, 1:2 for scratch coat and 1:3 for brown coat.	See Note 19	2 hrs.		1		19, 20, 21, 22	2
F/C-S-42	1''	Floor: 2'', 1:2:4 Portland cement concrete; Ceiling: 1'' Portland cement-sand plaster with 10 lbs. of hydrated lime for @ bag of cement 1:2 for scratch coat and $1:2^1/_2''$ for brown coat.	See Note 19	2 hrs.		1		19, 20, 21, 22	2
F/C-S-43	$1^1/_2''$	Floor: 2'', 1:2:4 Portland cement concrete; Ceiling: $1^1/_2''$, 1:2 sanded gypsum plaster on ribbed metal lath.	See Note 19	2 hrs. 30 min.		1		19, 20, 21, 22	$2^1/_2$
F/C-S-44	$1^1/_8''$	Floor: 2'', 1:2:4 Portland cement concrete; Ceiling: $1^1/_8''$, 1:1 sanded gypsum plaster.	See Note 19	2 hrs. 30 min.		1		19, 20, 21, 22	$2^1/_2$
F/C-S-45	1''	Floor: $2^1/_2''$, 1:2:4 Portland cement concrete; Ceiling: 1'', 1:2 sanded gypsum plaster.	See Note 19	2 hrs. 30 min.		1		19, 20, 21, 22	$2^1/_2$
F/C-S-46	$3/4''$	Floor: $2^1/_2''$, 1:2:4 Portland cement concrete; Ceiling: 1'' neat gypsum plaster or $3/4''$ gypsum-vermiculite plaster, ratio of gypsum to vermiculite 2:1 to 3:1.	See Note 19	3 hrs.		1		19, 20, 21, 22	3
F/C-S-47	$1^1/_8''$	Floor: $2^1/_2''$, 1:2:4 Portland cement, sand and cinder concrete plus $1/2''$, $1:2^1/_2''$ cement-sand finish; total thickness 3''; Ceiling: $1^1/_8''$, 1:1 sanded gypsum plaster.	See Note 19	3 hrs.		1		19, 20, 21, 22	3
F/C-S-48	$1^1/_8''$	Floor: $2^1/_2''$, gas expanded Portland cement-sand concrete plus $1/2''$, 1:2.5 cement-sand finish; total thickness 3''; Ceiling: $1^1/_8''$, 1:1 sanded gypsum plaster.	See Note 19	3 hrs. 30 min.		1		19, 20, 21, 22	$3^1/_2$
F/C-S-49	1''	Floor: $2^1/_2''$, 1:2:4 Portland cement concrete; Ceiling: 1'' gypsum-vermiculite plaster; ratio of gypsum to vermiculite 2:1 to 3:1.	See Note 19	4 hrs.		1		19, 20, 21, 22	4
F/C-S-50	$2^1/_2''$	Floor: 2'', 1:2:4 Portland cement concrete; Ceiling: 2'' interlocking gypsum tile supported on upper face of lower flanges of beams, $1/2''$ 1:3 sanded gypsum plaster.	See Note 19	2 hrs.		1		19, 20, 21, 22	2
F/C-S-51	$2^1/_2''$	Floor: 2'', 1:2:4 Portland cement concrete; Ceiling: 2'' precast metal reinforced gypsum tile, $1/2''$ 1:3 sanded gypsum plaster (tile clipped to channels which are clipped to lower flanges of beams).	See Note 19	4 hrs.		1		19, 20, 21, 22	4

For SI: 1 inch = 25.4 mm, 1 foot = 305 mm, 1 pound per square inch = 0.00689 MPa, 1 pound per square foot = 47.9 N/m^2.

Notes:
1. No protective membrane over structural steel.
2. Performance time indicates first endpoint reached only several tests were continued to points where other failures occurred.
3. Load failure.

(continued)

TABLE 3.2—continued
FLOOR/CEILING ASSEMBLIES—STEEL STRUCTURAL ELEMENTS

4. Thermal failure.
5. This is an estimated time to load bearing failure. The same joist and deck specimen was used for a later test with different membrane protection.
6. Test stopped at 3 hours to reuse specimen; no endpoint reached.
7. Test stopped at 1 hour to reuse specimen; no endpoint reached.
8. All plaster used = gypsum.
9. Specimen size - 18 feet by $13^{1}/_{2}$ inches. Floor deck - base material - $^{1}/_{4}$-inch by 18-foot steel plate welded to "I" beams.
10. "I" beams - 24 inches o.c.
11. "I" beams - 48 inches o.c.
12. Apply to open web joists, pressed steel joists or rolled steel beams, which are not stressed beyond 18,000 lbs./in.^{2} in flexure for open-web pressed or light rolled joists, and 20,000 lbs./in.^{2} for American standard or heavier rolled beams.
13. Ratio of weight of Portland cement to fine and coarse aggregates combined for floor slabs shall not be less than $1:6^{1}/_{2}$.
14. Plaster for ceiling shall be applied on metal lath which shall be tied to supports to give the equivalent of single No. 18 gage steel wires 5 inches o.c.
15. Load: maximum fiber stress in steel not to exceed 16,000 psi.
16. Prefabricated units 2 feet wide with length equal to the span, composed of two pieces of No. 18 gage formed steel welded together to give four longitudinal cells.
17. Depth not less than 3 inches and distance between cells no less than 2 inches.
18. Ceiling: metal lath tied to furring channels secured to runner channels hung from cellular steel.
19. Load: rolled steel supporting beams and steel plate base shall not be stressed beyond 20,000 psi in flexure. Formed steel (with wide upper flange) construction shall not be stressed beyond 16,000 psi.
20. Some type of expanded metal or woven wire shall be embedded to prevent cracking in concrete flooring.
21. Ceiling plaster shall be metal lath wired to rods or channels which are clipped or welded to steel construction. Lath shall be no smaller than 18 gage steel wire and not more than 7 inches o.c.
22. The securing rods or channels shall be at least as effective as single $^{3}/_{16}$-inch rods with 1-inch of their length bent over the lower flanges of beams with the rods or channels tied to this clip with 14 gage iron wire.

RESOURCE A—GUIDELINES ON FIRE RATINGS OF ARCHAIC MATERIALS AND ASSEMBLIES

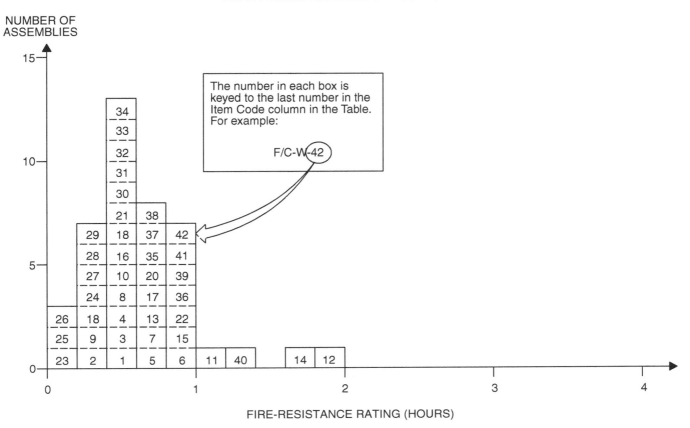

FIGURE 3.3
FLOOR/CEILING ASSEMBLIES—WOOD JOIST

TABLE 3.3
FLOOR/CEILING ASSEMBLIES—WOOD JOIST

ITEM CODE	MEMBRANE THICKNESS	CONSTRUCTION DETAILS	PERFORMANCE		REFERENCE NUMBER			NOTES	REC. HOURS
			LOAD	TIME	PRE-BMS-92	BMS-92	POST-BMS-92		
F/C-W-1	$3/8''$	12′ clear span - 2″ × 9″ wood joists; 18″ o.c.; Deck: 1″ T&G; Filler: 3″ of ashes on $1/2''$ boards nailed to joist sides 2″ from bottom; 2″ air space; Membrane: $3/8''$ gypsum board.	60 psf	36 min.			7	1, 2	$1/2$
F/C-W-2	$1/2''$	12′ clear span - 2″ × 7″ joists; 15″ o.c.; Deck: 1″ nominal lumber; Membrane: $1/2''$ fiber board.	60 psf	22 min.			7	1, 2, 3	$1/4$
F/C-W-3	$1/2''$	12′ clear span - 2″ × 7″ wood joists; 16″ o.c.; 2″ × $1 1/2''$ bridging at center; Deck: 1″ T&G; Membrane: $1/2''$ fiber board; 2 coats "distemper" paint.	30 psf	28 min.			7	1, 3, 15	$1/3$
F/C-W-4	$3/16''$	12′ clear span - 2″ × 7″ wood joists; 16″ o.c.; 2″ × $1 1/2''$ bridging at center span; Deck: 1″ nominal lumber; Membrane: $1/2''$ fiber board under $3/16''$ gypsum plaster.	30 psf	32 min.			7	1, 2	$1/2$
F/C-W-5	$5/8''$	As per previous F/C-W-4 except membrane is $5/8''$ lime plaster.	70 psf	48 min.			7	1, 2	$3/4$
F/C-W-6	$5/8''$	As per previous F/C-W-5 except membrane is $5/8''$ gypsum plaster on 22 gage $3/8''$ metal lath.	70 psf	49 min.			7	1, 2	$3/4$

(continued)

TABLE 3.3—continued
FLOOR/CEILING ASSEMBLIES—WOOD JOIST

ITEM CODE	MEMBRANE THICKNESS	CONSTRUCTION DETAILS	PERFORMANCE LOAD	PERFORMANCE TIME	REFERENCE NUMBER PRE-BMS-92	REFERENCE NUMBER BMS-92	REFERENCE NUMBER POST-BMS-92	NOTES	REC. HOURS
F/C-W-7	$1/2''$	As per previous F/C-W-6 except membrane is $1/2''$ fiber board under $1/2''$ gypsum plaster.	60 psf	43 min.			7	1, 2, 3	$2/3$
F/C-W-8	$1/2''$	As per previous F/C-W-7 except membrane is $1/2''$ gypsum board.	60 psf	33 min.			7	1, 2, 3	$1/2$
F/C-W-9	$9/16''$	12′ clear span - 2″ × 7″ wood joists; 15″ o.c.; 2″ × $1 1/2''$ bridging at center; Deck: 1″ nominal lumber; Membrane: $3/8''$ gypsum board; $3/16''$ gypsum plaster.	60 psf	24 min.			7	1, 2, 3	$1/3$
F/C-W-10	$5/8''$	As per F/C-W-9 except membrane is $5/8''$ gypsum plaster on wood lath.	60 psf	27 min.			7	1, 2, 3	$1/3$
F/C-W-11	$7/8''$	12′ clear span - 2″ × 9″ wood joists; 15″ o.c.; 2″ × $1 1/2''$ bridging at center span; Deck: 1″ T&G; Membrane: original ceiling joists have $3/8''$ plaster on wood lath; 4″ metal hangers attached below joists creating 15″ chases filled with mineral wool and closed with $7/8''$ plaster (gypsum) on $3/8''$ S.W.M. metal lath to form new ceiling surface.	75 psf	1 hr. 10 min.			7	1, 2	1
F/C-W-12	$7/8''$	12′ clear span - 2″ × 9″ wood joists; 15″ o.c.; 2″ × $1 1/2''$ bridging at center; Deck: 1″ T&G; Membrane: 3″ mineral wood below joists; 3″ hangers to channel below joists; $7/8''$ gypsum plaster on metal lath attached to channels.	75 psf	2 hrs.			7	1, 4	2
F/C-W-13	$7/8''$	12′ clear span - 2″ × 9″ wood joists; 16″ o.c.; 2″ × $1 1/2''$ bridging at center span; Deck: 1″ T&G on 1″ bottoms on $3/4''$ glass wool strips on $3/4''$ gypsum board nailed to joists; Membrane: $3/4''$ glass wool strips on joists; $3/8''$ perforated gypsum lath; $1/2''$ gypsum plaster.	60 psf	41 min.			7	1, 3	$2/3$
F/C-W-14	$7/8''$	12′ clear span - 2″ × 9″ wood joists; 15″ o.c.; Deck: 1″ T&G; Membrane: 3″ foam concrete in cavity on $1/2''$ boards nailed to joists; wood lath nailed to 1″ × $1 1/4''$ straps 14 o.c. across joists; $7/8''$ gypsum plaster.	60 psf	1 hr. 40 min.			7	1, 5	$1 2/3$
F/C-W-15	$7/8''$	12′ clear span - 2″ × 9″ wood joists; 18″ o.c.; Deck: 1″ T&G; Membrane: 2″ foam concrete on $1/2''$ boards nailed to joist sides 2″ from joist bottom; 2″ air space; 1″ × $1 1/4''$ wood straps 14″ o.c. across joists; $7/8''$ lime plaster on wood lath.	60 psf	53 min.			7	1, 2	$3/4$
F/C-W-16	$7/8''$	12′ clear span - 2″ × 9″ wood joists; Deck: 1″ T&G; Membrane: 3″ ashes on $1/2''$ boards nailed to joist sides 2″ from joist bottom; 2″ air space; 1″ × $1 1/4''$ wood straps 14″ o.c. ; $7/8''$ gypsum plaster on wood lath.	60 psf	28 min.			7	1, 2	$1/3$
F/C-W-17	$7/8''$	As per previous F/C-W-16 but with lime plaster mix.	60 psf	41 min.			7	1, 2	$2/3$
F/C-W-18	$7/8''$	12′ clear span - 2″ × 9″ wood joists; 18″ o.c.; 2″ × $1 1/2''$ bridging at center; Deck: 1″ T&G; Membrane: $7/8''$ gypsum plaster on wood lath.	60 psf	36 min.			7	1, 2	$1/2$
F/C-W-19	$7/8''$	As per previous F/C-W-18 except with lime plaster membrane and deck is 1″ nominal boards (plain edge).	60 psf	19 min.			7	1, 2	$1/4$

(continued)

RESOURCE A—GUIDELINES ON FIRE RATINGS OF ARCHAIC MATERIALS AND ASSEMBLIES

TABLE 3.3—continued
FLOOR/CEILING ASSEMBLIES—WOOD JOIST

ITEM CODE	MEMBRANE THICKNESS	CONSTRUCTION DETAILS	PERFORMANCE LOAD	PERFORMANCE TIME	REFERENCE NUMBER PRE-BMS-92	BMS-92	POST-BMS-92	NOTES	REC. HOURS
F/C-W-20	$7/8''$	As per F/C-W-19, except deck is 1″ T&G boards.	60 psf	43 min.			7	1, 2	$2/3$
F/C-W-21	1″	12′ clear span - 2″ × 9″ wood joists; 16″ o.c.; 2″ × $1^1/_2$″ bridging at center; Deck: 1″ T&G; Membrane: $3/8$″ gypsum base board; $5/8$″ gypsum plaster.	70 psf	29 min.			7	1, 2	$1/3$
F/C-W-22	$1^1/_8$″	12′ clear span - 2″ × 9″ wood joists; 16″ o.c.; 2″ × 2″ wood bridging at center; Deck: 1″ T&G; Membrane: hangers, channel with $3/8$″ gypsum baseboard affixed under $3/4$″ gypsum plaster.	60 psf	1 hr.			7	1, 2, 3	1
F/C-W-23	$3/8$″	Deck: 1″ nominal lumber; Joists: 2″ × 7″; 15″ o.c.; Membrane: $3/8$″ plasterboard with plaster skim coat.	60 psf	$11^1/_2$ min.			12	2, 6	$1/6$
F/C-W-24	$1/2$″	Deck: 1″ T&G lumber; Joists: 2″ × 9″; 16″ o.c.; Membrane: $1/2$″ plasterboard.	60 psf	18 min.			12	2, 7	$1/4$
F/C-W-25	$1/2$″	Deck: 1″ T&G lumber; Joists: 2″ × 7″; 16″ o.c.; Membrane: $1/2$″ fiber insulation board.	30 psf	8 min.			12	2, 8	$2/15$
F/C-W-26	$1/2$″	Deck: 1″ nominal lumber; Joists: 2″ × 7″; 15″ o.c.; Membrane: $1/2$″ fiber insulation board.	60 psf	8 min.			12	2, 9	$2/15$
F/C-W-27	$5/8$″	Deck: 1″ nominal lumber; Joists: 2″ × 7″; 15″ o.c.; Membrane: $5/8$″ gypsum plaster on wood lath.	60 psf	17 min.			12	2, 10	$1/4$
F/C-W-28	$5/8$″	Deck: 1″ T&G lumber; Joists: 2″ × 9″; 16″ o.c.; Membrane: $1/2$″ fiber insulation board; $1/2$″ plaster.	60 psf	20 min.			12	2, 11	$1/3$
F/C-W-29	No Membrane	Exposed wood joists.	See Note 13	15 min.		1		1, 12, 13, 14	$1/4$
F/C-W-30	$3/8$″	Gypsum wallboard: $3/8$″ or $1/2$″ with $1^1/_2$″ No. 15 gage nails with $3/16$″ heads spaced 6″ centers with asbestos paper applied with paperhangers' paste and finished with casein paint.	See Note 13	25 min.		1		1, 12, 13, 14	$1/2$
F/C-W-31	$1/2$″	Gypsum wallboard: $1/2$″ with $1^3/_4$″ No. 12 gage nails with $1/2$″ heads, 6″ o.c., and finished with casein paint.	See Note 13	25 min.		1		1, 12, 13, 14	$1/2$
F/C-W-32	$1/2$″	Gypsum wallboard: $1/2$″ with $1^1/_2$″ No. 12 gage nails with $1/2$″ heads, 18″ o.c., with asbestos paper applied with paperhangers' paste and secured with $1^1/_2$″ No. 15 gage nails with $3/16$″ heads and finished with casein paint; combined nail spacing 6″ o.c.	See Note 13	30 min.		1		1, 12, 13, 14	$1/2$
F/C-W-33	$3/8$″	Gypsum wallboard: two layers $3/8$″ secured with $1^1/_2$″ No. 15 gage nails with $3/8$″ heads, 6″ o.c.	See Note 13	30 min.		1		1, 12, 13, 14	$1/2$
F/C-W-34	$1/2$″	Perforated gypsum lath: $3/8$″, plastered with $1^1/_8$″ No. 13 gage nails with $5/16$″ heads, 4″ o.c.; $1/2$″ sanded gypsum plaster.	See Note 13	30 min.		1		1, 12, 13, 14	$1/2$
F/C-W-35	$1/2$″	Same as F/C-W-34, except with $1^1/_8$″ No. 13 gage nails with $3/8$″ heads, 4″ o.c.	See Note 13	45 min.		1		1, 12, 13, 14	$3/4$

(continued)

RESOURCE A—GUIDELINES ON FIRE RATINGS OF ARCHAIC MATERIALS AND ASSEMBLIES

TABLE 3.3—continued
FLOOR/CEILING ASSEMBLIES—WOOD JOIST

ITEM CODE	MEMBRANE THICKNESS	CONSTRUCTION DETAILS	PERFORMANCE		REFERENCE NUMBER			NOTES	REC. HOURS
			LOAD	TIME	PRE-BMS-92	BMS-92	POST-BMS-92		
F/C-W-36	$1/2''$	Perforated gypsum lath: $3/8''$, nailed with $1^{1}/_{8}''$ No. 13 gage nails with $3/8''$ heads, 4" o.c.; joints covered with 3" strips of metal lath with $1^{3}/_{4}''$ No. 12 nails with $1/2''$ heads, 5" o.c.; $1/2''$ sanded gypsum plaster.	See Note 13	1 hr.	1			1, 12, 13, 14	1
F/C-W-37	$1/2''$	Gypsum lath: $3/8''$ and lower layer of $3/8''$ perforated gypsum lath nailed with $1^{3}/_{4}''$ No. 13 nails with $5/16''$ heads, 4" o.c.; $1/2''$ sanded gypsum plaster or $1/2''$ Portland cement plaster.	See Note 13	45 min.	1			1, 12, 13, 14	$3/4$
F/C-W-38	$3/4''$	Metal lath: nailed with $1^{1}/_{4}''$ No. 11 nails with $3/8''$ heads or 6d common driven 1" and bent over, 6" o.c.; $3/4''$ sanded gypsum plaster.	See Note 13	45 min.	1			1, 12, 13, 14	$3/4$
F/C-W-39	$3/4''$	Same as F/C-W-38, except nailed with $1^{1}/_{2}''$ No. 11 barbed roof nails with $7/16''$ heads, 6" o.c.	See Note 13	1 hr.	1			1, 12, 13, 14	1
F/C-W-40	$3/4''$	Same as F/C-W-38, except with lath nailed to joists with additional supports for lath 27" o.c.; attached to alternate joists and consisting of two nails driven $1^{1}/_{4}''$, 2" above bottom on opposite sides of the joists, one loop of No. 18 wire slipped over each nail; the ends twisted together below lath.	See Note 13	1 hr. 15 min.	1			1, 12, 13, 14	$1^{1}/_{4}$
F/C-W-41	$3/4''$	Metal lath: nailed with $1^{1}/_{2}''$ No. 11 barbed roof nails with $7/16''$ heads, 6 o.c., with $3/4''$ Portland cement plaster for scratch coat and 1:3 for brown coat, 3 lbs. of asbestos fiber and 15 lbs. of hydrated lime/94 lbs. bag of cement.	See Note 13	1 hr.	1			1, 12, 13, 14	1
F/C-W-42	$3/4''$	Metal lath: nailed with 8d, No. $11^{1}/_{2}$ gage barbed box nails, $2^{1}/_{2}''$ driven, $1^{1}/_{4}''$ on slant and bent over, 6" o.c.; $3/4''$ sanded gypsum plaster, 1:2 for scratch coat and 1:3 for below coat.	See Note 13	1 hr.	1			1, 12, 13, 14	1

For SI: 1 inch = 25.4 mm, 1 foot = 305 mm, 1 pound per square inch = 0.00689 MPa, 1 pound per square foot = 47.9 N/m².

Notes:
1. Thickness indicates thickness of first membrane protection on ceiling surface.
2. Failure mode—flame thru.
3. Failure mode—collapse.
4. No endpoint reached at termination of test.
5. Failure imminent—test terminated.
6. Joist failure—11.5 minutes; flame thru—13 minutes; collapse—24 minutes.
7. Joist failure—17 minutes; flame thru—18 minutes; collapse—33 minutes.
8. Joist failure—18 minutes; flame thru—8 minutes; collapse—30 minutes.
9. Joist failure—12 minutes; flame thru—8 minutes; collapse—22 minutes.
10. Joist failure—11 minutes; flame thru—17 minutes; collapse—27 minutes.
11. Joist failure—17 minutes; flame thru—20 minutes; collapse—43 minutes.
12. Joists: 2-inch by 10-inch southern pine or Douglas fir; No. 1 common or better. Subfloor: $3/4$-inch wood sheathing diaphragm of asbestos paper, and finish of tongue-and-groove wood flooring.
13. Loadings: not more than 1,000 psi maximum fiber stress in joists.
14. Perforations in gypsum lath are to be not less than $3/4$-inch diameter with one perforation for not more than 16/in.² diameter.
15. "Distemper" is a British term for a water-based paint such as white wash or calcimine.

RESOURCE A—GUIDELINES ON FIRE RATINGS OF ARCHAIC MATERIALS AND ASSEMBLIES

FIGURE 3.4
FLOOR/CEILING ASSEMBLIES—HOLLOW CLAY TILE WITH REINFORCED CONCRETE

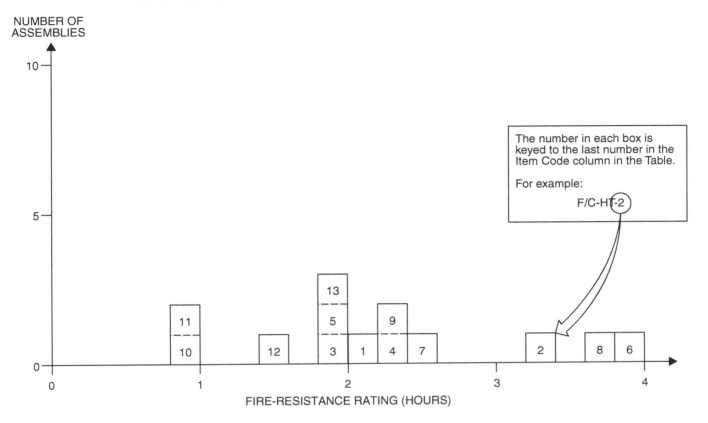

TABLE 3.4
FLOOR/CEILING ASSEMBLIES—HOLLOW CLAY TILE WITH REINFORCED CONCRETE

ITEM CODE	ASSEMBLY THICKNESS	CONSTRUCTION DETAILS	PERFORMANCE		REFERENCE NUMBER			NOTES	REC. HOURS
			LOAD	TIME	PRE-BMS-92	BMS-92	POST-BMS-92		
F/C-HT-1	6″	Cover: $1^1/_2$″ concrete (6080 psi); three cell hollow clay tiles, 12″ × 12″ × 4″; $3^1/_4$″ concrete between tiles including two $1/_2$″ rebars with $3/_4$″ concrete cover; $1/_2$″ plaster cover, lower.	75 psf	2 hrs. 7 min.			7	1, 2, 3	2
F/C-HT-2	6″	Cover: $1^1/_2$″ concrete (5840 psi); three cell hollow clay tiles, 12″ × 12″ × 4″; $3^1/_4$″ concrete between tiles including two $1/_2$″ rebars each with $1/_2$″ concrete cover and $5/_8$″ filler tiles between hollow tiles; $1/_2$″ plaster cover, lower.	61 psf	3 hrs. 23 min.			7	3, 4, 6	$3^1/_3$
F/C-HT-3	6″	Cover: $1^1/_2$″ concrete (6280 psi); three cell hollow clay tiles, 12″ × 12″ × 4″; $3^1/_4$″ concrete between tiles including two $1/_2$″ rebars with $1/_2$″ cover; $1/_2$″ plaster cover, lower.	122 psf	2 hrs.			7	1, 3, 5, 8	2
F/C-HT-4	6″	Cover: $1^1/_2$″ concrete (6280 psi); three cell hollow clay tiles, 12″ × 12″ × 4″; $3^1/_4$″ concrete between tiles including two $1/_2$″ rebars with $3/_4$″ cover; $1/_2$″ plaster cover, lower.	115 psf	2 hrs. 23 min.			7	1, 3, 7	$2^1/_3$
F/C-HT-5	6″	Cover: $1^1/_2$″ concrete (6470 psi); three cell hollow clay tiles, 12″ × 12″ × 4″; $3^1/_4$″ concrete between tiles including two $1/_2$″ rebars with $1/_2$″ cover; $1/_2$″ plaster cover, lower.	122 psf	2 hrs.			7	1, 3, 5, 8	2

(continued)

RESOURCE A—GUIDELINES ON FIRE RATINGS OF ARCHAIC MATERIALS AND ASSEMBLIES

TABLE 3.4—continued
FLOOR/CEILING ASSEMBLIES—HOLLOW CLAY TILE WITH REINFORCED CONCRETE

ITEM CODE	ASSEMBLY THICKNESS	CONSTRUCTION DETAILS	PERFORMANCE LOAD	PERFORMANCE TIME	REFERENCE NUMBER PRE-BMS-92	REFERENCE NUMBER BMS-92	REFERENCE NUMBER POST-BMS-92	NOTES	REC. HOURS
F/C-HT-6	8"	Floor cover: $1\frac{1}{2}$" gravel cement (4300 psi); three cell, $12" \times 12" \times 6"$; $3\frac{1}{2}$" space between tiles including two $\frac{1}{2}$" rebars with 1" cover from concrete bottom; $\frac{1}{2}$" plaster cover, lower.	165 psf	4 hrs.				7	1, 3, 9, 10	4
F/C-HT-7	9" (nom.)	Deck: $\frac{7}{8}$" T&G on $2" \times 1\frac{1}{2}$" bottoms (18" o.c.) $1\frac{1}{2}$" concrete cover (4600 psi); three cell hollow clay tiles, $12" \times 12" \times 4"$; 3" concrete between tiles including one $\frac{3}{4}$" rebar $\frac{3}{4}$" from tile bottom; $\frac{3}{4}$" plaster cover.	95 psf	2 hrs. 26 min.				7	4, 11, 12, 13	$2\frac{1}{3}$
F/C-HT-8	9" (nom.)	Deck: $\frac{7}{8}$" T&G on $2" \times 1\frac{1}{2}$" bottoms (18" o.c.) $1\frac{1}{2}$" concrete cover (3850 psi); three cell hollow clay tiles, $12" \times 12" \times 4"$; 3" concrete between tiles including one $\frac{3}{4}$" rebar $\frac{3}{4}$" from tile bottoms; $\frac{1}{2}$" plaster cover.	95 psf	3 hrs. 28 min.				7	4, 11, 12, 13	
F/C-HT-9	9" (nom.)	Deck: $\frac{7}{8}$" T&G on $2" \times 1\frac{1}{2}$" bottoms (18" o.c.) $1\frac{1}{2}$" concrete cover (4200 psi); three cell hollow clay tiles, $12" \times 12" \times 4"$; 3" concrete between tiles including one $\frac{3}{4}$" rebar $\frac{3}{4}$" from tile bottoms; $\frac{1}{2}$" plaster cover.	95 psf	2 hrs. 14 min.				7	3, 5, 8, 11	
F/C-HT-10	$5\frac{1}{2}$"	Fire clay tile (4" thick); $1\frac{1}{2}$" concrete cover; for general details, see Note 15.	See Note 14	1 hr.			43	15		1
F/C-HT-11	8"	Fire clay tile (6" thick); 2" cover.	See Note 14	1 hr.			43	15		1
F/C-HT-12	$5\frac{1}{2}$"	Fire clay tile (4" thick); $1\frac{1}{2}$" cover; $\frac{5}{8}$" gypsum plaster, lower.	See Note 14	1 hr. 30 min.			43	15		$1\frac{1}{2}$
F/C-HT-13	8"	Fire clay tile (6" thick); 2" cover; $\frac{5}{8}$" gypsum plaster, lower.	See Note 14	2 hrs.			43	15		$1\frac{1}{2}$

For SI: 1 inch = 25.4 mm, 1 foot = 305 mm, 1 pound per square inch = 0.00689 MPa, 1 pound per square foot = 47.9 N/m².

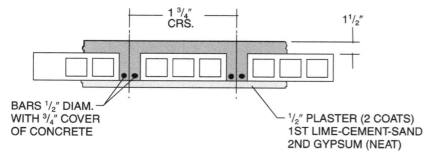

Notes:
1. A generalized cross section of this floor type follows:
2. Failure mode - structural.
3. Plaster: base coat—lime-cement-sand; top coat—gypsum (neat).

(continued)

TABLE 3.4—continued
FLOOR/CEILING ASSEMBLIES—HOLLOW CLAY TILE WITH REINFORCED CONCRETE

4. Failure mode—collapse.
5. Test stopped before any endpoints were reached.
6. A generalized cross section of this floor type follows:

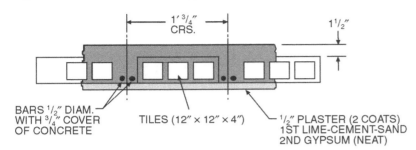

7. Failure mode—thermal—back face temperature rise.
8. Passed hose stream test.
9. Failed hose stream test.

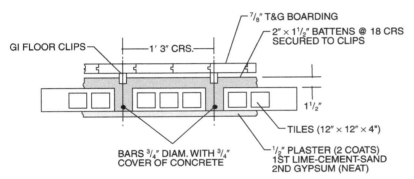

10. Test stopped at 4 hours before any endpoints were reached.
11. A generalized cross section of this floor type follows:
12. Plaster: base coat—retarded hemihydrate gypsum-sand; second coat—neat gypsum.
13. Concrete in Item 7 is P.C. based but with crushed brick aggregates while in Item 8 river sand and river gravels are used with the P.C.
14. Load - unspecified.
15. The 12-inch by 12-inch fire-clay tiles were laid end to end in rows spaced $2\frac{1}{2}$ inches or 4 inches apart. The reinforcing steel was placed between these rows and the concrete cast around them and over the tile to form the structural floor.

SECTION IV
BEAMS

TABLE 4.1.1
REINFORCED CONCRETE BEAMS
DEPTH 10″ TO LESS THAN 12″

ITEM CODE	DEPTH	CONSTRUCTION DETAILS	PERFORMANCE LOAD	PERFORMANCE TIME	REFERENCE NUMBER PRE-BMS-92	REFERENCE NUMBER BMS-92	REFERENCE NUMBER POST-BMS-92	NOTES	REC. HOURS
B-11-RC-1	11″	24″ wide × 11″ deep reinforced concrete "T" beam (3290 psi); Details: see Note 5 figure.	8.8 tons	4 hrs. 2 min.			7	1, 2, 14	4
B-10-RC-2	10″	24″ wide × 10″ deep reinforced concrete "T" beam (4370 psi); Details: see Note 6 figure.	8.8 tons	1 hr. 53 min.			7	1, 3	$1^{3}/_{4}$
B-10-RC-3	$10^{1}/_{2}″$	24″ wide × $10^{1}/_{2}″$ deep reinforced concrete "T" beam (4450 psi); Details: see Note 7 figure.	8.8 tons	2 hrs. 40 min.			7	1, 3	$2^{2}/_{3}$
B-11-RC-4	11″	24″ wide × 11″ deep reinforced concrete "T" beam (2400 psi); Details: see Note 8 figure.	8.8 tons	3 hrs. 32 min.			7	1, 3, 14	$3^{1}/_{2}$
B-11-RC-5	11″	24″ wide × 11″ deep reinforced concrete "T" beam (4250 psi); Details: see Note 9 figure.	8.8 tons	3 hrs. 3 min.			7	1, 3, 14	3
B-11-RC-6	11″	Concrete flange: 4″ deep × 2′ wide (4895 psi) concrete; Concrete beam: 7″ deep × $6^{1}/_{2}″$ wide beam; "I" beam reinforcement; 10″ × $4^{1}/_{2}″$ × 25 lbs. R.S.J.; 1″ cover on flanges; Flange reinforcement: $^{3}/_{8}″$ diameter bars at 6″ pitch parallel to "T"; $^{1}/_{4}″$ diameter bars perpendicular to "T"; Beam reinforcement: 4″ × 6″ wire mesh No. 13 SWG; Span: 11′ restrained; Details: see Note 10 figure.	10 tons	6 hrs.			7	1, 4	6
B-11-RC-7	11″	Concrete flange: 6″ deep × 1′ $6^{1}/_{2}″$ wide (3525 psi) concrete; Concrete beam: 5″ deep × 8″ wide precast concrete blocks $8^{3}/_{4}″$ long; "I" beam reinforcement; 7″ × 4″ × 16 lbs. R.S.J.; 2″ cover on bottom; $1^{1}/_{2}″$ cover on top; Flange reinforcement: two rows $^{1}/_{2}″$ diameter rods parallel to "T"; Beam reinforcement: $^{1}/_{8}″$ wire mesh perpendicular to 1″ ; Span: 1′ 3″ simply supported; Details: see Note 11 figure.	3.9 tons	4 hrs.			7	1, 2	4
B-11-RC-8	11″	Concrete flange: 4″ deep × 2′ wide (3525 psi) concrete; Concrete beam 7″ deep × $4^{1}/_{2}″$ wide; (scaled from drawing); "I" beam reinforcement; 10″ × $4^{1}/_{2}″$ × 25 lbs. R.S.J.; no concrete cover on bottom; Flange reinforcement: $^{3}/_{8}″$ diameter bars at 6 pitch parallel to "T"; $^{1}/_{4}″$ diameter bars perpendicular to "T"; Span: 11′ restricted.	10 tons	4 hrs.			7	1, 2, 12	4
B-11-RC-9	$11^{1}/_{2}″$	24″ wide × $11^{1}/_{2}″$ deep reinforced concrete "T" beam (4390 psi); Details: see Note 12 figure.	8.8 tons	3 hrs. 24 min.			7	1, 3	$3^{1}/_{3}$

For SI: 1 inch = 25.4 mm, 1 foot = 305 mm, 1 pound = 0.004448 kN, 1 pound per square inch = 0.00689 MPa, 1 ton = 8.896 kN.

Notes:
1. Load concentrated at mid span.
2. Achieved 4 hour performance (Class "B," British).
3. Failure mode—collapse.
4. Achieved 6 hour performance (Class "A," British).

(continued)

RESOURCE A—GUIDELINES ON FIRE RATINGS OF ARCHAIC MATERIALS AND ASSEMBLIES

TABLE 4.1.1—continued
REINFORCED CONCRETE BEAMS
DEPTH 10″ TO LESS THAN 12″

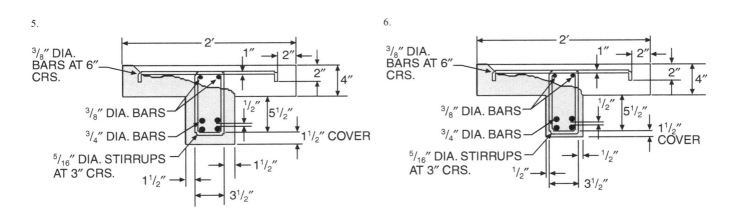

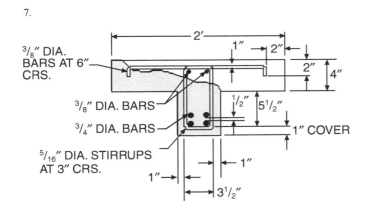

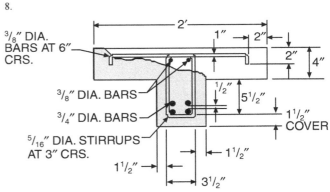

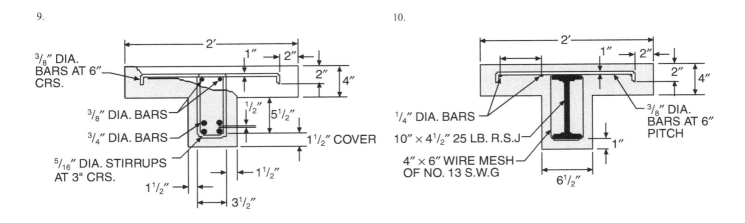

(continued)

TABLE 4.1.1—continued
REINFORCED CONCRETE BEAMS
DEPTH 10″ TO LESS THAN 12″

11.

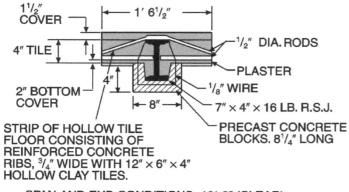

SPAN AND END CONDITIONS:-10′-3″ (CLEAR). SIMPLY SUPPORTED.

12.

13.

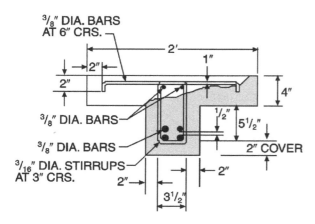

14. The different performances achieved by B-11-RC-1, B-11-RC-4 and B-11-RC-5 are attributable to differences in concrete aggregate compositions reported in the source document but unreported in this table. This demonstrates the significance of material composition in addition to other details.

RESOURCE A—GUIDELINES ON FIRE RATINGS OF ARCHAIC MATERIALS AND ASSEMBLIES

TABLE 4.1.2
REINFORCED CONCRETE BEAMS
DEPTH 12″ TO LESS THAN 14″

ITEM CODE	DEPTH	CONSTRUCTION DETAILS	PERFORMANCE		REFERENCE NUMBER			NOTES	REC. HOURS
			LOAD	TIME	PRE-BMS-92	BMS-92	POST-BMS-92		
B-12-RC-1	12″	12″ × 8″ section; 4160 psi aggregate concrete; Reinforcement: 4-⅞″ rebars at corners; 1″ below each surface; ¼″ stirrups 10″ o.c.	5.5 tons	2 hrs.			7	1	2
B-12-RC-2	12″	Concrete flange: 4″ deep × 2′ wide (3045 psi) concrete at 35 days; Concrete beam: 8″ deep; "I" beam reinforcement: 10″ × 4½″ × 25 lbs. R.S.J.; 1″ cover on flanges; Flange reinforcement: ⅜″ diameter bars at 6″ pitch parallel to "T"; ¼″ diameter bars perpendicular to "T"; Beam reinforcement: 4″ × 6″ wire mesh No. 13 SWG; Span: 10′ 3″ simply supported.	10 tons	4 hrs.			7	2, 3, 5	4
B-13-RC-3	13″	Concrete flange: 4″ deep × 2′ wide (3825 psi) concrete at 46 days; Concrete beam: 9″ deep × 8½″ wide; (scaled from drawing); "I" beam reinforcement: 10″ × 4½″ × 25 lbs. R.S.J.; 3″ cover on bottom flange; 1″ cover on top flange; Flange reinforcement: ⅜″ diameter bars at 6″ pitch parallel to "T"; ¼″ diameter bars perpendicular to "T"; Beam reinforcement: 4″ × 6″ wire mesh No. 13 SWG; Span: 11′ restrained.	10 tons	6 hrs.			7	2, 3, 6, 8, 9	4
B-12-RC-4	12″	Concrete flange: 4″ deep × 2′ wide (3720 psi) concrete at 42 days; Concrete beam: 8″ deep × 8½″ wide; (scaled from drawing); "I" beam reinforcement: 10″ × 4½″ × 25 lbs. R.S.J.; 2″ cover bottom flange; 1″ cover top flange; Flange reinforcement: ⅜″ diameter bars at 6″ pitch parallel to "T"; ¼″ diameter bars perpendicular to "T"; Beam reinforcement: 4″ × 6″ wire mesh No. 13 SWG; Span: 11′ restrained.	10 tons	6 hrs.			7	1, 3, 4, 7, 8, 9	4

For SI: 1 inch = 25.4 mm, 1 foot = 305 mm, 1 pound = 0.004448 kN, 1 pound per square inch = 0.00689 MPa, 1 ton = 8.896 kN.

Notes:
1. Qualified for 2 hour use. (Grade "C," British) Test included hose stream and reload at 48 hours.
2. Load concentrated at mid span.
3. British test.
4. British test—qualified for 6 hour use (Grade "A").

(continued)

TABLE 4.1.2—continued
REINFORCED CONCRETE BEAMS
DEPTH 12″ TO LESS THAN 14″

5.

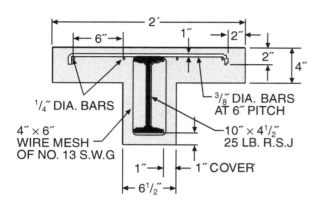

6.

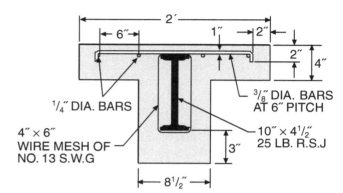

7.

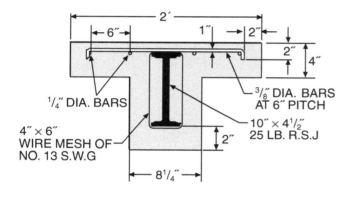

8. See Table 4.1.3, Note 5.

9. Hourly rating based upon B-12-RC-2 above.

TABLE 4.1.3
REINFORCED CONCRETE BEAMS
DEPTH 14″ TO LESS THAN 16″

ITEM CODE	DEPTH	CONSTRUCTION DETAILS	PERFORMANCE		REFERENCE NUMBER			NOTES	REC. HOURS
			LOAD	TIME	PRE-BMS-92	BMS-92	POST-BMS-92		
B-15-RC-1	15″	Concrete flange: 4″ deep × 2′ wide (3290 psi) concrete; Concrete beam: 10″ deep × $8\frac{1}{2}$″ wide; "I" beam reinforcement: 10″ × $4\frac{1}{2}$″ × 25 lbs. R.S.J.; 4″ cover on bottom flange; 1″ cover on top flange; Flange reinforcement: $\frac{3}{8}$″ diameter bars at 6″ pitch parallel to "T"; $\frac{1}{4}$″ diameter bars perpendicular to "T"; Beam reinforcement: 4″ × 6″ wire mesh No. 13 SWG; Span: 11′ restrained.	10 tons	6 hrs.			7	1, 2, 3, 5, 6	4
B-15-RC-2	15″	Concrete flange: 4″ deep × 2′ wide (4820 psi) concrete; Concrete beam: 10″ deep × $8\frac{1}{2}$″ wide; "I" beam reinforcement: 10″ × $4\frac{1}{2}$″ × 25 lbs. R.S.J.; 1″ cover over wire mesh on bottom flange; 1″ cover on top flange; Flange reinforcement: $\frac{3}{8}$″ diameter bars at 6″ pitch parallel to "T"; $\frac{1}{4}$″ diameter bars perpendicular to "T"; Beam reinforcement: 4″ × 6″ wire mesh No. 13 SWG; Span: 11′ restrained.	10 tons	6 hrs.			7	1, 2, 4, 5, 6	4

For SI: 1 inch = 25.4 mm, 1 foot = 305 mm, 1 pound = 0.004448 kN, 1 pound per square inch = 0.00689 MPa, 1 ton = 8.896 kN.

Notes:
1. Load concentrated at mid span.
2. Achieved 6 hour fire rating (Grade "A," British).
3.
4.

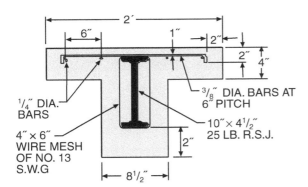
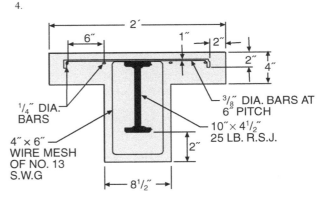

5. Section 43.147 of the 1979 edition of the *Uniform Building Code Standards* provides:
   "A restrained condition in fire tests, as used in this standard, is one in which expansion at the supports of a load-carrying element resulting from the effects of the fire is resisted by forces external to the element. An unrestrained condition is one in which the load-carrying element is free to expand and rotate at its support."
   "Restraint in buildings is defined as follows: Floor and roof assemblies and individual beams in buildings shall be considered restrained when the surrounding or supporting structure is capable of resisting the thermal expansion throughout the range of anticipated elevated temperatures. Construction not complying . . . is assumed to be free to rotate and expand and shall be considered as unrestrained."
   "Restraint may be provided by the lateral stiffness of supports for floor and roof assemblies and intermediate beams forming part of the assembly. In order to develop restraint, connections must adequately transfer thermal thrusts to such supports. The rigidity of adjoining panels or structures shall be considered in assessing the capability of a structure to resist therm expansion."
   Because it is difficult to determine whether an existing building's structural system is capable of providing the required restraint, the lower hourly ratings of a similar but unrestrained assembly have been recommended.
6. Hourly rating based upon Table 4.2.1, Item B-12-RC-2.

RESOURCE A—GUIDELINES ON FIRE RATINGS OF ARCHAIC MATERIALS AND ASSEMBLIES

**TABLE 4.2.1
REINFORCED CONCRETE BEAMS—UNPROTECTED DEPTH
10″ TO LESS THAN 12″**

ITEM CODE	DEPTH	CONSTRUCTION DETAILS	PERFORMANCE		REFERENCE NUMBER			NOTES	REC. HOURS
			LOAD	TIME	PRE- BMS-92	BMS-92	POST-BMS-92		
B-SU-1	10″	10″ × 4$\frac{1}{2}$″ × 25 lbs. "I" beam.	10 tons	39 min.			7	1	$\frac{1}{3}$

For SI: 1 inch = 25.4 mm, 1 pound = 0.004448 kN, 1 ton = 8.896 kN.

Notes:
1. Concentrated at mid span.

**TABLE 4.2.2
STEEL BEAMS—CONCRETE PROTECTION DEPTH
10″ TO LESS THAN 12″**

ITEM CODE	DEPTH	CONSTRUCTION DETAILS	PERFORMANCE		REFERENCE NUMBER			NOTES	REC. HOURS
			LOAD	TIME	PRE- BMS-92	BMS-92	POST- BMS-92		
B-SC-1	10″	10″ × 8″ rectangle; aggregate concrete (4170 psi) with 1″ top cover and 2″ bottom cover; No. 13 SWG iron wire loosely wrapped at approximately 6″ pitch about 7″ × 4″ × 16 lbs. "I" beam.	3.9 tons	3 hrs. 46 min.			7	1, 2, 3	3$\frac{3}{4}$
B-SC-1	10″	10″ × 8″ rectangle; aggregate concrete (3630 psi) with 1″ top cover and 2″ bottom cover; No. 13 SWG iron wire loosely wrapped at approximately 6″ pitch about 7″ × 4″ × 16 lbs. "I" beam.	5.5 tons	5 hrs. 26 min.			7	1, 4, 5, 6, 7	3$\frac{3}{4}$

For SI: 1 inch = 25.4 mm, 1 pound = 0.004448 kN, 1 pound per square inch = 0.00689 MPa, 1 ton = 8.896 kN.

Notes:
1. Load concentrated at mid span.
2. Specimen 10-foot 3-inch clear span simply supported.
3. Passed Grade "C" fire resistance (British) including hose stream and reload.
4. Specimen 11-foot clear span—restrained.
5. Passed Grade "B" fire resistance (British) including hose stream and reload.
6. See Table 4.1.3, Note 5.
7. Hourly rating based upon B-SC-1 above.

RESOURCE A—GUIDELINES ON FIRE RATINGS OF ARCHAIC MATERIALS AND ASSEMBLIES

SECTION V
DOORS

**FIGURE 5.1
RESISTANCE OF DOORS TO FIRE EXPOSURE**

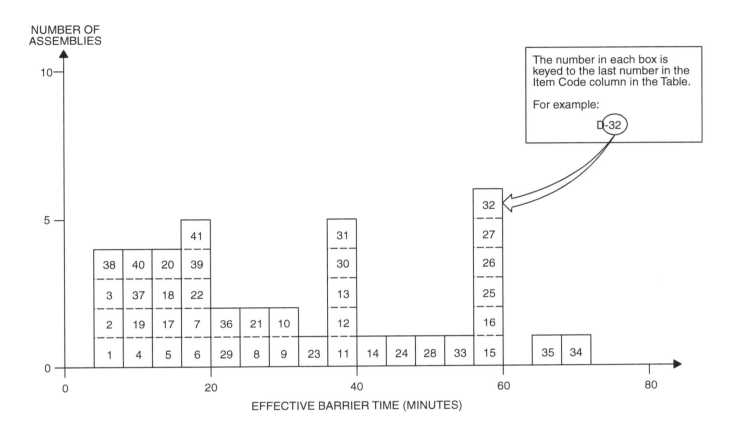

**TABLE 5.1
RESISTANCE OF DOORS TO FIRE EXPOSURE**

| ITEM CODE | DOOR MINIMUM THICKNESS | CONSTRUCTION DETAILS | PERFORMANCE || REFERENCE NUMBER ||| NOTES | REC. (MIN.) |
			EFFECTIVE BARRIER	EDGE FLAMING	PRE-BMS-92	BMS-92	POST-BMS-92		
D-1	$3/8''$	Panel door; pine perimeter ($1^{3}/_{8}''$); painted (enamel).	5 min. 10 sec.	N/A			90	1, 2	5
D-2	$3/8''$	As above, with two coats U.L. listed intumescent coating.	5 min. 30 sec.	5 min.			90	1, 2, 7	5
D-3	$3/8''$	As D-1, with standard primer and flat interior paint.	5 min. 55 sec.	N/A			90	1, 3, 4	5
D-4	$2^{5}/_{8}''$	As D-1, with panels covered each side with $1/2''$ plywood; edge grouted with sawdust filled plaster; door faced with $1/8''$ hardboard each side; paint see (5).	11 min. 15 sec.	3 min. 45 sec.			90	1, 2, 5, 7	10

(continued)

TABLE 5.1—continued
RESISTANCE OF DOORS TO FIRE EXPOSURE

ITEM CODE	DOOR MINIMUM THICKNESS	CONSTRUCTION DETAILS	PERFORMANCE		REFERENCE NUMBER			NOTES	REC. (MIN.)
			EFFECTIVE BARRIER	EDGE FLAMING	PRE-BMS-92	BMS-92	POST-BMS-92		
D-5	$3/8''$	As D-1, except surface protected with glass fiber reinforced intumescent fire retardant coating.	16 min.	N/A			90	1, 3, 4, 7	15
D-6	$1\,5/8''$	Door detail: As D-4, except with $1/8''$ cement asbestos board facings with aluminum foil; door edges protected by sheet metal.	17 min.	10 min. 15 sec.			90	1, 3, 4	15
D-7	$1\,5/8''$	Door detail with $1/8''$ hardboard cover each side as facings; glass fiber reinforced intumescent coating applied.	20 min.	N/A			90	1, 3, 4, 7	20
D-8	$1\,5/8''$	Door detail same as D-4; paint was glass reinforced epoxy intumescent.	26 min.	24 min. 45 sec.			90	1, 3, 4, 6, 7	25
D-9	$1\,5/8''$	Door detail same as D-4 with facings of $1/8''$ cement asbestos board.	29 min.	3 min. 15 sec.			90	1, 2	5
D-10	$1\,5/8''$	As per D-9.	31 min. 30 sec.	7 min. 20 sec.			90	1, 3, 4	6
D-11	$1\,5/8''$	As per D-7; painted with epoxy intumescent coating including glass fiber roving.	36 min. 25 sec.	N/A			90	1, 3, 4	35
D-12	$1\,5/8''$	As per D-4 with intumescent fire retardant paint.	37 min. 30 sec.	24 min. 40 sec.			90	1, 3, 4	30
D-13	$1\,1/2''$ (nom.)	As per D-4, except with 24 ga. galvanized sheet metal facings.	39 min.	39 min.			90	1, 3, 4	39
D-14	$1\,5/8''$	As per D-9.	41 min. 30 sec.	17 min. 20 sec.			90	1, 3, 4, 6	20
D-15	—	Class C steel fire door.	60 min.	58 min.			90	7, 8	60
D-16	—	Class B steel fire door.	60 min.	57 min.			90	7, 8	60
D-17	$1\,3/4''$	Solid core flush door; core staves laminated to facings but not each other; Birch plywood facings $1/2''$ rebate in door frame for door; $3/32''$ clearance between door and wood frame.	15 min.	13 min.	37			11	13

(continued)

TABLE 5.1—continued
RESISTANCE OF DOORS TO FIRE EXPOSURE

ITEM CODE	DOOR MINIMUM THICKNESS	CONSTRUCTION DETAILS	PERFORMANCE		REFERENCE NUMBER			NOTES	REC. (MIN.)
			EFFECTIVE BARRIER	EDGE FLAMING	PRE- BMS-92	BMS-92	POST- BMS-92		
D-18	$1\frac{3}{4}''$	As per D-17.	14 min.	13 min.			37	11	13
D-19	$1\frac{3}{4}''$	Door same as D-17, except with 16 ga. steel; $\frac{3}{32}''$ door frame clearance.	12 min.	—			37	9, 11	10
D-20	$1\frac{3}{4}''$	As per D-19.	16 min.				37	10, 11	10
D-21	$1\frac{3}{4}''$	Doors as per D-17; intumescent paint applied to top and side edges.	26 min.	—			37	11	25
D-22	$1\frac{3}{4}''$	Door as per D-17, except with $\frac{1}{2}'' \times \frac{1}{8}''$ steel strip set into edges of door at top and side facing stops; matching strip on stop.	18 min.	6 min.			37	11	18
D-23	$1\frac{3}{4}''$	Solid oak door.	36 min.	22 min.			15	13	25
D-24	$1\frac{7}{8}''$	Solid oak door.	45 min.	35 min.			15	13	35
D-25	$1\frac{7}{8}''$	Solid teak door.	58 min.	34 min.			15	13	35
D-26	$1\frac{7}{8}''$	Solid (pitch) pine door.	57 min.	36 min.			15	13	35
D-27	$1\frac{7}{8}''$	Solid deal (pine) door.	57 min.	30 min.			15	13	30
D-28	$1\frac{7}{8}''$	Solid mahogany door.	49 min.	40 min.			15	13	45
D-29	$1\frac{7}{8}''$	Solid poplar door.	24 min.	3 min.			15	13, 14	5
D-30	$1\frac{7}{8}''$	Solid oak door.	40 min.	33 min.			15	13	35
D-31	$1\frac{7}{8}''$	Solid walnut door.	40 min.	15 min.			15	13	20
D-32	$2\frac{5}{8}''$	Solid Quebec pine.	60 min.	60 min.			15	13	60
D-33	$2\frac{5}{8}''$	Solid pine door.	55 min.	39 min.			15	13	40
D-34	$2\frac{5}{8}''$	Solid oak door.	69 min.	60 min.			15	13	60
D-35	$2\frac{5}{8}''$	Solid teak door.	65 min.	17 min.			15	13	60
D-36	$1\frac{1}{2}''$	Solid softwood door.	23 min.	8.5 min.			15	13	10
D-37	$\frac{3}{4}''$	Panel door.	8 min.	7.5 min.			15	13	5
D-38	$\frac{5}{16}''$	Panel door.	5 min.	5 min.			15	13	5
D-39	$\frac{3}{4}''$	Panel door, fire retardant treated.	$17\frac{1}{2}$ min.	3 min.			15	13	8
D-40	$\frac{3}{4}''$	Panel door, fire retardant treated.	$8\frac{1}{2}$ min.	$8\frac{1}{2}$ min.			15	13	8
D-41	$\frac{3}{4}''$	Panel door, fire retardant treated.	$16\frac{3}{4}$ min.	$11\frac{1}{2}$ min.			15	13	8

For SI: 1 inch = 25.4 mm, 1 foot = 305 mm.

Notes:
1. All door frames were of standard lumber construction.
2. Wood door stop protected by asbestos millboard.
3. Wood door stop protected by sheet metal.
4. Door frame protected with sheet metal and weather strip.
5. Surface painted with intumescent coating.
6. Door edge sheet metal protected.
7. Door edge intumescent paint protected.
8. Formal steel frame and door stop.
9. Door opened into furnace at 12 feet.
10. Similar door opened into furnace at 12 feet.
11. The doors reported in these tests represent the type contemporaries used as 20-minute solid-core wood doors. The test results demonstrate the necessity of having wall anchored metal frames, minimum cleaners possible between door, frame and stops. They also indicate the utility of long throw latches and the possible use of intumescent paints to seal doors to frames in event of a fire.
12. Minimum working clearance and good latch closure are absolute necessities for effective containment for all such working door assemblies.
13. Based on British tests.
14. Failure at door-frame interface.

BIBLIOGRAPHY

1. Central Housing Committee on Research, Design, and Construction; Subcommittee on Fire Resistance Classifications, "Fire-Resistance Classifications of Building Constructions," *Building Materials and Structures,* Report BMS 92, National Bureau of Standards, Washington, Oct. 1942. (Available from NTIS No. COM-73-10974)

2. Foster, H. D., Pinkston, E. R., and Ingberg, S. H., "Fire Resistance of Structural Clay Tile Partitions," *Building Materials and Structures, Report* BMS 113, National Bureau of Standards, Washington, Oct. 1948.

3. Ryan, J. V., and Bender, E.W., "Fire Endurance of Open-Web Steel-Joist Floors with Concrete Slabs and Gypsum Ceilings," *Building Materials and Structures,* Report BMS 141, National Bureau of Standards, Washington, Aug. 1954.

4. Mitchell, N. D., "Fire Tests of Wood-Framed Walls and Partitions with Asbestos-Cement Facings," *Building Materials and Structures,* Report BMS 123, National Bureau of Standards, Washington, May 1951.

5. Robinson, H. E., Cosgrove, L. A., and Powell, F. J., "Thermal Resistance of Airspace and Fibrous Insulations Bounded by Reflective Surfaces," *Building Materials and Structures,* Report BMS 151, National Bureau of Standards, Washington, Nov. 1957.

6. Shoub, H., and Ingberg, S. H., "Fire Resistance of Steel Deck Floor Assemblies," *Building Science Series,* 11, National Bureau of Standards, Washington, Dec. 1967.

7. Davey, N., and Ashton, L. A., "Investigations on Building Fires, Part V: Fire Tests of Structural Elements," *National Building Studies,* Research Paper, No. 12, Dept. of Scientific and Industrial Research (Building Research Station), London, 1953.

8. National Board of Fire Underwriters, *Fire Resistance Ratings of Beam, Girder, and Truss Protections, Ceiling Constructions, Column Protections, Floor and Ceiling Constructions, Roof Constructions, Walls and Partitions,* New York, April 1959.

9. Mitchell, N.D., Bender, E.D., and Ryan, J.V., "Fire Resistance of Shutters for Moving-Stairway Openings," *Building Materials and Structures,* Report BMS 129, National Bureau of Standards, Washington, March 1952.

10. National Board of Fire Underwriters, *National Building Code; an Ordinance Providing for Fire Limits, and Regulations Governing the Construction, Alteration, Equipment, or Removal of Buildings or Structures,* New York, 1949.

11. Department of Scientific and Industrial Research and of the Fire Offices' Committee, Joint Committee of the Building Research Board, "Fire Gradings of Buildings, Part I: General Principles and Structural Precautions," *Post-War Building Studies,* No. 20, Ministry of Works, London, 1946.

12. Lawson, D. I., Webster, C. T., and Ashton, L. A., "Fire Endurance of Timber Beams and Floors," *National Building Studies,* Bulletin, No. 13, Dept. of Scientific and Industrial Research and Fire Offices' Committee (Joint Fire Research Organization), London, 1951.

13. Parker, T. W., Nurse, R. W., and Bessey, G. E., "Investigations on Building Fires. Part I: The Estimation of the Maximum Temperature Attained in Building Fires from Examination of the Debris, and Part II: The Visible Change in Concrete or Mortar Exposed to High Temperatures," *National Building Studies,* Technical Paper, No. 4, Dept. of Scientific and Industrial Research (Building Research Station), London, 1950.

14. Bevan, R. C., and Webster, C. T., "Investigations on Building Fires, Part III: Radiation from Building Fires," *National Building Studies,* Technical Paper, No. 5, Dept. of Scientific and Industrial Research (Building Research Station), London, 1950.

15. Webster, D. J., and Ashton, L. A., "Investigations on Building Fires, Part IV: Fire Resistance of Timber Doors," *National Building Studies,* Technical Paper, No. 6, Dept. of Scientific and Industrial Research (Building Research Station), London, 1951.

16. Kidder, F. E., *Architects' and Builders' Handbook: Data for Architects, Structural Engineers, Contractors, and Draughtsmen,* comp. by a Staff of Specialists and H. Parker, editor-in-chief, 18th ed., enl., J. Wiley, New York, 1936.

17. Parker, H., Gay, C. M., and MacGuire, J. W., *Materials and Methods of Architectural Construction,* 3rd ed., J. Wiley, New York, 1958.

18. Diets, A. G. H., *Dwelling House Construction,* The MIT Press, Cambridge, 1971.

19. Crosby, E. U., and Fiske, H. A., *Handbook of Fire Protection,* 5th ed., The Insurance Field Company, Louisville, Ky., 1914.

20. Crosby, E. U., Fiske, H. A., and Forster, H.W., *Handbook of Fire Protection,* 8th ed., R. S. Moulton, general editor, National Fire Protection Association, Boston, 1936.

21. Kidder, F. E., *Building Construction and Superintendence,* rev. and enl., by T. Nolan, W. T. Comstock, New York, 1909-1913, 2 vols.

22. National Fire Protection Association, Committee on Fire-Resistive Construction, *The Baltimore Conflagration,* 2nd ed., Chicago, 1904.

23. Przetak, L., *Standard Details for Fire-Resistive Building Construction,* McGraw-Hill Book Co., New York, 1977.

24. Hird, D., and Fischl, C. F., "Fire Hazard of Internal Linings," *National Building Studies,* Special Report, No. 22, Dept. of Scientific and Industrial Research and Fire Offices' Committee (Joint Fire Research Organization), London, 1954.

25. Menzel, C. A., *Tests of the Fire-Resistance and Strength of Walls Concrete Masonry Units,* Portland Cement Association, Chicago, 1934.

26. Hamilton, S. B., "A Short History of the Structural Fire Protection of Buildings Particularly in England," *National Building Studies,* Special Report, No. 27, Dept. of Scientific and Industrial Research (Building Research Station), London, 1958.

27. Sachs, E. O., and Marsland, E., "The Fire Resistance of Doors and Shutters being Tabulated Results of Fire Tests Conducted by the Committee," *Journal of the British Fire Prevention Committee,* No. VII, London, 1912.

28. Egan, M. D., *Concepts in Building Firesafety,* J. Wiley, New York, 1978.

29. Sachs, E. O., and Marsland, E., "The Fire Resistance of Floors being Tabulated Results of Fire Tests Conducted by the Committee," *Journal of the British Fire Prevention Committee,* No. VI, London, 1911.

30. Sachs, E. O., and Marsland, E., "The Fire Resistance of Partitions being Tabulated Results of Fire Tests Conducted by the Committee," *Journal of the British Fire Prevention Committee,* No. IX, London, 1914.

31. Ryan, J. V., and Bender, E. W., "Fire Tests of Precast Cellular Concrete Floors and Roofs," *National Bureau of Standards Monograph,* 45, Washington, April 1962.

32. Kingberg, S. H., and Foster, H. D., "Fire Resistance of Hollow Load-Bearing Wall Tile," *National Bureau of Standards* Research Paper, No. 37, (Reprint from *NBS Journal of Research,* Vol. 2) Washington, 1929.

33. Hull, W. A., and Ingberg, S. H., "Fire Resistance of Concrete Columns," *Technologic Papers of the Bureau of Standards,* No. 272, Vol. 18, Washington, 1925, pp. 635-708.

34. National Board of Fire Underwriters, *Fire Resistance Ratings of Less than One Hour,* New York, Aug. 1956.

35. Harmathy, T. Z., "Ten Rules of Fire Endurance Rating," *Fire Technology,* Vol. 1, May 1965, pp. 93-102.

36. Son, B. C., "Fire Endurance Test on a Steel Tubular Column Protected with Gypsum Board," *National Bureau of Standards,* NBSIR, 73-165, Washington, 1973.

37. Galbreath, M., "Fire Tests of Wood Door Assemblies," *Fire Study,* No. 36, Div. of Building Research, National Research Council Canada, Ottawa, May 1975.

38. Morris, W. A., "An Investigation into the Fire Resistance of Timber Doors," *Fire Research Note,* No. 855, Fire Research Station, Boreham Wood, Jan. 1971.

39. Hall, G. S., "Fire Resistance Tests of Laminated Timber Beams," *Timber Association Research Report,* WR/RR/1, High Sycombe, July 1968.

40. Goalwin, D. S., "Fire Resistance of Concrete Floors," *Building Materials and Structures,* Report BMS 134, National Bureau of Standards, Washington, Dec. 1952.

41. Mitchell, N. D., and Ryan, J. V., "Fire Tests of Steel Columns Encased with Gypsum Lath and Plaster," *Building Materials and Structures,* Report BMS 135, National Bureau of Standards, Washington, April 1953.

42. Ingberg, S. H., "Fire Tests of Brick Walls," *Building Materials and Structures,* Report BMS 143, National Bureau of Standards, Washington, Nov. 1954.

43. National Bureau of Standards, "Fire Resistance and Sound-Insulation Ratings for Walls, Partitions, and Floors," *Technical Report on Building Materials,* 44, Washington, 1944.

44. Malhotra, H. L., "Fire Resistance of Brick and Block Walls," *Fire Note,* No. 6, Ministry of Technology and Fire Offices' Committee Joint Fire Research Organization, London, HMSO, 1966.

45. Mitchell, N. D., "Fire Tests of Steel Columns Protected with Siliceous Aggregate Concrete," *Building Materials and Structures,* Report BMS 124, National Bureau of Standards, Washington, May 1951.

46. Freitag, J. K., *Fire Prevention and Fire Protection as Applied to Building Construction; a Handbook of Theory and Practice,* 2nd ed., J. Wiley, New York, 1921.

47. Ingberg, S. H., and Mitchell, N. D., "Fire Tests of Wood and Metal-Framed Partition," *Building Materials and Structures,* Report BMS 71, National Bureau of Standards, Washington, 1941.

48. Central Housing Committee on Research, Design, and Construction, Subcommittee on Definitions, "A Glossary of Housing Terms," *Building Materials and Structures,* Report BMS 91, National Bureau of Standards, Washington, Sept. 1942.

49. Crosby, E. U., Fiske, H. A., and Forster, H.W., *Handbook of Fire Protection,* 7th ed., D. Van Nostrand Co., New York 1924.

50. Bird, E. L., and Docking, S. J., *Fire in Buildings,* A. & C. Black, London, 1949.

51. American Institute of Steel Construction, Fire Resistant Construction in Modern Steel-Framed Buildings, New York, 1959.

52. Central Dockyard Laboratory, "Fire Retardant Paint Tests—a Critical Review," CDL Technical Memorandum, No. P87/73, H. M. Naval Base, Portsmouth, Dec. 1973.

53. Malhotra, H. L., "Fire Resistance of Structural Concrete Beams," *Fire Research Note,* No. 741, Fire Research Station, Borehamwood, May 1969.

54. Abrams, M. S., and Gustaferro, A. H., "Fire Tests of Poke-Thru Assemblies," *Research and Development Bulletin,* 1481-1, Portland Cement Association, Skokie, 1971.

55. Bullen, M. L., "A Note on the Relationship between Scale Fire Experiments and Standard Test Results," *Building Research Establishment Note,* N51/75, Borehamwood, May 1975.

56. The America Fore Group of Insurance Companies, Research Department, *Some Characteristic Fires in Fire Resistive Buildings,* Selected from twenty years record in the files of the N.F.P.A. "Quarterly," New York, c. 1933.

57. Spiegelhalter, F., "Guide to Design of Cavity Barriers and Fire Stops," *Current Paper,* CP 7/77, Building Research Establishment, Borehamwood, Feb. 1977.

58. Wardle, T. M. "Notes on the Fire Resistance of Heavy Timber Construction," *Information Series,* No. 53, New Zealand Forest Service, Wellington, 1966.

59. Fisher, R. W., and Smart, P. M. T., "Results of Fire Resistance Tests on Elements of Building Construction," *Building Research Establishment Report,* G R6, London, HMSO, 1975.

60. Serex, E. R., "Fire Resistance of Alta Bates Gypsum Block Non-Load Bearing Wall," Report to Alta Bates Community Hospital, *Structural Research Laboratory Report,* ES-7000, University of Calif., Berkeley, 1969.

61. Thomas, F. G., and Webster, C. T., "Investigations on Building Fires, Part VI: The Fire Resistance of Reinforced Concrete Columns," *National Building Studies,* Research Paper, No. 18, Dept. of Scientific and Industrial Research (Building Research Station), London, HMSO, 1953.

62. Building Research Establishment, "Timber Fire Doors," *Digest,* 220, Borehamwood, Nov. 1978.

63. *Massachusetts State Building Code; Recommended Provisions,* Article 22: Repairs, Alterations, Additions, and Change of Use of Existing Buildings, Boston, Oct. 23, 1978.

64. Freitag, J. K., *Architectural Engineering; with Especial Reference to High Building Construction, Including Many Examples of Prominent Office Buildings,* 2nd ed., rewritten, J. Wiley, New York, 1906.

65. Architectural Record, *Sweet's Indexed Catalogue of Building Construction for the Year 1906,* New York, 1906.

66. Dept. of Commerce, Building Code Committee, "Recommended Minimum Requirements for Fire Resistance in Buildings," *Building and Housing,* No. 14, National Bureau of Standards, Washington, 1931.

67. British Standards Institution, "Fire Tests on Building Materials and Structures," *British Standards,* 476, Pt. 1, London, 1953.

68. Löberg-Holm, K., "Glass," *The Architectural Record,* Oct. 1930, pp. 345-357.

69. Structural Clay Products Institute, "Fire Resistance," *Technical Notes on Brick and Tile Construction,* 16 rev., Washington, 1964.

70. Ramsey, C. G., and Sleeper, H. R., *Architectural Graphic Standards for Architects, Engineers, Decorators, Builders, and Draftsmen,* 3rd ed., J. Wiley, New York, 1941.

71. Underwriters' Laboratories, *Fire Protection Equipment List,* Chicago, Jan. 1957.

72. Underwriters' Laboratories, *Fire Resistance Directory; with Hourly Ratings for Beams, Columns, Floors, Roofs, Walls, and Partitions,* Chicago, Jan. 1977.

73. Mitchell, N. D., "Fire Tests of Gunite Slabs and Partitions," *Building Materials and Structures,* Report BMS 131, National Bureau of Standards, Washington, May 1952.

74. Woolson, I. H., and Miller, R. P., "Fire Tests of Floors in the United States," *Proceedings International Association for Testing Materials,* VIth Congress, New York, 1912, Section C, pp. 36-41.

75. Underwriters' Laboratories, "An Investigation of the Effects of Fire Exposure upon Hollow Concrete Building Units, Conducted for American Concrete Institute, Concrete Products Association, Portland Cement Association, Joint Submittors," *Retardant Report,* No. 1555, Chicago, May 1924.

76. Dept. of Scientific & Industrial Research and of the Fire Offices' Committee, Joint Committee of the Building Research Board, "Fire Gradings of Buildings. Part IV: Chimneys and Flues," *Post-War Building Studies,* No. 29, London, HMSO, 1952.

77. National Research Council of Canada. Associate Committee on the National Building Code, *Fire Performance Ratings,* Suppl. No. 2 to the National Building Code of Canada, Ottawa, 1965.

78. Associated Factory Mutual Fire Insurance Companies, The National Board of Fire Underwriters, and the Bureau of Standards, *Fire Tests of Building Columns; an Experimental Investigation of the Resistance of Columns, Loaded and Exposed to Fire or to Fire and Water, with Record of Characteristic Effects,* Jointly Conducted at Underwriters. Laboratories, Chicago, 1917-19.

79. Malhotra, H. L., "Effect of Age on the Fire Resistance of Reinforced Concrete Columns," *Fire Research Memorandum,* No. 1, Fire Research Station, Borehamwood, April 1970.

80. Bond, H., ed., *Research on Fire; a Description of the Facilities, Personnel and Management of Agencies*

Engaged in Research on Fire, a Staff Report, National Fire Protection Association, Boston, 1957.

81. *California State Historical Building Code,* Draft, 1978.

82. Fisher, F. L., et al., "A Study of Potential Flashover Fires in Wheeler Hall and the Results from a Full Scale Fire Test of a Modified Wheeler Hall Door Assembly," *Fire Research Laboratory Report,* UCX 77-3; UCX-2480, University of Calif., Dept. of Civil Eng., Berkeley, 1977.

83. Freitag, J. K., *The Fireproofing of Steel Buildings,* 1st ed., J. Wiley, New York, 1906.

84. Gross, D., "Field Burnout Tests of Apartment Dwellings Units," *Building Science Series,* 10, National Bureau of Standards, Washington, 1967.

85. Dunlap, M. E., and Cartwright, F. P., "Standard Fire Tests for Combustible Building Materials," *Proceedings of the American Society for Testing Materials,* vol. 27, Philadelphia, 1927, pp. 534-546.

86. Menzel, C. A., "Tests of the Fire Resistance and Stability of Walls of Concrete Masonry Units," *Proceedings of the American Society for Testing Materials,* vol. 31, Philadelphia, 1931, pp. 607-660.

87. Steiner, A. J., "Method of Fire-Hazard Classification of Building Materials," *Bulletin of the American Society for Testing and Materials,* March 1943, Philadelphia, 1943, pp. 19-22.

88. Heselden, A. J. M., Smith, P. G., and Theobald, C. R., "Fires in a Large Compartment Containing Structural Steelwork; Detailed Measurements of Fire Behavior," *Fire Research Note,* No. 646, Fire Research Station, Borehamwood, Dec. 1966.

89. Ministry of Technology and Fire Offices' Committee Joint Fire Research Organization, "Fire and Structural Use of Timber in Buildings; Proceedings of the Symposium Held at the Fire Research Station, Borehamwood, Herts on 25th October, 1967," *Symposium,* No. 3, London, HMSO, 1970.

90. Shoub, H., and Gross, D., "Doors as Barriers to Fire and Smoke," *Building Science Series,* 3, National Bureau of Standards, Washington, 1966.

91. Ingberg, S. H., "The Fire Resistance of Gypsum Partitions," *Proceedings of the American Society for Testing and Materials,* vol. 25, Philadelphia, 1925, pp. 299-314.

92. Ingberg, S.H., "Influence of Mineral Composition of Aggregates on Fire Resistance of Concrete," *Proceedings of the American Society for Testing and Materials,* vol. 29, Philadelphia, 1929, pp. 824-829.

93. Ingberg, S. H., "The Fire Resistive Properties of Gypsum," *Proceedings of the American Society for Testing and Materials,* vol. 23, Philadelphia, 1923, pp. 254-256.

94. Gottschalk, F.W., "Some Factors in the Interpretation of Small-Scale Tests for Fire-Retardant Wood," *Bulletin of the American Society for Testing and Materials,* October 1945, pp. 40-43.

95. Ministry of Technology and Fire Offices' Committee Joint Fire Research Organization, "Behaviour of Structural Steel in Fire; Proceedings of the Symposium Held at the Fire Research Station Borehamwood, Herts on 24th January, 1967," *Symposium,* No. 2, London, HMSO, 1968.

96. Gustaferro, A. H., and Martin, L. D., *Design for Fire Resistance of Pre-cast Concrete,* prep. for the Prestressed Concrete Institute Fire Committee, 1st ed., Chicago, PCI, 1977.

97. "The Fire Endurance of Concrete; a Special Issue," *Concrete Construction,* vol. 18, no. 8, Aug. 1974, pp. 345-440.

98. The British Constructional Steelwork Association, "Modern Fire Protection for Structural Steelwork," *Publication,* No. FPl, London, 1961.

99. Underwriters' Laboratories, "Fire Hazard Classification of Building Materials," *Bulletin,* No. 32, Sept. 1944, Chicago, 1959.

100. Central Housing Committee on Research, Design, and Construction, Subcommittee on Building Codes, "Recommended Building Code Requirements for New Dwelling Construction with Special Reference to War Housing; Report," *Building Materials and Structures,* Report BMS 88, National Bureau of Standards, Washington, Sept. 1942.

101. De Coppet Bergh, D., *Safe Building Construction; a Treatise Giving in Simplest Forms Possible Practical and Theoretical Rules and Formulae Used in Construction of Buildings and General Instruction,* new ed., thoroughly rev. Macmillan Co., New York, 1908.

102. *Cyclopedia of Fire Prevention and Insurance; a General Reference Work on Fire and Fire Losses, Fireproof Construction, Building Inspection...,* prep. by architects, engineers, underwriters and practical insurance men. American School of Correspondence, Chicago, 1912.

103. Setchkin, N. P., and Ingberg, S. H., "Test Criterion for an Incombustible Material," *Proceedings of the American Society for Testing Materials,* vol. 45, Philadelphia, 1945, pp. 866-877.

104. Underwriters' Laboratories, "Report on Fire Hazard Classification of Various Species of Lumber," *Retardant,* 3365, Chicago, 1952.

105. Steingiser, S., "A Philosophy of Fire Testing," Journal of Fire & Flammability, vol. 3, July 1972, pp. 238-253.

106. Yuill, C. H., Bauerschlag, W. H., and Smith, H. M., "An Evaluation of the Comparative Performance of 2.4.1 Plywood and Two-Inch Lumber Roof Decking under Equivalent Fire Exposure," *Fire Protection Section, Final Report,* Project No. 717A-3-211, Southwest Research Institute, Dept. of Structural Research, San Antonio, Dec. 1962.

107. Ashton, L. A., and Smart, P.M. T., *Sponsored Fire-Resistance Tests on Structural Elements,* London, Dept. of Scientific and Industrial Research and Fire Offices. Committee, London, 1960.

108. Butcher, E. G., Chitty, T. B., and Ashton, L. A., "The Temperature Attained by Steel in Building Fires," *Fire Research Technical Paper,* No. 15, Ministry of Technology and Fire Offices. Committee, Joint Fire Research Organization, London, HMSO, 1966.

109. Dept. of the Environment and Fire Offices' Committee, Joint Fire Research Organization, "Fire-Resistance Requirements for Buildings—a New Approach; Proceedings of the Symposium Held at the Connaught Rooms, London, 28 September 1971," *Symposium,* No. 5, London, HMSO, 1973.

110. Langdon Thomas, G. J., "Roofs and Fire," *Fire Note,* No. 3, Dept. of Scientific and Industrial Research and Fire Offices' Committee, Joint Fire Research Organization, London, HMSO, 1963.

111. National Fire Protection Association and the National Board of Fire Underwriters, *Report on Fire the Edison Phonograph Works,* Thomas A. Edison, Inc., West Orange, N.J., December 9, 1914, Boston, 1915.

112. Thompson, J. P., *Fire Resistance of Reinforced Concrete Floors,* Portland Cement Association, Chicago, 1963.

113. Forest Products Laboratory, "Fire Resistance Tests of Plywood Covered Wall Panels," Information reviewed and reaffirmed, *Forest Service Report,* No. 1257, Madison, April 1961.

114. Forest Products Laboratory, "Charring Rate of Selected Woods—Transverse to Grain," Forest Service *Research Paper,* FLP 69, Madison, April 1967.

115. Bird, G. I., "Protection of Structural Steel Against Fire," *Fire Note,* No. 2, Dept. of Scientific and Industrial Research and Fire Offices' Committee, Joint Fire Research Organization, London, HMSO, 1961.

116. Robinson, W. C., *The Parker Building Fire,* Underwriters' Laboratories, Chicago, c. 1908.

117. Ferris, J. E., "Fire Hazards of Combustible Wallboards," *Commonwealth Experimental Building Station Special Report,* No. 18, Sydney, Oct. 1955.

118. Markwardt, L. J., Bruce, H. D., and Freas, A. D., "Brief Description of Some Fire-Test Methods Used for Wood and Wood Base Materials," *Forest Service Report,* No. 1976, Forest Products Laboratory, Madison, 1976.

119. Foster, H. D., Pinkston, E. R., and Ingberg, S. H., "Fire Resistance of Walls of Gravel-Aggregate Concrete Masonry Units," *Building Materials and Structures,* Report, BMS 120, National Bureau of Standards, Washington, March 1951.

120. Foster, H. D., Pinkston, E.R., and Ingberg, S. H., "Fire Resistance of Walls of Lightweight-Aggregate Concrete Masonry Units," *Building Materials and Structures,* Report BMS 117, National Bureau of Standards, Washington, May 1950.

121. Structural Clay Products Institute, "Structural Clay Tile Fireproofing," *Technical Notes on Brick & Tile Construction,* vol. 1, no. 11, San Francisco, Nov. 1950.

122. Structural Clay Products Institute, "Fire Resistance Ratings of Clay Masonry Walls—I," *Technical Notes on Brick & Tile Construction,* vol. 3, no. 12, San Francisco, Dec. 1952.

123. Structural Clay Products Institute, "Estimating the Fire Resistance of Clay Masonry Walls—II," *Technical Notes on Brick & Tile Construction,* vol. 4, no. 1, San Francisco, Jan. 1953.

124. Building Research Station, "Fire: Materials and Structures," *Digest,* No. 106, London, HMSO, 1958.

125. Mitchell, N. D., "Fire Hazard Tests with Masonry Chimneys," *NFPA Publication,* No. Q-43-7, Boston, Oct. 1949.

126. Clinton Wire Cloth Company, *Some Test Data on Fireproof Floor Construction Relating to Cinder Concrete, Terra Cotta and Gypsum,* Clinton, 1913.

127. Structural Engineers Association of Southern California, Fire Ratings Subcommittee, "Fire Ratings, a Report," part of *Annual Report,* Los Angeles, 1962, pp. 30-38.

128. Lawson, D. I., Fox, L. L., and Webster, C. T., "The Heating of Panels by Flue Pipes," *Fire Research, Special Report,* No. 1, Dept. of Scientific and Industrial Research and Fire Offices' Committee, London, HMSO, 1952.

129. Forest Products Laboratory, "Fire Resistance of Wood Construction," Excerpt from 'Wood Handbook—Basic Information on Wood as a Material of Construction with Data for its Use in Design and Specification,' *Dept. of Agriculture Handbook,* No. 72, Washington, 1955, pp. 337-350.

130. Goalwin, D. S., "Properties of Cavity Walls," *Building Materials and Structures,* Report BMS 136, National Bureau of Standards, Washington, May 1953.

131. Humphrey, R. L., "The Fire-Resistive Properties of Various Building Materials," *Geological Survey Bulletin,* 370, Washington, 1909.

132. National Lumber Manufacturers Association, "Comparative Fire Test on Wood and Steel Joists," *Technical Report,* No. 1, Washington, 1961.

133. National Lumber Manufacturers Association, "Comparative Fire Test of Timber and Steel Beams," *Technical Report,* No. 3, Washington, 1963.

134. Malhotra, H. L., and Morris, W. A., "Tests on Roof Construction Subjected to External Fire," *Fire Note,* No. 4, Dept. of Scientific and Industrial Research and Fire Offices' Committee, Joint Fire Research Organization, London, HMSO, 1963.

135. Brown, C. R., "Fire Tests of Treated and Untreated Wood Partitions," *Research Paper,* RP 1076, part of *Journal of Research of the National Bureau of Standards,* vol. 20, Washington, Feb. 1938, pp. 217-237.

136. Underwriters' Laboratories, "Report on Investigation of Fire Resistance of Wood Lath and Lime Plaster Interior Finish," *Publication,* SP. 1.230, Chicago, Nov. 1922.

137. Underwriters' Laboratories, "Report on Interior Building Construction Consisting of Metal Lath and Gypsum Plaster on Wood Supports," *Retardant,* No. 1355, Chicago, 1922.

138. Underwriters' Laboratories, "An Investigation of the Effects of Fire Exposure upon Hollow Concrete Building Units," *Retardant,* No. 1555, Chicago, May 1924.

139. Moran, T. H., "Comparative Fire Resistance Ratings of Douglas Fir Plywood," *Douglas Fir Plywood Association Laboratory Bulletin,* 57-A, Tacoma, 1957.

140. Gage Babcock & Association, "The Performance of Fire-Protective Materials under Varying Conditions of Fire Severity," Report 6924, Chicago, 1969.

141. International Conference of Building Officials, *Uniform Building Code* (1979 ed.), Whittier, CA, 1979.

142. Babrauskas, V., and Williamson, R. B., "The Historical Basis of Fire Resistance Testing, Part I and Part II," *Fire Technology,* vol. 14, no. 3 & 4, Aug. & Nov. 1978, pp. 184-194, 205, 304-316.

143. Underwriters' Laboratories, "Fire Tests of Building Construction and Materials," 8th ed., *Standard for Safety,* UL263, Chicago, 1971.

144. Hold, H. G., *Fire Protection in Buildings,* Crosby, Lockwood, London, 1913.

145. Kollbrunner, C. F., "Steel Buildings and Fire Protection in Europe," *Journal of the Structural Division,* ASCE, vol. 85, no. ST9, Proc. Paper 2264, Nov. 1959, pp. 125-149.

146. Smith, P., "Investigation and Repair of Damage to Concrete Caused by Formwork and Falsework Fire," *Journal of the American Concrete Institute,* vol. 60, Title no. 60-66, Nov. 1963, pp. 1535-1566.

147. "Repair of Fire Damage," 3 parts, *Concrete Construction,* March-May, 1972.

148. National Fire Protection Association, *National Fire Codes; a Compilation of NFPA Codes, Standards, Recommended Practices and Manuals,* 16 vols., Boston, 1978.

149. Ingberg, S. H. "Tests of Severity of Building Fires," *NFPA Quarterly,* vol. 22, no. 1, July 1928, pp. 43-61.

150. Underwriters' Laboratories, "Fire Exposure Tests of Ordinary Wood Doors," *Bulletin of Research,* no. 6, Dec. 1938, Chicago, 1942.

151. Parson, H., "The Tall Building under Test of Fire," *Red Book,* no. 17, British Fire Prevention Committee, London, 1899.

152. Sachs, E. O., "The British Fire Prevention Committee Testing Station," *Red Book,* no. 13, British Fire Prevention Committee, London, 1899.

153. Sachs, E. O., "Fire Tests with Unprotected Columns," *Red Book,* no. 11, British Fire Prevention Committee, London, 1899.

154. British Fire Prevention Committee, "Fire Tests with Floors a Floor by the Expended Metal Company," *Red Book,* no. 14, London, 1899.

155. *Engineering News,* vol. 56, Aug. 9, 1906, pp. 135-140.

156. *Engineering News,* vol. 36, Aug. 6, 1896, pp. 92-94.

157. Bauschinger, J., *Mittheilungen de Mech.-Tech. Lab. der K. Tech. Hochschule, Müchen,* vol. 12, 1885.

158. *Engineering News,* vol. 46, Dec. 26, 1901, pp. 482-486, 489-490.

159. *The American Architect and Building News,* vol. 31, March 28, 1891, pp. 195-201.

160. British Fire Prevention Committee, First International Fire Prevention Congress, *Official Congress Report,* London, 1903.

161. American Society for Testing Materials, *Standard Specifications for Fire Tests of Materials and Construction (C19-18),* Philadelphia, 1918.

162. International Organization for Standardization, *Fire Resistance Tests on Elements of Building Construction (R834),* London, 1968.

163. *Engineering Record,* vol. 35, Jan. 2, 1897, pp. 93-94; May 29, 1897, pp. 558-560; vol. 36, Sept. 18, 1897, pp. 337-340; Sept. 25, 1897, pp. 359-363; Oct. 2, 1897, pp. 382-387; Oct. 9, 1897, pp. 402-405.

164. Babrauskas, Vytenis, "Fire Endurance in Buildings," PhD Thesis. *Fire Research Group,* Report, No. UCB FRG 76-16, University of California, Berkeley, Nov. 1976.

165. The Institution of Structural Engineers and The Concrete Society, *Fire Resistance of Concrete Structures,* London, Aug. 1975.

INDEX

A

ACCESSIBILITY . . . . 107.2, 306, 801.1, 901.2, 1101.2, 1508, Appendix B

ADDITIONS . . . . 101.2, 101.3, 101.4, 104.2, 104.2.1, 104.10.1, 106.2.3, 109.3.3, 113.2, 115.5, 202, 301.1, 301.3, 302.2, 302.2.1, 303.2, 304.1, 304.2, 306.1, 306.3, 306.6, 307.1, 308.1, 309, 501.1, 502, 601.1, 606, Chapter 11, 1301.1, 1301.1.1, 1301.2, 1301.2.3, 1301.3, 1301.4.1, 1301.4.3, 1501.3, 1505.2

ADMINISTRATION . . . . Chapter 1

ALTERATIONS . . . . 101.2, 101.3, 101.4, 101.7, 104.2, 104.2.1, 104.2.2, 104.10.1, 105.1.1, 105.1.2, 105.2.2, 105.2.3, 106.2.3, 106.2.6, 108.5, 110.2, 113.2, 115.5, 202, 301.1, 301.3, 302.2, 302.2.1, 302.4, 304, 306.1, 306.3, 306.3.1, 306.4, 306.5, 306.7, 306.7.1, 306.7.3, 306.7.4, 306.7.5, 306.7.16, 306.7.16.6, 307.1, 308.1, 309.1, 501.1, 501.1.1, 502.1, 502.4, 502.5, 503, 504.1, 601.1, 602, 603, 604, Chapter 7, Chapter 8, Chapter 9, 1101.3, 1103.1, 1103.2, 1201.1, 1201.2, 1201.4, 1203.1, 1301.1, 1301.1.1, 1301.2, 1301.2.4, 1301.2.6, 1301.3, 1301.4.1, 1301.4.3, 1301.9.1, 1401.2, 1501.3, 1504.1, 1505.2, A105.2, A304.1.1, A402.1, A403.2

Level 1 104.2.2, 602, 603.2, 604.2, Chapter 7, 801.2
Level 2 . . . . 106.2.3, 505.2, 603, Chapter 8
Level 3 . . . . 106.2.3, 604, Chapter 9, 1011.4

ALTERNATIVE MATERIALS, DESIGN AND METHODS OF CONSTRUCTION 104.11

APPEALS 104.8, 112, 116.6

APPROVED 202

ARCHAIC MATERIALS AND ASSEMBLIES . . . . Resource A

ARCHITECT (see REGISTERED DESIGN PROFESSIONAL)

AREA (see BUILDING)

AUTOMATIC SPRINKLER SYSTEM . . . 110.2, 501.2, 504.5, 802.2.1, 802.6, 803.1.1, 803.2, 803.2.1.1, 803.2.3, 803.2.5, 803.2.6, 803.3, 803.4, 804.4.1.1, Table 804.4.1.1(1), 804.4.1.2.1, 804.4.1.2.2, 804.4.2, 804.6.1, 804.7, 902.2, 904.1, 1011.2.1, 1011.6.1, 1011.6.1.1, 1011.7.3, 1011.8.3, 1203.4, 1203.8, 1204.4, 1204.9, 1301.5.1, Table 1301.6.9, 1301.6.10.1, 1301.6.17, Table 1301.6.17, 1301.6.17.1, 1301.6.19, Table 1301.6.19, 1507

B

BOARD OF APPEALS . . . . . . 104.8, 112, Appendix D

BUILDING

Area . . . 503.5, 503.7, 503.8, 503.9, 503.10, 503.11, 503.16, 503.18, 506.5.2, 506.5.3, 506.6, 603.1, 604.1, 1006.2, 1006.3, 1011.6, Table 1011.6, 1011.6.1, 1011.6.1.1, 1011.6.2, 1102, 1204.2, 1301.2.3, 1301.6.2

Dangerous 104.6, 114.1, 117.1, 202, 302.1, 1203.9, 1205.2

Height . . . . 202, 502.1, 804.4.1.1, 1011.6, Table 1011.6, 1011.7.1, 1011.7.3, 1102, 1301.2.3, 1301.6.1, Table 1301.7, A301.2

Historic . . . . 202, 306.1, 306.7.16, 401.1, 501.1, 507, 601.1, 607, 701.1, 1001.2.1, Chapter 12, 1301.1, B101

Relocatable 202, 1401.1
Relocated . . . . 301.4, 1206, Chapter 14
Relocation of 101.2, 101.3, 101.4, 104.2, 301.1
Unsafe 104.6, 115, 116, 117.1, 202

C

CARBON MONOXIDE 308
CARPETING 105.2, 702.2
CEILING 702.1, 801.4, 802.2.1, 802.4, 903.2.1, 1011.3, 1011.5.1, 1203.7, 1204.10, 1301.6.3.3, A105.3, A113.1
CERTIFICATE OF OCCUPANCY 110, 506.2, 1001.2, 1001.3
CHANGE OF OCCUPANCY . . . . . 101.2, 101.3, 101.4, 104.2, 104.2.2, 106.2.6, 110.1, 115.5, 202, 301.1, 301.3, 306.1, 306.5, 307.1, 308.1, 501.1, 505.3, 506, 601.1, 605, 702.5, Chapter 10, 1201.1, 1201.2, 1201.4, 1204, 1301.1, 1301.2, 1301.2.6, 1301.4.1, 1301.4.3, 1301.6.17, 1301.9.1, 1401.2, B101.2, B101.3, B101.4, C101.1, C201.1
CHANGE OF USE 202
CLASSROOM ACOUSTICS 502.6, 503.16, 506.6, 903.4, 1011.4, 1101.4
CODE OFFICIAL . . . . . . Chapter 1, 202, 301.3, 302.1, 302.3, 303.2.1, 304.2, 405.2.3.1, 501.2, 506.1, 506.1.1, 507.2, 507.4, 802.6, 803.2.6, 804.2, 1001.2, 1011.5.1, 1201.2, 1201.3, 1201.5, 1203.2, 1203.3, 1203.11, 1203.12, 1204.6, 1204.7, 1204.11, 1204.12, 1204.14, 1205.1, 1205.2, 1301.3, 1301.3.1, 1301.4.2, 1301.4.3, 1402.7, 1501.6.7, 1509.1, A102.2, A107.1, A108.1, A113.7, A205.3.2, A301.1, A301.2, A301.3, A302.1, A303.1, A304.2.1, A304.2.2, A304.2.3, Table A304.2.3(1), Table A304.2.3(2), A304.5, A403.8, A403.9.3.2, A404.1, A405.3, A405.3.2

INDEX

COMPARTMENTATION . . . 1301.6.3, Table 1301.6.3, 1301.6.20, Table 1301.6.20, Table 1301.7

COMPLIANCE METHODS . . . . Chapter 3
   Performance compliance method. . . . . 301.3.3, Chapter 13
   Prescriptive compliance method. . . . . 301.3.1, Chapter 5
   Work area compliance method. . . . . 301.3.2, Chapter 6, Chapter 7, Chapter 8, Chapter 9, Chapter 10, Chapter 11, Chapter 12

CONFLICT 102.1, 102.4.1, 102.4.2, 104.10.1, 113.1, 302.2, 1301.2.2

CONSTRUCTION DOCUMENTS. 104.2, 105.3, 105.3.1, 105.4, 106, 113.4, 202, 601.2, A105.4, A205.3, A301.1, A406, A407.1

CONSTRUCTION SAFEGUARDS . . . . 101.5, Chapter 15

CORRIDOR
   Dead-end. . . . . . . 801.4, 804.7, 1011.5.1, 1301.6.12, 1301.6.12.1, Table 1301.7
   Doors. 804.6.1, 1011.5.1
   Exit access 801.4, 802.2.1, 802.4, 802.4.1, 803.2.1, 803.2.2, 803.2.5, 803.3, 804.1, 804.4.1, 901.2, 1011.8.2, 1203.3, 1204.6
   Openings. 804.6, 804.6.1, 804.6.2, 804.6.3, 804.6.3.1, 804.6.4, 1011.5.1, 1011.8.2, 1203.4, 1204.8
   Projections. 704.1.1
   Rating . . . . 803.1.1, 1011.5.1, 1301.6.5, Table 1301.6.5, 1301.6.5.1, Table 1301.7

D

DANGEROUS
   Nonstructural. 115.1, 116.1
   Structural. 104.6, 114.1, 117.1, 202, 302.1, 1203.9, 1205.2

DEFERRED SUBMITTAL 106.3.4, 106.6, 202

DEFINITIONS. 202

DEMOLITION 101.5, 104.2, 106.2.6, 108.5, 113.2, 115.5, 117, 1501.6, 1501.6.1, 1501.7, 1502.1, 1503.1, 1504.1, 1505.2

DEPARTMENT OF BUILDING SAFETY 104.10, 105.3, 109.3.8

DISPROPORTIONATE EARTHQUAKE DAMAGE 202, 405.2.2, 502.2

E

EGRESS (see MEANS OF EGRESS)

ELECTRICAL. 105.1, 105.1.1, 105.2, 108.3, 109.3.4, 202, 306.7.1, 406, 801.3, 801.4, 806, 1007, 1101.2, A102.1

ELEVATOR . . . . . . 306.7.7, 902.1.2, 905.4, 1011.8.3, 1102.2, 1301.6.14, Table 1301.6.14, 1301.6.14.1, Table 1301.7

EMERGENCY ESCAPE AND RESCUE OPENINGS 202, 505, 506.4, 702.4, 702.5, 702.5.1, 702.6

EMERGENCY POWER 804.5.5, 1301.6.15.1

ENERGY. 302.2, 702.7, 708, 809, 907, 1104

ENGINEER (see REGISTERED DESIGN PROFESSIONAL)

EQUIPMENT OR FIXTURE 202

EXISTING (see BUILDING)

EXIT . . . . . . 504.1.4, 802.2.1, 802.4, 802.4.1, 803.2.1, 803.2.2, 803.2.5, 803.3, 804, 903.1, 903.3, 905.2, 905.3, 1011.5, Table 1011.5, 1011.5.1, 1011.8.2, 1102.2, 1203.2, 1203.3, 1203.6, 1204.6, 1204.7, 1204.12, 1204.13, 1301.6.3.2, 1301.6.6, 1301.6.10.1, 1301.6.11, 1301.6.11.1, 1301.6.12, 1301.6.13, Table 1301.6.15, 1301.6.15.1, Table 1301.7, A402

EXIT, HORIZONTAL 503.15, 804.11, 1301.6.3.1

EXTERIOR WALL COVERING 202
   Addition and replacement 309

EXTERIOR WALL ENVELOPE 202
   Addition and replacement 309

F

FIRE ESCAPES 504, 802.2.1, 804.4.1.2, 1301.6.11, 1301.6.11.1

FIRE PROTECTION. 106.2.2, 202, 306.7.1, 403, 703, 801.3, 803, 804.6.1, 804.6.2, 904, 1001.2, 1004, 1011.1, 1011.2, 1011.2.1, 1011.2.2, 1011.8.4, 1102.3, 1501.2.1, 1501.3, 1509

FIRE RATINGS . . . . 804.4.1.2.1, 804.6.1, 804.6.2, 1011.6.1, 1011.6.1.1, 1011.6.3, 1011.7, 1011.7.1, 1011.7.2, 1011.7.3, 1011.8.2, 1011.8.3, 1011.8.4, 1203.6, 1203.7, 1203.8, 1204.4, 1204.10, 1301.2.2, 1301.6.3.2, 1301.6.3.3, 1301.6.4, 1301.6.4.1, 1301.6.5, 1301.6.5.1, 1301.6.6, 1301.6.16.1, Table 1301.7, Resource A

FIRE SAFETY . . . 101.7, 104.10, 107.2, 1101.2, 1203, 1204.10, 1301.5, 1301.5.1, 1301.5.3, 1301.6.1, 1301.6.2, 1301.6.3, 1301.6.4, 1301.6.5, 1301.6.6, 1301.6.7, 1301.6.8, 1301.6.9, 1301.6.14, 1301.6.16, 1301.6.17, 1301.6.18, 1301.6.19, 1301.6.20, Table 1301.7, Table 1301.8, Table 1301.9, A102.1

INDEX

FIRE-RESISTANCE RATING........802.2.1, 802.6, 803.1.1, 803.2.2.1, 902.2, 1011.6.1.1, 1011.6.3, 1011.7, 1011.8.2, 1203.4, 1203.6, 1203.7, 1203.8, 1204.3, 1204.8, 1204.10, 1301.2.2, 1301.5.1, 1301.6.4, 1301.6.4.1, 1301.6.5, 1301.6.5.1, 1301.6.6, 1301.6.16, 1301.6.16.1, 1301.7, Resource A
FLAME SPREAD................Resource A
FLOOD HAZARD AREA........104.2.1, 104.10.1, 109.3.3, 109.3.10, 202, 301.3, 401.3, 405.2.6, 502.3, 503.2, 507.3, 701.3, 1103.3, 1201.4, 1301.3.3, 1402.6
FUEL GAS................302.2, 702.7.1

G

GLASS......402.1, 406, 804.6.1, 804.6.2, 1011.8.2, 1202.2, 1204.8, 1301.6.10.1, Resource A
GRAVITY LOADS (see STRUCTURAL LOADS/FORCES)
GUARDS........802.5, 802.5.1, 802.5.2, 804.12, 804.12.1, 804.12.2, 1203.9, 1203.10
GUIDELINES FOR STRUCTURAL RETROFIT
  Seismic.........................Appendix A
  Wind...........................Appendix C

H

HANDRAILS............503.1, 804.10, 1011.5.1, 1011.5.4, 1203.9
HAZARD CATEGORIES......1011.5, Table 1011.5, 1011.6, Table 1011.6, Table 1011.7
HEIGHT (see BUILDING, HEIGHT)
HIGH-RISE BUILDING......803.2.1, 902.1, 904.1.1, 1011.6.1, 1301.6.17, 1301.6.19
HISTORIC BUILDING (see BUILDING, HISTORIC)

I

INCIDENTAL USE AREAS.................1002.2
INSPECTION.....104.4, 104.6, 104.7, 105.2, 106.1, 106.3.1, 106.6, 109, 1402.7, A105.3, A105.4, A107.1, A107.4, A113.7, A205.3.1, A205.4, A304.4.3, A304.5, A406.3.4, A407.3
INTERIOR FINISHES........702.1, 702.2, 1011.3, 1203.5, 1203.9
INTERIOR TRIM.........................702.3

L

LIVE LOAD (see STRUCTURAL LOADS/FORCES)
LOAD-BEARING ELEMENT........105.2.2, 202, Resource A

M

MAINTENANCE.....105.2, 306.3, 1501.3, 1501.6.6
MEANS OF EGRESS........101.2, 105.2.2, 106.2.3, 107.2, 115.1, 202, 306.3, 306.7.2, 306.7.16, 404, 501.2, 504.1.1, 504.1.2, 704, 802.2.1, 802.2.3, 802.6, 803.2.4, 804, 903.1, 905, 1005, 1011.5, Table 1011.5, 1011.5.1, 1011.5.2, 1101.2, 1201.3, 1203.3, 1204.6, 1301.5, 1301.5.2, 1301.5.3, 1301.6.1, 1301.6.2, 1301.6.3, 1301.6.4, 1301.6.5, 1301.6.6, 1301.6.7, 1301.6.8, 1301.6.9, 1301.6.10, 1301.6.11, 1301.6.12, 1301.6.14, 1301.6.15, 1301.6.17, 1301.6.18, 1301.6.19, 1301.6.20, 1301.6.21.1, 1301.6.21.2, 1301.6.21.3, Table 1301.7, Table 1301.8, Table 1301.9, 1505
MECHANICAL......105.1, 105.1.1, 105.2, 105.2.2, 108.3, 302.2, 306.7.1, 308.1, 407, 503.4, 702.7, 801.3, 807, 902.1.1, 1008, 1101.2, 1301.6.7.1, 1301.6.8, 1301.6.8.1, 1301.6.10, 1301.6.10.1, 1501.2.1, A102.1
MOVED BUILDINGS (see BUILDING, RELOCATED)

N

NONCOMBUSTIBLE MATERIAL........202, 504.3, 804.4.1.2.2, Resource A

O

OPENINGS...106.2.4, 116.2, 202, 504.5, 505, 506.4, 702.4, 702.5, 702.6, 802.2, 804.4.1.2.1, 804.5.3, 804.6, 804.6.3, 804.6.3.1, 804.6.4, 903.1, 1011.5.1, 1011.5.6, 1011.7.1, 1011.7.2, 1011.7.3, 1011.8, 1011.8.2, 1011.8.3, 1011.8.4, 1102.2, 1203.3, 1203.8, 1204.3, 1204.6, 1301.6.6, Table 1301.6.6(1), 1301.6.6.1, 1301.6.10.1, Table 1301.7, 1501.6.4

P

PERMITS........101.4.1, 104.2, 104.2.2, 104.7, 105, 106.1, 106.2.6, 106.3.1, 106.3.2, 106.3.3, 106.6, 107.1, 107.4, 108.1, 108.2, 108.3, 108.4, 108.5, 109.1, 109.2, 109.3.11, 109.5, 109.6, 110.1, 110.2, 110.3, 111.1, 113.2, 113.4, 503.6, 706.1, 706.3, 706.3.1, A105.3
PLUMBING.....105.1, 105.1.1, 105.2, 108.3, 109.3.4, 202, 302.2, 308.1, 408, 702.7, 808, 1009, 1101.2, 1301.2.6, 1501.7, A102.1, A304.1.3, A304.3.2, A304.4.2
PRIMARY FUNCTION..........202, 306.6, 306.7.1

R

RAMPS..............306.7.6, 306.7.16.1, 802.2.1, 804.3, 1301.6.6, B101.3, B101.4
REFERENCED STANDARDS..........Chapter 16
REFUGE AREAS..................503.15, 804.11

INDEX

REGISTERED DESIGN PROFESSIONAL 104.2.2.1, 106.1, 106.3.4, 106.6, 202, 405.2.3.1, 1201.2, A106.2.3.4, A301.1, A301.2, A301.3, A304.2.2, A304.2.3, A304.4.1, A405.3

REGISTERED DESIGN PROFESSIONAL IN RESPONSIBLE CHARGE 106.3.4, 106.6

REHABILITATION 104.2.1, 202, 405.2.4

RELOCATABLE BUILDINGS (see BUILDING, RELOCATABLE)

RELOCATED BUILDINGS (see BUILDING, RELOCATED)

REPAIR . . . . 101.2, 101.3, 101.4, 101.7, 104.2, 104.2.1, 104.2.2, 104.10.1, 105.1, 105.2, 105.2.1, 105.2.2, 105.2.3, 106.2.6, 113.1, 113.2, 115.3, 115.5, 116.1, 116.4, 116.5, 117.1, 202, 301.1, 301.2, 302.2, 302.2.1, 302.4, 306.1, 306.3, 308.1, 309.1, Chapter 4, 1101.3, 1201.1, 1201.4, 1202, 1205.1, 1301.3, 1401.2, 1402.7, 1501.3, 1501.6.6, A105.2, A105.3, A106.1, A106.2.1, A106.2.3.9, A403.2, C101.1, C201.1

REROOFING 705

RISK CATEGORY . . . . . . 202, 304.3.1, Table 304.3.1, 304.3.2, Table 304.3.2, 506.5.2, 506.5.3, 506.5.4, 1006.2, 1006.3, 1006.4, A102.2, A401.2, A403.3, C201.2

ROOF 308.1, 405.2.5, 503.3, 503.4, 503.7, 503.8, 504.3, 705, 906.4, 906.5, 1204.5, 1501.2.1, 1501.6.5, 1502.1
  Diaphragms 503.12, 706.3.2, 906.4
  Permit 706.1, 706.3, 706.3.1
  Recover 202, 705.2.1, 705.2.1.1, 705.3
  Repair 202
  Replacement 202, 705.1, 705.2
  Reroofing 202, 503.6, 705, 706.3.1

ROOF COATING 705.2.1

S

SAFEGUARDS DURING CONSTRUCTION (see CONSTRUCTION SAFEGUARDS)

SAFETY PARAMETERS . . . . 1301.5.3, 1301.6, Table 1301.7

SEISMIC FORCES (see STRUCTURAL LOADS/FORCES)

SEISMIC LOADS (see STRUCTURAL LOADS/FORCES)

SEISMIC RETROFIT . . . . Appendix A

SHAFT ENCLOSURES 802.2.2, 1011.8, 1301.6.6

SMOKE ALARMS 307

SMOKE COMPARTMENTS . . . 503.14, 503.15, 802.3, 803.2.3, 904.1.5, 1301.6, 1301.6.20, Table 1301.6.20, 1301.6.20.1, 1301.6.21, 1301.6.21.1, 1301.6.21.2, 1301.6.21.2.1, Table 1301.7

SMOKE CONTROL . . . . . 1301.6.10, Table 1301.6.10, 1301.6.10.1, Table 1301.7

SMOKE DETECTORS . . . 803.4, 1011.8.4, 1301.6.8.1

SNOW LOAD (see STRUCTURAL LOADS/FORCES)

SPECIAL USE AND OCCUPANCY . . . 802, 902, 1002

SPRINKLER SYSTEM (see AUTOMATIC SPRINKLER SYSTEM)

STAIRWAY 306.7.9, 503.1, 504.1.3, 504.2, 504.4, 504.5, 506.3, 802.2.1, 802.2.2, 802.2.3, 803.1.1, 804.4.1.1, 804.4.1.2.1, 804.4.1.2.3, 804.5.3, 804.5.3.1, 804.10.1, 804.12.1, 806.4.5, 903.1, 1011.5.1, 1011.5.2, 1011.5.4, 1011.8.2, 1011.8.3, 1011.8.4, 1102.2, 1203.3, 1203.6, 1203.9, 1204.6, 1204.11, 1204.13, 1301.6.3, 1301.6.6, 1301.6.10.1, 1504.1, 1505.1, 1506.1

STANDPIPE SYSTEMS . . . . . 105.2.2, 803.2.1, 803.3, 1301.6.18, Table 1301.6.18, 1301.6.18.1, Table 1301.7, 1506, 1503.2

STORM SHELTER 303

STRUCTURAL 304.1, 405, 502.4, 502.5, 503.3, 503.4, 503.5, 503.6, 503.7, 503.8, 503.9, 503.10, 503.11, 503.12, 503.13, 506.5, 507.4, 706, 805, 906, 1006, 1103, 1205, 1301.4.1, 1402

STRUCTURAL LOADS/FORCES
  Gravity loads 202, 304.1, 405.2.4, 405.2.4.1, 502.4, 503.3, 706.2, 805.2, 1007.1, 1103.1, A403.5, A403.10.1, A403.10.2
  International Building Code-level . . . 506.5.3, 506.5.4, 1006.3, 1103.2, 1402.4
  Live loads 202, 304.1, 405.2.4, 502.4, 503.3, 504.3, 506.5.1, 507.4, 706.2, 804.4.1.2.2, 805.2, 1006.1, 1103.1, 1204.13, 1205.1, 1501.6.1, 1501.6.5, A104.1, A106.2.3.6, A108.3, A108.6, A403.7
  Reduced . . . . 304.3.2, Table 304.3.2, 405.2.3.1, 405.2.3.3, 503.4, 503.5, 503.6, 503.7, 503.8, 503.9, 503.10, 503.11, 506.5.3, 706.3.1, 805.3, 906.2, 906.3, 906.4, 906.5, 906.6, 906.7, 1006.3
  Seismic loads . . . . . 202, 304.3, 405.2.3, 405.2.4.1, 502.2, 502.5, 503.4, 503.5, 503.6, 503.7, 503.8, 503.9, 503.10, 503.11, 503.13, 506.5.3, 506.5.4, 706.3.1, 805.3, 906.2, 906.3, 906.4, 906.5, 906.6, 906.7, 1006.3, 1006.4, 1103.2, 1402.4, Appendix A
  Snow loads 304.2, 405.2.1.1, 405.2.4, 405.2.5, 502.4, 503.3, 506.5.2, 706.2, 805.2, 1006.2, 1103.1, 1402.5
  Wind loads . . . . 303.1, 303.2, 405.2.3, 405.2.4.1, 503.12, 506.5.2, 706.3.2, 1006.2, 1402.3, 1501.6.7, Appendix C

SUBSTANTIAL DAMAGE 202, 507.3, 1103.3

SUBSTANTIAL IMPROVEMENT . . . 104.2.1, 104.10.1, 109.3.3, 202, 401.3, 502.3, 503.2, 507.3, 701.3, 1103.3, 1201.4, 1301.3.3

INDEX

**SUBSTANTIAL STRUCTURAL
    ALTERATION** 202, 503.11, 906.2
**SUBSTANTIAL STRUCTURAL
    DAMAGE** 202, 405.2.1, 405.2.2, 405.2.3,
        405.2.4, 405.2.4.1, 405.2.5,
        502.2, 507.4, 1205.1

T

TECHNICALLY INFEASIBLE 202, 306.7,
    306.7.11, 306.7.12, 306.7.14
TEMPORARY STRUCTURE 107
TESTING 104.11.2, 111.2, 305, 1301.6.10.1,
    1507.1, A104.1, A105.3, A105.4, A106.2.1,
    A106.2.3, A107.2, A107.3, A107.4, A107.5, A108.2,
    A108.5, A113.1.3, A114.1, A205.3.2, A205.4,
    A206.2, A304.2.2, A403.9.2.1, A405.3, A407.3

U

UNSAFE 104.6, 105.2, 114.1, 114.4, 115,
    116.1, 116.2, 116.3, 116.5, 117.1, 202,
    302.3, 302.4, 1007.2, 1201.5, 1301.3.1
UTILITIES 111

V

VERTICAL OPENING PROTECTION 802.2,
    903.1, 1011.8,
    1301.6.6, Table 1301.7
VIOLATIONS 101.7, 104.6, 105.2, 105.4, 105.6,
    109.1, 110.1, 110.2, 110.4, 113, 114.4

W

WIND LOAD (see STRUCTURAL LOADS/FORCES)
WINDOWS
    Emergency escape and
        rescue openings 505.3, 505.3.1, 505.4,
            702.5, 702.5.1, 702.6
    Glazing 402.1, 804.6, 804.6.2,
        1202.2, 1203.8, 1204.8
    Opening control devices 505.2, 505.3.1,
        702.4, 702.5.1
    Replacement 505.1
WORK AREA 202, 301.3.2, 306.5, 306.7.4,
    306.7.10.3, 503.5, 503.7, 503.8, 503.9, 503.10,
    503.11, 503.16, 503.18, 506.6, 601.2, 603.1, 604.1,
    801.3, 802.1, 802.2.1, 802.2.2, 802.2.3, 802.3,
    802.4, 802.4.1, 802.5, 803.1, 803.2.1, 803.2.1.1,
    803.2.2, 803.2.2.1, 803.2.4, 803.2.5, 803.3, 803.4.1,
    803.4.1.1, 803.4.1.2, 803.4.1.4, 803.4.1.6, 803.4.2,
    804.1, 804.2, 804.4.1, 804.4.2, 804.5, 804.5.1,
    804.5.1.1, 804.5.1.2, 804.5.2, 804.5.2.1, 804.5.3,
    804.5.3.1, 804.5.4, 804.5.4.1, 804.6, 804.6.1,
    804.6.2, 804.6.3, 804.6.3.1, 804.6.4, 804.7, 804.8,
    804.8.1, 804.8.2, 804.9.1, 804.9.2, 804.10, 804.10.1,
    804.12, 806.1, 806.2, 807.1, 901.2, 902.1.2, 903.1,
    903.2.1, 903.3, 903.4, 904.1, 904.1.1, 904.1.2,
    904.1.3, 904.1.4, 904.1.5, 904.1.6, 904.1.7, 904.2.1,
    904.2.2, 905.2, 905.3, 906.7

Valuable Guides to Changes in the 2021 I-Codes®

SIGNIFICANT CHANGES TO THE 2021 INTERNATIONAL CODES®

Practical resources that offer a comprehensive analysis of the critical changes made between the 2018 and 2021 editions of the codes. Authored by ICC code experts, these useful tools are "must-have" guides to the many important changes in the 2021 International Codes.

Key changes are identified then followed by in-depth, expert discussion of how the change affects real world application. A full-color photo, table or illustration is included for each change to further clarify application.

SIGNIFICANT CHANGES TO THE IBC®, 2021 EDITION
#7024S21

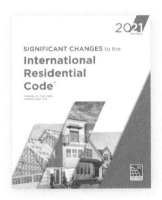

SIGNIFICANT CHANGES TO THE IRC®, 2021 EDITION
#7101S21

SIGNIFICANT CHANGES TO THE IFC®, 2021 EDITION
#7404S21

SIGNIFICANT CHANGES TO THE IPC®/IMC®/IFGC®, 2021 EDITION
#7202S21

SIGNIFICANT CHANGES TO THE IECC®, 2021 EDITION
#7808S21

New to the series!

Order Your Helpful Guides Today! 1-800-786-4452 | www.iccsafe.org/books

 ICC EVALUATION SERVICE

☑ **Specify** and
☑ **Approve** *with* **Confidence**

When facing new or unfamiliar materials, look for an ICC-ES Evaluation Report or Listing before approving for installation.

ICC-ES® **Evaluation Reports** are the most widely accepted and trusted technical reports for code compliance.

ICC-ES **Building Product Listings** and **PMG Listings** show product compliance with applicable standard(s) referenced in the building and plumbing codes as well as other applicable codes.

When you specify or approve products or materials with an ICC-ES report, building product listing or PMG listing, you avoid delays on projects and improve your bottom line.

ICC-ES is a subsidiary of ICC®, the publisher of the codes used throughout the U.S. and many global markets, so you can be confident in their code expertise.

www.icc-es.org | 800-423-6587

Ever wonder why contractors, architects and other building professionals rely on ICC plan reviews?

- **Experience** – Our I-Code experts have expertise in ALL the International Codes® (I-Codes®)
- **Detailed Report** – identifies code compliance and violations
- **Complimentary** re-review of reissued plans*

Additionally, ICC Plan Review Services has years of experience in code application, certified design experts on board, and International Code Council Certifications to ensure we deliver the industry's most comprehensive and reliable plan reviews.

Find Out More!
888-422-7233, x4337
www.iccsafe.org/PLR3

*Applies to "Complete Plan Review Services". Contact ICC Plan Review staff for details.

Telework with ICC Digital Codes Premium
An essential online platform for all code users

ICC's Digital Codes is the most trusted and authentic source of model codes and standards, which conveniently provides access to the latest code text from across the United States. With a Premium subscription, users can further enhance their code experience with powerful features such as team collaboration and access sharing, bookmarks, notes, errata and revision history, and more.

- Never miss a code update
- Available anywhere 24/7
- Use on any mobile or digital device
- View hundreds of code titles and standards

Go Beyond the Codes with a Premium Subscription
Start Your 14-Day Premium trial at *codes.iccsafe.org/trial*